'AN IMMENSE AND EXCEEDINGLY COMMODIOUS GOODS STATION'

Published by Pre-Construct Archaeology Limited
Copyright © Pre-Construct Archaeology Limited 2016
All rights reserved. No part of this publication may be reproduced, stored in a retrieval system or transmitted, in any form or by any means, electronic, mechanical, photocopying, recording or otherwise, without prior permission of the copyright owner.
ISBN 978-0-9926672-6-9
Edited by Victoria Ridgeway and Frank Meddens
Typeset by Cate Davies
Printed by Henry Ling Ltd, The Dorset Press

Cover artwork: Internal view of the Granary building before regeneration.

'AN IMMENSE AND EXCEEDINGLY COMMODIOUS GOODS STATION'

The Archaeology and History of the Great Northern Railway's
Goods Yard at King's Cross, 1849 to the Present Day

By Rebecca Haslam and Guy Thompson

with contributions from
Kevin Hayward, Chris Jarrett, Tim Smith and Berni Sudds

PCA Monograph Series

1   Excavations at Hunt's House, Guy's Hospital, London Borough of Southwark
By Robin Taylor-Wilson, 2002

2   Tatberht's Lundenwic: Archaeological excavations in Middle Saxon London
By Jim Leary with Gary Brown, James Rackham, Chris Pickard and Richard Hughes, 2004

3   Iwade: Occupation of a North Kent Village from the Mesolithic to the Medieval period
By Barry Bishop and Mark Bagwell, 2005

4   Saxons, Templars & Lawyers in the Inner Temple: Archaeological excavations in Church Court & Hare Court
By Jonathan Butler, 2005

5   Unlocking the Landscape: Archaeological excavations at Ashford Prison, Middlesex
By Tim Carew, Barry Bishop, Frank Meddens and Victoria Ridgeway, 2006

6   Reclaiming the Marsh: Archaeological excavations at Moor House, City of London 1998–2004
By Jonathan Butler, 2006

7   From Temples to Thames Street – 2000 Years of Riverside Development: Archaeological excavations at the Salvation Army International Headquarters
By Timothy Bradley and Jonathan Butler, 2008

8   A New Millennium at Southwark Cathedral: Investigations into the first two thousand years
By David Divers, Chris Mayo, Nathalie Cohen and Chris Jarrett, 2009

9   On the Boundaries of Occupation: Excavations at Burringham Road, Scunthorpe and Baldwin Avenue, Bottesford, North Lincolnshire
By Peter Boyer, Jennifer Proctor and Robin Taylor-Wilson, 2009

10   The Sea and the Marsh: The Medieval Cinque Port of New Romney revealed through archaeological excavations and historical research
By Gillian Draper and Frank Meddens, 2009

11   Pegswood Moor, Morpeth: A Later Iron Age and Romano-British Farmstead settlement
By Jennifer Proctor, 2009

12   A Roman Settlement and Bath House at Shadwell: Excavations at Tobacco Dock, London
By Alistair Douglas, James Gerrard and Berni Sudds, 2011

13   Settlement, Ceremony and Industry on Mousehold Heath: Excavations at Laurel Farm (Phase II), Broadland Business Park, Thorpe St Andrew, Norfolk
By Barry Bishop and Jennifer Proctor, 2011

14   Roman Archaeology in the Upper Reaches of the Walbrook Valley: Excavations at 6–8 Tokenhouse Yard, London EC2
By Jim Leary and Jonathan Butler, 2012

15   Faverdale, Darlington: Excavations at a major settlement in the northern frontier zone of Roman Britain
By Jennifer Proctor, 2012

16   Friars, Quakers, Industry and Urbanisation: The Archaeology of the Broadmead Expansion Project, Cabot Circus, Bristol, 2005–2008
Edited by Victoria Ridgeway and Martin Watts, 2013

17   Roman Burials in Southwark: Excavations at 52–56 Lant Street and 56 Southwark Bridge Road, London SE1
By Victoria Ridgeway, Kathelen Leary and Berni Sudds, 2013

18   Temples and Suburbs: Excavations at Tabard Square, Southwark
By Douglas Killock, John Shepherd, James Gerrard, Kevin Hayward, Kevin Rielly and Victoria Ridgeway, 2015

Contributors

Principal authors   Rebecca Haslam and Guy Thompson
Volume editors   Victoria Ridgeway and Frank Meddens
Academic advisor   Mike Neville
Project managers   Helen Hawkins and Charlotte Matthews
Post-excavation managers   Victoria Ridgeway and Frank Meddens
Graphics   Mark Roughley
Photography   Strephon Duckering
Pottery   Berni Sudds
Glass   Chris Jarrett
Building materials   Kevin Hayward
Hydraulics   Tim Smith
Archaeological reconstruction Chris Mitchell
Series editor   Victoria Ridgeway

Contents

Contributors	v
Figures	x
Summary	xx
Acknowledgements	xxi

Chapter 1 Introduction — 1

1.1	Organisation of the report	4
1.2	Geology and topography	6
1.3	Methodology and limitations of the study	6

Chapter 2 King's Cross before the Great Northern Railway — 9

2.1	The early history of King's Cross	9
2.2	Urbanisation and industrialisation between the mid-18th and the mid-19th centuries	10
2.3	The Construction of the Regent's Canal and the Imperial Gasworks at St Pancras	15
2.4	Synopsis	17

Railway politics and the formation of the Great Northern Railway — 18

Chapter 3 The Construction of the King's Cross Goods Station: 1847–1852 — 23

3.1	Preliminary ground works	27
3.2	The architect's vision	29
3.3	Construction work begins	30
3.4	The Regent's Canal (Landscape Feature 1) and the Retaining Wall (Landscape Feature 4) by 1852	31
3.5	Canal infrastructure: the Granary Basin (Landscape Feature 5), the Coal and Stone Basin (Landscape Feature 6) and associated structures (The Canal Branch, Canal Tunnels 1 to 4 and Docks 1 to 4) by 1852	33
3.6	The Main Goods Station: The Western and Eastern Transit Sheds (Landscape Features 7 and 8), the Western and Eastern Stables (Landscape Features 9 and 10), the Train Assembly Shed (Landscape Feature 11) and the Granary (Landscape Feature 12) by 1852	37
3.7	The Western Transit Shed (Landscape Feature 7) by 1852	38
3.8	The Eastern Transit Shed (Landscape Feature 8) by 1852	45
3.9	The Western and Eastern Transit Shed Stables (Landscape Features 9 and 10) by 1852	47
3.10	The Train Assembly Shed (Landscape Feature 11) by 1852	51
3.11	The Granary (Landscape Feature 12) by 1852	54
3.12	Railway infrastructure in and around the Main Goods Shed (Landscape Features 7–11) and the Granary (Landscape Feature 12) by 1852	67
3.13	The Lamp Room and Coffee Shop (Landscape Feature 13) by 1852	72
3.14	The Temporary Passenger Station (Landscape Feature 14) and the Great Northern Carriage Shed (Landscape Feature 15) by 1852	73
3.15	The Hydraulic Station (Landscape Feature 16), hydraulic machinery and the hydraulic network by 1852	76
3.16	The Coal Drops (Landscape Feature 17), the Coal Offices (Landscape Feature 18) and Wharf Road Viaduct (Landscape Feature 19) by 1852	81
3.17	Hard landscaping and weighbridges by 1852	86
3.18	The building materials and the evolution of the Great Northern's track network by 1852	87
3.19	Synopsis	88

Building the Goods Yard: the human cost — 90

Chapter 4 Competition and Congestion at the King's Cross Goods Station: c.1853–1879   93

4.1   Hard landscaping throughout the yard: the mid-1850s to the 1870s   98
4.2   The Granary Basin (Landscape Feature 5): the mid-1850s to the 1870s   99
4.3   The Inward (Eastern) and Outward (Western) Goods Offices (Landscape Features 20 and 21): the mid-1850s to the 1870s   101
4.4   The Western and Eastern Transit Sheds (Landscape Features 7 and 8): the mid-1850s to the 1870s   107
4.5   The Train Assembly Shed (Landscape Feature 11): the mid-1850s to the 1870s   114
4.6   The Granary (Landscape Feature 12): the mid-1850s to the 1870s   115
4.7   The Lamp Room and Coffee Shop, later the Refreshment Room (Landscape Feature 13): the mid-1850s to the 1870s   116
4.8   The Potato Market (Landscape Feature 14): the mid-1850s to the 1870s   118
4.9   The construction of new facilities between the Eastern Transit Shed and Midland Goods Shed: the mid-1850s to the 1870s   119
4.10   The temporary Smiths' shed (Landscape Feature 15) c.1852–8   120
4.11   The Midland Railway Goods Shed and Midland Offices (Landscape Feature 15): the late 1850s to the 1870s   120
4.12   The expansion of the coal trade and the evolution of the Coal Offices (Landscape Feature 18): the mid-1850s to the 1870s   126
4.13   The Western Coal Drops (Landscape Feature 22): the mid-1850s to the 1870s   131
4.14   The Cambridge Street Coal Drops and the Plimsoll Viaduct (Landscape Feature 23): the mid-1850s to the 1870s   132
4.15   The Gasworks Viaduct (Landscape Feature 24): the mid-1850s to the 1870s   133
4.16   The Hydraulic Station (Landscape Feature 16), hydraulic infrastructure and the hydraulic network: the mid-1850s to the 1870s   133
4.17   The Horse Provender Store (Landscape Feature 25): the mid-1850s to the 1870s   134
4.18   Synopsis   136

The shed that never sleeps: how King's Cross fed the city and fuelled the North   138

Chapter 5 The Introduction of Capstan Shunting in the early 1880s   143

5.1   The Main Goods Shed and Granary Complex (Landscape Features 7–12): the early 1880s   144
5.2   The Hydraulic Station (Landscape Feature 16): the early 1880s   150
5.3   Building materials and the evolution of the Great Northern's track network: the early 1880s   154
5.4   Synopsis   154

Innovation, bankruptcy and reinvention: Samuel Plimsoll and the Great Northern Railway   156

Chapter 6 Overcrowding at the King's Cross Goods Station during the 1880s and 1890s   159

6.1   Road vehicle infrastructure: the mid-1880s to the 1890s   161
6.2   The Eastern Transit Shed (Landscape Feature 8): the mid-1880s to the 1890s   161
6.3   The Potato Market (Landscape Feature 14) and the West and East Handyside Canopies (Landscape Features 26 and 27): the mid-1880s to the 1890s   161
6.4   The Granary (Landscape Feature 12): the mid-1880s to the 1890s   164
6.5   The Refreshment Club (Landscape Feature 13): the mid-1880s to the 1890s   165
6.6   Improvements to stables (Landscape Features 9, 10 and 25) in the Goods Yard: the mid-1880s to the 1890s   167
6.7   The relocation of the coal trade to the south of the canal (Landscape Feature 18) between the mid-1880s to the 1890s   171
6.8   The Midland Goods Shed (Landscape Feature 15): the mid-1880s to the 1890s   172
6.9   Synopsis   173

Vital cogs in a vast machine: the labour force remembered through history and archaeology   174

Chapter 7 The King's Cross Goods Station at the turn of the 20th Century: Investment and Enlargement — 179

7.1 The Inwards Goods Shed (Landscape Features 7–11): the late 1890s to the early 1900s — 181
7.2 The Granary (Landscape Feature 12): the late 1890s to the early 1900s — 185
7.3 The Western and Eastern Goods Offices (Landscape Features 20 and 21): the late 1890s to the early 1900s — 186
7.4 The Hydraulic Station (Landscape Feature 16): the late 1890s to the early 1900s — 187
7.5 The Gasworks Viaduct (Landscape Feature 24): the late 1890s to the early 1900s — 195
7.6 The sources of the building materials used at King's Cross during the late 1890s and the early 1900s — 196
7.7 Synopsis — 197

Finds from the Refreshment Club and the extension of 1908 — 200

Chapter 8 The King's Cross Goods Station during the First World War: 1914–1918 — 203

8.1 The construction of Goods Way (Landscape Feature 29) and the infilling of the Granary Basin (Landscape Feature 5): 1914–8 — 206
8.2 The Hydraulic Station (Landscape Feature 16): 1914–8 — 208
8.3 The Midland Goods Shed (Landscape Feature 15): 1914–8 — 208
8.4 Alterations to the Inwards Goods Shed (Landscape Features 7–11) and the Granary (Landscape Feature 12): 1914–8 — 210
8.5 Staff welfare facilities, 1914–8 — 210
8.6 Synopsis — 211

A short history of the horse at King's Cross Goods Yard — 212

Chapter 9 The King's Cross Goods Station during the Interwar Period: Grouping, Depression and Transposition — 219

9.1 The Granary (Landscape Feature 12) and the modernisation of goods handling: 1919–38 — 223
9.2 The Midland Goods Shed (Landscape Feature 15): 1919–38 — 226
9.3 The Hydraulic Station (Landscape Feature 16): 1919–38 — 227
9.4 The Coal, Fish and Cartage Department Offices (Landscape Feature 18) and the Road Motor Engineering Garage (Landscape Feature 30): 1919–38 — 228
9.5 The East and West Handyside Canopies (Landscape Features 26 and 27) and the fish trade: 1919–38 — 230
9.6 The Eastern Goods Shed (Landscape Features 7–11) and the Granary (Landscape Feature 12): 1919–38 — 230
9.7 Improvements to staff welfare facilities in the Western Goods Offices (Landscape Feature 20) and the Inwards Staff Messroom (later the Outwards Staff Messroom; Landscape Feature 31): 1919–38 — 237
9.8 Building materials, trade and the evolution of the LNER's track network: 1919–38 — 239
9.9 Synopsis — 239

'No human activity can be carried on without paper': the working lives of the goods station's clerks — 242

Chapter 10 The King's Cross Goods Station during the Second World War: 1939–1945 — 245

10.1 ARP structures in the Outwards Goods Shed (Landscape Features 7–11): 1939–45 — 248
10.2 The vulnerability of the Hydraulic Station (Landscape Feature 16): 1939–45 — 250
10.3 New messrooms for a changing workforce: 1939–45 — 251
10.4 Air Raid Damage at King's Cross Goods Station: 1940–45 — 251
10.5 Repairing bomb damage: 1940–45 — 254
10.6 Synopsis — 256

Chapter 11 Nationalisation and Austerity: 1946 to the mid-1950s   257

11.1	The Eastern Goods Shed (Landscape Features 7–11) and the Granary (Landscape Feature 12): 1946 to the 1950s	259
11.2	The Unclaimed Goods Warehouse (Landscape Feature 15): 1946 to the 1950s	265
11.3	The Western and Eastern Goods Offices (Landscape Features 20 and 21): 1946 to the 1950s	266
11.4	Staff welfare in the Outwards Goods and Granary Complex (Landscape Features 7–12) and the Goods Offices (Landscape Features 20 and 21): 1946 to the 1950s	266
11.5	The London Hydraulic Power Company Metering Station (Landscape Feature 32): 1946 to the 1950s	266
11.6	The RME Garage and associated messroom (Landscape Feature 30): 1946 to the 1950s	267
11.7	The Goods Department Offices (Landscape Feature 18): 1946 to the 1950s	268
11.8	Synopsis	268

King's Cross Goods Yard and the bottle trade   270

Chapter 12 The Decline of the Former Eastern Goods Yard in the late 20th Century   277

12.1	The Eastern Goods Shed and Granary Complex as a British Rail Parcels Depot (Landscape Features 7–12, 20 and 21): 1966–81	279
12.2	The Road Distribution Terminal (Landscape Features 7–12, 20 and 21): the 1980s and 1990s	280
12.3	The former Coal and Fish Offices (Landscape Feature 18): the 1960s to the 1990s	285
12.4	Re-use, neglect and abandonment: King's Cross Goods Yard: the late 1990s to the early 21st century	285
12.5	Synopsis	287

Hydraulic power at King's Cross Goods Yard   288

Chapter 13 Conclusions   297

13.1	A reconstruction of goods handling, use of space and traffic flow over time and an exploration of the driving forces of change at King's Cross	297
13.2	Britain's industrial boom. The causes and effects of infrastructure and working practice changes at King's Cross Goods Station: 1849–1913	298
13.3	Conflict and crisis. The causes and effects of infrastructure and working practice changes at King's Cross Goods Station: 1914–45	312
13.4	The industrial decline of post-war Britain. The causes and effects of infrastructure and working practice changes at King's Cross Goods Station: 1945–90	317
13.5	Developments at King's Cross Goods Yard compared and contrasted with two major 'rivals' in the capital	321
13.6	The value of a multidisciplinary approach when investigating industrial sites: case studies from King's Cross Goods Yard	325
13.7	The importance of King's Cross Goods Yard both past and present	327

Chapter 14 Epilogue   329

Appendix 1: The archaeological sites and built heritage surveys that are included in this document and an inventory of the relevant interventions   330

Appendix 2: An inventory of features (buildings, Landscape Features and infrastructure networks) discussed in this publication (arranged alphabetically)   332

Appendix 3: Bricks used at King's Cross, their fabrics and origins   333

Glossary   334

Bibliography   336

Index   348

Figures

Fig. 1.1	Site location	1
Fig. 1.2	Detailed site and Landscape Feature locations detailing the whereabouts of the built heritage surveys	2
Fig. 1.3	Detailed site and Landscape Feature locations detailing the whereabouts of the archaeological investigations	3
Fig. 1.4	Graphical conventions used to illustrate modifications to standing buildings	5
Fig. 2.1	'The great dust heap at King's Cross' by E.H. Dixon 1837	11
Fig. 2.2	The Rocque map of 1769	12
Fig. 2.3	Davies's map of 1834	13
Fig. 2.4	An elevation of the monument to King George IV	14
Fig. 2.5	North facing elevation of the south wall of the Regent's Canal (Landscape Feature 1)	16
Fig. 3.1	Parliamentary plan of 1846	24
Fig. 3.2	The locations of the Landscape Features present in the Goods Yard from *c.*1850–2	25
Fig. 3.3	The locations of the significant archaeological features dated to *c.*1850–2	26
Fig. 3.4	The layer of burnt clay ballast that was deposited in 1849–51 revealed in section	28
Fig. 3.5	Lewis Cubitt's March 1850 proposal plan of the iron roofing of the Train Assembly Shed and the Temporary Passenger Terminus	29
Fig. 3.6	Lewis Cubitt's 1851 watercolour of the Granary	30
Fig. 3.7	Captain Galton's sketch plan of October 1852 depicting the Goods Yard and its environs	32
Fig. 3.8	The Humber Plan of 1866 depicting the Goods Yard and its environs	34
Fig. 3.9	Elevations showing the entrances to Canal Tunnels 1 and 2 in the north wall of the Granary Basin	35
Fig. 3.10	Southwest facing elevation showing the former entrance to the Coal and Stone Basin (Landscape Feature 6) integral to the north wall of the Regent's Canal (Landscape Feature 1) constructed in 1850–1	36
Fig. 3.11	Cross-sectional elevation through the southern end of the Western Transit Shed (Landscape Feature 7) showing the internal face of the south wall and the original 1850 masonry	39
Fig. 3.12	External elevation of the south wall of the Western Transit Shed (Landscape Feature 7) and the original 1850 masonry	40
Fig. 3.13	External elevation of the north wall of the Western Transit Shed (Landscape Feature 7) showing the original 1850 masonry	40
Fig. 3.14	External elevation of the west wall of the Western Transit Shed (Landscape Feature 7) showing the original 1850 masonry	41
Fig. 3.15	Cross-sectional elevation through the Western Transit Shed (Landscape Feature 7) and the Western Transit Shed Stables (Landscape Feature 9) showing the internal faces of their west walls and the original 1850–1 masonry	41
Fig. 3.16	Cross-sectional elevation through the Train Assembly Shed (Landscape Feature 11) and the Granary (Landscape Feature 12) showing the external face of the east wall of the Western Transit Shed (Landscape Feature 7), the internal face of the west wall of the Granary and the original 1850–1 masonry	42

Fig. 3.17	Cross-sectional elevation through the Western Transit Shed (Landscape Feature 7) and Western Transit Shed Stables (Landscape Feature 9) showing the internal faces of their east walls and the original 1850–1 masonry	42
Fig. 3.18	The 1850–1 foundations of Platform 1 to the immediate east of the Western Transit Shed Stables Ramp	43
Fig. 3.19	The 1850–1 foundations of Platform 1 in the Western Transit Shed	43
Fig. 3.20	An engraving from *The Illustrated London News* of 28th May 1853 (Anon 1853) showing the Western Transit Shed (Landscape Feature 7) in operation shortly after the yard fully opened for business	44
Fig. 3.21	External elevation of the east walls of the Eastern Transit Shed (Landscape Feature 8) and the Granary (Landscape Feature 12) showing the original 1850–1 masonry	45
Fig. 3.22	Cross-sectional elevation through the Train Assembly Shed (Landscape Feature 11) showing the external face of the west wall of the Eastern Transit Shed (Landscape Feature 8), the internal face of the east wall of the Granary (Landscape Feature 12) and the 1850–1 masonry	46
Fig. 3.23	Circular niche incorporating an 1850–1 turntable base in the Eastern Transit Shed (Landscape Feature 8)	46
Fig. 3.24	Semi-ovoid niche incorporating three 1850–1 turntable bases in the Eastern Transit Shed (Landscape Feature 8)	47
Fig. 3.25	Basement plan of the Western Stables (Landscape Feature 9) *c*.1851 reconstructed from archaeological remains, built heritage evidence and cartographic sources	48
Fig. 3.26	Ramp, constructed in 1850–1, providing a link between the Western Transit Shed (Landscape Feature 7) and the underground stables (Landscape Feature 9)	49
Fig. 3.27	Tunnel, constructed in 1850–1, connecting the Western Transit Shed Stables (Landscape Feature 9) with the low ground on the western side of the Retaining Wall (Landscape Feature 4)	49
Fig. 3.28	The Western Transit Shed Stables (Landscape Feature 9)	49
Fig. 3.29	Original 1850–1 horse trough against the east wall at the south end of the Western Transit Shed Stables (Landscape Feature 9)	49
Fig. 3.30	Tack room of 1850–1 at the north end of the Western Transit Shed Stables (Landscape Feature 9)	49
Fig. 3.31	The foundations of a turntable support of 1850–1 in the east wall of the Western Transit Shed Stables (Landscape Feature 9)	49
Fig. 3.32	Basement plan of the Eastern Transit Shed Stables (Landscape Feature 10) *c*.1851 reconstructed from archaeological remains, built heritage evidence and cartographic sources	50
Fig. 3.33	The Train Assembly Shed (Landscape Feature 11) looking towards the Granary (Landscape Feature 12)	52
Fig. 3.34	External elevation of the north sides of the Western Transit Shed (Landscape Feature 7), the Eastern Transit Shed (Landscape Feature 8), the Train Assembly Shed (Landscape Feature 11) and the Granary (Landscape Feature 12) showing the original 1850–1 masonry	53
Fig. 3.35	The Train Assembly Shed (Landscape Feature 11) after the removal of the roof; note six of the eight 1850–1 hollow cast iron columnar roof supports in the right (west) side of the image and the three masonry turntable supports of the same date in the foreground	53
Fig. 3.36	Maker's mark on an early 1850s column within the Train Assembly Shed (Landscape Feature 11)	54
Fig. 3.37	The Granary (Landscape Feature 12), flanked by the Eastern and Western Goods Offices (both later additions)	55
Fig. 3.38	The timber chutes of 1850–1 that were used to convey sacks of grain from the upper floors of the Granary (Landscape Feature 12) through the easternmost ground floor window in the south wall to waiting carts	56

Fig. 3.39 External elevation of the south wall of the Granary (Landscape Feature 12) showing the original 1850–1 masonry   57

Fig. 3.40 Cross-sectional elevation through the Granary (Landscape Feature 12) showing the internal face of its south wall and the original 1850–1 masonry   57

Fig. 3.41 An example of the 1850–1 hoist mechanisms and hipped canopies that surmounted the Granary's (Landscape Feature 12) fifth floor loading bays   58

Fig. 3.42 External elevation of the north wall of the Granary (Landscape Feature 12) showing the original 1850–1 masonry   58

Fig. 3.43 Southeast stairwell in the Granary (Landscape Feature 12)   59

Fig. 3.44 Iron framework around the base of a column in the Granary (Landscape Feature 12), which was used as a fairlead   59

Fig. 3.45 Roofing plan of the Granary (Landscape Feature 12) showing the original 1850–1 elements   60

Fig. 3.46 Central roof valley of the Granary (Landscape Feature 12) surmounted by infrastructure associated with the sack hoists' hydraulic winches of 1850–1   60

Fig. 3.47 Housing for the sack hoists' hydraulic winches of 1850–1 on the roof of the Granary (Landscape Feature 12)   60

Fig. 3.48 Gate on the second floor of the southwest stairwell of the Granary (Landscape Feature 12)   61

Fig. 3.49 Cross-sectional elevation through the Granary (Landscape Feature 12) showing the internal face of its northern wall and 1850–1 masonry   61

Fig. 3.50 Roof trusses of 1850–1 in the Granary (Landscape Feature 12)   62

Fig. 3.51 Cross-sectional elevation through the eastern stairs of 1850–1 that led to the central roof valley of the Granary (Landscape Feature 12) showing their southern wall   63

Fig. 3.52 Schematic diagram showing the distribution of trapdoors and chutes across all floors of the Granary (Landscape Feature 12) based on the scarring that was observed in the floors and ceilings of the building   64

Fig. 3.53 Reconstruction of the rail and canal infrastructure and the trapdoor and chute network inside the Granary (Landscape Feature 12) *c.*1851 (Landscape Feature 12)   65

Fig. 3.54 An engraving from The Illustrated London News of 28th May 1853 showing horse-drawn vehicles lining up against the southern wall of the Granary (Landscape Feature 12) to receive sacks of grain; two barges sail into the building via Canal Tunnels 2 and 3 (Anon 1853)   66

Fig. 3.55 Turntable A of 1850–2   69

Fig. 3.56 Turntable B of 1850–2   69

Fig. 3.57 Internal view of the turning mechanism of Turntable B of 1850–2   69

Fig. 3.58 The ground floor of the Main Goods Station (i.e. Landscape Features 7–11) and the Granary (Landscape Feature 12)   71

Fig. 3.59 Plan of the Lamp Room and Coffee Shop (Landscape Feature 13) *c.*1852   72

Fig. 3.60 Queen Victoria and Prince Albert at the Temporary Passenger Station (Landscape Feature 14) in August 1851   73

Fig. 3.61 External elevation of the east wall of the Carriage Shed (Landscape Feature 15) showing the original 1850 masonry   75

Fig. 3.62 Crane Base 1 of 1851 exposed by excavation   78

Fig. 3.63 Plan of the Hydraulic Station (Landscape Feature 16) *c.*1852   79

Fig. 3.64 The south elevation of the Coal Offices (Landscape Feature 18) as seen from the Regent's Canal   81

Fig. 3.65	Floor plans of the Coal Offices (Landscape Feature 18) *c.*1851	83
Fig. 3.66	External elevation of the north wall of the Coal Offices (Landscape Feature 18) showing original 1850–1 masonry	84
Fig. 3.67	External elevation of the south wall of the Coal Offices (Landscape Feature 18) showing original 1850–1 masonry	84
Fig. 3.68	External elevation of the east wall of the Coal Offices (Landscape Feature 18) showing original 1850–1 masonry	85
Fig. 3.69	Arched entranceway of 1851 (since infilled) integral to the north wall of the Regent's Canal (Landscape Feature 1) leading to the stables in the vaults below the Wharf Road Viaduct (Landscape Feature 19)	86
Fig. 4.1	The recently constructed junction between the Great Northern and Metropolitan railways at King's Cross, which enabled the former company to carry their goods to and from the heart of the City (Anon 1868), taken from The Illustrated London News of 1868	95
Fig. 4.2	The locations of the Landscape Features present in the Goods Yard from the mid 1850s to the 1870s	96
Fig. 4.3	The locations of the significant archaeological features dated to the mid 1850s to the 1870s	97
Fig. 4.4	The Humber Plan of 1866 detailing the southern end of the Main Goods Shed (Landscape Features 7–11), the Granary (Landscape Feature 12) and the Granary Basin (Landscape Feature 5)	99
Fig. 4.5	The Ordnance Survey Map of 1871 depicting the Goods Yard and its environs	100
Fig. 4.6	The Crockett Plan of 1873 detailing the southern end of the Main Goods Station (Landscape Features 7–11) and the Granary (Landscape Feature 12)	101
Fig. 4.7	Plan of the Western Goods Offices (Landscape Feature 20) by 1866	102
Fig. 4.8	External elevation of the south wall of the Western Goods Offices (Landscape Feature 20) showing the changes that took place from 1858–66	104
Fig. 4.9	Plan of the Eastern Goods Offices (Landscape Feature 21) by 1873	105
Fig. 4.10	External elevation of the south wall of the Eastern Goods Offices (Landscape Feature 21) showing the changes that took place from 1858–73	106
Fig. 4.11	External elevation of the east wall of the Eastern Goods Offices (Landscape Feature 21) showing the changes that took place from 1865–71	107
Fig. 4.12	A diagonal scar was just visible running down the west face of the Western Goods Offices (Landscape Feature 20), caused by a structure that was built against it and the southern face of the Western Transit Shed (Landscape Feature 7) at some point after 1858	109
Fig. 4.13	The Great Northern Goods Station Plan of 1882 depicting the Goods Yard and its environs	109
Fig. 4.14	External elevation of the north wall of the Western Transit Shed (Landscape Feature 20)	110
Fig. 4.15	The clock mechanism inside its housing in the north wall of the Western Transit Shed (Landscape Feature 7)	110
Fig. 4.16	Cross-sectional elevation through the Train Assembly Shed (Landscape Feature 11)	112
Fig. 4.17	Cross-sectional elevation through the Eastern Transit Shed (Landscape Feature 8) showing the changes that took place from 1858–73	112
Fig. 4.18	External elevation of the north wall of the Eastern Transit Shed (Landscape Feature 8) showing the changes that took place from *c.*1853–72	113
Fig. 4.19	External elevation of the south wall of the Granary (Landscape Feature 12) and the Goods Offices (Landscape Features 20 and 21) showing the changes that took place from 1858–73	115

Fig. 4.20	Plan of the Refreshment Room by 1872 (Landscape Feature 13)	117
Fig. 4.21	Porters moving sacks of potatoes in the Potato Market (Landscape Feature 14) in 1951	118
Fig. 4.22	An engraving of the Potato Market taken from the Illustrated Times of October 1st 1864	119
Fig. 4.23	The Ordnance Survey Map of 1871 detailing the Midland Goods Shed (Landscape Feature 15)	121
Fig. 4.24	External elevation of the west wall of the Midland Goods Shed (Landscape Feature 15) showing the changes of 1858–77	123
Fig. 4.25	External elevation of the north wall of the Midland Goods Shed (Landscape Feature 15) showing the changes of 1858–77	124
Fig. 4.26	External elevation of the east wall of the Midland Goods Shed (Landscape Feature 15) showing the changes of 1858–77	124
Fig. 4.27	The Ordnance Survey Map of 1894–6 detailing the Midland Goods Shed (Landscape Feature 15)	126
Fig. 4.28	Ground, first and second floor Great Northern Railway plan of the Coal Offices (Landscape Feature 18) printed in 1860	127
Fig. 4.29	The extension to the Coal Offices (Landscape Feature 18) seen from the south bank of the Regent's Canal	128
Fig. 4.30	External elevation of the south wall of the Coal Offices (Landscape Feature 18) showing Blocks 3 to 5 and the changes of 1860	129
Fig. 4.31	External elevation of the north wall of the Coal Offices (Landscape Feature 18) showing Blocks 3 to 5 and the changes of 1860	130
Fig. 4.32	The Coal Offices (Landscape Feature 18), with the west (side) elevation of Block 5 in the foreground	131
Fig. 4.33	Stanford's Map of 1862 detailing the Hydraulic Station (Landscape Feature 16) and the Horse Provender Store (Landscape Feature 25)	134
Fig. 4.34	Plan of the western end of the Horse Provender Store (Landscape Feature 25) by 1854	135
Fig. 5.1	The locations of the landscape features present in the Goods Yard from 1881–3	144
Fig. 5.2	The locations of the significant archaeological features dated 1881–3	145
Fig. 5.3	Great Northern Railway Plan of 1882, detailing the Main Goods Station (Landscape Features 7–11), the Granary (Landscape Feature 12) and the Granary Basin (Landscape Feature 5)	147
Fig. 5.4	Ordnance Survey Second Edition 1894–6 detailing the Main Goods Station (Landscape Features 7–11), the Granary (Landscape Feature 12) and the Granary Basin (Landscape Feature 5)	147
Fig. 5.5	External elevation of the north wall of the Granary (Landscape Feature 12)	148
Fig. 5.6	Hydraulic Capstan 6 with *in situ* mechanism, unearthed in the northern end of the Train Assembly Shed (Landscape Feature 11)	148
Fig. 5.7	Capstanmen in 1951, photographed in the process of turning a wagon in the King's Cross Potato Market (Landscape Feature 14) through the use of a hydraulic capstan and snatch head	149
Fig. 5.8	Accumulator bases in the 1881–2 Accumulator Tower of the Hydraulic Station (Landscape Feature 16)	152
Fig. 5.9	Plan of the Hydraulic Station (Landscape Feature 16) by 1882	153
Fig. 6.1	The locations of the Landscape Features that were present in the Goods Yard from the mid 1880s to the mid 1890s	160
Fig. 6.2	The West Handyside Canopy (Landscape Feature 26) abutting the eastern wall of the Eastern Transit Shed (Landscape Feature 8)	163
Fig. 6.3	The West Handyside Canopy (Landscape Feature 26) after the removal of the roof covering abutting the western wall of the Midland Goods Shed (Landscape Feature 15)	163

xv

Fig. 6.4	Wooden gable added to the east wall of the Eastern Transit Shed (Landscape Feature 8) to enable it to support the upper reaches of the West Handyside Canopy (Landscape Feature 26)	163
Fig. 6.5	Internal elevation of the east side of the partition that separates the West and East Handyside Canopies (Landscape Features 26 and 27) showing one of the telegraph poles that was probably instated in 1897	164
Fig. 6.6	Cross-sectional elevation through the Granary (Landscape Feature 12) showing the internal face of its southern wall and the changes of 1892–3	165
Fig. 6.7	Great Northern Railway Plan of June 1888 showing proposed alterations to the ground and first floors of the Refreshment Club (Landscape Feature 13)	166
Fig. 6.8	The Ordnance Survey Map of 1894–6 detailing the modifications that were made to the Refreshment Club (Landscape Feature 13) in order to accommodate the new smoking rooms	167
Fig. 6.9	Reconstruction of the Refreshment Club (Landscape Feature 13) by 1888	167
Fig. 6.10	Plan of 1906 showing the Horse Provender Store (Landscape Feature 25) after its conversion to stables	168
Fig. 6.11	Plan of the ground floor of the Horse Provender Store and Stables (Landscape Feature 25) by *c*.1891	168
Fig. 6.12	Plan of the Western and Eastern Stables (Landscape Features 9 and 10) by 1898	170
Fig. 6.13	External elevation of the north wall of Block 5 of the Coal Offices (Landscape Feature 18) showing the changes of 1898	171
Fig. 7.1	The locations of the landscape features present in the Goods Yard from 1897–1913	180
Fig. 7.2	The western end of the newly constructed 'Western Goods Shed' (Landscape Feature 28) fronting the north bank of the Regents Canal	182
Fig. 7.3	The Great Northern Railway Plan of 1905 detailing the Eastern Goods Shed (Landscape Features 7 to 11), the Granary (Landscape Feature 12) and the Granary Basin (Landscape Feature 5)	183
Fig. 7.4	Cross-sectional elevation through the Train Assembly Shed (Landscape Feature 11) showing the external face of the east wall of the Western Transit Shed (Landscape Feature 7) and the changes of 1898–9	183
Fig. 7.5	External elevation of the north wall of the Western Transit Shed (Landscape Feature 7) showing the changes of 1898–9	184
Fig. 7.6	External elevation of the western side of the north wall of the Granary (Landscape Feature 12) showing the changes of 1898–9	185
Fig. 7.7	Cross-sectional elevation (looking west) showing the changes of 1898–9	186
Fig. 7.8	Great Northern Railway Plan of 1905 detailing the Hydraulic Station (Landscape Feature 16)	188
Fig. 7.9	The 1906 Plan detailing the Hydraulic Station (Landscape Feature 16)	188
Fig. 7.10	The remodelled Hydraulic Station (Landscape Feature 16) by 1899	189
Fig. 7.11	The Goad Fire Insurance Plan of 1921 detailing the Hydraulic Station (Landscape Feature 16)	191
Fig. 7.12	East facing section showing the partially rebuilt external wall of the Hydraulic Station (Landscape Feature 16) in the location of the door and stairwell showing the changes of 1898–9	194
Fig. 7.13	The Hydraulic Station's Boiler Room (Landscape Feature 16) of 1898–9 under excavation	195
Fig. 7.14	The Hydraulic Station (Landscape Feature 16) and associated structures in plan by 1906 reconstructed from archaeological and cartographic evidence	195
Fig. 7.15	The largely redundant Granary Basin standing empty	197
Fig. 8.1	Wagon used for carriage of gunpowder and explosives	204
Fig. 8.2	The locations of the Landscape Features that were present in the Goods Yard from 1914–8	205

Fig. 8.3	Great Northern Railway Plan of August 1915 detailing the proposed alterations to the water supply from the Regent's Canal	207
Fig. 8.4	Great Northern Railway Plan of October 1916 detailing alterations to the water supply from the Regent's Canal	208
Fig. 8.5	External elevation of the west wall of the former Midland Goods Shed (Landscape Feature 15) showing the changes of 1915	209
Fig. 8.6	External elevation of the east wall of the former Midland Goods Shed (Landscape Feature 15) showing the changes of 1915	209
Fig. 9.1	The locations of the Landscape Features that were present in the Goods Yard from 1919–38	220
Fig. 9.2	The locations of the significant archaeological features dated from 1919–38	221
Fig. 9.3	A Ransomes and Rapier mobile crane in operation at King's Cross Goods Yard in 1949	224
Fig. 9.4	A sugar beet lineside storage yard	225
Fig. 9.5	External elevation of the eastern side of the north wall of the Granary (Landscape Feature 12) showing the windows that were modified when the lift was inserted in 1926	226
Fig. 9.6	Plan of the Hydraulic Station (Landscape Feature 16) after electrification in 1927	227
Fig. 9.7	LNER Plan of King's Cross Goods Station of 1933 detailing the RME Garage (Landscape Feature 30)	229
Fig. 9.8	LNER Plan of proposed alterations to the Outwards Goods Shed of 1942 illustrating the track arrangement in the Eastern Goods Shed (Landscape Features 7 to 11) and the Granary (Landscape Feature 12) in the late 1930s	231
Fig. 9.9	LNER Plan of the Cowans Sheldon electric traverser for the Outwards Goods Shed (Landscape Features 7–11) of September 1936	232
Fig. 9.10	One of the Cowans Sheldon Electric Traversers in operation in the Train Assembly Shed (Landscape Feature 11) in 1938	233
Fig. 9.11	Cross-sectional elevation through the northern end of the Western Transit Shed (Landscape Feature 7) showing the internal face of the east wall and the changes of 1935–6	233
Fig. 9.12	Remnants of the traverser support of 1935–6 exposed by excavation	234
Fig. 9.13	The 1935–6 electric motor that powered the traversers in the Western Transit Shed (Landscape Feature 7)	234
Fig. 9.14	Cross-sectional elevation through the southern end of the Western Transit Shed (Landscape Feature 7) and the Western Transit Shed Stables (Landscape Feature 9) showing their east walls and the changes of 1935–6	235
Fig. 9.15	LNER Plan of King's Cross Goods Station of 1933 detailing the Inwards Staff Messroom (Landscape Feature 31)	236
Fig. 9.16	External elevation of the western wall of the Inwards Staff Messroom, later the Outwards Staff Messroom (Landscape Feature 31) showing the changes 1924–36	237
Fig. 9.17	External elevation of the eastern wall of the Inwards Staff Messroom, later the Outwards Staff Messroom (Landscape Feature 31) showing the changes of 1924–36	237
Fig. 9.18	External elevation of the north wall of the Inwards Staff Messroom, later the Outwards Staff Messroom (Landscape Feature 31) showing the changes of 1924–36	238
Fig. 9.19	External elevation of the south wall of the Inwards Staff Messroom, later the Outwards Staff Messroom (Landscape Feature 31) showing the changes of 1924–36	238
Fig. 10.1	A naval gun is conveyed by special train from Catterick Bridge to Woolwich on 24th July 1946	246

Fig. 10.2	The locations of the Landscape Features and ARP structures that were present in the Goods Yard from 1939–45	247
Fig. 10.3	External elevation of the south wall of the Western Transit Shed (Landscape Feature 7) showing the changes of 1939–42 that included the construction of an air raid shelter	248
Fig. 10.4	The Western and Eastern Transit Shed Shelters (Landscape Features 9 and 10) in plan *c.*1942	249
Fig. 10.5	The damaged walls of the Potato Market following the air raid in Delhi Street, taken October 1940	251
Fig. 10.6	The Goods Yard in the immediate aftermath of the air raid that took place on 9th November 1940	251
Fig. 10.7	The remains of the Eastern Goods Offices (Landscape Feature 8) in the wake of the air raid of 9th November 1940	252
Fig. 10.8	The damage that was inflicted upon No. 1 Office on 9th November, 1940	252
Fig. 10.9	Composite elevation incorporating a rectified photograph of the external side of the east wall of the Granary (Landscape Feature 12) showing shrapnel damage from the 1940 bomb blast	252
Fig. 10.10	Bomb damage to the RME Garage (Landscape Feature 30) and the Granary Basin car park incurred on 9th November 1940	253
Fig. 10.11	External elevation of the south wall of the Eastern Goods Offices (Landscape Feature 21) showing the repair work that was carried out in the wake of the bomb blast of 9th November 1940	254
Fig. 10.12	External elevation of the east wall of the Eastern Goods Offices (Landscape Feature 21) showing the repair work that was carried out in the wake of the bomb blast of November 1940	255
Fig. 11.1	The newly introduced 'door to door' express services for modern containerised traffic, one of the few improvements made to freight transportation services in the 1950s	258
Fig. 11.2	The locations of the landscape features present in the Goods Yard from 1946 to the mid 1950s	259
Fig. 11.3	A collection of LNER plans and sections from 1946 entitled 'proposed reconstruction of the Outwards Goods Shed Roofs (A and G Banks)'	261
Fig. 11.4	Detail of a roof truss taken from a collection of LNER plans and sections from 1946 entitled 'proposed reconstruction of the Outwards Goods Shed Roofs (A and G Banks)'	261
Fig. 11.5	A collection of LNER plans and sections from 1952 entitled 'King's Cross Granary provision of lift general drawing for builders work'	262
Fig. 11.6	Cross-sectional elevation through the Western Transit Shed (Landscape Feature 7) showing the internal face of the south wall and the changes of 1948	263
Fig. 11.7	External elevation of the south wall of the Western Transit Shed (Landscape Feature 7) showing the changes of 1948	263
Fig. 11.8	External (left) and internal (right) elevations of the north wall of the Granary (Landscape Feature 12) showing the changes of 1952	264
Fig. 11.9	LNER plan of *c.*1955 detailing the ground floor of the Granary (Landscape Feature 12) and the lift shafts that were inserted in 1952 and 1955	264
Fig. 11.10	The London Hydraulic Power Company Metering Station (Landscape Feature 32) exposed by excavation	267
Fig. 12.1	The locations of the landscape features present in the Goods Yard in the latter half of the 20th century	278
Fig. 12.2	NCL Trailers parked by the Western Transit Shed (Landscape Feature 7) in 1975	280
Fig. 12.3	External elevation of the west wall of the Western Transit Shed (Landscape Feature 7) showing the modifications that were made during the late 20th century	281

Fig. 12.4 External elevation of the north wall of the Western Transit Shed (Landscape Feature 7) showing the modifications that were made during the late 20th century — 282

Fig. 12.5 External elevation of the south walls of the Western Transit Shed (Landscape Feature 7) and the Western Goods Offices (Landscape Feature 20) showing the modifications that were made during the late 20th century — 282

Fig. 12.6 Cross-sectional elevation through the southern end of the Western Transit Shed (Landscape Feature 7) showing the internal face of the west wall and the changes of the late 20th century — 283

Fig. 12.7 Cross-sectional elevation through the Western Transit Shed (Landscape Feature 7) and the Western Transit Shed Stables (Landscape Feature 9) — 283

Fig. 12.8 External elevation of the east wall of the Eastern Transit Shed (Landscape Feature 8) showing late 20th century modifications — 284

Fig. 12.9 External elevation of the north wall of the Eastern Transit Shed (Landscape Feature 8) showing later 20th century modifications — 284

Fig. 13.1 Incoming and outgoing Commodities in the Goods Yard *c.*1852 overlain on Captain Galton's Sketch Plan of October 1852 — 299

Fig. 13.2 Incoming and Outgoing Commodities in the Goods Yard *c.*1866 overlain on the Humber Plan of 1866 — 300

Fig. 13.3 Incoming and Outgoing Commodities in the Goods Yard *c.*1871 overlain on the Ordnance Survey Plan of 1871 — 301

Fig. 13.4 Incoming and Outgoing Commodities in the Goods Yard *c.*1882 overlain on the Great Northern Goods Station Plan of 1882 — 302

Fig. 13.5 Incoming and Outgoing Commodities in the Goods Yard *c.*1894–6 overlain on the Ordnance Survey Map of 1894–6 — 303

Fig. 13.6 Incoming and Outgoing Commodities in the Goods Yard *c.*1905 overlain on the Great Northern Railway Station Plan of 1905 — 304

Fig. 13.7 Reconstruction of the probable goods management procedures and goods flow through the Main Goods Shed (Landscape Features 7–11) and the Granary (Landscape Feature 12) *c.*1852 — 306

Fig. 13.8 Reconstruction of the probable goods management procedures and goods flow through the Main Goods Shed (Landscape Features 7–11) and the Granary (Landscape Feature 12) *c.*1879 — 307

Fig. 13.9 Reconstruction of the probable goods management procedures and goods flow through the Main Goods Shed (Landscape Features 7–11) and the Granary (Landscape Feature 12) *c.*1882 — 308

Fig. 13.10 Reconstruction of the probable goods management procedures and goods flow through the Eastern Goods Shed (Landscape Features 7–11) and the Granary (Landscape Feature 12) *c.*1899 — 309

Fig. 13.11 Reconstruction of the Granary — 310

Fig. 13.12 Incoming and outgoing commodities in the Goods Yard *c.*1921 overlain on the Goad Fire Insurance Plan of 1921 — 313

Fig. 13.13 Incoming and outgoing commodities in the Goods Yard *c.*1950 overlain on the British Rail Plan of 1950 — 314

Fig. 13.14 Reconstruction of the probable goods management procedures and goods flow through the Eastern Goods Shed (Landscape Features 7–11) and the Granary (Landscape Feature 12) *c.*1935 — 315

Fig. 13.15 Incoming and outgoing commodities in the Goods Yard *c.*1953 overlain on the Ordnance Survey Map of 1953 — 318

Fig. 13.16 Incoming and outgoing commodities in the Goods Yard *c.*1982–3 overlain on the Ordnance Survey Map of 1982–3 — 319

Fig. 13.17 Reconstruction of the probable Goods Management procedures and goods flow through the Eastern Goods Shed (Landscape Features 7–11) and the Granary (Landscape Feature 12) *c*.1953   320

Fig. 13.18 Reconstruction of the probable Goods Management procedures and goods flow through the Parcels Depot (Landscape Features 7–12) *c*.1983   321

Fig. 13.19 The granary after regeneration. Water features in front of the renovated Granary building mirror the location of the former granary basin   328

Fig. 14.1 The King's Cross regeneration area by night, seen from the south, with the Regent's Canal in the foreground and the Granary building at the centre   329

Summary

The decision in the early 2000s to redevelop a swathe of the former railway lands to the north of King's Cross Station presented a unique opportunity to undertake an in-depth archaeological, built heritage and historical investigation of this important suburb of London. Forming the first monograph in a series of three, this volume targets the former site of the Great Northern Railway's first goods terminus in the capital. Archaeological and built heritage work pertaining to that area of investigation was carried out between November 2006 and July 2009 by Pre-Construct Archaeology Ltd (PCA).

The below-ground investigations comprised a series of interventions that included open area excavations, trial pits and archaeological monitoring (i.e. watching briefs), the extent of which was dictated by the needs of the redevelopment project. Those works were carried out in tandem with a series of built heritage surveys that recorded to the highest standard all of the extant structures that were situated within the confines of the former Goods Station prior to their demolition or alteration. In combination with an extensive programme of historic research this has enabled a detailed chronology of the archaeology and history of the site to be produced.

The archaeological record relating to the area of investigation commences in 1849 when the Great Northern Railway began to shape the future location of their London goods railhead. That construction project demanded that an extensive programme of landscaping work be carried out, which resulted in the creation of an entirely man-made topography. A side effect of that undertaking was the concealment or obliteration of any earlier archaeological horizons. As a consequence, the research that is detailed herein primarily focuses upon the archaeology and history of the area from the mid 19th century onwards.

This volume begins by presenting the circumstances, methodologies, aims and objectives of the fieldwork, the layout and themes that are presented herein, a summary of the geology and topography of the King's Cross area and a discussion of the strengths and weaknesses of the study (Chapter 1). It goes on to present a brief history of the development and industrialisation of King's Cross and the surrounding areas and explores why that district was chosen by the Great Northern Railway as the future site of their goods and passenger termini (Chapter 2). A detailed chronology of the area of investigation is then presented, from the foundation of the Goods Station in 1849–52 (Chapter 3), its expansion during the latter half of the 19th century and the first decade of the 20th (Chapters 4 to 7), the impact of the two World Wars and the period of austerity that separated them (Chapters 8 to 10), and the causes and effects of its gradual demise in the latter half of the 20th century (Chapters 11 and 12). The concluding chapter presents a summary of how, when and why the yard evolved in the ways that it did and the effects that this had upon traffic and goods flow through the complex before going on to compare those findings with what is known about two of the yard's major competitors, the London and North Western's yard in Camden (formerly the London and Birmingham) and the Midland Railway's depot in Somers Town (Chapter 13). The results of that exercise are then used to examine wider trends in goods handling in London and to explore the value of a combined archaeological and historical approach when investigating industrial sites of this nature. Concluding remarks focus upon the significance of this historically important complex not only to the capital but to the nation at large.

Acknowledgements

The high level of archaeological and built heritage recording that was undertaken at King's Cross was made possible thanks to the input of several organisations. First and foremost, thanks must be awarded to King's Cross Partnership, who generously funded both the fieldwork and the post-excavation analysis that culminated in the publication of this book. They also masterminded the sympathetic redevelopment of the King's Cross area and are therefore responsible for the preservation and restoration of the Goods Station for posterity. Particular thanks are awarded to Rebecca Bennett and Nick Foster of Argent for their help and feedback during the compilation of this monograph and for the provision of a selection of archive images. Thanks also to Bridget Evans, Joanna Facer, Richard Meier and Ken Trew of Argent and King's Cross Partnership for facilitating the on-site work.

Pre-Construct Archaeology Ltd would like to thank Richard Hughes, Michael Bussell and Ela Palmer of International Heritage Conservation and Management (IHCM Ltd) for commissioning a swathe of the excavation and built heritage work on behalf of Kings Cross Partnership. Thanks in particular to Michael Bussell for sharing his extensive knowledge regarding the industrial heritage of King's Cross and for his helpful feedback throughout. Kim Stabler of the Greater London Archaeology Service (GLAAS) is to be thanked for monitoring the fieldwork on behalf of English Heritage and the London Borough of Camden as is Michelle O'Doherty, Senior Conservation Officer at the London Borough of Camden, for her guidance regarding the built heritage element of the redevelopment.

The authors would like to thank Helen Hawkins, Charlotte Matthews and Alex Rose-Deacon of Pre-Construct Archaeology for their project management and Victoria Ridgeway and Frank Meddens, also of Pre-Construct Archaeology, for their post-excavation management and editing. This publication text was prepared by archaeologist Rebecca Haslam, who was primarily responsible for undertaking the stratigraphic and built heritage analysis and historian Guy Thompson, who carried out an extensive programme of primary archival research. Particular thanks are awarded to Mark Roughley, who devised the innovative figures that are presented herein and Cate Davies, who prepared the schematic diagrams for the hydraulics chapter and typeset this document. Thanks are also due to Pre-Construct Archaeology's team of specialist contributors, without whom such a well-rounded and well-presented account of the Goods Station's history would not have been possible. Strephon Duckering (finds photography), Marit Gaimster (small finds), Kevin Hayward (brick and worked stone), Chris Jarrett (post-medieval glass) and Berni Sudds (post-medieval pottery) are therefore acknowledged for their input. External specialists Tim Smith and Andy Guy are awarded thanks for providing hydraulic and railway advice, the acquisition of which was vital to the successful interpretation of the industrial remains that were encountered, as is artist Chris Mitchell for his wonderful reconstruction of the Granary (Fig. 13.11). The authors would also like to thank the staff of the Camden Local Studies and Archives Centre, the Nation Archives at Kew, the British Library, the British Museum, the Corporation of London Guildhall Library, the Institution of Civil Engineers Archive, the London Metropolitan Archive, the London School of Economics, the National Portrait Gallery, the Network Rail Record Group at York, the Science and Society Picture Library, the Tyne and Wear County Archives and the Wellcome Library.

The on-site archaeological work was undertaken by an enormous team of practitioners from Pre-Construct Archaeology who are too numerous to name in full here, however this publication could not have come to fruition without their hard work and dedication. Particular thanks are due to archaeological fieldwork director Shane Maher and assistant supervisor Tomasz Mazurkiewicz, who oversaw the KXI07 and KXP08 phases of the archaeological works; their post-excavation assistant, Richard Archer, was responsible for checking and ordering the associated archives. Co-author Rebecca Haslam directed the KXO08 excavation and watching brief and checked the resulting archive. Built heritage recording was overseen by Kari Bower, Malcolm Gould, Daniel Graham, Amanda Hayhurst, Daniel Jackson, Paul McGarrity, Adam Garwood, Tom O'Gorman, Rhiannon Rhys, Helen Robertson and Tudor Skinner (KXC06, KXD07, KXE08, KXF07, KXK08 and KXM08). Thanks are also due to Strephon Duckering and Edwin Baker for the onsite photography, Nathalie Barrett, Phil Frickers, Aidan Turner and Jeremy Rogers for the surveying and Lisa Lonsdale for her technical and logistical support.

Finally, many thanks to Mike Neville, for reading and producing useful feedback and comment on previous versions of this text.

Aerial photograph of the railway lands surrounding King's Cross and St Pancras Passenger Stations, during redevelopment works, looking east, with the former Great Northern Railway Goods Station centre left

CHAPTER 1

Introduction

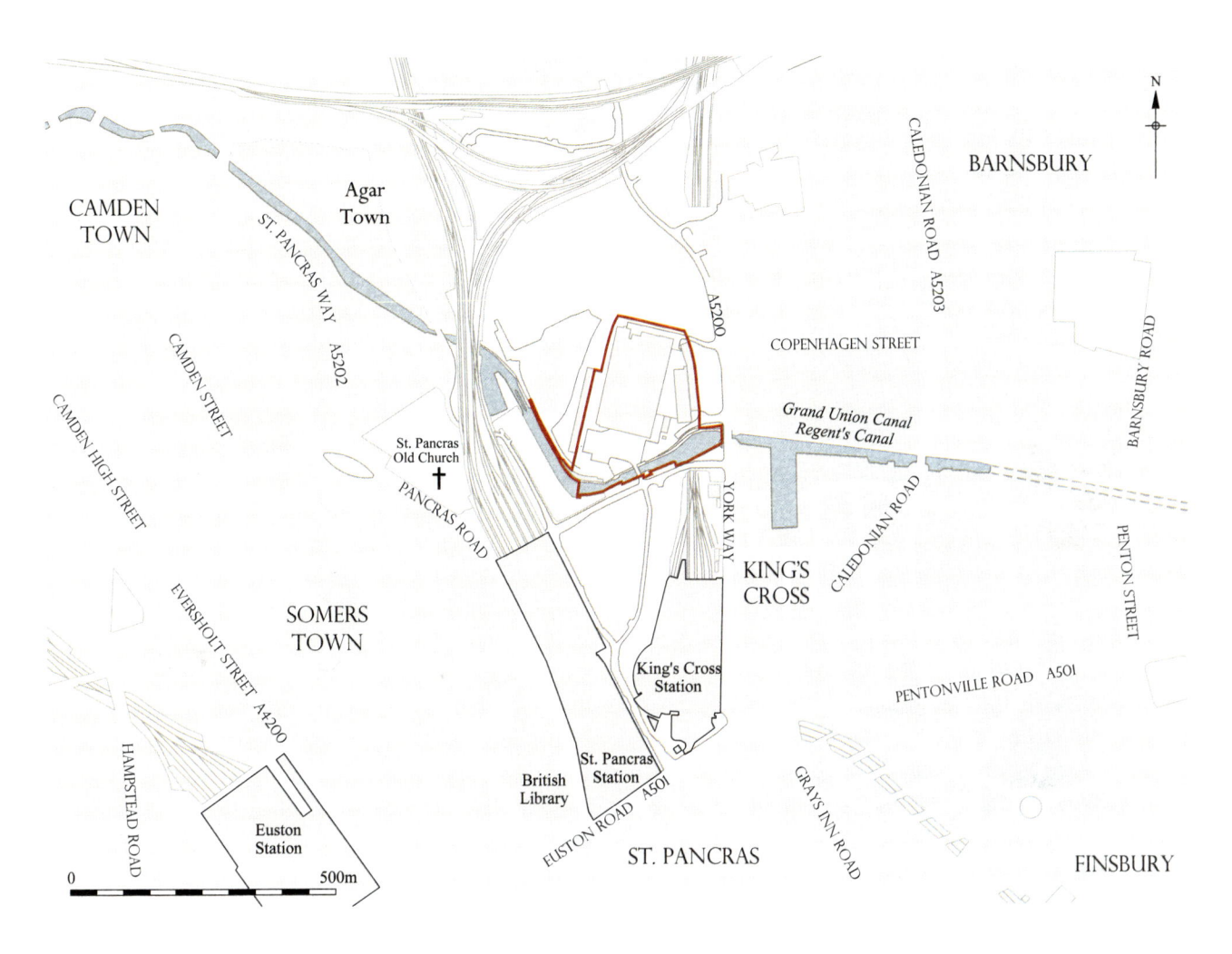

Fig. 1.1   Site location, scale 1:12,500
© Crown copyright 2016 Ordnance Survey 100020795

Between November 2006 and July 2009 extensive work was undertaken to transform the area around the Great Northern Railway Company's (GNR) former Goods Station at King's Cross providing a mixed development of commercial units, buildings forming part of the University of the Arts London and public open spaces. Many extant buildings falling within the site boundaries were renovated and modified and some new structures were erected. Pre-Construct Archaeology (PCA) monitored all the subterranean ground works that were carried out during the redevelopment so that any archaeology exposed in the process could be recorded prior to its destruction, a mode of working that is known as a 'watching brief'. A series of detailed archaeological excavations were also undertaken in advance of the redevelopment, which targeted sample areas of particular interest and sensitivity. The nature and appearance of the extant structures that occupied the site were documented before any modifications were made to them through a programme of historic building recording. This publication combines the results of these studies with historical research in order to reconstruct the design and function of the Goods Station as completely as possible.

2 *Introduction*

The depot is incredibly well preserved and is one of the earliest, largest and most complex examples in the country. It was constructed not only as a goods terminus but also as an interchange hub between the railway, the canal and the road network. Whilst one rival company, the London & North Western, had already constructed a London goods interchange of a similar nature in Camden, the Great Northern bravely sought to gain an advantage through the sheer scale of their operation as well as through the use of a pioneering network of hydraulic machinery (Thorne *et al.* 1990: 93, 105). Although the earliest design of their goods station was unique and innovative, it did not function as was originally envisaged for long. Its survival hinged on its dynamism and it remained competitive through the introduction of new technologies and modes of

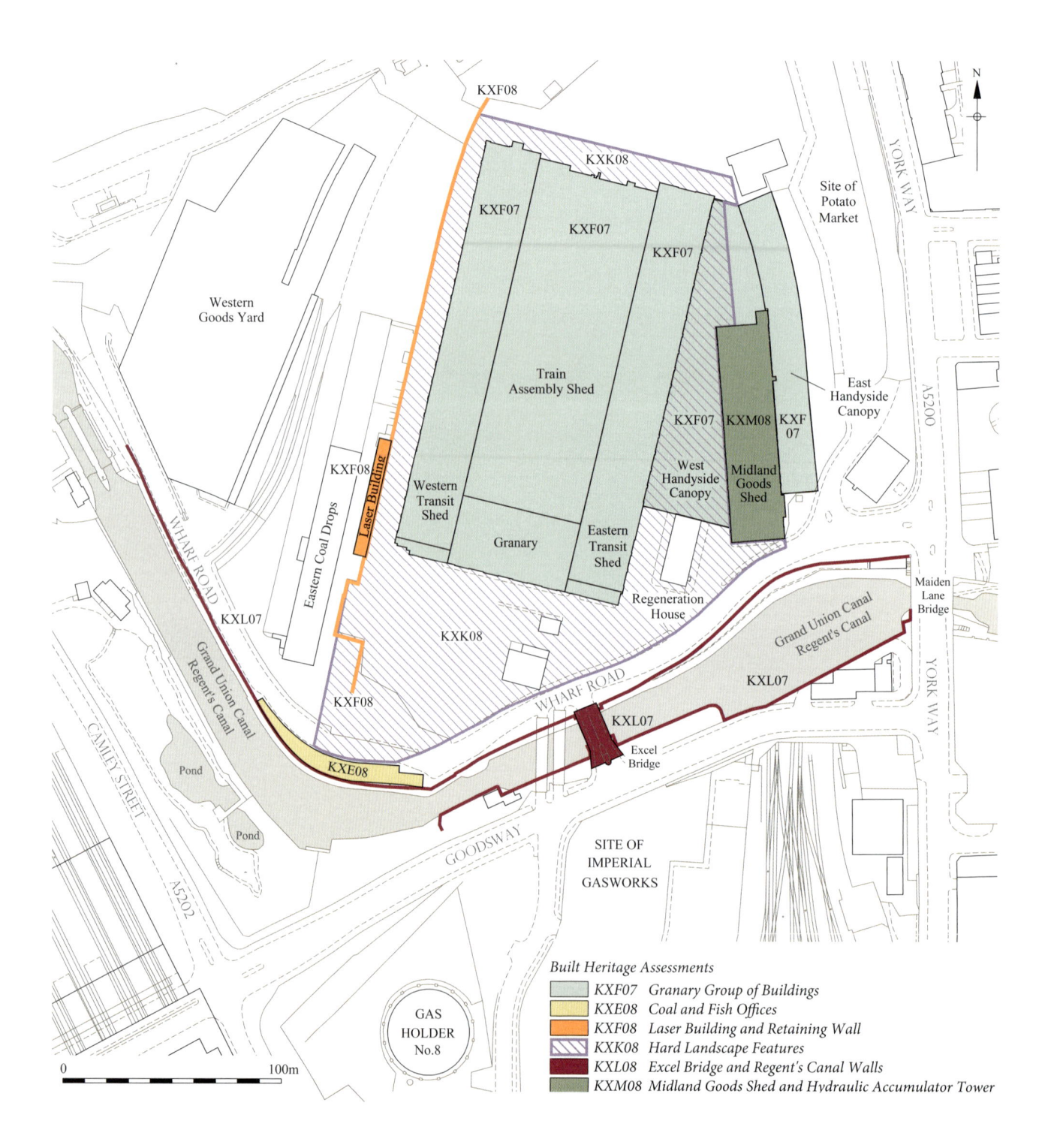

Fig. 1.2 Detailed site and Landscape Feature locations detailing the whereabouts of the built heritage surveys, scale 1:2,500 © Crown copyright 2016 Ordnance Survey 100020795

operation. Although its fortunes rose and fell over the decades, its adaptability facilitated its survival under the ownership of various private and public concerns into the late 20th century. Redevelopment work has presented an unrivalled opportunity to mount a detailed investigation into its design, workings and its eventual demise. Of the handful of extant, comparable yards of this nature, none have been recorded or researched in such a comprehensive manner.

The sites were situated to the north of the extant Passenger Terminus, bounded by the Regent's Canal to the south, York Way to the east and former and standing railway infrastructure and brown field land to the north and west (Fig. 1.1; Fig. 1.2; Fig. 1.3). Together they formed an area that was roughly pentagonal in plan, bounded by National Grid References TQ 30315 83667, TQ 30092 83725, TQ 3000 83524, TQ 30100 83421, TQ 30233 83488 and TQ 30323 83551.

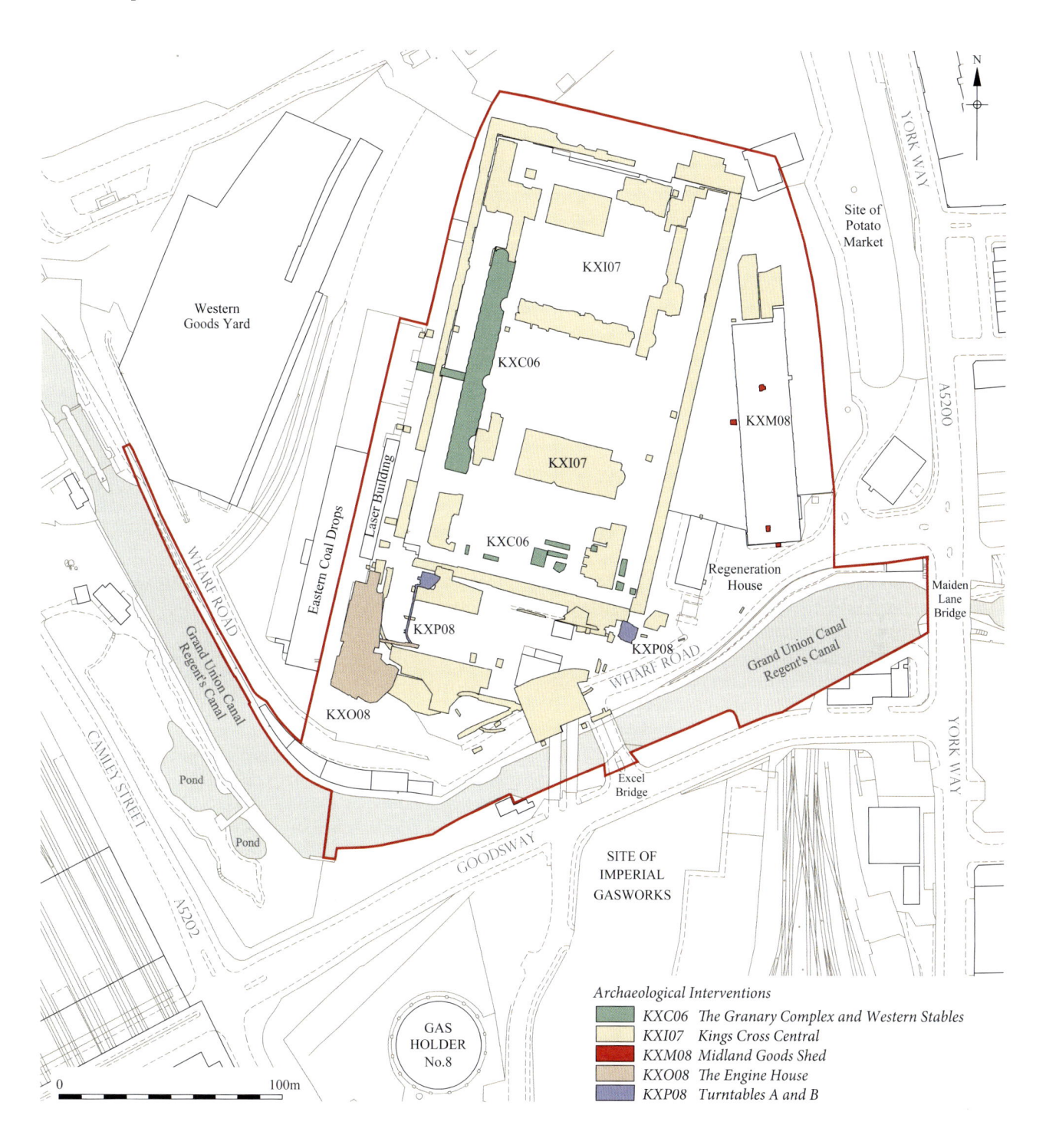

Fig. 1.3　Detailed site and Landscape Feature locations detailing the whereabouts of the archaeological investigations, scale 1:2,500 © Crown copyright 2016 Ordnance Survey 100020795

1.1 Organisation of the Report

Terminology and scope

The scope of this document, encompassing five archaeological sites and six historic building recording surveys, is detailed in Appendix 1; the area covered by the works is cumulatively referred to as the 'Area of Investigation'.

The Grade II listed status of the Goods Yard's buildings necessitated that they be investigated in accord with English Heritage's Level 4 guidelines, the highest and most comprehensive tier of historic building recording (English Heritage 2006; Fig. 1.2). The results of the archaeological monitoring exercises and the detailed excavations were recorded using methodologies developed for deeply stratified urban sites including the single context recording system (PCA 1999) with individual descriptions of all archaeological strata and features excavated and exposed being entered onto *pro-forma* recording sheets.

In-depth analyses of the results of these studies form a series of archaeological assessments and standing buildings reports, (or 'grey literature': Archer 2009; Fairman 2007; Haslam *et al.* 2011; Hawkins 2011; O'Gorman 2007; Thompson and Gould 2010; Thompson *et al.* 2011; Thompson and Matthews 2011; Thompson and Matthews 2012; Thompson and O'Gorman 2009). Assessments of all eleven sites, including context indices, matrices, detailed finds reports and methodologies, have been deposited with the project archives at the Museum of London Archaeological Archive and Research Centre (LAARC), Eagle Wharf Road, where they can be consulted by prior arrangement.

A total of 172 archaeological interventions, consisting of 23 excavation areas and 149 watching brief trenches, were excavated and monitored on the five archaeological sites. For clarity and brevity, only those that yielded positive results are made use of here. All trenches considered in this document are listed in Appendix 1 and their locations illustrated in Fig. 1.3. Each intervention was given a unique alphanumeric reference code (the site code, see Appendix 1); where referred to below, trench names are suffixed by the relevant site code in parentheses.

It is common practice to split the archaeological record into component parts known as contexts, each of which is assumed to have been created by a 'single action' that leaves a positive record, such as the construction of a wall or the dumping of a layer, or a negative imprint such as the removal of spoil to create a rubbish pit (MOLAS 1994: 1.2). Unique numbers are awarded to each, so that individual 'actions' can be identified, located and tied to any associated records or finds. This system enables the results of an archaeological investigation to be organised temporally and spatially, thus facilitating interpretation. Over 5,000 contexts were assigned during fieldwork for this project, and these were vital at the assessment stage. In contrast, complete descriptions of all contexts and finds catalogues are not included in this document as this would produce an overly complex and confusing account. In order to present the data coherently, they have been grouped into landscape features (i.e. buildings, infrastructure networks, bridges, roadways etc.). Various, often fluid terms have been used to describe these assorted landscape features over time, so for ease of reference and continuity each has been awarded a specific identifier (i.e. an unchanging landscape feature [LF] number), which is used throughout the ensuing text (given after the applicable historical name in parentheses; see Table 1.1, Appendix 2, Fig. 1.2; Fig. 1.3). The data is also discussed holistically by landscape feature rather than by archaeological trench or historic building recording survey, an approach that has prevented the results from being fragmented into arbitrary blocks of information.

Specific elements of the landscape features detailed above (such as railway lines, turntables, capstans, engine beds, machine beds and boiler supports) are also referred to throughout this publication using unique identifiers (descriptive terms followed by a number where appropriate; for example Engine Bed 2, Crane Base 1 etc.). The detailed record of these encompasses a myriad of context numbers which have been forsaken here in order to create a broader contextual approach and maintain coherence. Similarly the original nine archaeological phases that were used to structure the assessment documents and analyses, whilst heavily relied upon during that process, are not referred to directly here. Instead, local, regional and national historical events that influenced the phasing are referred to.

Layout and themes

A synopsis of the underlying geology and topography of the Area of Investigation is given at the end of this chapter, followed by a critical consideration of the methodologies that were used to collect the data upon which this study is based. The archaeological sequence and the historical background are then discussed in a series of chapters, each of which correspondes to a specific chronological phase. They are arranged sequentially from the earliest period (Chapter 2) to the latest (Chapter 12).

Chapter 2 summarises the history of the area before the coming of the railway, charting its transition from a semi-rural backwater to an industrial, urban landscape. It explores why it was chosen by the Great Northern Railway as the site of their future goods depot. Chapters 3 to 12 detail the history of the Goods Station by presenting the archaeological results against a backdrop that is provided by documentary sources. These lines of evidence are compared, contrasted and integrated to create a chronological account. The historical background is presented in an introductory section given at the start of each of these chapters that details the various local, regional and national events that impacted upon the Goods Station. When and where appropriate, these include relevant aspects of the economic and political situation in Britain, the affairs of the various companies that were directly or indirectly involved with the yard and the antics of rival concerns at local, regional and national levels. The core texts of Chapters 3 to 12 detail the responses to these events by the Goods Station's owners through an analysis of the archaeological record and any relevant historical documentation. In order to optimise clarity within the narrative and maximise integration between the archaeological, built heritage and historic information, these sections are arranged by landscape feature. Pertinent historical information relating to that feature is presented first, followed by a synopsis of the contents of the archaeological and built heritage records relating to it. The latter borrows from the former in order to increase the power of the ensuing interpretation whilst simultaneously critiquing the reliability of the historical information. As shall be demonstrated, these lines of evidence in combination demonstrate in a thorough and comprehensive fashion how the different elements of this complex functioned at various points in its history and how, when and why they evolved in the ways that they did.

Since the complex could not have functioned without the individuals that were employed to administrate, construct, run and maintain it, the changing roles and working conditions of the employees are considered where relevant.

Synopses, included at the end of Chapters 3 to 12, consider whether or not the responses of the Goods Station's various owners to the internal and external problems that they were confronted with were effective. The ultimate goal of the interchange was to move goods between the canal, the railway and the road as rapidly as possible, thus facilitating trade. As a consequence, the goals of these chapters are to consider how this was done during each temporal period, to explore the efficiency of the process and to draw conclusions concerning the complex's importance to trade and exchange at local, regional and national scales. Consequently, the ultimate aim of each synopsis is to determine whether or not the Goods Station was successful during the various phases of its existence.

A series of short vignettes, placed at appropriate positions within the text, explore issues of particular interest in more detail. The last of these summarises the evolution of the yard's pioneering hydraulic network and considers its mechanical intricacies from an engineering perspective.

The concluding chapter pulls the various strands of evidence together to discuss how, when and why the function and efficiency of the Goods Yard altered with time. In doing so, its changing fortunes and those of the various concerns that owned it can be charted whilst considering the causes that lay behind the variances. In doing so the interplay between events at the yard and the wider socioeconomics of 19th and 20th-century Britain are drawn out. This chapter goes on to compare developments at King's Cross with two rival London goods stations in order to see whether events at the depot fit within wider trends in railway goods handling. It then evaluates the power of the multidisciplinary approach that has been used to compile this account (i.e. the use of archaeological, built heritage and historical data in combination) and asks whether this method is pertinent to the investigation of other goods stations and industrial sites. The text concludes by looking towards the yard's bright future.

Graphical conventions used in the text

The text is illustrated through contemporary and modern photographs, maps and other images. These are accompanied by overall schematic plans illustrating major landscape features and changes to the complex of buildings by period, combined with occasional, more detailed illustrations of particular archaeological details; individual keys accompany these drawings. A series of annotated drawn building elevations also accompany the following text, illustrating the major modifcations to individual buildings by period. These have been colour coded using the conventions set out below. On occasion the phasing can be more closely refined within the period in question and in these instances a key to conventions used accompanies the drawing.

Chapter 3
1847-1852

Chapter 4
Early 1850s?
1855
1858
Post January 1858
Post June 1858
1860
1865-66
1867
1866-1882
1871
1872
1873
1875
1877

Chapter 5
1881-1882

Chapter 6
1895-1893
1883-1899
1898

Chapter 7
1897-1899

Chapter 8
1900-1914

Chapter 9
1924
1926
1935
1935-1936

Chapter 10
1939-1945
Post 1940

Chapter 11
1948
1952-1955

Chapter 12
Late C20th
Retained

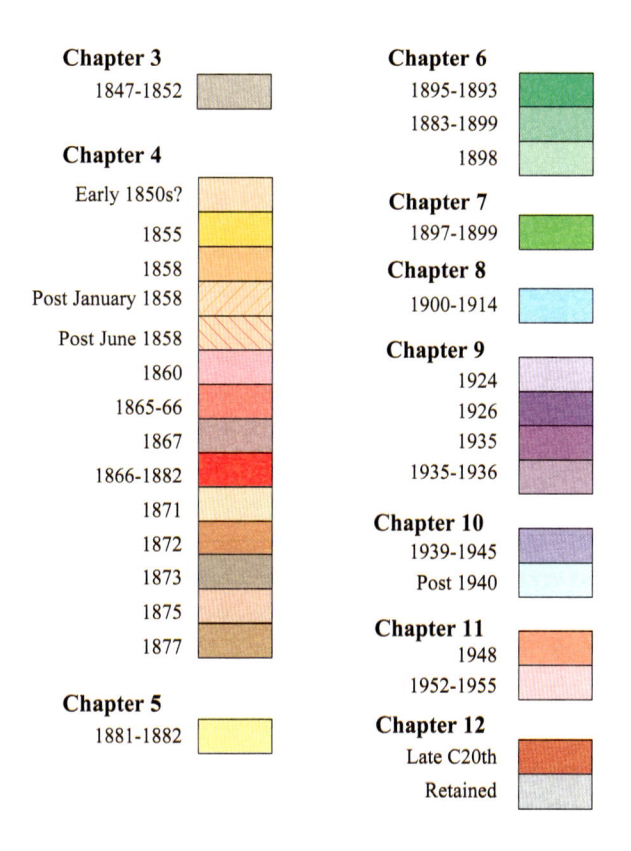

Fig. 1.4   Graphical conventions used to illustrate modifications to standing buildings

1.2 Geology and Topography

In order to understand the significant landscaping works that were undertaken as part of the historical development of the site, it is necessary to consider the pre-urban natural landscape and how that differs from its modern counterpart. The natural topography of the area, and indeed that of north London in general, would prove to be somewhat of a challenge to early locomotives, particularly those frequenting the Permanent Passenger Station in the mid 19th century (see Chapter 3).

The Geological Survey of England and Wales (Sheet 260) indicates that Eocene London Clay underlies the entire site. It was rarely definitively encountered during the project, as it is buried at depth below a thick layer of 19th-century made ground. Interpretive problems were exacerbated by the similarities that existed between the natural clay and the overlying deposits, which were formed of virtually identical but redeposited material. London Clay was only conclusively identified during a watching brief in the west central section of the site where it was observed at a depth of 23.56m OD in Trial Pit 4 (KXC06).

As was the case in this region more generally, in antiquity the Area of Investigation most probably occupied a tract of land that exhibited a pronounced slope from higher ground in the north to lower ground in the south. Although this is no longer apparent within the confines of the former Goods Yard, York Way, which runs along its eastern boundary, still exhibits such a gradient. Whilst this road has been artificially landscaped (its original topography must have been altered in order to carry it over the Regent's Canal via Maiden Lane Bridge) this was not as extreme as the modifications that were made within the confines of the Area of Investigation. As a consequence, the original topography has been approximated from the lie of York Way and the towpath of the Regent's Canal.

The modern topography is entirely the result of landscaping work that was undertaken in the early to mid 19th century. Between 1817 and 1820, a cutting for the Regent's Canal was dug along the southern boundary where it forms a distinctive 'U' shape that deviates markedly from its more general course (Faulkner 1990: 48–9). The creators of the canal originally proposed a more direct route, the meander being the product of a dispute between them and a local landowner (see Section 2.3 for further details). The presence of the meander made the site an ideal location for a rail, road and canal interchange and was no doubt instrumental in attracting the Great Northern Railway to the area.

The remaining sections of the Area of Investigation were extensively modified in the early 1850s, when the Great Northern built its London Goods Station. Archaeological work has demonstrated that the flank of the hill was levelled through the dumping of made ground obtained from elsewhere in order to create a flat surface upon which the bulk of the Goods Yard could be constructed. This was important, since early locomotives found it difficult to operate at low speeds on strong gradients, whilst the cart horses that would be primarily responsible for moving wagons through the Main Goods Station would similarly struggle to control those heavy vehicles on anything other than a level surface. More material therefore had to be deposited at the southern end of the site at the bottom of the hill than the northern section at the top of the incline. The end result was a fairly level terrace, the top of which was found to be between 24.00m OD and 24.30m OD at the southern end and 23.95m OD and 24.10m OD at the northern end. To the east and north, this terrace merged seamlessly with the topography of the wider area. In contrast retaining walls and buildings, constructed for the purpose of shoring the high ground, flanked the western and southern edges of the raised area. This produced vertical height differentials of 3.70m and 2.65m respectively in the southern and western edges of the Area of Investigation, the naturally occuring high ground being found to the north and east of the retaining structures. The course of the Regent's Canal was unaltered by these landscaping works since it still runs along the base of the southern terrace today. Given the depth of the redeposited material that was observed in the southern end of the site, the results suggest that the towpath of the feature represents the approximate level of the pre-1849 natural ground surface.

1.3 Methodology and Limitations of the Study

The combined archaeological, built heritage and documentary approach that was utilised in this study proved to be a great boon to the quality of the results and conclusions that are presented herein. When and where appropriate this methodology has enabled the validity of certain historic records to be verified, refined or disproved and has elevated the level of interpretation that has been applied to the remains that were recorded in the field. The archaeology and built heritage data revealed entirely new information that could not have been gleaned from the archives, and occasionally disproved certain historical accounts. Likewise, the historical research that was undertaken yielded a plethora of facts that would have been unobtainable by any alternative route. This approach has also enabled certain biographic details of the Goods Yard's buildings and the people that worked in them to be told, which has gone some way to bringing the complex back to life in the mind of the reader.

It was important to present the data within a chronological framework so that the findings could be placed in a wider socioeconomic and political context that encompasses developments at local, regional, national and, when necessary, international scales. This has enabled the reasoning behind the various changes that occurred at different points in time to be better understood.

The data collection methodologies that were employed on the King's Cross site (as outlined in Section 1.1) involved recording features to established archaeological guidelines, using the metric system, to levels of accuracy that are within ten millimetres. This is normal practice on commercial archaeological sites and complies with the standards that are set by the industry's regulatory bodies. These commonly used techniques are adequate for the successful recording and interpretation of pre-industrial sites, however most specialist industrial archaeologists would agree that they are sometimes insufficient when recording 'modern' structures like those that were found within the Goods Yard at King's Cross. A thorough review and overhaul of commercial archaeological guidance and practice on sites of this nature is therefore overdue, so it is worth pausing to consider how the recording methods that were deployed at King's Cross could be improved so that better and more versatile data can be routinely collected in the future.

The importance of accurate Imperial measurements to the interpretation of a site of this nature cannot be underestimated. It is therefore important to highlight the slight but unnecessary margin of error that is introduced when converting Metric to Imperial, which could have been avoided entirely had the latter been obtained directly in the field. Having access to accurate Imperial measurements can be extremely useful as an interpretive tool on industrial sites. Taking an example from King's Cross to illustrate this point, the collection of such measurements would have aided the interpretation of the various machine bases that were encountered archaeologically, perhaps enabling the makes, models or parts of the different machines that they upheld at various points in time to be reconstructed with greater consistency.

Issues of precision measurement are further compounded on an archaeological site of this age and type because most component parts of 19th to 20th-century machines were built to extremely high tolerances of accuracy. As such recording them with conventional archaeological methods that produce margins of error in the region of 10mm becomes far less appropriate (such as hand planning from a grid or baseline with tape measures, use of the Global Positioning System or a Total Station). As the bulk of recording is generally undertaken and indeed independently monitored by non-industrialists, there is also a danger that small but vital components of buildings and machines go unnoticed and undocumented. The following case study from King's Cross, which is given for illustrative purposes, offers a clear example of the kinds of questions that cannot be addressed when these modes of recording are used in isolation.

One of several issues that were encountered during post-excavation work on King's Cross was the successful interpretation, dating and evolution of the yard's hydraulic pipe network. As the thicknesses of the pipes changed by several millimetres over time (Tim Smith pers. comm.) it should have been possible to phase the network by obtaining approximate typological dates for the various elements based upon very precise measurements of their thicknesses. However, this was not known at the time of recording and the methods that were generally deployed, coupled with the fact that many of the pipes were corroded, did not permit this level of accuracy to be reliably achieved. Rather than take the risk of producing an inaccurate reconstruction, this task has therefore not been attempted in this document. However, had an alternative recording methodology been available to the excavation team from the outset, it should have been possible to shed far more light on this pioneering, innovative and therefore historically important infrastructure network. This interpretative problem represents just one of many that will be routinely encountered on virtually any 19th or 20th-century industrial site if it is exclusively recorded using conventional commercial archaeological methods. Rather than viewing these problems in an entirely negative light, the lessons that were learned at King's Cross should be used to advance archaeological practice in the future.

There are a number of cost effective, high accuracy recording options that are available to commercial archaeological contractors that mitigate these issues. The first is both cheap and easy: Imperial as well as Metric measurements should be compiled on site as and when necessary and conventional hand planning methods should be substituted for high accuracy techniques when appropriate (e.g. recording component parts with callipers etc.). Contract archaeologists also need to explore the potential of alternative technologies. For example, techniques such as photogrammetry, rectified photography and laser scanning offer the opportunity to capture the intricacies of an industrial site in minute and accurate detail, meaning that small but important components that may be of great importance to an expert interpreter but of no apparent significance to a layman do not go unrecorded.

Documenting entire archaeological or built heritage sites using these more expensive or time consuming methods is rarely possible given the temporal and budgetary constraints that are almost always imposed upon a rescue excavation by a construction programme. A degree of pragmatism and direction is therefore required so that money and time can be spent appropriately. High level recording should therefore be undertaken when deemed appropriate by a consulting industrial specialist. This individual should regularly visit the site throughout the project to offer advice concerning interpretation and recording techniques.

It is this author's opinion (R. Haslam) that the limitations in the standard methodology that have been

8 Introduction

highlighted here should be used to inform and guide future industrial archaeological rescue excavations at the design stage. Such an undertaking would require the assistance of the archaeological and built heritage monitoring bodies so that appropriate time and funding for the increased level of work that would be required in order to obtain a sufficiently high level of accuracy can be sourced from clients in advance. A review of the guidelines that are commonly issued to archaeological contractors for the excavation of industrial sites is therefore needed and should be undertaken in the near future so that recording and interpretation can be improved at national level. If these suggestions are implemented then increasingly accurate and versatile data sets should be generated in the future so that more refined engineering based questions can be addressed (for further information on suggested methodological improvements see Haslam 2013).

Table 1.1 Landscape feature numbers used in this publication (for details of the interventions see Appendix 2)

Landscape Feature no.	Original function	Part of / later uses	First constructed by (date)
LF1	Regent's Canal, Towpath and Canal Walls		1818–1819
LF2	Maiden Lane Bridge		1819
LF3	Gasworks Basin		1824
LF4	Retaining Wall		1852
LF5	Granary Basin and Canal Tunnels		1851
LF6	Coal and Stone Basin		1852
LF7	Western Transit Shed		1852
LF8	East Transit Shed		1852
LF9	Western Stables	Part of the original Goods Shed complex, later known as the Eastern Goods Yard	1852
LF10	Eastern Stables		1851
LF11	Train Assembly Shed		1852
LF12	Granary		1851
LF13	The Lamp Room and Coffee Shop	The Refreshment Club	1852
LF14	The Temporary Passenger Station	The Potato Market	1851
LF15	The Carriage Shed	The Midland Goods Shed	1852
LF16	Hydraulic Station		1852
LF17	The Coal Drops	Eastern Coal Drops	1852
LF18	Coal and Fish Offices		1851
LF19	Wharf Road Viaduct		1852
LF20	Western Goods Offices	Part of Goods Shed complex	1866
LF21	Eastern Goods Offices		1858
LF22	Western Coal Drops		1860
LF23	The Plimsoll Viaduct and the Cambridge Street Coal Drops		1866
LF24	Gasworks Viaduct		1867
LF25	Horse Provender Store / Stables		1860
LF26	West Handyside Canopy		1888
LF27	East Handyside Canopy		1888
LF28	Western Goods Shed		1899
LF29	Goods Way and Excel Bridge		1918
LF30	RME Garage		1933
LF31	Laser Building		1933
LF32	Hydraulic Metering Station		1958

CHAPTER 2

King's Cross before the Great Northern Railway

'What strange mutations does the hand of 'public improvement' work in our metropolis. Less than a score of years have rolled away since a very anomalous pile was reared at the point where meet the New-road, Maiden-lane, Pentonville-hill, the Gray's Inn-road, &c.; the spot receiving the somewhat grandiloquent name of 'King's Cross.' The building boasted, however, of correspondent pretension; the lower story was classically embellished…the upper stories were less ornate; but, if the expression be allowable, the structure was crowned with a composition statue of the Fourth George- and a very sorry representative of one who was every inch a king. The pennyworths of artistical information, doled out from week to week, soon taught the people that the above was a very uncomplimentary effigy of majesty; even the very cabmen grew critical; the watermen jeered; and the omnibus drivers ridiculed royalty in so parlous a state, at length the statue was removed in toto, *or rather by piecemeal...The time to pull down at length arrived; the strange pile has been cleared away; and lest a future generation should ask where the fabric stood we have consigned its whereabouts to our columns'.*

A description of the widely derided monument to George IV (from which the place name 'King's Cross' originates), taken from ***The Illustrated London News*** of January 1842

2.1 The Early History of King's Cross

Any evidence of human activity pre-dating the early 1850s was destroyed within the confines of the Area of Investigation by the extensive landscaping works that immediately preceded the construction of the Goods Station. Whilst the possibility of earlier exploitation in this location cannot be ruled out, archaeological and historical evidence strongly suggest that activity predating the 18th century was not intensive. The underlying geology of the King's Cross area consists of poorly drained London Clay which is prone to waterlogging. Consequently, these claylands are thought to have been more sparsely occupied during the prehistoric and early historic periods than the free draining gravel terraces.

No evidence of prehistoric activity has been found in the area, the earliest feature of note being a possible Roman road that may have run along the course of York Way (formerly known as Maiden Lane) to the Roman city of *Londinium* to the south (Robertson 1977: 255). A fragment of a tombstone dedicated to the memory of a Roman legionary was reportedly found on the eastern side of Maiden Lane in 1842 (Godfrey and Marcham, 1952: 102). A Saxon settlement may have developed around the site of St Pancras Old Church, where an exceptionally early Christian altar, dating to 600 AD, was reputedly unearthed (Weinreb and Hibbert 1993: 774). At some point before the Norman Conquest the entire ancient parish of St Pancras came into the possession of the Canons of St Paul's Cathedral (Lovell and Marcham, 1938: 1–31). In the 12th century, this estate was divided into manors (known as prebends) for the endowment of the Cathedral Canons. It is likely that the prebendal manor of St Pancras, in which both the King's Cross Goods and Passenger Stations lie, formed an original endowment of the church of St Pancras. The church was first mentioned in 1183, and was subsequently rebuilt in the mid-14th century. A small hamlet which developed around the church in the medieval period had been largely abandoned by the early 14th century owing to flooding from the River Fleet. The Area of Investigation itself lay in a parcel of three fields known as 'Alfrichbury' or 'Allenbury', which was demised in the late 12th century by the Prebendary of St Pancras to the Hospital of St Bartholomew, West Smithfield (*ibid*.).

The countryside to the west and south of the church (the future location of Somers Town and Agar Town) remained predominantly rural until the second half of the 18th century. To the southeast, in the area that would eventually be called King's Cross, the ancient hamlet of Battle Bridge developed around a crossing of the River Fleet. A 16-acre close that lay on the south side of the hamlet and the Fleet and east of Gray's Inn Road was known as Battlebridge Field in the 16th century (*ibid*.). There is no credible evidence to support the legend that the settlement owes its name to a skirmish fought between the Romans and the *Iceni* during the Boudiccan revolt, nor the equally fanciful claim that Boudicca herself lies buried beneath Platform 8 of King's Cross Passenger Station. It is thought that the place name is ultimately derived from a 'Broad Ford' in the Fleet, which gave its name to Bradford Bridge and thence Battle Bridge (Godfrey & Marcham 1952: 102).

London began to expand during the post-medieval period, its population quadrupling in number between the late 15th and the mid-18th centuries. Settlement spread beyond the medieval city walls into outlying districts like Islington, Shoreditch and Clerkenwell, which were absorbed into the urban sprawl. In contrast, the remoteness of the area between St Pancras and Battle Bridge persisted into the post-medieval period, prompting one writer to describe the church in 1593 as "all alone, utterly forsaken, old and weather-beaten" (Stamp 1990: 19). During the first half of the 18th century however, the Battle Bridge area began to gain a reputation for respite and healing amongst the residents of the encroaching city. Several nearby springs that fed the River Fleet, including Bagnigge Wells and St Chad's Well, became popular spa resorts by the middle of the century (Thornbury 1878: 296–8; Barker *et al*. 2008: 59). The promoters of the latter claimed that it attracted between 800 and 900 visitors per day, who were drawn by the reputedly therapeutic waters. The area also became renowned for its clean air. In 1763 the proprietors of the Middlesex County Hospital for Smallpox in Clerkenwell purchased a property in St Pancras where they built a new hospital which opened in 1767 on the future site of King's Cross Passenger Station (Godfrey and Marcham 1952: 114–117 fn. C1). The adjacent Fever Hospital was erected in the gardens between the Smallpox Hospital and St Pancras Road in 1802, followed in 1805 by St Pancras Hospital on St Pancras Way.

As late as 1777, the locality was described as being almost entirely rural, with St Pancras Old Church commanding unrivalled views of the open countryside to the north and the growing city to the south from Tottenham Court Road to Highgate (Walford 1878: 340). The Area of Investigation remained in the possession of St Bartholomew's Hospital at the end of the 18th century, the pastures of the Allenbury estate still untouched by the encroaching metropolis.

2.2 Urbanisation and Industrialisation between the mid-18th and mid-19th Centuries

The earliest shoots of urbanisation emerged to the south of the Area of Investigation during the mid-18th century. By the turn of the 19th century the environs of Battle Bridge and St Pancras had been transformed from an essentially rural environment to a mosaic of residential areas, semi-industrial zones and brick fields, interspersed with pockets of agricultural land. The principal catalyst for development was the construction of the 'the New Road from Paddington to Islington', which was sanctioned by Act of Parliament in 1756 (now known as Marylebone Road, Euston Road and Pentonville Road). John Rocque's map of 1769 (Fig. 2.2) shows the road skirting the fashionable suburb of Bloomsbury to the south, separating it from the fields to the north. Originally designed to allow cattle to be driven to Smithfield Market whilst bypassing the city's western suburbs, the highway greatly improved access to the Battle Bridge area. The Act of Parliament stipulated that new buildings should be set back from the road, which encouraged the development of residential properties with long gardens in front (Godfrey and Marcham 1952: 114–117). The north side of the stretch of the Euston Road between Tottenham Court Road and Battle Bridge in particular became a magnet for developers and by the 1770s was lined with residential properties.

Somers Town, which lay to the west of the Area of Investigation and to the north of the New Road, was developed by the speculator Jacob Leroux on land belonging to Baron Somers from 1784. Although Leroux set out to build a suburb for the middle classes, his scheme did not unfold as planned and the development fell into a rapid decline after 'some unforeseen cause checked the fervour of building, and many carcasses of houses were sold for less than the value of the building materials' (Walford 1878: 340). By the end of the 18th century, the value of property in the development had fallen considerably and Somers Town had become associated with the poorer end of society. Émigrés fleeing the French Revolution moved in from the 1790s, followed in the 1820s by Spanish exiles (Weinreb and Hibbert 1993: 817). A number of famous artists and writers lived in the area, including Mary Wollstonecraft, who lived in the Polygon in Clarendon Square, which was also briefly the boyhood home of Charles Dickens.

During the three decades that followed the construction of the Smallpox Hospital, the strip of land to the north of the building became increasingly built up, as builders erected rows of houses along a number of narrow passages that linked St Pancras Road and Pancras Walk. By 1800 Paradise Row, Wellers Place, Duke of Clarence Passage and Red Lion Passage were all packed with high density, poor quality housing (Thompson 1804 not illustrated). In 1806 building leases were granted on a triangular plot of land between Pancras Walk and Maiden Lane, which was subsequently developed as a residential district known as the Drakefield Estate (TNA RAIL 1189/5). The area was densely built-up by the mid-1830s (Fig. 2.3), although part of it was subsequently cleared to make way for the King's Cross Passenger Station.

In order to meet the demand of developers for construction materials, local landowners began to let large areas of former pasture and market gardens in the vicinity to contractors such as Thomas Cubitt for brick making. Soon 'one vast brick-field' peppered with clay extraction quarries developed between Battle Bridge and Somers Town (Walford 1878: 340). Local brick makers included John Smith and William and Henry Hickman, who were described as brick makers of Battle Bridge in early 19th-century insurance documents (CCRO 148/2/839; CCRO 148/2/844). A field on the west side of Gray's Inn Road was home to a substantial brickworks owned by the Harrison family by the 1740s (Godfrey and Marcham, 1952: 70). A short distance to the north of Harrison's premises lay Smith's 'Dust Ground', in which a huge heap of ash stood in the north-east corner of Battle Bridge Field at the junction of Euston Road and Gray's Inn Road. The ashes were mixed with the local brickearth to make the bricks necessary for urban development. The heap was finally removed in 1826 and the site sold for development (Stamp 1990: 17; Fig 2.1).

Tile making also became established in the fields on either side of Maiden Lane in the early decades of the 19th century. In 1810 Adams' tile kilns were making garden and chimney pots at Belle Isle; by 1829 the company's premises had grown to include 8 acres of land, tile kilns, sheds and cottages (Baker and Elrington 1985: 3–8). Belle Isle also attracted refuse collectors, knackers' yards, manure processing factories and chemical works. In 1828 Randell's tile kilns moved from Bagnigge Wells Road, Clerkenwell to premises on the east side of Maiden Lane.

The district's already unsavoury reputation declined even further following the arrival of a variety of noxious industrial processes, which flocked to the sparsely

Fig. 2.1   'The great dust heap at King's Cross' by E.H. Dixon 1837 (WL 38712i © Wellcome Library)

inhabited vicinity of Maiden Lane during the early decades of the 19th century. Local directories listed a feather dresser, a mustard manufactory and a brewery at the southern end of Maiden Lane in 1811, while a varnish factory, a patent yellow paint factory and a business that boiled animal bones were trading from premises at Battle Bridge in 1829 (*ibid*.: 3–8).

The development of the Regent's Canal further contributed to the increasingly industrial character of the area. Completed in 1820, the course of the canal is shown on B.R. Davies's Map of 1834 (Fig. 2.3). Whilst its construction was beset by problems, it greatly improved the transport of raw materials and finished goods, making the district ripe for further commercial and industrial development. Owing to the importance of the canal in attracting the railway to King's Cross in the mid-19th century, its development is discussed in greater detail below.

A little over a year after the canal opened the Imperial Gas Light and Coke Company purchased a plot of land on the south bank of the waterway, upon which it set about building a new gasworks to serve the northern districts of London (Fig. 2.3). The St Pancras Gasworks opened in 1824 and remained the largest in the capital until the development of Beckton in 1869. The architect William Bardwell described gasworks as the 'most offensive and pestilential nuisances in London' and the St Pancras works became notorious for the 'mephitic vapours' it emitted (Bardwell 1854: 29). The establishment of the gasworks is elaborated upon below.

In 1830, a monument to King George IV was erected at the crossroads created by Euston Road, Pentonville Road, St Pancras Road and Gray's Inn Road in the centre of Battle Bridge (Fig. 2.4). Designed by Stephen Geary, the 60 ft-high structure consisted of an octagonal base adorned with Doric columns and the four patron saints of Britain, surmounted by a statue of the King. The design of the memorial was widely derided by contemporaries and proved to be exceedingly unpopular with the general public. One commentator recalled that '*Little boys used to chalk their political opinion freely on the pedestal, accompanied by rough cartoons of their parents and guardians, their pastors and masters; omnibus drivers and conductors pointed the finger of hilarity at it, as*

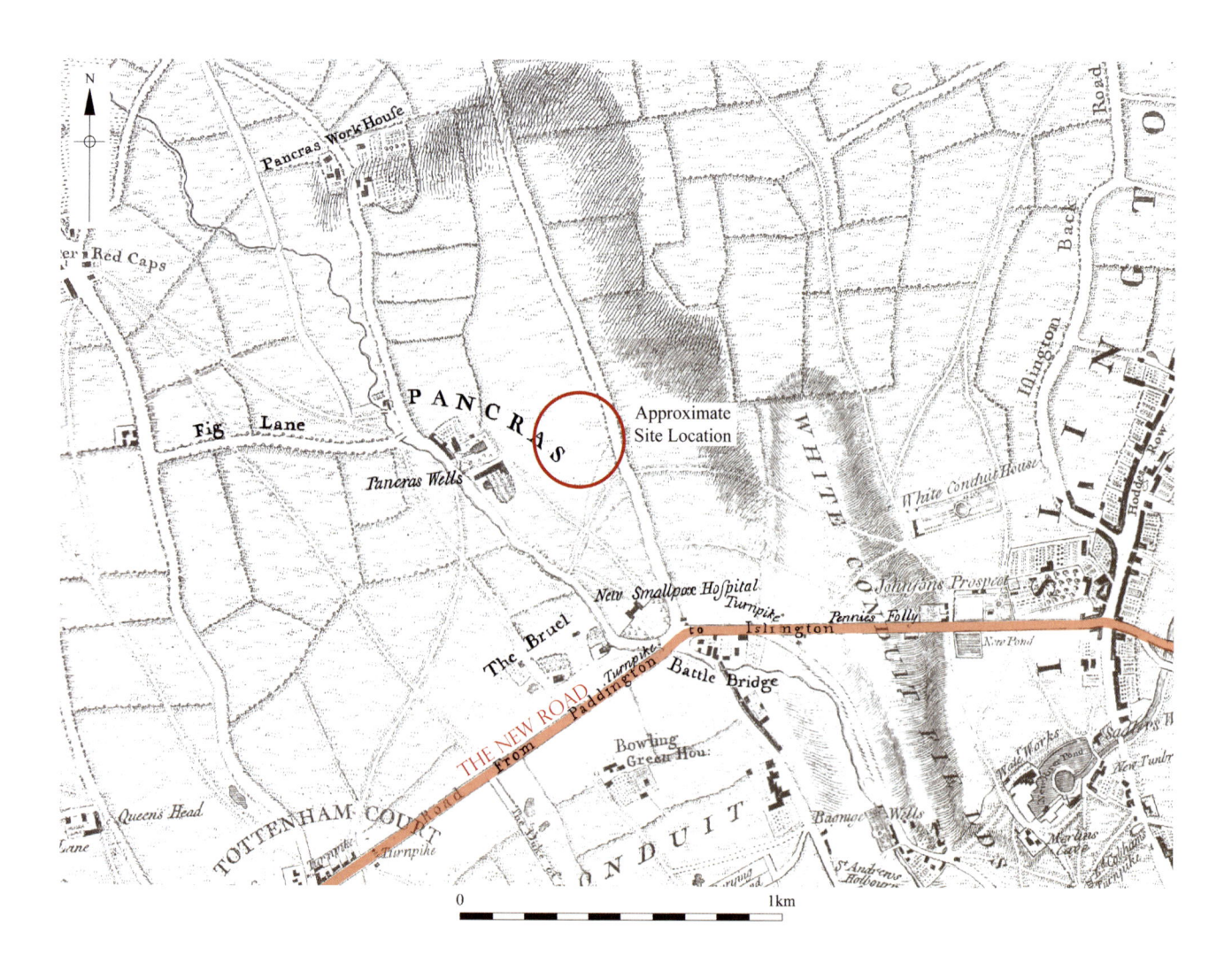

Fig. 2.2   The Rocque map of 1769 illustrating the course of the New Road from Paddington to Islington and the environs of Battle Bridge indicating the approximate location of the Area of Investigation, scale 1:20,000

they passed by' (Sala 1859: 56). Public derision led to the removal of the monarch's likeness in 1842 followed by the remainder of the structure three years later, after it had been used variously as an exhibition space, a police station and even a beer shop (Anon 1842; Stamp 1990:19).

Despite the monument's short life, developers preferred to use the name 'King's Cross' in favour of Battle Bridge, the reputation of which had been irreparably damaged over generations. Ultimately, the new name superseded Battle Bridge, which fell into obscurity.

The railways first arrived in the district in July 1837, when the London & Birmingham Railway opened London's first passenger terminus north of the river at Euston. The site of the station, to the west of Battle Bridge and to the immediate north of the New Road, was occupied by market gardens prior to development. At first, six trains per day ran to Harrow, Watford and Boxmoor, following which services to Birmingham commenced in September 1838. The London & Birmingham would also be the first company to establish a London goods interchange depot, which was built on the Regent's Canal at Camden. Soon afterwards the company merged with the Grand Junction and the Manchester and Birmingham to become the London & North Western Railway Company, which inherited the Camden facility in 1846. The latter depot was to be the main rival in the capital to the Great Northern Railway Goods Yard at King's Cross.

In contrast to the increasingly built-up area around Battle Bridge and Somers Town, one part of the district remained relatively unaffected by development in the early decades of the 19th century. This was an estate of approximately 70 acres which surrounded Elm Lodge, the mansion house of the Prebendary of St Pancras. The estate occupied a tract of land bounded by the Imperial Gasworks to the west, Battle Bridge to the north and west, the New Road to the south, Somers Town to the east and the growing suburb of Camden Town to the north. In 1810 the lawyer William Agar purchased the lease of the house, where he settled with his family (Lovell and Marcham 1938: 60–62). A

Fig. 2.3 Davies's map of 1834 showing the course of the Regent's Canal, the location of the newly constructed Imperial Gasworks and further urbanisation in the Battle Bridge area overlain with the location of the Area of Investigation, scale 1:12,500

couple of years after Agar's death in 1838, his widow Louisa divided part of the estate (now largely covered by the northern approaches to St Pancras Station) into small plots that were let out under 21-year sub leases to labourers, in order that they might build their own houses. The unforeseen consequence of this decision was the rapid and entirely uncoordinated development of a district which came to be known as Agar Town. The district soon gained a notorious reputation amongst contemporary commentators as the worst kind of slum, where families lived in extreme poverty alongside a criminal underclass (Hollingshead 1861: 127–9). It was most famously described in a short story by W.M. Thomas, published in Charles Dickens' magazine *Household Words* under the title 'A Suburban Connemara'. According to Thomas' fictionalised account, the district was 'a complete bog of mud and filth with deep cart-ruts… wretched hovels… doors blocked up with mud and filth … heaps of ashes, oyster shells and decayed vegetables…The stench of a rainy morning is enough to knock down a bullock' (Thomas 1851: 310–1). In contrast to the catalogue of horrors described by contemporaries, a modern study of the district, based upon analysis of census returns and the records of the St Pancras Vestry has found that the population was not particularly poor by the standards of the time and included many members of respectable trades (Denford 1995: 21). Relatively few householders received poor relief and the streets of the district were paved and lit by 1860.

Fig. 2.4 An elevation of the monument to King George IV at the road junction that would come to be known as 'King's Cross' in its honour (BM Crace Collection Main Series 1880 1113.4759 © Trustees of the British Museum)

2.3 The Construction of the Regent's Canal and the Imperial Gasworks at St Pancras

Plans to build a canal connecting the east London docks and the Grand Junction Canal terminus at Paddington were first mooted in 1802 (Faulkner 1990: 41). Although that scheme failed to attract much interest from potential investors, eight years later a barge owner named Thomas Homer proposed to build a canal from Paddington across London to join the Limehouse Cut, a navigable channel built and maintained by the Trustees of the Navigation of the River Lea. Homer's co-sponsor, the architect John Nash, sought patronage for the scheme from the banker Sir Thomas Bernard and the Prince of Wales, later the Prince Regent. In summer 1811 the promoters submitted a Bill to Parliament seeking authorization for the scheme. After a long and disputed passage through the Commons, the Regent's Canal Act received Royal Assent the following July.

The proprietors of the new company held their first meeting in August 1812, when Nash's assistant, John Morgan, was appointed Engineer, Architect and Land Surveyor to the project, whilst Homer was appointed Superintendent (TNA RAIL 860/1: 5). Following the ceremonial commencement of works in October, construction started in earnest that December, when the excavation of the Maida Hill tunnel began.

By the end of 1814 almost £180,000 had been spent constructing the canal as far as Hampstead Road (*ibid*.: 42; Faulkner 1990: 44). With the total cost of construction estimated to be nearly £250,000 the company set out to raise a further £45,900 by subscription. Although this had been achieved by the following June, the company's precarious finances were further depleted by Homer, who absconded in April 1815 after having misappropriated funds (TNA RAIL 860/1: 50). The financial crisis was further exacerbated by the expenditure of substantial sums on an unsuccessful hydro-pneumatic canal lift designed by Major General Sir William Congreve. By the end of March 1816, the company had come to realise that Congreve's lock had been little more than an 'expensive experiment' although money continued to be wasted on it for another year.

The company encountered a further obstacle to progress in the form of William Agar, whose 'pertinacious opposition' to the construction of the waterway through his estate at St Pancras resulted in a series of legal disputes that lasted for decades (*ibid*.: 103). Despite numerous attempts to mediate between the two parties, the ensuing stalemate had yet to be resolved by the end of 1816, when the company's deteriorating financial situation forced it to apply for a loan from the Government.

At the beginning of December 1817 the Commissioners for the Issue of Exchequer Bills granted a loan to the proprietors subject to the condition that construction resumed immediately in order to provide work for the 'labouring poor' at a time of rising unemployment following the end of the Napoleonic Wars (*ibid*.: 212). The company duly complied and negotiations with Agar resumed whilst tenders for the excavation of the canal, the supply of bricks and the construction of bridges were invited.

In May 1817 the parties finally reached agreement on the course of the canal through Agar's estate (TNA RAIL 860/14: 91). At the end of that month the canal company paid him a total of £15,750 in compensation and purchase costs, following which possession was granted that June.

The lines of the new roads were laid out that August, while excavations and bridge building proceeded throughout the autumn. Contracts for the brickwork of the new bridges on Agar's land were awarded to Richardson and Want (*ibid*.: 196, 217), whilst William Harkom was paid a total of £417 for providing ballast for both these bridges and the Maiden Lane Bridge (TNA RAIL 860/16: 38, 52, 82). A contract to cast and install pipes to carry the public water supply across the bridges was awarded to another contractor in November 1820 (TNA RAIL 860/18: 195).

In December 1818, the Directors of the canal company were informed that the 'heaviest part of the Excavation through Mr Agar's land is done and the several bridges to be erected on that line are far advanced in their execution' (TNA RAIL 860/1: 233). Six months later, it was reported that in addition to the completion of canal wharves at Maiden Lane, 'a considerable extent' of the banks of the canal behind St Pancras Church had also been laid out for wharves (*ibid*.: 252). At the same time a large canal basin was created at Battle Bridge on land belonging to William Horsfall (Faulkner, 1990: 50). While construction of the canal wharves, towing paths and bridges had largely been completed by the summer of 1819, Agar appears to have reverted to type the following spring when he halted work on the construction of the western compartment of the Pancras Lock and the ensuing court case was not resolved for another five years (TNA RAIL 860/18: 48). Despite the fact that Agar continued to instigate proceedings against the company for a further twelve years, the canal finally opened to through traffic at the beginning of August 1820.

The development of the Imperial Gas Light and Coke Company's St Pancras Gasworks on the south bank of the Regent's Canal commenced little more than a year after the canal opened. In October 1821 the directors of the gas company approached the Governors of St Bartholomew's Hospital with a view to acquiring part of the latter's land on the south bank of the Regent's Canal westward of Maiden Lane Bridge (LMA B/IMP/GLC/1: 12/10/1821). The hospital consented to the sale the following year, shortly after which the gas company made separate arrangements with the parish of St Pancras for the purchase of an adjacent plot on the same side of the canal (*ibid*.: 15/02/22, 26/02/1822). Unfortunately for the gas company, the transaction attracted the unwelcome attention of William Agar over whose land contractors were obliged to cross in order to gain access to the site of the new works (*ibid.*: 01/11/1822). According to the records of the gas company, Agar placed obstructions at the entrance to their premises and interfered with the work of the contractors who brought building materials to the site.

The gas company finally began to clear these hurdles the following year. The works at St Pancras gained renewed momentum during the spring and early summer of 1823, when contracts for the construction of the main buildings and the gasholder tanks were awarded (B/IMP/GLC/1: 30/05/1823). On Thursday 24th July, a "multitude of spectators attracted by the interesting ceremony and the fineness of the weather" assembled to watch Sir William Congreve lay the foundation stone of the retort house (*ibid.*: 24/06/1823).

The following February the Imperial held a second ceremony at St Pancras, on this occasion celebrating the completion of the 'indent', a short branch of the canal that enabled barges to unload cargoes of coal to be burnt in the retort houses of the new gasworks (LMA B/IMP/GLC/2: 30/02/1824; TNA RAIL 1189/1423). This short passage subsequently came to be known as the Gasworks Basin (Fig. 3.2). The St Pancras gasworks was formally opened on Wednesday 25th August.

In 1825 the Regent's Canal Company granted the Imperial permission to construct a dock approximately 300ft long on the adjacent plot between the Gasworks Basin and the Maiden Lane Bridge, on land leased from St Bartholomew's Hospital (TNA RAIL 1189/1423). This section of the southern wall of the canal was therefore rebuilt further to the south, effectively widening the waterway in that location. The new structure was multipurpose, serving both as a dock for the gasworks and as a lay-by that allowed barges to draw up so that vessels approaching from the opposite direction could pass. These wharves were leased to the gas company in January 1826.

Archaeological evidence of the construction, appearance, footprint and function of the Regent's Canal Walls in the early 19th century

Landscape Feature 1

A stretch of the Regent's Canal (Landscape Feature 1), from Maiden Lane Bridge in the east to St Pancras Lock in the west, ran along the southern boundary of the Area of Investigation (Fig. 3.2). It was studied in detail during an historic building recording survey under the Site Code KXD07 and a synopsis of the results, taken from Thompson and Matthews 2011, is given below.

The canal was lined with early 19th-century stock moulded purple bricks in fabric 3032 (see Appendix 3 for an explanation of fabric codes), which were presumably laid between 1818 and 1819 when this section was constructed. Its southern bank consisted of a vertical brick structure, the earliest sections of which were made of identical materials to the lining, suggesting that they were built as one and were contemporary. The wall had the dual function of retaining both the canal water and the slightly higher ground on its landward side. Successive rebuilding from 1825 onwards had destroyed the bulk of the initial structure so that only three courses of it were usually visible above water level (Fig. 2.5). In contrast, the taller north wall that bounded the towpath to the north is almost certainly a later addition. The earliest elements of the structure were held together with Portland cement, indicating a construction date of *c.*1830 or later, which strongly suggests that it was erected after the canal opened to traffic in 1820. It was therefore constructed when the Great Northern's Goods Station was built in 1850–2 (see Chapter 2).

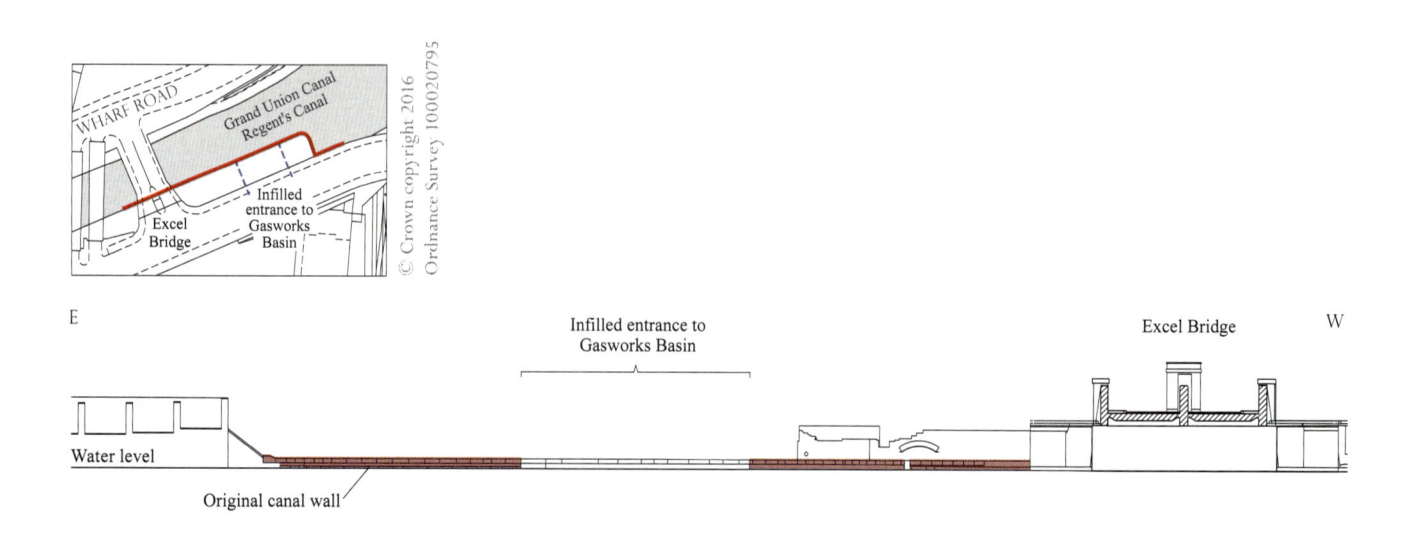

Fig. 2.5   North facing elevation of the south wall of the Regent's Canal (Landscape Feature 1) illustrating the entrance to the Gasworks Basin of 1824, scale 1:500

The hump-backed Somers Bridge to the west and Maiden Lane Bridge (Landscape Feature 2) to the east, which carried York Way (Maiden Lane), had been erected by 1819. Neither survived in their original state; the former was removed and replaced in 1920, whilst the latter was rebuilt on several occasions. The original locations of these structures have therefore been extrapolated from historic maps (Fig. 3.2).

After the Imperial Gas Company constructed its gasworks on the southern bank of the canal, two significant modifications were made to the south wall of the waterway, both of which were identified archaeologically. The earliest of these, the 1824 'Indent' or 'Gasworks' Basin' (Landscape Feature 3; Fig. 3.2), was marked by two straight joints in the south canal wall (Fig. 2.5). Above the waterline, this modification consisted of three courses of early 19th-century red stock bricks in English bond, capped by ashlar blocks of Millstone Grit. Cast iron rings for mooring boats, inserted into the top of some of the stone blocks, were also observed. This section of the wall was raised repeatedly during later periods (Fig. 2.5).

The canal was widened in 1825 in order to construct the Imperial Gasworks' dock and lay-by and this modification was also identified. Curved walls running southwards, deviating from the original course of the canal, marked the extremities of the rebuild (Fig. 3.2). Early 19th-century red bricks surmounted by an Aberdeen granite ashlar capping were used to construct the edge of the new dock and cast iron plates were bolted to it at regular intervals to protect it from damage by boats moored alongside. At the time of the historic building recording survey about five courses of brickwork relating to this phase of construction were visible above the waterline.

2.4 Synopsis

The development of the Battle Bridge area between the mid-18th and mid-19th centuries was inextricably linked with improvements made to the local transport network. The construction of the New Road in the 1750s initiated the district's transformation from a rural hinterland into a thriving suburb and centre of industry. The opening of the Regent's Canal in 1820 was equally crucial to the area's development, attracting a number of noxious industries drawn both by the transport opportunities offered by the canal and by the remoteness of Battle Bridge. The development of the Imperial Gasworks on the south bank of the canal in the early 1820s further stimulated the local economy, creating new jobs for the residents of St Pancras, who in turn demanded goods and services that local businesses were eager to supply. The relative efficiency of water over road transport enabled raw materials and finished products to be imported and exported with comparative ease, giving this peripheral area on the very edge of the growing metropolis an important competitive advantage over less favoured areas. Unlike the suburbs to the south, it still contained plenty of investment opportunities in the form of relatively undeveloped land close to these aforementioned transport arteries as well as, from 1837 onwards, north London's first railway passenger station at Euston. It would not be long before the potential of the undeveloped fields of the Allenbury estate would be spotted by the promoters of the largest railway project in Britain at the time; an ambitious scheme to build a railway between London and York.

Railway Politics and the Formation of the Great Northern Railway

Sir Edmund Denison QC, Chairman of the Great Northern Railway, after the artist Henry William Pickersgill, 1846 (NPG D35018 © National Portrait Gallery, London)

The Great Northern's nemesis George Hudson, 'the Railway King', by George Raphael Ward, 1848 (NPG D9521 © National Portrait Gallery, London)

In the years preceding the formation of the Great Northern Railway Company, few rail connections existed between London and the north of Britain. The principal trunk route connecting the metropolis with the north of the country was the line from Euston to Birmingham, which had been opened in 1838 by the London & Birmingham Railway Company. Connections with lines owned by the Grand Junction and North Union Companies enabled travellers to proceed as far north as Preston by 1839, beyond which a steamer could be taken from the port of Fleetwood to Ardrossan, where it was possible to join the Ayr railway to Glasgow (Grinling, 1898: 6). The formation of the Midland Railway in May 1844 by the railway tycoon George Hudson created a network that connected with the London & Birmingham Company's line from Euston at Rugby, which ran northwards via Leicester, Derby and Leeds, with a branch line to Nottingham (Wrottesley, 1979: 11). Another concern owned by Hudson, The York & North Midland, provided a connection from a junction with the Midland's line at Altofts to York, from where it was possible to continue northwards via the Great North of England Railway, the Stockton and Darlington and the Darlington, Newcastle & Gateshead line. The circuitous nature of this route, coupled with the lack of rival lines on the eastern side of the country represented a lucrative opening for an ambitious newcomer to exploit.

Although the development of a direct rail link between London and York had been proposed by the engineer Joseph Gibbs as early as 1835, nothing had come of this or any other east coast lines promoted during that decade. The mania for railway construction that briefly raged during the mid-1830s came to an abrupt end when the economy faltered in 1837, before entering a downturn which lasted until 1843. As the economy revived, investors' confidence in railway shares also began to rise. By the spring of 1844, three rival concerns had emerged, each of which sought to be the first to establish a major east coast trunk line. They included a revived version of Gibbs' scheme, a scheme devised by the engineer James Walker in the 1830s called the Cambridge & York line, and a scheme called the Direct Northern, which boasted the services of the engineer John Rennie. The prospectus of the Cambridge & York was issued on 22 February, the Direct Northern's was published on 4th April, while Gibbs' scheme was advertised on 16th April 1844 as 'The Great Northern Railway' (Railway Chronicle 1844: 644).

Within days of the announcement of The Direct Northern and Great Northern schemes, the promoters of the Cambridge and York scheme voted to extend their line to London via Peterborough, abandoning Cambridge altogether. Hastily renamed the 'London and York Committee', the promoters proposed to establish a London terminus 'near Pentonville' (ibid.: 220). On the 17th May 1844 representatives of the London & York and Great Northern

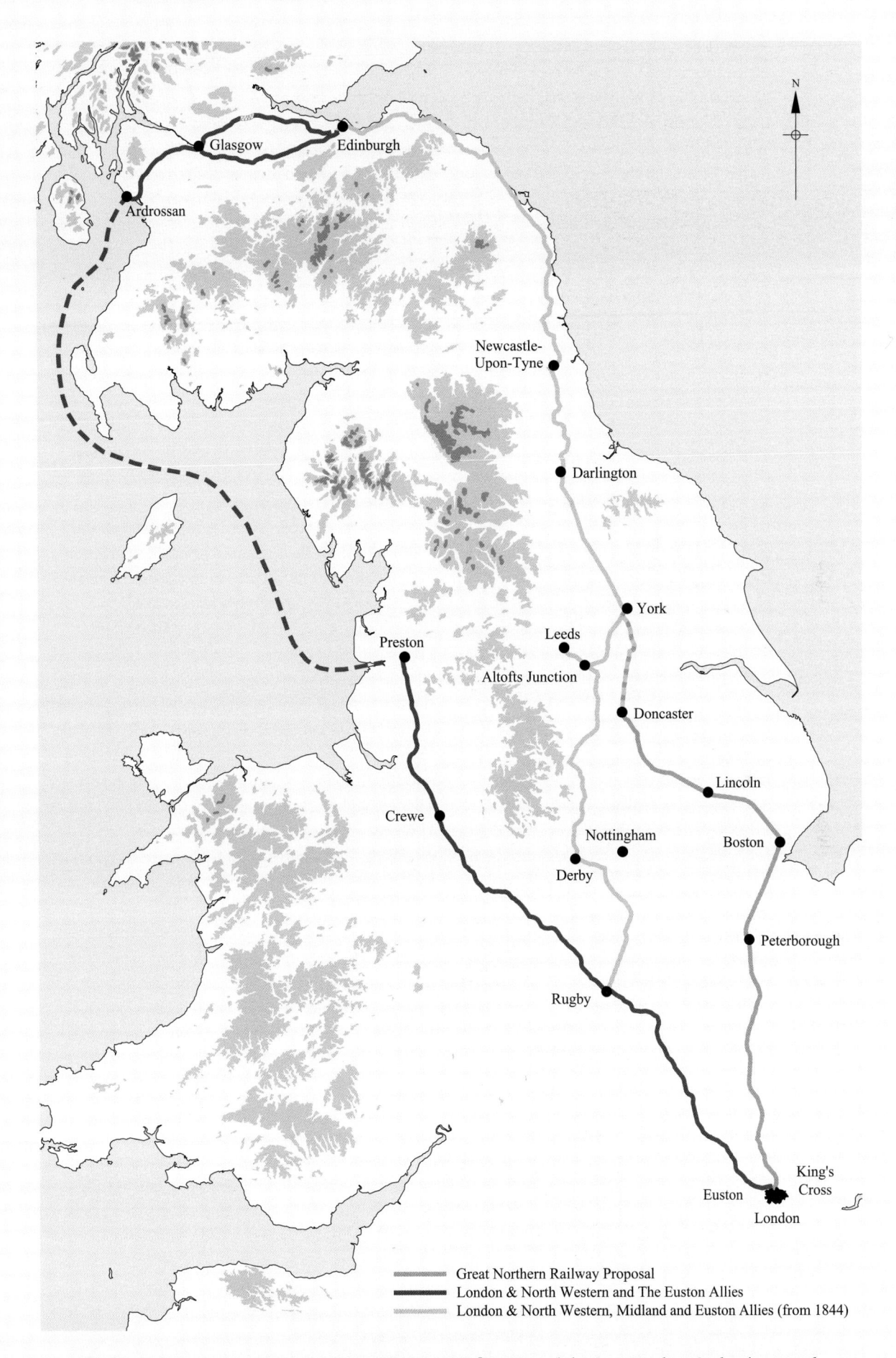

Pre-1850 railway communication between London and the north of Britain and the Great Northern Railway's proposed scheme, scale 1:2,500,000

concerns met at the offices of the former's solicitors, where they voted to merge their businesses under the banner of the London & York Railway. A 'Committee of Direction' was established under the Chairmanship of William Astell, MP for Bedfordshire, while Francis Mowatt, a director of the East India Company and Edmund Denison, MP for the West Riding, were appointed Vice-Chairmen. Within days of the formation of the new committee, James Walker resigned the post of chief engineer. Joseph Locke, the engineer of the Grand Junction Railway, accepted an invitation to serve in Walker's place.

The promoters of the London & York published their prospectus on 11th June 1844. This promised to establish a direct connection between the markets of London and Yorkshire, each of which served a population of more than 1.5 million people (Railway Chronicle, 1844: 220). It was claimed that the line would afford the coalfields of Yorkshire, Derbyshire and Nottinghamshire and the manufacturers of Yorkshire and London opportunities to 'find a ready market along the whole Line'. The prospectus offered farmers and graziers in the East Midlands 'daily opportunities of sending their fat cattle to Smithfield in a few hours', while the 'Market Gardeners of Biggleswade will be enabled to send their produce fresh to the London Market in about two hours'. The proprietors estimated that the scheme would generate a net return of more than 9% on capital of £4,500,000.

The launch of the scheme coincided with the start of the 'Railway Mania' of 1844–6, a period when an unprecedented number of railway schemes were promoted by a mixture of regional concerns such as the London & York, by companies formed by towns bypassed by existing trunk lines, and by the occasional unscrupulous speculator eager to turn a quick profit. In order to obtain the right to the land over which a railway was to be built, promoters put their schemes before Parliament in the form of Private Bills. Although many of the schemes submitted to Parliament during the height of Railway Mania simply duplicated other railway routes, MPs were often reluctant to choose between rival schemes, for fear of showing undue favour to one town over another. When MPs failed to come down in favour of one of a number of competing schemes, promoters were encouraged to merge their interests, stimulating cooperation and potentially encouraging schemes that benefitted all parties (Casson 2009: 17–18). At the end of August Edmund Denison orchestrated a merger with the promoters of the Wakefield, Lincoln & Boston line, which crossed the proposed London & York line at Doncaster. The arrangement enabled the London & York to gain access to the West Riding of Yorkshire, added a further £500,000 of capital to the company's coffers and brought five former directors of the smaller company into the fold, including its head promoter Captain Laws (Wrottesley 1979: 13–14). In October the promoters of the Gainsborough, Sheffield & Chesterfield line were also persuaded to throw in their lot with the London & York. The absorption of the smaller concern completed a scheme which encompassed a total distance of 327½ miles of new track, making the scheme the largest railway project in Britain at that time (Grinling 1898: 25).

George Hudson appreciated the threat that the new company represented to his own interests. Should the London & York's proposals come to fruition, Hudson's empire faced losing a significant share of traffic to and from the metropolis. Hudson therefore pursued a twofold strategy aimed at undermining the interloper. Firstly he set about forging new alliances with smaller companies and promoters, in order that the London & York's 'territory' in the east of England might be penetrated. Secondly, Hudson used his position as an MP to instigate a programme of legal action, vociferous criticism and filibustering in order to disrupt the passage of the London & York Bill through Parliament (Wrottesley 1979: 13). The machinations of Hudson's Euston allies against the London & York in the summer of 1844 were cited as the reason behind Joseph Locke's surprise resignation from the latter's service on 20th September (Nock 1974: 5). The loss of their chief engineer infuriated the committee, who feared the effect that it might have upon shareholder confidence. Fortunately for them, Locke's letter of resignation arrived late on a Friday, which meant that his departure was not reported until the following Monday. In the meantime Denison acted quickly, visiting William Cubitt, the chief engineer of the Wakefield, Lincoln & Boston line at his home in Clapham in order to offer him the post vacated by Locke. On the morning of Monday 23rd September the London & York committee announced the simultaneous resignation of Locke and appointment of Cubitt as chief engineer to their undertaking (Grinling 1898: 23).

Meanwhile the London & York came under attack from the Direct Northern committee, which published its full prospectus at the end of September. The promoters accused the London & York of attempting 'to appropriate upwards of 70 miles' of the route earmarked for their line, as well as copying their plan to establish a London terminus at King's Cross (Railway Journal, 1844: 622). The London & York committee hit back on 3rd October, claiming that the disputed line had been identified by Joseph Gibbs several years before Rennie and his company had even surveyed their route. With respect to the allegation that the London & York had copied the proposal to establish a terminus at King's Cross there was no answer; unsurprising perhaps, given that the company's decision to build there had only been announced by Joseph Locke as recently as August (Grinling, 1898: 21).

By the middle of August the London & York was ready to submit its Bill to the forthcoming session of Parliament. So concerned was Parliament with the anticipated deluge of railway bills, however, that it was decided to screen schemes before they reached the committees responsible for scrutinising them. To this end the Board of Trade was instructed to assess the viability of competing projects, in order that likely 'winners' could be picked at an early stage. On 20th August the Board announced that it intended to consider separately three groups of proposed railway schemes: firstly the various London to York projects, secondly the proposed east–west lines between Lincolnshire and the West Riding and finally the schemes for the rail connection between England and Scotland. Since the London & York committee's scheme was placed in a group containing only north–south schemes, members were unable to contest the various east-west schemes which were being promoted by Hudson with the sole aim of frustrating the committee's plans (Wrottesley, 1979: 15).

The London & York scheme was one of 224 railway bills submitted to Parliament during the 1844 session. Scrutiny of the committee's proposals by Parliament took place in parallel with the enquiry carried out by the Board of Trade. By the beginning of March 1845 the Bill had passed the 'Standing Orders' stage in the House of Commons. Although the Board of Trade had yet to announce its preferred bidder for the London to York line, it was widely assumed that the London & York committee had the strongest case. However opinions began to change during the first week of the month, when it was reported that the promoters of two north–south schemes, the Direct Northern and the Cambridge & Lincoln, were showing renewed confidence in their chances of success. On Saturday 8th March a rumour swept the City of London that the Board would shortly announce its rejection of the London & York's bid. When the markets reopened the following Monday, investors dumped London & York scrip and rushed to buy into the two rival schemes.

The following day the Board formally published its verdict, which as anticipated, rejected the London & York's bid. All was not lost however for the scheme's promoters, who vowed to press ahead. Fortunately the London & York Bill had passed its second reading in the Commons shortly before the publication of the report. Denison and his associates in Parliament questioned the Board's neutrality, asking how its findings had come to be leaked to the City four days before they were scheduled for publication. Much was made of the Board's acceptance of a statement made by George Hudson that cast doubt on the costs of the London & York's schemes, which the Board published almost verbatim in its report. It was also pointed out that neither Locke nor his successor Cubitt had been given an opportunity to explain their figures.

On 25th March the London & York committee published a notice which highlighted the most egregious errors contained in the report. Shortly afterwards the committee's solicitor, Robert Baxter, prepared a comprehensive response to the Board's report, which called into question many of its conclusions. The committee also received the support of the towns through which the line would pass, many of which were bypassed by the Direct Northern and Cambridge & Lincoln schemes. The Bill received support from several leading Parliamentarians, including the Earl of Lincoln, Lord Worsley and William Gladstone. During the Commons Committee hearings into the London & York Bill, the colliery owner Joseph Pease testified about the benefits the line would bring, most notably by reducing the cost of carrying coal to London (Wrottesley 1979: 15).

Another attempt to derail the Bill arose in July 1845. Inspired by a fraud involving the Dublin & Galway Railway, Hudson instructed one of his employees, a Mr Croucher, to canvass local postmasters in an attempt to ascertain the solvency of the London & York's investors. Croucher's findings suggested that numerous subscription holders were classified either as 'needy' or as 'paupers' and would therefore be unable to meet their financial obligations. The list was so long that the deficit amounted to nearly half a million pounds (Nock 1974: 8). The 'offending' individuals were summoned to attend the forthcoming Parliamentary investigation, revealing that £30,000 worth of London & York shares had been signed for by missing persons whilst another £45,000 had been allocated to individuals who could not afford to pay for the stock that they wished to buy. There was insufficient time to investigate further before the summer recess, so the only option was to further delay the passage of the London & York Bill until the following year.

The political wranglings that continued into the next Parliamentary Session are too lengthy to detail here, suffice to say that the passage of the Bill took many months, degenerating into the longest Parliamentary contest on record (Wrottersley 1979: 15). The promoters of the London & York eventually got their Act and in the process Hudson's machinations and questionable practices were exposed (Nock 1974: 8). Ultimately the passage of the Bill triggered a sequence of events that led to Hudson's downfall (Wolmar 2007: 102).

Events finally began to draw to a close on 5th May 1846 when the London & York Bill came before the Lords. On the 30th of that month the promoters dropped their former name in favour of 'The Great Northern Railway', the name coined by Gibbs a decade earlier. On the 8th of June, a Lords committee finally found in favour of the Great Northern. Although wrangling over branch lines continued, Royal Assent was granted on 26th June 1846. Only then could work begin on the construction of the Great Northern Railway's trunk route between Yorkshire and the capital.

CHAPTER 3

The Construction of the King's Cross Goods Station: 1847–1852

'The station now being formed at King's Cross promises to be one of the largest and most important near London. Indeed, in respect to actual area, we believe it excels all the rest. It will be the terminus of the Great Northern Railway, by which the shortest route will be obtained (and consequently the route) to Yorkshire and Newcastle'

Knight's Cyclopedia of London 1851: 846

Following an economic downturn that began in the late 1830s, by 1843 the British economy was on the road to recovery. Positive economic indicators included a series of good harvests, low interest rates and a high gross domestic product (Campbell and Turner 2010: 6). As economic fortunes revived, investors' confidence began to grow, not least in the profits that could be made from the ownership of shares in railway companies. The prolific growth in new railway companies during this period came to be known as the 'Railway Mania' of 1844–6, a period during which would-be shareholders invested vast amounts of money in a hitherto unprecedented number of railway schemes. In order to obtain the right to the land over which a railway was to be built, promoters were obliged to put their schemes before Parliament in the form of Private Bills, the merits of which were considered by MPs. A total of 562 Railway Bills were submitted to the House of Commons in the year 1845 alone, illustrating the extent of the phenomenon (Campbell and Turner 2010: 5).

In order to become a major player in the industry, establishing a terminus in London afforded a considerable advantage over would-be competitors. Having decided to establish their goods and passenger terminals at King's Cross, the promoters of the London & York (as the Great Northern was originally known), submitted their Railway Bill during the 1844 Parliamentary session. After a slow, difficult and expensive passage through the Commons, predominantly due to the numerous legal challenges and assorted delaying tactics employed by rival railway companies, Royal Assent was granted on 26th June 1846. The Great Northern Act set the company's authorised capital at £5,600,000 and granted borrowing powers of £1,868,000 (Wrottesley 1979: 17). The Act also granted the company powers to acquire and demolish the Fever and Smallpox Hospitals and a small number of residential properties at the London end of the proposed line (Stamp 1990: 23; Biddle 1990: 62).

Although the promoters of the London & York originally intended to establish their London terminus somewhere 'near Pentonville', the eventual choice of King's Cross was an astute one. The development of the New Road nearly 90 years earlier had placed this hitherto remote district within easy reach of the heart of the metropolis. Notwithstanding the area's excellent road links, the land on either side of Maiden Lane was sufficiently underdeveloped to permit the construction of extensive railway goods and passenger facilities. Furthermore, the nearby working-class districts of Somers Town, Agar Town and the Drakefield Estate contained sufficient residential accommodation to house the workforce necessary to operate the stations (Fig. 2.3). More prosaically, the Great Northern was prohibited from building on land to the south of the Euston Road by the judgement of the 1846 Royal Commission on Railway Termini, which forbade encroachment on the area known as the 'London quadrilateral' (bounded by the New Road and the Thames) meaning that the chosen site was actually as close as the Great Northern were able to get to the centre of the city (Biddle 1990: 59). Like the Imperial

Gas Light and Coke Company twenty years earlier, the Great Northern was also doubtless attracted by the proximity of the Regent's Canal, which offered an artery for the onward transport of goods and coals received by rail. The establishment of a canal and rail interchange at King's Cross offered the company a ready opportunity to compete with its main rival, the London & North Western, whose yard in Camden was also equipped to handle canal traffic. The Great Northern eventually acquired the 40 acre site for the company's London goods yard from St Bartholomew's Hospital in May 1847 (Wrottesley, 1979: 26). The company's Chief Engineer, Joseph Cubitt, subsequently commended the site, which he maintained offered 'an excellent position, readily accessible by Train Waggons [sic] and communicating conveniently with the Canal' (TNA RAIL 236/273: 27/09/1849).

In contrast to the undeveloped brick fields on the north bank of the canal, the site selected for the Permanent Passenger Station contained a number of existing properties, which meant that development there could not proceed as quickly as the Goods Station. In October 1846 negotiations to purchase the Fever and Smallpox Hospitals broke down, owing to the disparity between the Great Northern's valuation of the site and the sum demanded by the hospitals' owners (Wrottesley 1979: 21). Alternative accommodation for the hospitals then had to be found before the site could be cleared and the matter eventually had to be dragged before a jury resulting in an additional delay of six months (*ibid.*). Of equally pressing concern to the company was the sudden, but inevitable end to 'railway mania' in 1847. Britain's economy entered into a phase of depression as food prices began to inflate after the Corn Laws were

Fig. 3.1 Parliamentary plan of 1846 showing the intended trajectories of the Great Northern Railway's approaches to their Permanent Goods and Passenger Stations at King's Cross, scale 1:8,000

repealed, a situation that caused the unsustainable railway bubble to burst (Wolmar 2007: 88, 106). In fact, by early 1850 when work on the Goods Station had just begun, railway shares had collectively plummeted to their lowest value (Cambell and Turner 2010: 29–33). Although the Great Northern's scrip was not the worst affected, the company's finances were straitened. The slow passage of the Great Northern Act necessitated a huge and unanticipated financial outlay to cover legal fees that amounted to half a million pounds, whilst the construction cost of the project, which was already extremely high by 19th-century standards, had been inflated by a nationwide labour shortage (Biddle 1990: 62; Wolmar 2007: 101). To compound the company's growing financial woes, the topography of the approaches to the proposed site of the Permanent Passenger Station were found to be particularly unforgiving: an outbound steam locomotive would have to go from a standing start in the station and rapidly gain sufficient speed to enable its passage under the Regent's Canal before immediately beginning a tortuous ascent to the top of the Northern Heights (Biddle 1990: 60; Nock 1974:12–3). The construction of the necessary track and tunnel infrastructure represented a challenging and expensive engineering task, which obliged the Great Northern to engage one of the most famous railway contractors of the day, Thomas Brassey, whose services came at a cost (Nock 1974: 13).

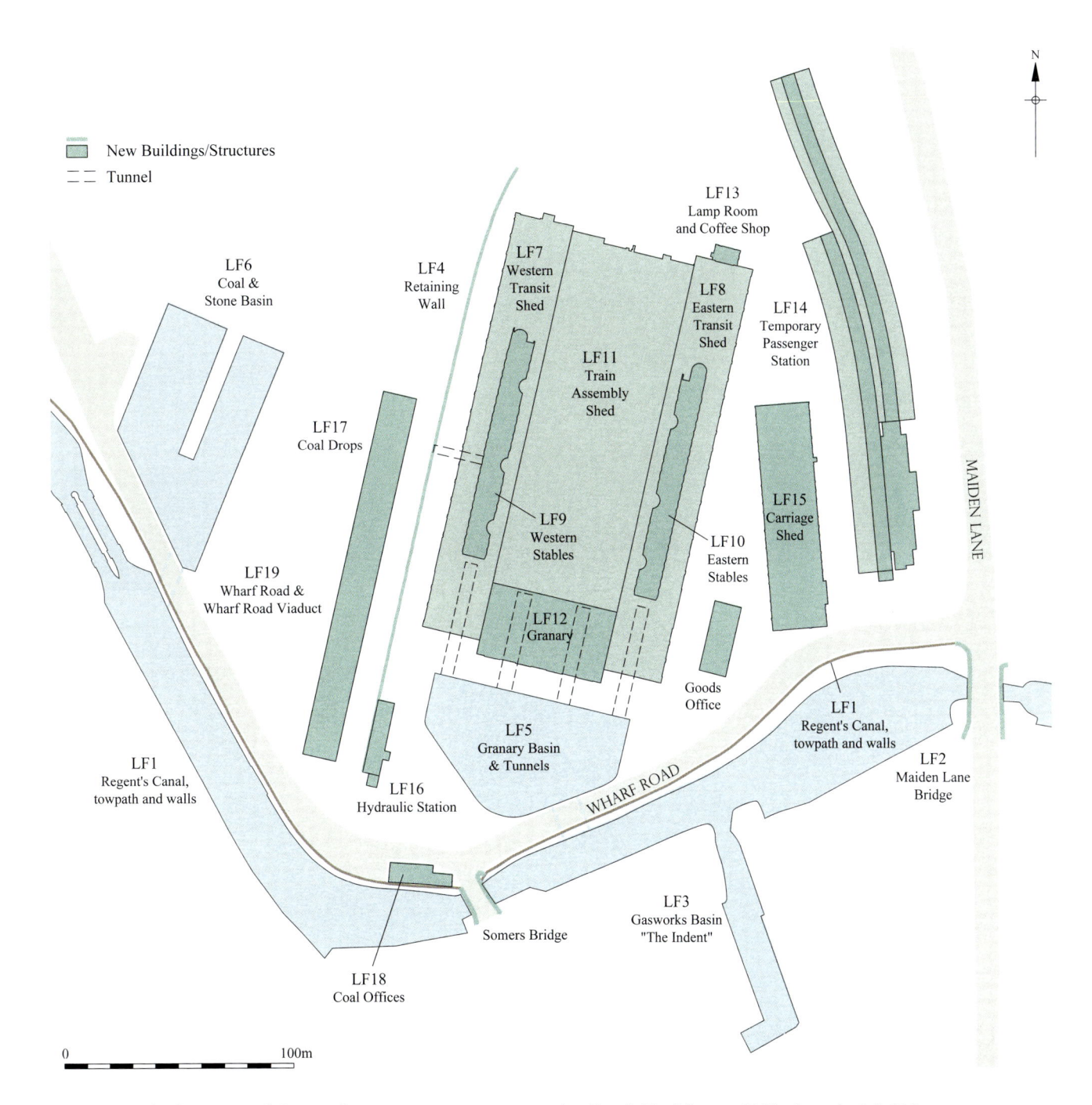

Fig. 3.2 The locations of the Landscape Features present in the Goods Yard from c.1850–2, scale 1:2,500

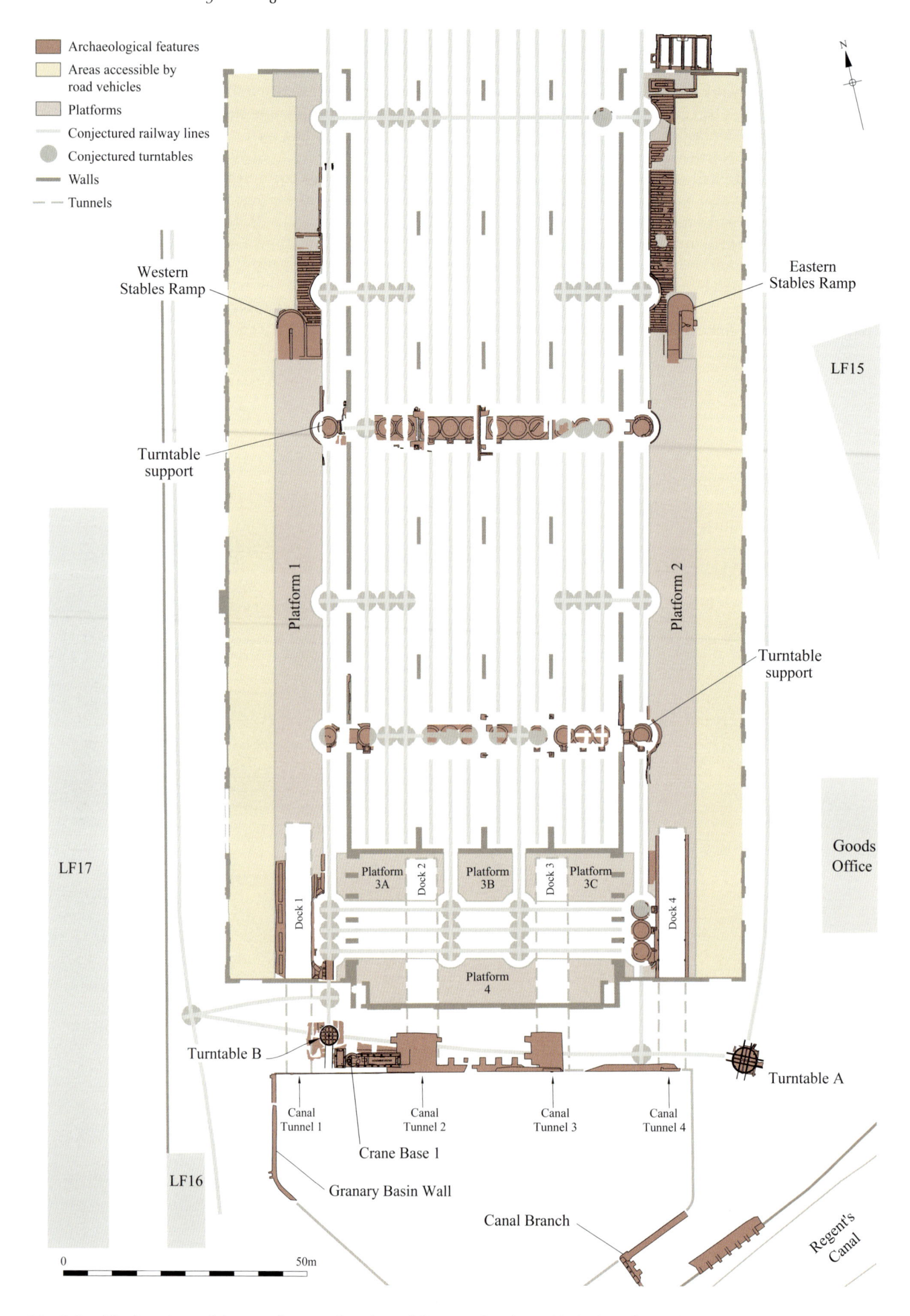

26 *The Construction of the King's Cross Goods Station: 1847–1852*

Fig. 3.3   The locations of the significant archaeological features dated to *c*.1850–2, scale 1:1,000

Although the Great Northern appointed a contractor for the construction of the southernmost section of the line in 1848, the company had still yet to gain full possession of the site earmarked for the Permanent Passenger Station by that date. By the beginning of 1849 a number of Directors had come to the conclusion that instead of continuing with the expensive and technically challenging task of proceeding south of the canal, it would be cheaper to build a terminus on the north side at Maiden Lane (Thorne *et al.* 1990: 94–6). Fortunately the Great Northern's Consulting Engineer William Cubitt managed to convince the Board that the original plan was the right one, and that if the company was to build a passenger terminus at Maiden Lane, it would only be a temporary measure. The Board subsequently authorised the construction of a temporary passenger terminus at Maiden Lane alongside the planned Goods Station (Biddle 1990: 60). This solution would enable passenger services to commence as rapidly as possible thus providing a much needed source of revenue as well as appeasing the company's shareholders, who wanted the London end of the line to open as quickly as possible.

The Parliamentary Plan of the southernmost stretch of the Great Northern mainline gives some idea of the impact that the new stations would have on the area earmarked for the construction of the termini (Fig. 3.1). In addition to the Smallpox and Fever Hospitals, a number of residential properties in the narrow streets to the north of the hospital, together with a number of houses in the eastern half of the Drakefield estate would need to be removed to make way for the Permanent Passenger Terminus and its approaches. The properties scheduled for demolition included the entirety of Paradise Row, Union Place, Essex Street, Lower Edmund Street and part of Norfolk Street. While a well-known and much reproduced engraving of houses in Paradise Row suggests that the street may have earned its reputation as 'that awful rookery at the back of St Pancras Road', the description of the rest of the district as a slum was probably unmerited (Weale 1851: 811). A degree of social disruption must have rippled through the neighbourhood when those evicted to make way for the station were forced to find alternative accommodation in an already crowded district. On the other hand, the construction of the Great Northern Railway's London termini must have generated significant opportunities for employment, which in turn must have furthered the expansion of neighbouring working class districts such as Islington, Camden Town, Agar Town and Somers Town, as Londoners and migrants alike were drawn to the area.

This chapter describes and analyses the construction of the various structures that were erected during the formative period of the Goods Yard and the Temporary Passenger Terminus between 1849 and 1852 (Fig. 3.2). It also sets out to reconstruct the way that the yard operated when it first opened to traffic in the early 1850s.

3.1 Preliminary Ground Works

Two years after the Great Northern Railway Act received Parliamentary Assent, the railway company awarded the contract to develop the site of its future London Goods Station to John Jay, a prolific building contractor based in the City of London. Jay's workforce were unable to commence preparatory works until the following spring owing to the slow progress of the negotiations to acquire the necessary land from a number of owners including the Governors of St Bartholomew's Hospital (TNA RAIL 236/273: 18/04/1849). In April 1849, the Great Northern's Chief Engineer Joseph Cubitt informed the Board that Jay was 'prepared and anxious to make a beginning, and [was] prepared to start work without delay' (*ibid.*). The following month, Jay gained access to part of the site and commenced 'levelling the same down for a station' although the railway company did not gain formal possession of the entire site until the end of August (*ibid.*: 31/05/1849; TNA RAIL 1189/1424: 29/08/1849).

By the following winter, site preparation works were well underway. On 17th January 1850 Joseph Cubitt reported to the Board of the Great Northern that nearly fifteen acres of the Goods Station site had been 'dressed-off to formation level, and 25,000 cubic yards of… clay [had] been burnt for ballast, ready for spreading as soon as the surface is prepared for it' (TNA RAIL 236/273: 17/01/1850). Less than a month later he noted that more than 30,000 cubic yards of 'good burnt clay ballast' was ready to be spread across the station ground once the necessary drains and sewers were complete (*ibid.*: 13/02/1850). Once 'formation level' had been reached and the drainage network finished, work on the foundations of the new Goods Station buildings was scheduled to begin (*ibid.*: 27/03/1850). These were 'well advanced' by the end of March, by which time preparations to attach the iron roofing to the Temporary Passenger Station were also underway (*ibid.*). Five weeks later, Cubitt reported that the Station Ground in the vicinity of the Goods Sheds and the Temporary Passenger Station was 'nearly formed…and a portion thereof [was] ballasted' (*ibid.*: 09/05/1850).

These ground works commenced in April 1849, several months before the Goods Station buildings had been designed. In fact, they began eight months before an architect had even been appointed. The fact that the Great Northern chose to press on in this way may give some indication of the importance that the Company placed upon the completion of the railhead at King's Cross.

Archaeological evidence relating to the nature and extent of the preliminary ground works that were undertaken prior to 1852

The topography of the Study Area took the form of a natural slope, which had to be modified when the Goods Station was built in the mid-19th century in order to make it suitable for railway use (see Section 1.2 for further details). A flat terrace was created through the re-deposition of London Clay so as to create a level area that could be easily traversed by early locomotives, which struggled to cope with anything but the most gradual of gradients. A great deal of wagon shunting was to be undertaken by horses rather than engines and they likewise would have found it difficult or impossible to control heavy vehicles on uneven terrain. In the first instance, waste from the excavation of the Gasworks and Copenhagen Tunnels[1] may have been used to level the area, along with material from the construction of the cuttings that ran towards the station's northern approaches. After May 1850, when work began on the yard's two canal basins, clay may also have been recycled from there. Comparing the height of the towpath of the Regent's Canal at the bottom of the southern edge of the terrace with the top of the made ground suggests that this landscaping work raised the pre-existing ground surface by up to 3.70m around the southern and western sides of the site. This work was presumably carried out by the contractor John Jay and his men from May 1849 onwards.

Joseph Cubitt's reports to the Great Northern's Board of Directors concerning the earliest phases of the Goods Station construction project paint a detailed picture of proceedings on site. In summary, they indicate that ground levelling work began in May 1849 and that this formed the very first stage of the construction process. Work on the foundations of the Main Goods Station buildings then seems to have commenced in the late winter or the early spring of 1850, after 'formation level' was reached in January. Shortly afterwards a layer of 'burnt clay ballast' was deposited across the entire area.

Had events proceeded in this neat order, with the site being fully elevated to 'formation level' before work on the footings of the Goods Yard began, then foundation trenches (termed 'construction cuts' in archaeological parlance) would have been dug into the newly raised ground before the walls of the earliest buildings and other structures were erected within them. These construction cuts should have been immediately obvious to a trained archaeological eye but were striking in their absence. Instead, it was clear that the made ground had been dumped against the sides of the earliest footings, strongly suggesting that these structures pre-dated at least some of the ground raising activity. This contrasts with Cubitt's account, thus demonstrating that early work on the Goods Yard was not as straightforward as he described. Indeed, Cubitt alluded to this himself, when he stated in March 1850 that 'a large part [i.e. not all] of the Station Ground is levelled and formed' whilst indicating that the foundations of the 'large Goods Buildings' (i.e. the Transit and Train Assembly Sheds) were in place (*ibid.*: 27/03/1850). However the archaeological record suggests that the overlap was even more extreme than this: neither the earliest phases of the canal infrastructure nor the Hydraulic Station possessed construction cuts, yet work on the former reportedly did not begin until May 1850, whilst the latter was constructed as late as the first few months of 1851 (see Sections 3.5 and 3.15 for further details). Either the foundations of those structures were erected earlier than documentary sources suggest or, more probably, ground raising works continued for far longer than Cubitt's account suggests.

An orange layer of small burnt clay nodules, no more than 0.65m thick, was found above the redeposited London Clay across most of the site (Fig. 3.4). It is highly likely that this deposit represents the 'good burnt clay ballast' that Cubitt describes in his report to the Great Northern's Board. The effort of manufacturing and dumping this material seems extreme, but impermeable London Clay would have formed an unforgiving construction surface for a major building project, being boggy, slippery and dangerous in wet conditions. The addition of the crushed, burnt clay would have mitigated these problems, providing a well drained working platform that would have enabled proceedings to continue quickly and efficiently in all weathers.

---

[1] These tunnels were excavated by the famous railway contractor Thomas Brassey, who served as engineer to a number of the lines built during the 'railway mania' of the 1840s (Nock 1974: 12; Wrottesley, 1979: 26; see also Section 0). The Copenhagen tunnel enabled the passage of all services through the elevated ground that can be found to the north of the Goods Yard, whilst the Gasworks Tunnel enabled passenger trains to pass below the Regent's Canal and the Imperial Gasworks. Without their insertion, a departing mid nineteenth century locomotive would not have been able to gather enough speed to successfully deal with the adverse inclines. Modern travellers arriving or departing from King's Cross Passenger Station still pass through both of Brassey's tunnels today.

Fig. 3.4 The layer of burnt clay ballast that was deposited in 1849–51 revealed in section

3.2 The Architect's Vision

At the beginning of December 1849 Lewis Cubitt joined his nephew, Joseph, and his brother, William, in the employ of the Great Northern Railway when he was made architect of the Temporary Passenger and Permanent Goods Stations at King's Cross (TNA RAIL 236/15: 199). Lewis set to work designing the complex immediately, inviting tenders for the 'execution of works or buildings necessary for the Permanent Goods Station' on Boxing Day and issuing the schedule of prices for the station buildings at the beginning of January 1850 (TNA RAIL 236/239: 77).

Central to Cubitt's blueprint was the Goods Shed, which would comprise two 'large Warehouses for Goods departing and Goods arriving; with stabling under each' (BL 8244f5: Cubitt 1850: 4). The eastern warehouse, historically known as the 'Eastern Transit Shed' or the 'arrivals', 'import', 'inwards' or 'up' shed, was to receive incoming goods for onward dispatch to London and the south. Its counterpart, the 'Western Transit Shed' or the 'departures', 'export' or 'down' shed, was to handle outbound freight destined for the north (*ibid.*). Once complete, the two Transit Sheds would frame the iron roofed central area known as the 'Train Assembly Shed' or the 'Arcade'. This area had a dual function: it would receive incoming trains, which were to be disassembled into individual wagons for unloading in the Eastern Transit Shed or the Granary and would also deal with the reassembly of loaded or empty wagons into trains for onward dispatch to the north (TNA RAIL 236/239: 4).

Cubitt submitted his original schedule for the Goods Station buildings to the Stations Committee of the Great Northern in January 1850. Although this schedule did not provide for a granary for the storage of grain, he subsequently added one to his designs, which he depicted on a set of 'coloured rough drawings' that he sent to the Board a few days (*ibid.*: 78). In a report accompanying the drawings, Cubitt stated that he had incorporated the structure in his plans at the instigation of the Great Northern's General Manager, Seymour Clarke, who was keen that it be placed near the south end of the Goods Warehouses (*ibid.*: 83).

Although Lewis Cubitt's earliest blueprints have not survived, there is no doubt that the connection between the goods sheds and the canal formed a central element of his design. A plan of the Train Assembly Shed and Temporary Passenger Station signed off by Cubitt in late March 1850 (ICE Cubitt 1850; Fig. 3.5) included an outline of a canal branch leading from the Regent's Canal into a basin (labelled 'Dock'), from which five subterranean tunnels extended beneath the warehouse complex. While only four tunnels were eventually built, these enabled barges to be brought into the heart of the Goods Station, in order that goods could be moved between the railway and the canal quickly and efficiently.

Stables also featured prominently in Cubitt's design. The principal stables were to be incorporated in the below ground foundations of the Eastern and Western Transit Sheds (TNA RAIL 236/239: 80). Horses were integral to the scheme as they were required to haul wagons, turn capstans and so forth. Not all of the yard's apparatus was to rely on human or animal force as partial mechanisation through the use of a pioneering hydraulic system was proposed. This made the construction of an onsite power house a necessity, with the southwest corner of the yard being set aside for the future hydraulic station.

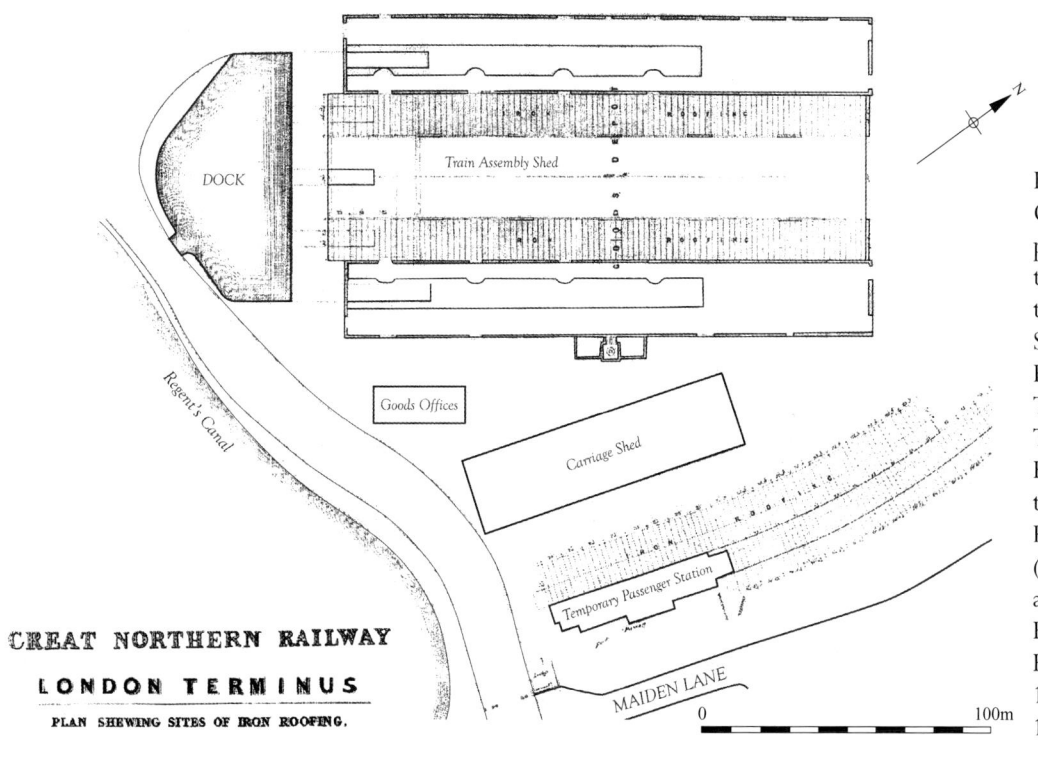

Fig. 3.5 Lewis Cubitt's March 1850 proposal plan of the iron roofing of the Train Assembly Shed (Landscape Feature 11) and the Temporary Passenger Terminus (Landscape Feature 14) showing the Western and Eastern Transit Sheds (Landscape Features 7 and 8) and the Canal Basin (Landscape Feature 5), scale 1:2,500 (ICE Cubitt 1850: drawing no. 1)

3.3 Construction Work Begins

Tenders for the construction of the Goods Station buildings were received during the third week of January 1850, following which the Stations Committee formally awarded the contract to build the Goods Shed complex to John Jay (*ibid.*: 96). Jay was also responsible for undertaking the necessary landscaping and groundworks which archaeological evidence confirms were still ongoing at this point. Two weeks after Jay's second appointment, Cubitt submitted a set of working plans and elevations of the Goods Offices, the Goods Warehouses and the Temporary Passenger Station to the Board, following which construction began rapidly.

By late March, Jay had already laid 'a considerable part of the foundations of the large Goods Building' (TNA RAIL 236/273: 27/03/1850). At the end of the month, Cubitt signed off a set of drawings of the iron roofing of the Train Assembly Shed, which represent the earliest surviving illustration of the two Transit Sheds as originally designed (ICE Cubitt 1850; Fig. 3.5).

Lewis Cubitt's vision of the Goods Shed was encapsulated in a watercolour, which he painted for exhibition at the Royal Academy in the summer of 1851 (Fig. 3.6). The viewer looks across his creation in a northwesterly direction at the Granary, flanked by the Transit Sheds, with the accumulator tower of the Hydraulic Station in the back left corner and the Granary Basin and associated canal tunnels in the foreground.

Fig. 3.6 Lewis Cubitt's 1851 watercolour of the Granary (Landscape Feature 12) and the West and East Transit Sheds (Landscape Features 7 and 8) with the Hydraulic Station Accumulator Tower (Landscape Feature 16) in the background (SSPL 1975-8523 © National Railway Museum and SSPL)

3.4 The Regent's Canal and the Retaining Wall by 1852

Landscape Features 1 & 4

The earliest phase of the extant North Wall of the Regent's Canal (Landscape Feature 1) was probably built at around the same time as the Retaining Wall (Landscape Feature 4) since both performed the same function, shoring up the raised ground that had been deposited to form a level surface in the location of the main Goods Shed buildings. Road access across the canal was also improved at this time since Somers Bridge in its pre-1850 state was too small to cope with the increased volume of traffic that the opening of the Goods Yard would generate. It was therefore rebuilt in 1850–1 in conjunction with the canal-side walls.

A sketch map of the Goods Yard drawn by Captain Douglas Galton of the Royal Engineers in 1852 (Fig. 3.7) illustrates the early arrangement of structures within the yard. Captain Galton drew the plan when he visited King's Cross in October of that year to inspect the signalling arrangements of the recently completed Permanent Passenger Station on behalf of the Board of Trade (TNA MT 6/10/38). Given the purpose of the sketch, it should not necessarily be considered to be an accurate survey of the area. Nevertheless, it is of importance as it represents one of the earliest surviving cartographic representations of the Goods Station. It depicts a curvilinear line running close to and parallel with the southern boundary of the Area of Investigation. This is the earliest illustration of the North Wall of the Regent's Canal (Landscape Feature 1). In contrast to Galton's sketch plan, the extant remains that were recorded demonstrated that the earliest incarnation of the feature continued along the entire southern boundary of the Goods Yard. The wall that retained the western side of the terrace (Landscape Feature 4) was also erected at an early stage, despite the fact that it does not appear on Galton's map.

Archaeological evidence relating to the construction, appearance, footprint and function of the North Wall of the Regent's Canal by 1852

Landscape Feature 1

A high wall, Landscape Feature 1, bounds the north side of the Regent's Canal towpath (Fig. 3.2). The earliest sections of it were built with mid to late 19th-century red and purple stock bricks in English bond held together with Portland cement, a bonding material that did not come into use until 1830. Consequently, it cannot be contemporary with the Regent's Canal, which opened to traffic a decade earlier.

The first incarnation of the structure was no doubt built in 1850–2 by the Great Northern Railway as part of the Goods Station. It functioned as a retaining wall for the raised ground that formed the Goods Yard to the north and as a parapet along the south side of one of the station's thoroughfares, Wharf Road. Whilst many repairs and modifications to the North Wall were made throughout its life, many original sections of it survive. These include the openings to the Coal and Stone Basin and the Granary Basin, the south wall of the basement of the Coal Offices and the south wall of the vaults under the Wharf Road Viaduct.

Archaeological evidence relating to the construction, appearance, footprint and function of the Retaining Wall by 1852

Landscape Feature 4

Landscape Feature 4 consisted of a long wall, orientated north-northeast–south-southwest retaining wall (Landscape Feature 4) (Fig. 3.2). With the exception of cartographic depictions, no documentary evidence concerning the creation of the Retaining Wall has been unearthed during the course of this investigation. The following reconstruction is therefore predominantly based on archaeological evidence. An historic building survey (KXF07) and a series of archaeological excavations (KXI07; KXO08; KXP08) demonstrated that the remains of the Retaining Wall were complex and multi-phase. Whilst the majority of the wall had been rebuilt or altered during later periods, original elements survived above ground level at the northern and southern extremes of the structure, along with a third fragment, visible in the west facing edifice. Below ground excavations revealed a fourth section, which had been truncated horizontally to a depth of 21.64m OD by later versions of the Hydraulic Station. The survival of these elements enabled an estimation of the wall's original appearance, age, method of construction, alignment and function to be reconstructed.

The investigations also revealed 19th-century cobbled paving on both sides of the Retaining Wall. The relative positions of these surfaces indicated that the 19th-century ground surface to the east had been raised by up to 2.63m to a height of 24.00m OD, whilst the land to the west was not built up to the same extent, being at a lower level of 21.37m OD. Logic dictates that the Retaining Wall was in place before the land that abutted its eastern face was raised, otherwise the newly formed terrace would have collapsed immediately. Historical documents do not discuss this explicitly; nevertheless, it is possible to use the available evidence

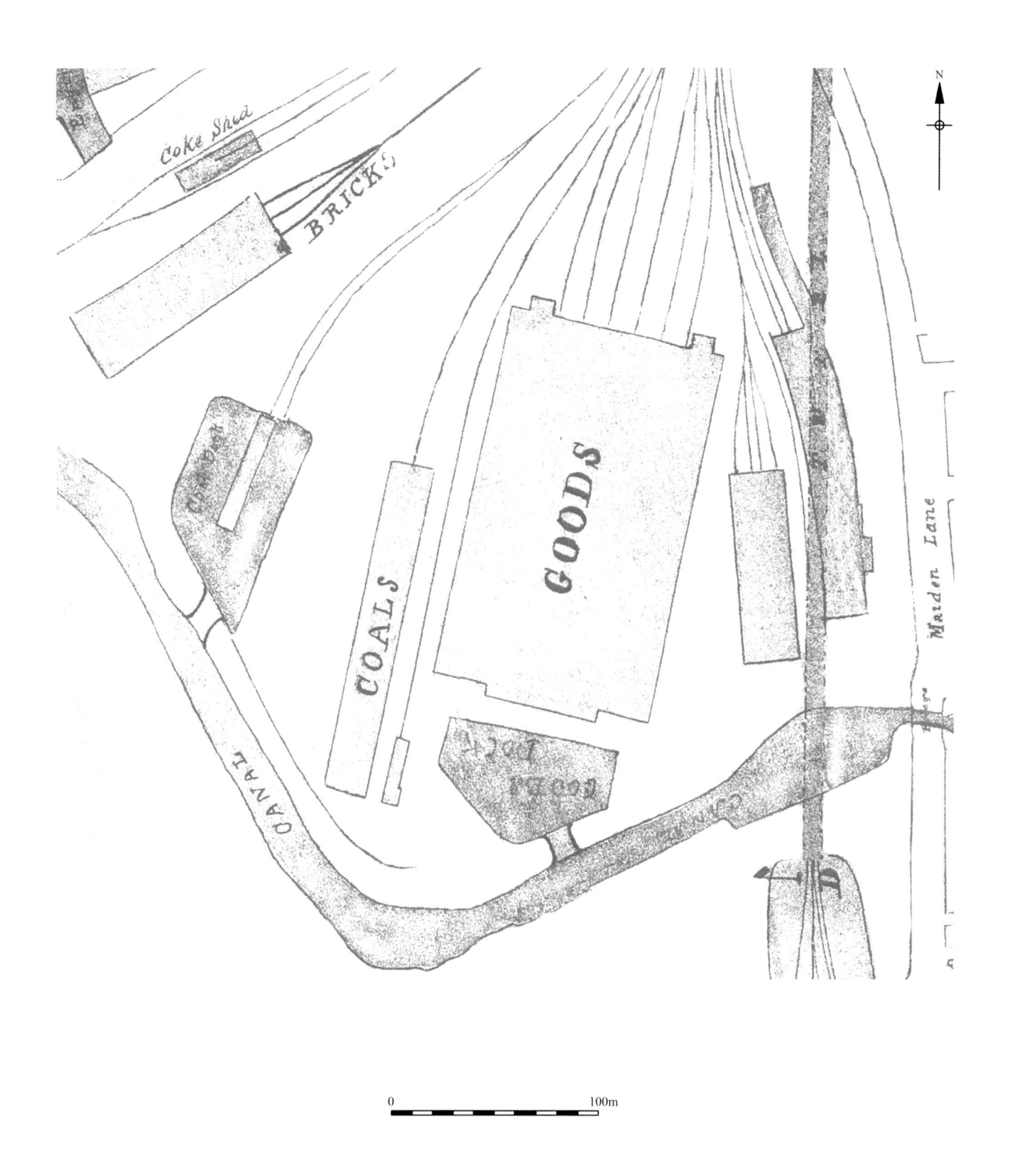

Fig. 3.7 Captain Galton's sketch plan of October 1852 depicting the Goods Yard and its environs, scale 1:2,500 (TNA MT 6/10/38)

in combination with known archaeological facts to deduce the wall's approximate date of construction. In mid February 1850, Cubitt reported that drains and sewers were being laid across the site of the Goods Station in advance of ballast spreading (TNA RAIL 236/273: 13/02/1850). Towards the end of March, he announced that 'a large part of the Station Ground is levelled and formed' and indicated that the foundations of the 'large Goods Buildings' (i.e. the Transit and Train Assembly Sheds) were in place (*ibid.*: 27/03/1850). Given the stratigraphic relationship between the Retaining Wall (Landscape Feature 4) and the raised ground that abutted its eastern face, it must have been constructed before the area occupied by the Granary complex was fully raised and ballasted, perhaps during the late winter or early spring of 1850. The northern end must have been completed later, however, as the entrance to the Western Transit Shed Stables, which archaeological evidence proves was an integral part of the original Retaining Wall rather than a later addition, was not constructed until February 1851 (*ibid.*: 18/02/1851). As the ground cannot have been fully raised across the entire west side of the yard until the Retaining Wall and the tunnel leading to the stables were finished in full, this supports the premise that the landscaping works that were begun by John Jay in 1849 were not finished until the late winter or spring of 1851.

The Retaining Wall was originally built with red stock bricks in English bond and was surmounted by a band of projecting brickwork that was itself capped by sandstone slabs. This capping probably marked the original apex of the wall. At its highest point, the structure was virtually flush with the top of the terrace to the east at 24.20m OD so it is reasonable to assume that it was crowned by a fence or railing to prevent falls from height.

The Retaining Wall and the Hydraulic Station were both original components of the earliest version of the Goods Yard and the fact that they were built at very similar times is not in doubt. However, archaeological evidence indicated that the former pre-dated all elements of the latter. The two structures were not one integral structure bonded together through their brickwork; rather the eastern face of the earlier Retaining Wall was abutted by the later Hydraulic Station.

3.5 Canal Infrastructure: The Granary Basin, the Coal and Stone Basin and Associated Structures by 1852

Landscape Features 5 & 6, the Canal Branch, Canal Tunnels 1 to 4 and Docks 1 to 4

A crucial component of the Great Northern's blueprint of its London Goods Station was the provision of an interchange between the railway and the canal network. This enabled the station to compete on an equal footing with the London & North Western's rival depot at Camden, which similarly incorporated an interchange with the Regent's Canal. The senior officers of the Great Northern were never in any doubt regarding the importance of this to the future success of their enterprise, the presence of the canal being vital to their choice of King's Cross as the future location of their Goods Station (*ibid.*: 27/09/1849).

Although the Directors of the Regent's Canal Company initially objected to the Great Northern's plans to build a goods station at King's Cross, by the beginning of 1849 they had reconciled themselves to the impending arrival of their new neighbour (TNA RAIL 860/42: 89). Shortly after New Year the two companies began to negotiate terms for the transport of goods to and from the station 'with a view to the modelling of traffic arrangements which may be mutually beneficial' (*ibid.*: 169). Whilst the Great Northern's Board had yet to finalise the final design of the goods handling facilities, Joseph Cubitt was given permission to push ahead with the construction of 'two large docks' for the exchange of goods and coal between the railway and the canal. In early May 1850 he met with William Radford, his counterpart at the canal company, in order to discuss the proposed works (*ibid.*: 224). Radford raised no objections, so on the 14th of that month the parties signed an agreement that permitted the railway company to construct two canal arms. The eastern arm would consist of 'a dock communicating with the canal opposite the Imperial Gas Works', which would handle general goods, whilst the western arm would consist of a second dock near St Pancras Lock, devoted to the transshipment of coal (TNA RAIL 236/469: 14/05/1850 A: iii, v). The articles of the agreement permitted the Great Northern to undertake an extensive programme of works above and beside the canal, some of which had already commenced before the agreement was signed and sealed (*ibid.*). In addition to the construction of the new basins, the company was also authorised to take down and rebuild Maiden Lane Bridge (Landscape Feature 2), place a culvert beneath the canal to drain King's Cross Passenger Station, rebuild Somers Bridge and create a third bridge to carry the canal towpath over the mouth of one of the entrances to the new basins (*ibid.*: 14/05/1850 A: i, ii, iv, v). The terms compelled the Great Northern to pay for the cost of all construction, make an annual payment of £350 to the Regent's Canal Company and to bear all responsibility for maintenance of the new structures 'for ever hereafter' (*ibid.*: 14/05/1850, B: i–iii).

Joseph Cubitt reported that the bridge over the dock connecting the canal with the Goods Station was 'in a forward state' by early May 1850 (*ibid.*: 09/05/1850). Cubitt subsequently indicated that work on the Granary Basin (Landscape Feature 5) began after the canal branch (i.e. the link with the Regent's Canal) was completed. Cubitt did not comment upon the excavation of the feature until late September 1850, when he noted that the undertaking was 'proceeding satisfactorily' (*ibid.*: 26/09/1850). Towards the end of the following month, Cubitt announced that the earthworks for the basin were 'well-advanced' and that the resulting spoil was being used to form 'the raised platforms or Embankment for the Coal Station' to the west of the Goods Station complex (*ibid.*: 24/10/1850).

Water had been admitted to the easternmost of the two canal arms by the beginning of January 1851 (Anon 1851a) and the entire installation was complete and fully operational by September of that year, when Lewis Cubitt supplied the canal company with a set of keys for the stop cocks that allowed the docks beneath the warehouses to be emptied in the event of an emergency (TNA RAIL 860/44: 81).

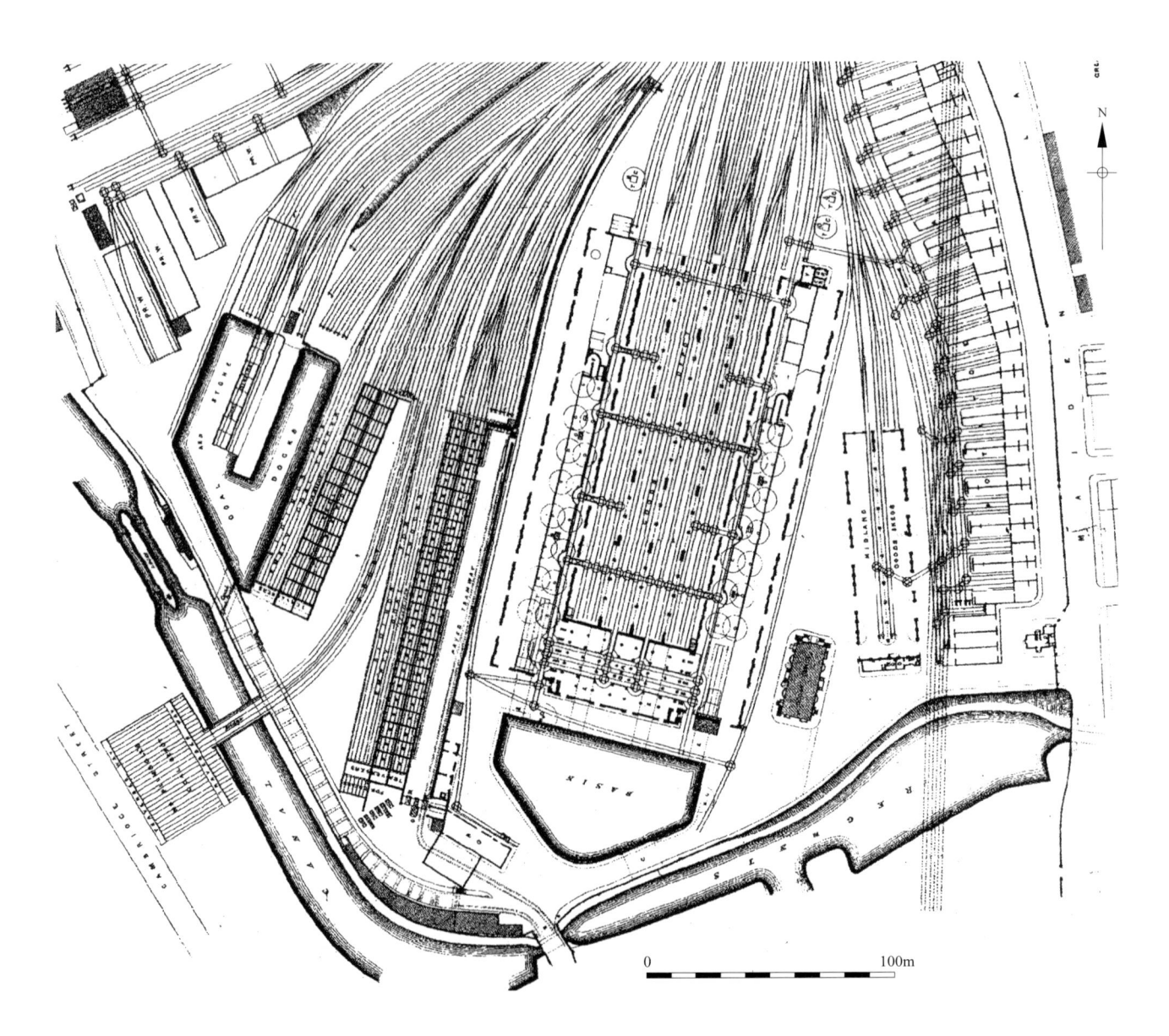

Fig. 3.8 The Humber Plan of 1866 depicting the Goods Yard and its environs, scale 1:2,500; see also Fig. 4.4 for detail (NNRG DMFP 00026266)

Archaeological evidence relating to the construction, appearance, footprint and function of the canal infrastructure by 1852

Lewis Cubitt's plan of March 1850 (ICE Cubitt 1850; Fig. 3.5) depicts the eastern canal arm as a five sided basin (labelled 'Dock'), with marked east–west symmetry, which was linked to the Goods Warehouses and Train Assembly area (Landscape Feature 7–11) by five canal tunnels (Fig. 3.5). The northernmost side of the basin was straight, whilst the southern side was broadly triangular in form, its apex gently rounded off in an almost apsidal form. This symmetry was interrupted on the southeastern side by the 'Canal Branch', which linked the basin with the Regent's Canal to the south, covered in part and surmounted by a roadway which provided access to the Goods Station from Maiden Lane to the east and Somers Bridge to the southwest (Fig. 3.5).

The subsequent decision of the Board of the Great Northern to authorise the construction of the Granary was probably the principal contributing factor that led Cubitt to modify his original arrangement of canal tunnels. Whereas his draft plan suggested that the outermost tunnels exiting the north side of the basin would continue in alignment with its east and west extremities (Fig. 3.5), their eventual locations were arranged a little further towards the centre of the basin, as indicated by William Humber's plan of the Goods Yard published in 1866 (Fig. 3.8) and confirmed by archaeological observation. Most significantly, the total number of tunnels was reduced from five to four, with one serving each Transit Shed and two serving the Granary. One of the earliest surviving depictions of the finished arrangements can be found in Lewis Cubitt's watercolour of 1851 (Fig. 3.6).

The canal tunnels and basin were not built within construction cuts. Instead, the ground was raised around the latter and over the former.

Archaeological evidence relating to the construction, appearance, footprint and function of the Granary Basin and Canal Tunnels by 1852

Landscape Feature 5

The Humber Plan of 1866 clearly shows that a total of four docks with corresponding tunnels ran from the Granary Basin (Landscape Feature 5) into the Goods Shed complex (Fig. 3.8). These will henceforth be differentiated from one another through the use of the terms 'Canal Tunnels 1, 2, 3 and 4', which will be applied sequentially from west to east (Fig. 3.3). The four docks are similarly referred to here as Dock 1, Dock 2 etc. (Fig. 3.3).

A brick wall capped with sandstone slabs, which formed part of the Granary Basin (Landscape Feature 5), was unearthed to the east of the Hydraulic Station (Landscape Feature 16; Fig. 3.2). This wall was 0.75m wide and over 0.59m deep, continuing beneath the vertical

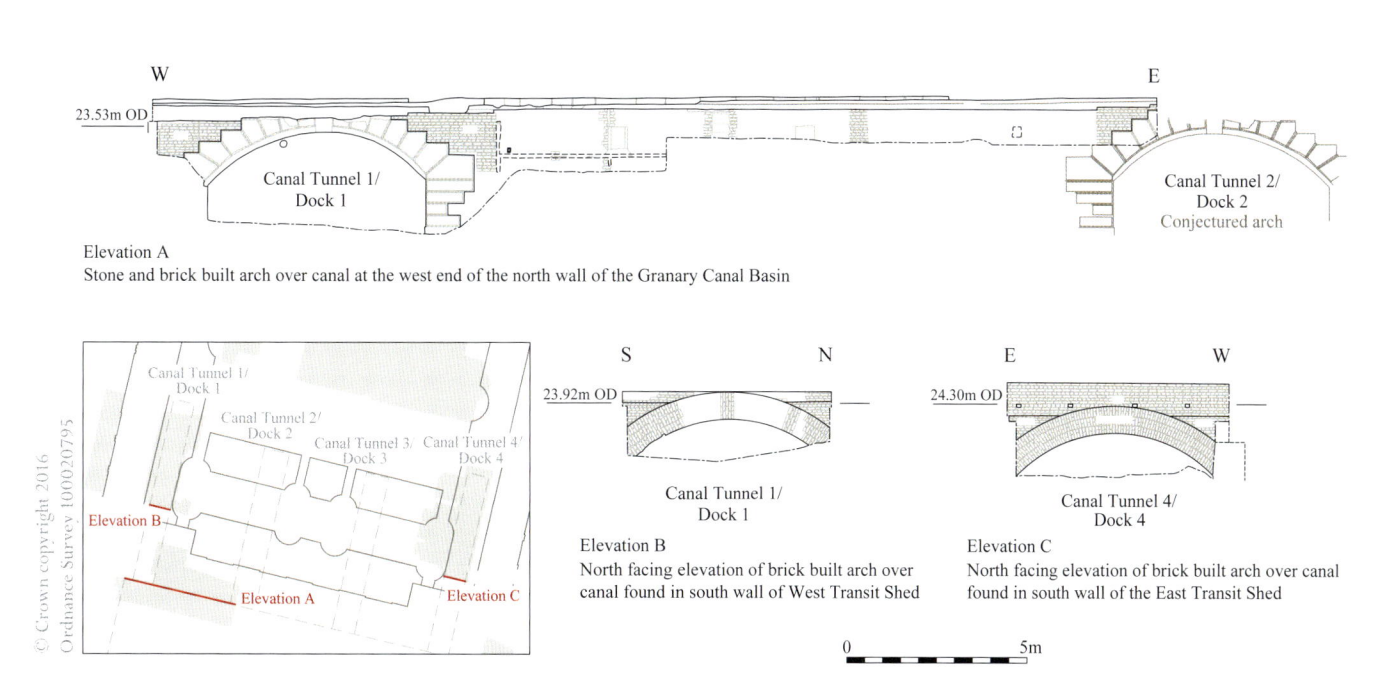

Fig. 3.9   Elevations showing the entrances to Canal Tunnels 1 and 2 in the north wall of the Granary Basin (Landscape Feature 5) and the south walls of the West and East Transit Sheds (Landscape Features 7 and 8) constructed in 1850–1, scale 1:200 (Areas of Excavation are grey in inset)

limit of the excavation, with the top of the stone capping being at a height of 24.10m OD. A second 'L' shaped section of virtually identical wall, over 29.60m in length and 3.40m in depth, was uncovered to the south of the Western Transit Shed (Landscape Feature 7). Parts of the Canal Branch and the Granary Basin wall were also uncovered to the south, which were found to be over 2.70m deep. They were constructed of frogged red, yellow and purple bricks (fabric codes 3032 and 3035) of mid to late 19th-century date set in English bond and capped with sandstone slabs (Fig. 3.2; Fig. 3.3). These cap stones were variously sized, being between 0.58m and 0.60m wide, 1.04m and 1.16m long and 95mm and 100mm thick. The positions of these structural elements correlates well with the location of the perimeter of the Granary Basin (Landscape Feature 5) as shown on the Humber Plan of 1866 (Fig. 3.8; Fig. 3.2).

An arch was incorporated in the western end of the southern face of the north wall of the Granary Basin (Landscape Feature 5; Fig. 3.9, Elevation A). It was over 3.50m tall and 5.95m wide, the opening being 0.89m below the apex of the masonry forming a segmented structure of sixteen five sided sandstone blocks, arranged either side of a central keystone, resting on two stone imposts. Internally, the intrados surrounded a cavity that was 5.96m in width at its widest point. This had been packed with clay and rubble, deposited after the structure fell out of use. The very edge of some identical masonry was observed in the same wall, 21.94m to the east, continuing beyond the eastern limit of the archaeological intervention (Fig. 3.9, Elevation A). This presumably formed the western edge of a second arch.

These two arches represent the external entrance and egress points that lead from the Granary Basin (Landscape Feature 5) to two of the four canal tunnels that could be found below the goods warehouses. These were striking in their similarity to the arched entrances depicted in the foreground of Lewis Cubitt's 1851 watercolour (Fig. 3.6) whilst their locations corresponded exactly with the anticipated positions of the entrances to Canal Tunnels 1 and 2. The remains described here therefore belong to those structures, which ran into the Western Transit Shed (Landscape Feature 7) and the western side of the Granary (Landscape Feature 12). The tunnel entrances had been built with frogged, stock moulded bricks (fabric 3032), indicative of a mid to late 19th-century date, some of which had been stamped with the initials 'JJ' (presumed to be the initials of the builder 'John Jay').

An arch also formed part of the below ground foundations of the internal face of the southern wall of Landscape Feature 7, providing entrance and egress to 'Canal Tunnel 1' from 'Dock 1' in the Western Transit Shed (Fig. 3.9, Elevation B). This lacked the ornamental and relatively expensive stone archivolts that characterised the external tunnel openings, presumably because it was internal to the warehouse complex and was therefore less visible. Instead, it was formed by three to six rows of red bricks, laid either in soldier bedding, rowlock bedding or some combination of these styles. An identical arch formed part of the internal face of the south wall of the Eastern Transit Shed, providing access between 'Dock 4' and 'Canal Tunnel 4' (Fig. 3.3; Fig. 3.9).

A sunken brick lined feature was discovered in front of the internal entrance to Canal Tunnel 1, which was integral to the earliest phase of the Western Transit Shed's platform foundation. It was over 23.30m long,

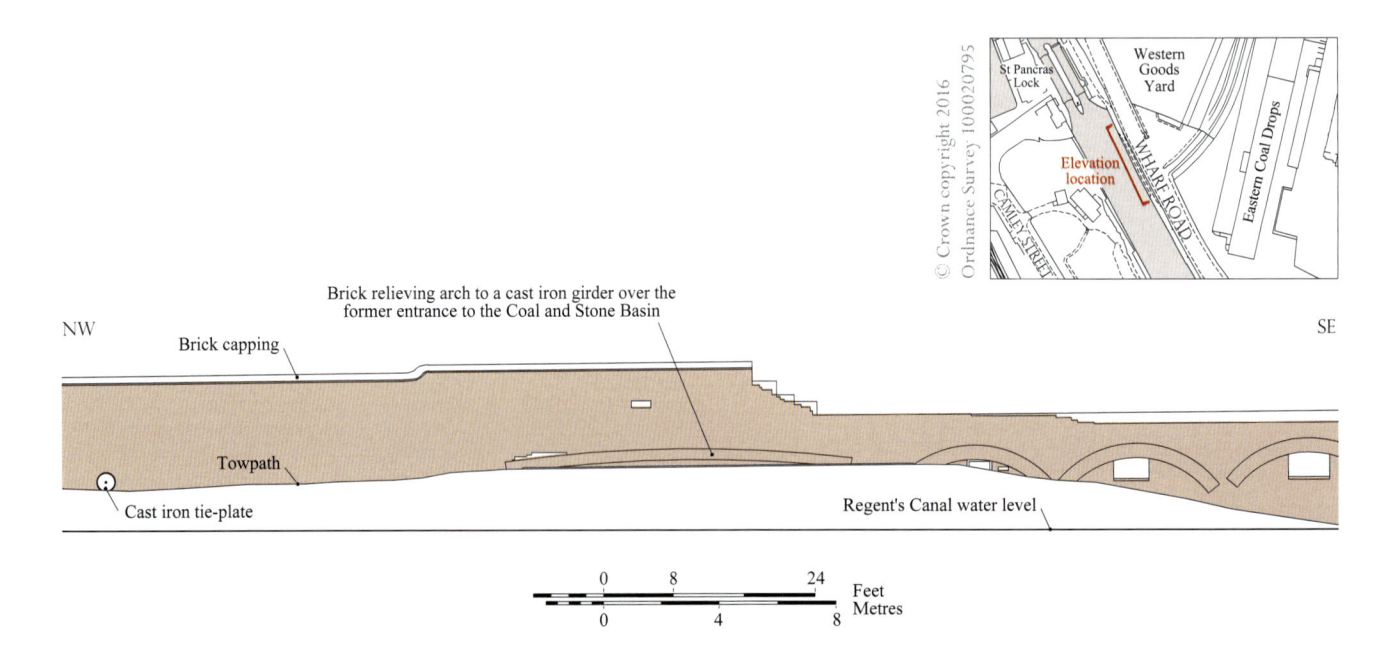

Fig. 3.10 Southwest facing elevation showing the former entrance to the Coal and Stone Basin (Landscape Feature 6) integral to the north wall of the Regent's Canal (Landscape Feature 1) constructed in 1850–1, scale 1:250

2.30m wide (approximately 7' 6") and over 1m deep and it indisputably represents a canal dock (henceforth termed 'Dock 1'; Fig. 3.3). Like the canal tunnels, the structure had also been constructed from frogged stock bricks (fabric 3032), corroborating a mid to late 19th-century construction date. In order to ensure sufficient clearance for the barges, the surface of the water in the dock and the tunnel must have been some way below the ceiling of the latter, which was found to be at a level of 24.24m OD. The barges would therefore have been considerably lower than the top of the West Transit Shed's platform, which was just 0.31m lower than the top of the canal tunnel at 23.93m OD. This would have made manual unloading of heavier items from a barge to the platform somewhat challenging, a difficulty that was presumably resolved through the use of hydraulic or hand operated cranes. The height of the platform would have allowed sufficient headroom for the use of these machines, which presumably off loaded goods from barges for onward transport to the north by train.

The associated canal tunnels were not directly observed, but calculating their lengths from the positions of the opposing tunnel entrances indicates that those that were identified in the Transit Sheds were 19.01m in length. The nature of the tunnels that entered the Granary (Landscape Feature 12) and their associated docks remain unknown, as the internal elements of those features were not encountered archaeologically, however it is reasonable to assume that they closely resembled those that were discovered in the Transit Sheds. If the Humber Plan is taken at face value, then they were shorter than those equivalents whilst their docks were covered by the Granary's platforms (Fig. 3.8). The available evidence therefore suggests that sacks of grain were manoeuvred into barges through trapdoors that were built into the aforementioned platforms (see Section 3.11 for further details).

Archaeological evidence relating to the construction, appearance, footprint and function of the Coal and Stone Basin by 1852

Landscape Feature 6

The Coal and Stone Basin (Landscape Feature 6) lay beyond the boundaries of the Area of Investigation, therefore the nature and location of this feature has been extrapolated from historic maps (Fig. 3.2). The only exception to this was the entrance to the canal arm that linked the Coal and Stone Basin with the Regent's Canal, which was identified during an historic building recording survey of the north wall of the waterway (Landscape Feature 1; Fig. 3.10). It consisted of an opening in the wall (now bricked up) that was surmounted by a relieving arch for a cast iron girder.

3.6 The Main Goods Station: The Western and Eastern Transit Sheds, the Western and Eastern Stables, the Train Assembly Shed and the Granary by 1852

Landscape Features 7 to 12

Lewis Cubitt's drawings, specifications and descriptions indicate that each element of the Main Goods Shed was carefully designed to fulfill a specific function. Cubitt's design ensured that these elements worked harmoniously together in order to deliver the most efficient railway goods handling facilities available in mid-19th-century Britain.

The loading and departures shed (the Western Transit Shed) dealt with all kinds of north-bound cargo. As a general goods shed, the bustling building handled a wide range of commodities, most of which presumably originated from London's manufacturing industries. These goods would have included considerable amounts of textiles, leather and finished clothing, metal products from foundries, chemicals and related substances like soap, starch and paint, wares from glass and pottery works and other specialist products such as rope, candles, matches, cigars and printed materials (Marshall 2013: 65–157). The shed also handled consignments of gunpowder until the late 1870s, and it is possible that cartridges and percussion caps manufactured by the famous firm of Eley Brothers in their factory in nearby St Pancras Road were distributed via the Goods Station (TNA RAIL 236/146: 365).

On a smaller scale, artisan goods produced by London based craftsmen, such as furniture, pianos, clocks and scientific instruments would have been a likely feature in the shed (Marshall 2013: 100–1). Railway companies would accept contracts of almost any size so one-off shipments of 'smalls' and special consignments from businesses and private individuals alike would have been welcome (Holden 1985: 97–100). For an additional charge, these items could be shipped door to door via railway cartage teams stationed at either end of the line (*ibid.*: 100).

The Eastern Transit Shed received commodities destined for London, the south of England and even further afield from all points on the line north of King's Cross. Along with a wide array of produce from the industrial heartlands of the north, such as textiles from the West Riding and steel from Sheffield, huge amounts of perishable foodstuffs destined for the London market flowed through the building, including vegetables, meat and fish.

London's appetite for necessities such as grain, coal and bricks was insatiable and, had the Eastern Transit Shed received these, it would doubtless have been overwhelmed as soon as it opened. To make matters worse, most goods that the city needed in bulk appear to have been stored in the yard for a slightly longer time than was common for more general commodities, presumably because one shipment of something like grain or coal, for example, might be split and collected by independent merchants or sold on in batches by representatives of the Great Northern Railway itself. This system would have been impossible to implement in the Eastern Transit Shed, where temporary storage took place on the platforms. Separate areas dedicated to the reception of commodities like grain, coal, bricks and stones were therefore needed elsewhere in the yard from the outset and were indeed created at an early stage of the yard's existence. With the exception of the Granary, these were separate from the main goods shed, which formed the centrepiece of a much larger goods handling depot.

Work on the foundations of the Transit Sheds began in the early months of 1850. John Jay's team made good progress and the bulk of the foundations were in place by late March (TNA RAIL 236/273: 27/03/1850). Work continued throughout the summer and early autumn of that year, during which time the construction of the yard's railway infrastructure had also commenced (*ibid.*: 04/07/1850). Whilst the Goods Shed could only be described as a work in progress when the writer John Weale visited in early 1851, enough had already been built to enable him to gain a clear impression of the 'magnificent scale' of Lewis Cubitt's creation (Weale 1851: 811). Weale observed that each of the two Transit Sheds comprised 'two side cart roads, running longitudinally and parallel to the rails', from which they were separated 'by a rather wide platform for the reception and classification of the goods' (*ibid.*: 812). Each was served by a single railway track, connected to the 'centre part of the shed' (i.e. the Train Assembly Shed) 'through a corresponding number of sliding doors, exactly opposite to other doors in the outer walls, communicating with access roads' (*ibid.*). This account suggests that the Goods Shed complex was being built in accordance with Cubitt's proposed design and gives no indication that it had been necessary to make any last minute alterations.

The main elements of the Goods Shed complex were recorded during a historic building recording survey (KXF07; Fig. 1.2) and an archaeological excavation and watching brief (KXI07; Fig. 1.3). The results of these investigations are considered in the context of the principal elements of the Goods Shed in the following paragraphs.

3.7 The Western Transit Shed by 1852

Landscape Feature 7

The Western Transit Shed was to be built in brick, 600' long by 75' wide, and covered with 'strong timber roofs, slated or glazed' (TNA RAIL 236/239: 80). Although partially obscured by the Granary, the Eastern Transit Shed can clearly be seen in Cubitt's 1851 watercolour. As these buildings were, with a few minor differences, mirror images of one another, this suggests that they both had three arches in their southern faces, two of which functioned as access points whilst the larger, central examples were blind and decorative.

According to Cubitt's specifications for the construction of the shed, an internal cart way was to run the length of the shed to the west of a 400' long, central platform that would receive goods for onward dispatch by rail. This was to be raised on a series of 'brick arches turned against Wrought Iron Girders' (*ibid.*: 80, 105), whilst the platform surface was to be formed of wooden planks as depicted in The Illustrated London News of May 1853 (Anon 1853). A single line of railway track was proposed along the east side of the platform that would be connected by turntables and transverse tracks to the Train Assembly Shed and, in later versions of Lewis Cubitt's design, the Granary (as shown on the Humber Plan of 1866, Fig. 3.8). Lewis Cubitt also planned to integrate subterranean stabling in the foundations of the Western Transit Shed's platform (TNA RAIL 236/239: 80).

Construction of the Western Transit Shed proceeded rapidly. Towards the end of October of 1850, Joseph Cubitt informed the Board that 'in the course of the next few days … the lines [will be] laid into the Western Goods Building [i.e. the Western Transit Shed] ready for traffic' (TNA RAIL 236/273: 24/10/1850), implying that the structure was nearing completion by that time. Cubitt subsequently reported that it was virtually finished by November of 1850, some four months earlier than its counterpart to the east (*ibid.*: 11/1850).

Built heritage and archaeological evidence relating to the construction, appearance, footprint and function of the Western Transit Shed by 1852

Landscape Feature 7

The Western Transit Shed is a brick building consisting of a single storey constructed in Flemish bond. Its roof has a gabled northern end and originally had a fully hipped southern end although the extant roof is now half hipped (a mid-20th century alteration). It is 182.88m (600') in length, 25m (82') in width and almost 7.5m (24' 6") high. The footprint of the building and its location relative to the other Landscape Features in the Goods Yard is shown in Fig. 3.2.

The south (i.e. front facing) elevation of the Western Transit Shed originally had two symmetrically placed openings of equal proportions arranged either side of a decorative blind arch (Fig. 3.6). The westernmost example admitted road vehicles to an internal cart way, whilst the eastern opening accommodated the shed's railway line as shown on the Humber Plan of 1866 (Fig. 3.8). Although the openings were infilled after they fell out of use, their locations can still be discerned in the extant structure. These remains demonstrate that they were originally surmounted by segmental arches that were composed of bricks that had been bedded 'on edge' (Fig. 3.11; Fig. 3.12), whilst on the internal side of the shed a secondary relieving arch was noted above the western opening which presumably provided additional structural support (Fig. 3.11). Externally, only the location of the western opening is visible today as its eastern equivalent was concealed when the Western Goods Offices (Landscape Feature 20) were constructed some years later (Fig. 3.12). Since little importance was attached to the aesthetics of the interior of the building, the central decorative blind arch was only present on the external face of the Western Transit Shed's southern wall. As such, most of it was concealed by the Western Goods Offices, although sections of it could be discerned from inside that later building.

The external face of the north (rear) wall of the Western Transit Shed (Fig. 3.13) was originally very similar to its southern equivalent, the main difference being the choice of a gabled rather than a hipped canopy. Mid 20th-century roof repairs reconstructed the upper reaches of the shed's walls as with the Eastern Transit Shed. Unlike its eastern counterpart the original parapet wall survived the rebuild on the western side, so the original stone corbels with cavetto and ovolo mouldings remain extant. On the eastern side, these original components were reinstated in the wake of late 19th-century building work.

The northern elevation was dominated by a decorative, blind segmental arch that was flanked by two smaller openings of identical sizes that were aligned with their counterparts in the southern façade. The most westerly of these therefore permitted the passage of road vehicles through the building, whilst the most easterly accommodated the shed's railway track. Given that the Western Transit Shed dealt with departures only, it is likely that the eastern opening was an exit that provided railway wagons with an egress point leading from the building, access being provided by the equivalent structure in the southern edifice as well as several openings in the western façade. No blind arches were observed on the plain internal side however an original timber beam was bolted to the wall around the eastern train wagon exit. It probably supported a sliding door, which suggests that these entrances could be shuttered off when necessary.

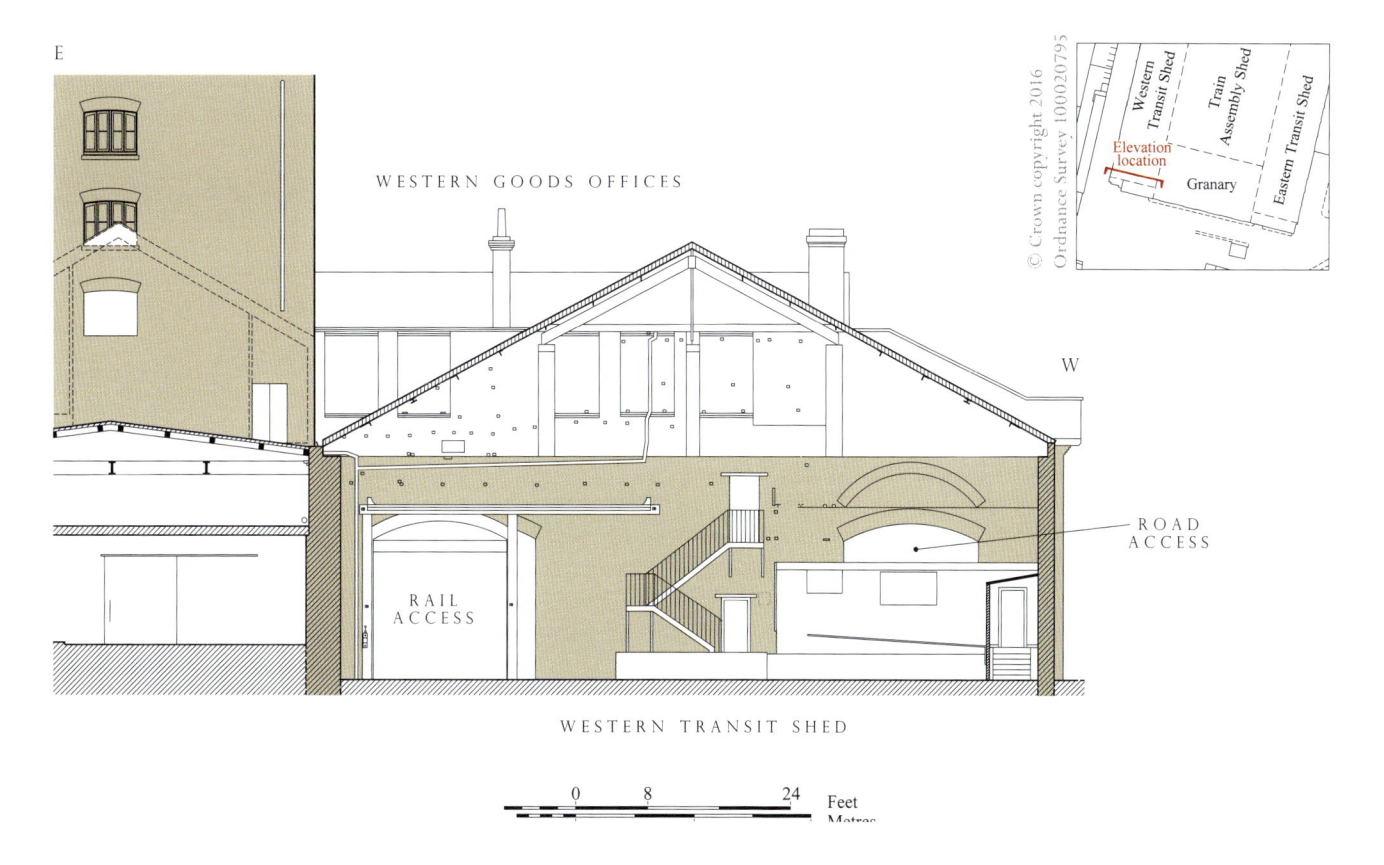

Fig. 3.11 Cross-sectional elevation through the southern end of the Western Transit Shed (Landscape Feature 7) showing the internal face of the south wall and the original 1850 masonry, scale 1:250

A thin scar formed by a structure with a pitched roof was visible on the northern exterior face of the Western Transit Shed (Fig. 3.13). Since the scar crossed the shed's ornamental blind arch, partially obscuring that decorative feature, it cannot have formed part of Cubitt's original plan for the building. Despite this it must represent a very early addition as it is shown on maps of the Goods Yard, dating back to 1852 (Fig. 3.7). Given its location at the northern end of the 'Outwards' Shed and its proximity to the terminus of a 'down' line and a heavy crane (Fig. 3.8), the structure probably represents a covered platform onto which goods for onward dispatch by rail were stockpiled. Its pitched canopy would have protected the commodities that were temporarily stored upon it from the elements before they were transferred to north-bound trains. In all probability this canopy was created after the shed opened for business between 1850 and 1852 to facilitate actual working practices.

The external west (side) elevation of the Western Transit Shed incorporated 23 decorative, evenly spaced recesses in the form of blind segmental arches. Six of these (the locations of which are illustrated in Fig. 3.14) originally had openings built into them, which provided road vehicles with access and egress points along the length of the internal cart way. On the internal side five of the six openings were delimited by wooden frames for sliding doors, demonstrating that they could be shuttered off when necessary (Fig. 3.15). The Humber Plan (Fig. 3.8) suggests that an additional five small entrances for pedestrians were also present, only two of which have survived into modern times (Fig. 3.14).

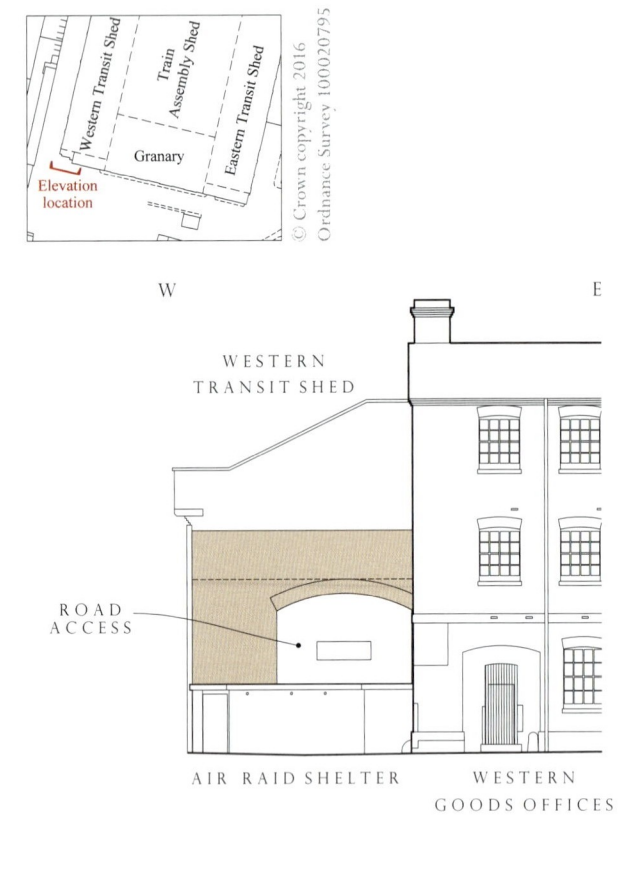

Fig. 3.12 External elevation of the south wall of the Western Transit Shed (Landscape Feature 7) and the original 1850 masonry, scale 1:250

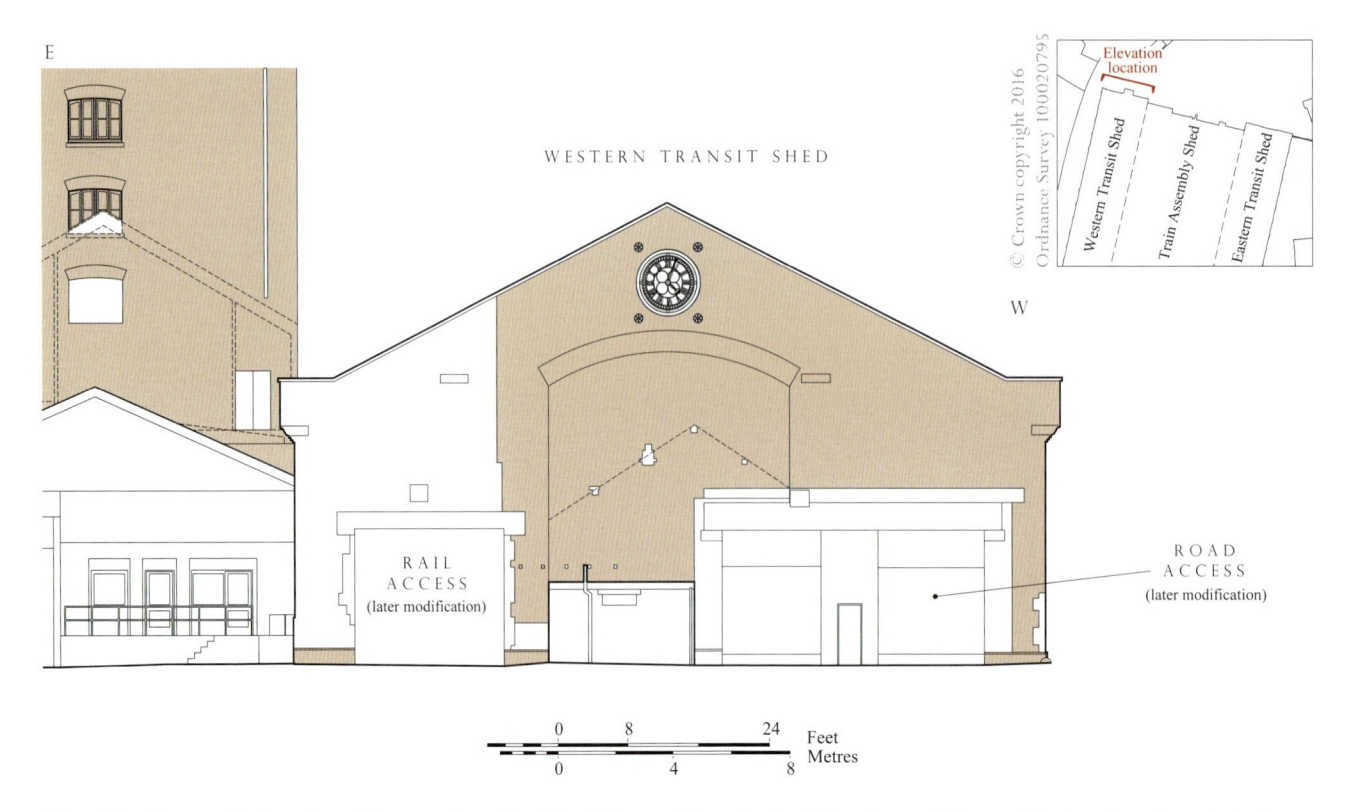

Fig. 3.13 External elevation of the north wall of the Western Transit Shed (Landscape Feature 7) showing the original 1850–1 masonry, scale 1:250

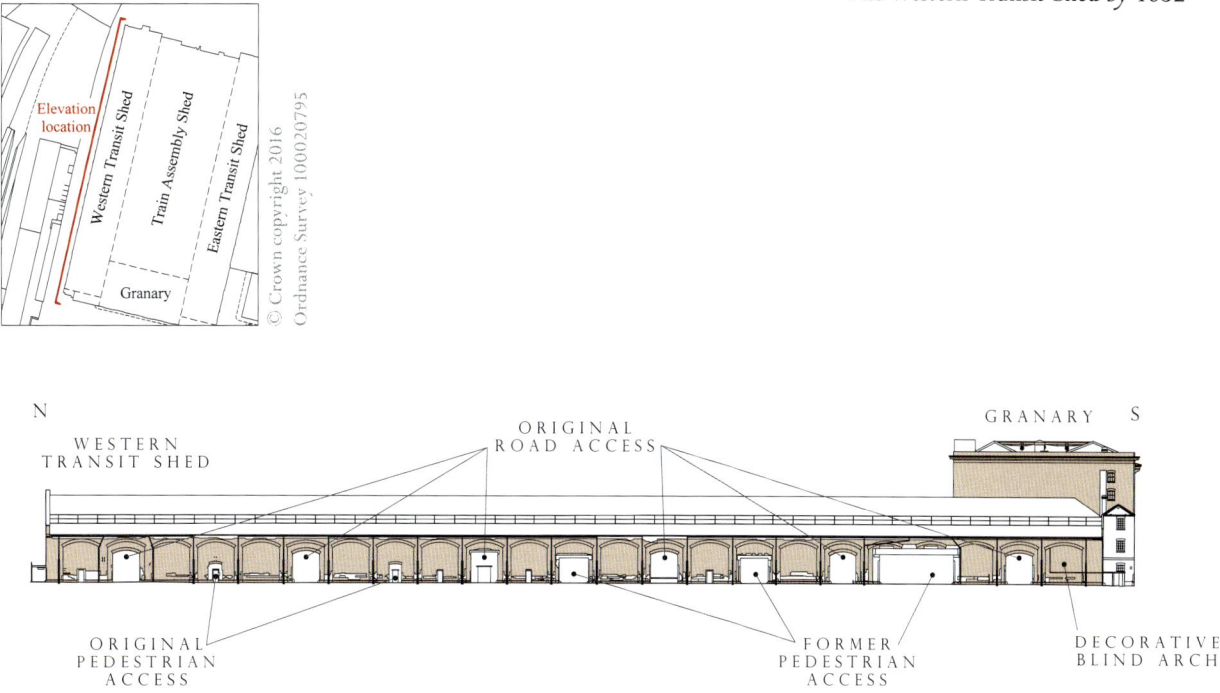

Fig. 3.14 External elevation of the west wall of the Western Transit Shed (Landscape Feature 7) showing the original 1850–1 masonry, scale 1:1,250

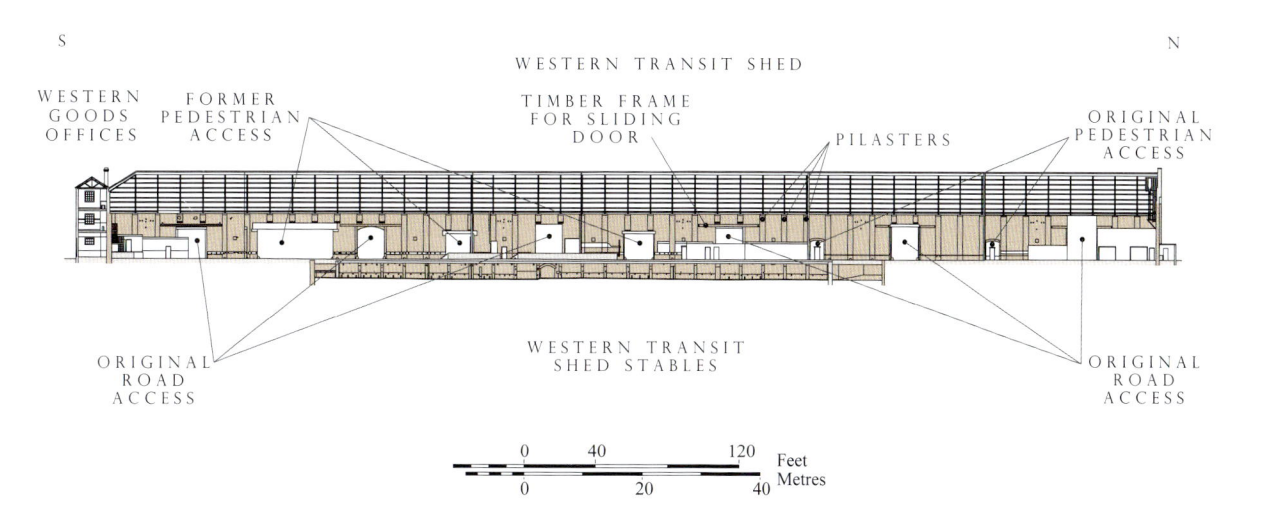

Fig. 3.15 Cross-sectional elevation through the Western Transit Shed (Landscape Feature 7) and the Western Transit Shed Stables (Landscape Feature 9) showing the internal faces of their west walls and the original 1850–1 masonry, scale 1:1,250

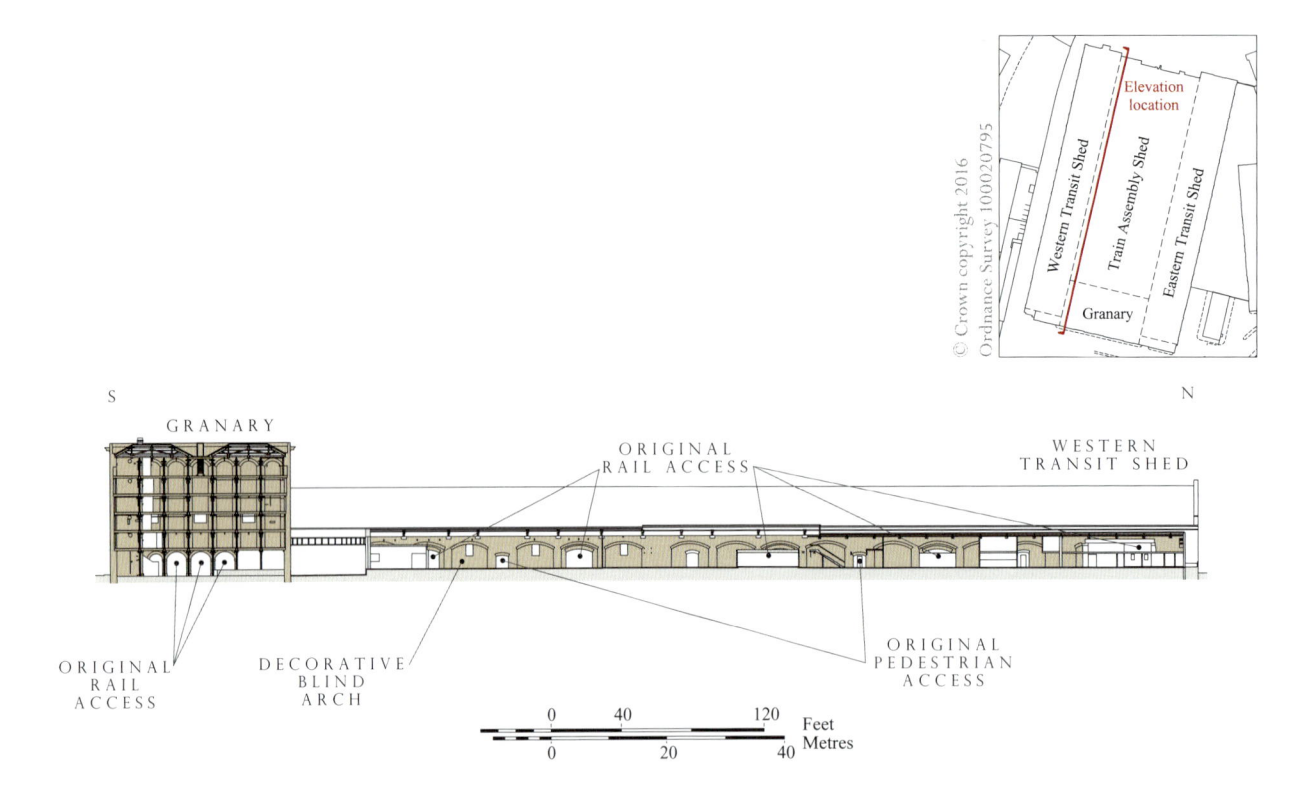

Fig. 3.16 Cross-sectional elevation through the Train Assembly Shed (Landscape Feature 11) and the Granary (Landscape Feature 12) showing the external face of the east wall of the Western Transit Shed (Landscape Feature 7), the internal face of the west wall of the Granary and the original 1850–1 masonry, scale 1:1,250

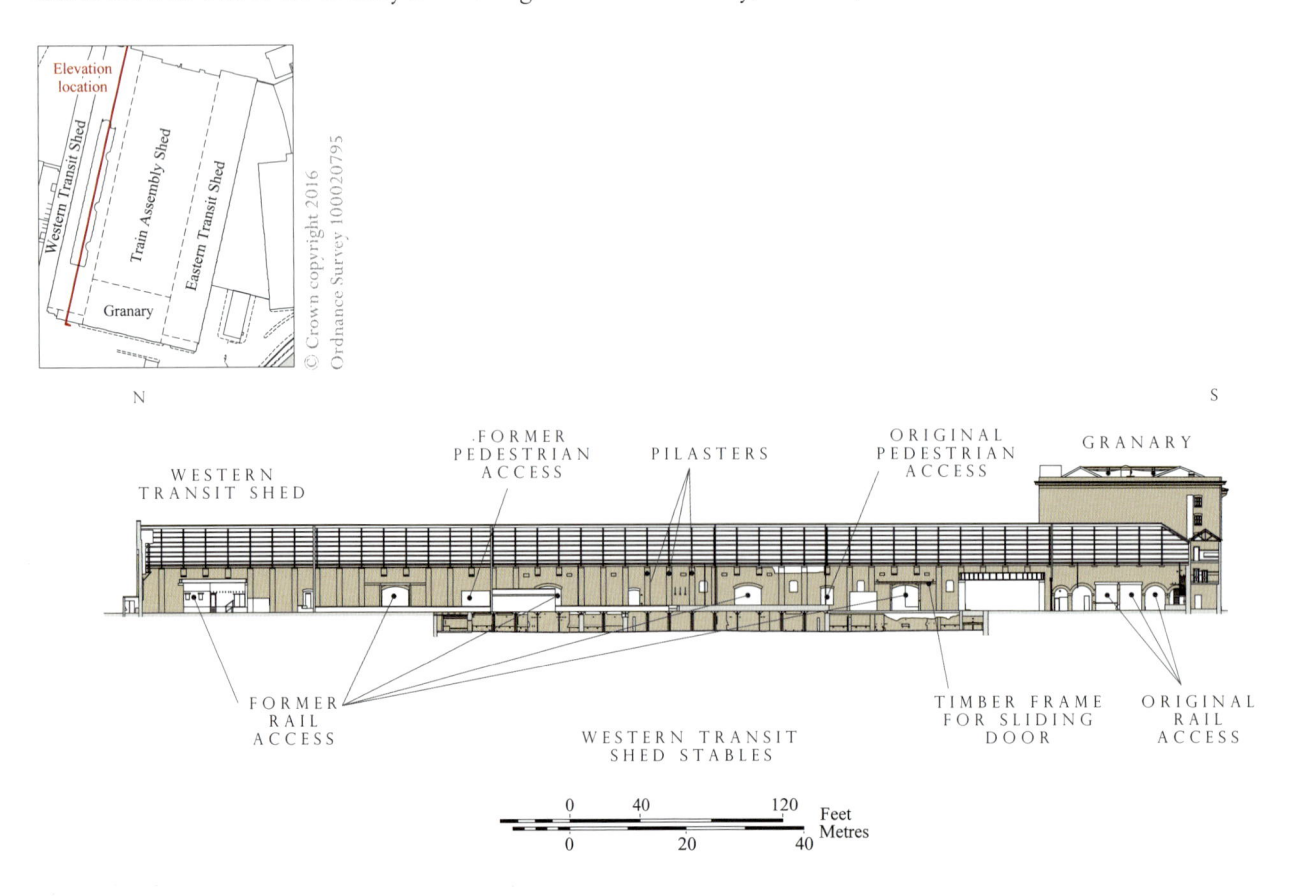

Fig. 3.17 Cross-sectional elevation through the Western Transit Shed (Landscape Feature 7) and Western Transit Shed Stables (Landscape Feature 9) showing the internal faces of their east walls and the original 1850–1 masonry, scale 1:1,250

The internal face of the west wall (Fig. 3.15) lacked the blind arches that were found on the opposing side in accordance with a decision to restrict superfluous and costly decorative elements to the exterior. Instead, this section of the building was originally characterised by 47 equally spaced brick pilasters (45 of which survive), arranged two to each bay. These were functional rather than decorative, stiffening the wall and supporting the roof trusses. Where vehicular openings were present, the pilasters did not extend to ground level; instead they terminated above the timber and cast iron supports for the sliding doors that originally shuttered these openings. Granite guard stones were placed at the base of some of the pilasters and at the sides of the vehicle openings, whilst horizontal timber beams were fixed to the length of the wall by a series of tie plates and bolts (Fig. 3.15). These features were presumably designed to protect the façade from vehicle strikes.

The eastern wall of the Western Transit Shed partitioned the building from the Train Assembly Shed and the Granary (Fig. 3.2). To the north of the Granary, a series of ornamental blind segmental arches graced the external façade (Fig. 3.16). Once again, the internal side of the wall was relatively plain, being characterised by the presence of 40 evenly spaced pilasters that directly faced their equivalents in the western wall (Fig. 3.17).

Four large, bricked up openings were noted in the opposing, eastern wall (Fig. 3.16; Fig. 3.17), along with a fifth example that was no doubt present at the northern end (this has been obscured from modern view by later alterations, its former presence being deduced from a run of shortened pilasters as shown in Fig. 3.17). The five openings were aligned with the positions of the transverse railway lines shown on the Humber Plan of 1866 (Fig. 3.8) thus enabling the passage of train wagons between the Western Transit Shed and the Train Assembly area. On the Western Transit Shed side, two were still surrounded by their original timber sliding door frames, demonstrating that they could be shuttered off in the same way as the cart road access points (Fig. 3.17). Two smaller openings were identified in the main body of the east wall (Bays B15 and B7 as shown in Fig. 3.16), which presumably represent doorways for pedestrian use.

The southern end of the eastern wall of the Western Transit Shed, which separated that structure from the Granary (Landscape Feature 12), was rather different to the rest of the façade. No blind segmental arches were present; instead six bays were identified, each of which once contained an opening as indicated by the Humber Plan of 1866 (Fig. 3.8) and confirmed by direct observation (Fig. 3.16; Fig. 3.17). The surviving remains demonstrated that the entrances consisted of six adjoining stilted segmental arches that were almost semi-circular in shape. They were formed by four courses of brickwork in 'brick on edge' positions, supported by brick piers that were surmounted by sandstone imposts. The six archways originally exhibited north–south symmetry with the four central examples being slightly taller and wider than their outer equivalents (Fig. 3.16; Fig. 3.17). Comparison with the 1866 Humber Plan (Fig. 3.8) demonstrates that three of the four large arches each accommodated one of the three transverse rail sidings that linked the Western Transit Shed with the Granary. The fourth and most northerly large archway did not accommodate a railway track and would also have been an unsuitable access point for road traffic due to obstructions in the form of the Western Transit Shed's platform. It is therefore reasonable to assume that the remaining openings were reserved for pedestrian use and that no internal road vehicular communication existed between the Western Transit Shed and the Granary.

The original ground floor surface of the Western Transit Shed was not found during the archaeological excavations. It is presumed to have been at an approximate level of 21.09m OD to 20.95m OD based on the height of the remains of a 19th-century floor that was revealed in the Eastern Transit Shed.

The substantial foundations of a rectangular red fabric brick structure were identified below the modern ground surface in the north-central section of the Western Transit Shed (Fig. 3.3; Fig. 3.18; Fig. 3.19) in the predicted location of the platform, which it must represent (henceforth termed 'Platform 1'). As illustrated by the Humber Plan, Platform 1 originally ran virtually the

Fig. 3.18 The 1850–1 foundations of Platform 1 to the immediate east of the Western Transit Shed Stables Ramp, photograph faces south

Fig. 3.19 The 1850–1 foundations of Platform 1 in the Western Transit Shed, photograph faces north

entire length of the Western Transit Shed (Fig. 3.8). The exposed remains of the platform were 37.90m in length, continuing beyond the northern and southern limits of the area of excavation, and between 5.20m and 9.90m in width. Two sections of identical masonry observed further south almost certainly represent additional sections of the same feature. The interior of the structure was spanned by brick beams that were orientated east–west, at a right angle to the outer masonry casing whilst seven masonry 'stubs' were situated to the east of a ramp that descended into a subterranean stable block. These presumably once supported the wooden surface of the platform. Beyond the ramp, within the footprint of the underground stables, the structure widened from 5.20m to 9.90m and the nature of its foundation design changed. Here it was formed by the outer walls of the stables that protruded above 19th-century ground level (Fig. 3.18).

The remains of a length of standard gauge railway track, comprising rails and sleepers, were uncovered to the immediate east of the southern end of Platform 1. The rails marked the location of the shed's 'down' line, upon which wagons were stationed whilst being loaded. As shown on the Humber plan (Fig. 3.8), the siding crossed eight wagon turntables, which were used to move vehicles at right angles to and from the transverse tracks that connected the Western Transit Shed with the Train Assembly area to the east (as shown in Fig. 3.58). Five were individually housed in a series of semi-circular niches that were integral to the eastern edge of Platform 1, whilst three more were found together in a semi-ovoid structure that was situated next to Dock 1 (Fig. 3.8). At 12' in diameter, the turntables were too small to be used by even the most modest tank engine. Given that steam locomotives were prohibited from entering goods sheds owing to the fire risk that they represented, the small size of the turntables is not surprising. Instead they moved single pieces of rolling stock through the complex after an incoming train had been broken up into its constituent parts. The turntables inside the Western Transit Shed would have been used to convey incoming empty wagons onto the shed's railway line for loading, whilst others would have been used to return them fully loaded to the Train Assembly Area for onward dispatch (Fig. 3.58).

The archaeological, documentary and cartographic sources detailed here have enabled the workings of the Western Transit Shed to be partially reconstructed. In summary, goods for onward dispatch to the north were brought into the building on barges or road vehicles via Dock 1 or the cart way, before being temporarily offloaded onto Platform 1 through the use of pairs of hydraulic or manually operated cranes, the locations of which are illustrated on the Humber Plan (Fig. 3.8). These items would then be transferred to train wagons on the western side of the platform for onward dispatch to the north as shown in an historic illustration of the Western Transit Shed (Fig. 3.20).

The railway infrastructure of the shed is further discussed in Section 3.12.

Fig. 3.20 An engraving from ***The Illustrated London News*** of 28th May 1853 (Anon 1853) showing the Western Transit Shed (Landscape Feature 7) in operation shortly after the yard fully opened for business. Looking northwards from Dock 1, it shows how goods were exchanged via Platform 1 between train wagons (shown on the rail siding to the right), barges (depicted in the central dock in the foreground) and horse-drawn carts (positioned on the roadway to the left)

3.8 The Eastern Transit Shed by 1852

Landscape Feature 8

From the outset the architect Lewis Cubitt intended that both Transit Sheds be built to identical specifications, resulting in two structures that were virtual mirror images of one another (TNA RAIL 236/239: 78). The Eastern Transit Shed, also known as the 'arrivals', 'import', 'inwards' or 'up' shed, performed the opposite function to its western twin, receiving goods from the north for onward dispatch by road or canal to London and the south.

Work began on the foundations of all the elements of the Goods Station complex simultaneously and a 'considerable part' of these had been laid by the end of March 1850 (TNA RAIL 236/273: 27/03/1850). By late October of that year, the rate of construction of the Eastern Transit Shed had fallen behind that of the Western Transit Shed, which was already approaching completion (*ibid*.: 24/10/1850). Towards the end of January 1851, Cubitt reported that 'the Arrival Goods Shed has made great progress' and that he anticipated that it would be completed within a month (*ibid*.: 28/01/1851). True to his word on the 18th February the architect announced that the shed would be ready for traffic a fortnight thence (*ibid*.: 18/02/1851).

Built heritage and archaeological evidence relating to the construction, appearance, footprint and function of the Eastern Transit Shed by 1852

Landscape Feature 8

The southern (front) elevation of the Eastern Transit Shed (Landscape Feature 8) is today almost completely obscured by the Eastern Goods Offices, which were built several years later (Landscape Feature 21). However, documentary evidence coupled with field observations confirmed the southern façade along with the other three sides of the building originally mirrored their equivalents in the Western Transit Shed (Fig. 3.2; Fig. 3.21; Fig. 3.22).

Removal of the modern ground surface of the Eastern Transit Shed revealed a 19th-century cobbled surface below, the top of which was found at a height of 20.95m OD to 21.09m OD. This probably represents the original ground floor of the building.

The well preserved remains of a large rectangular masonry structure were revealed during an archaeological excavation and watching brief (KXI07). It was strikingly similar in appearance to the remains of the Western Transit Shed's platform and it was therefore interpreted as an equivalent structure in the Eastern Transit Shed, henceforth termed Platform 2. Evidence

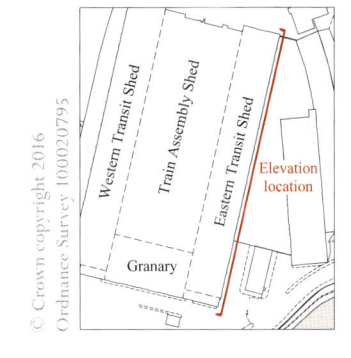

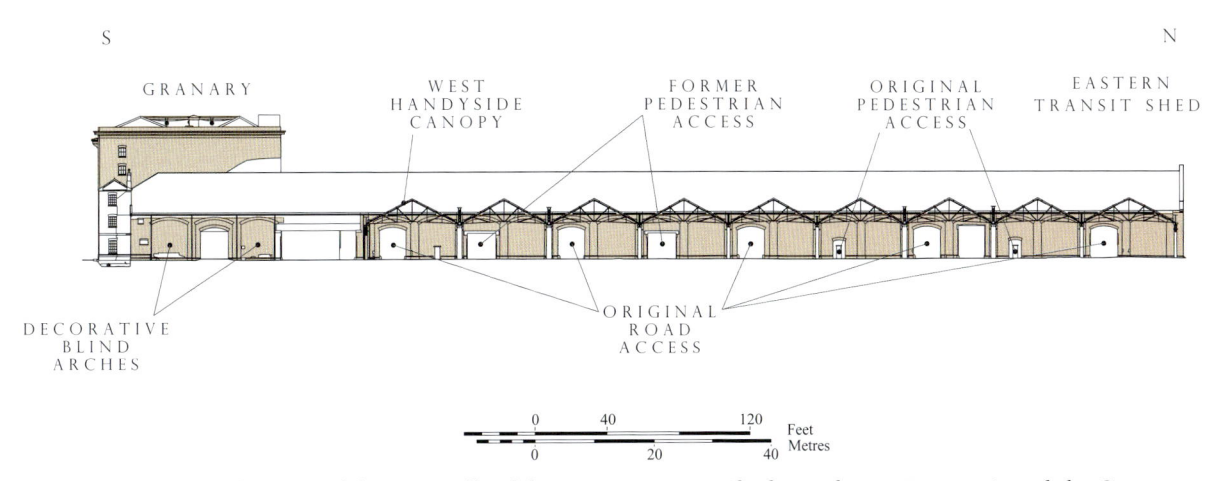

Fig. 3.21 External elevation of the east walls of the Eastern Transit Shed (Landscape Feature 8) and the Granary (Landscape Feature 12) showing the original 1850–1 masonry, scale 1:1,250

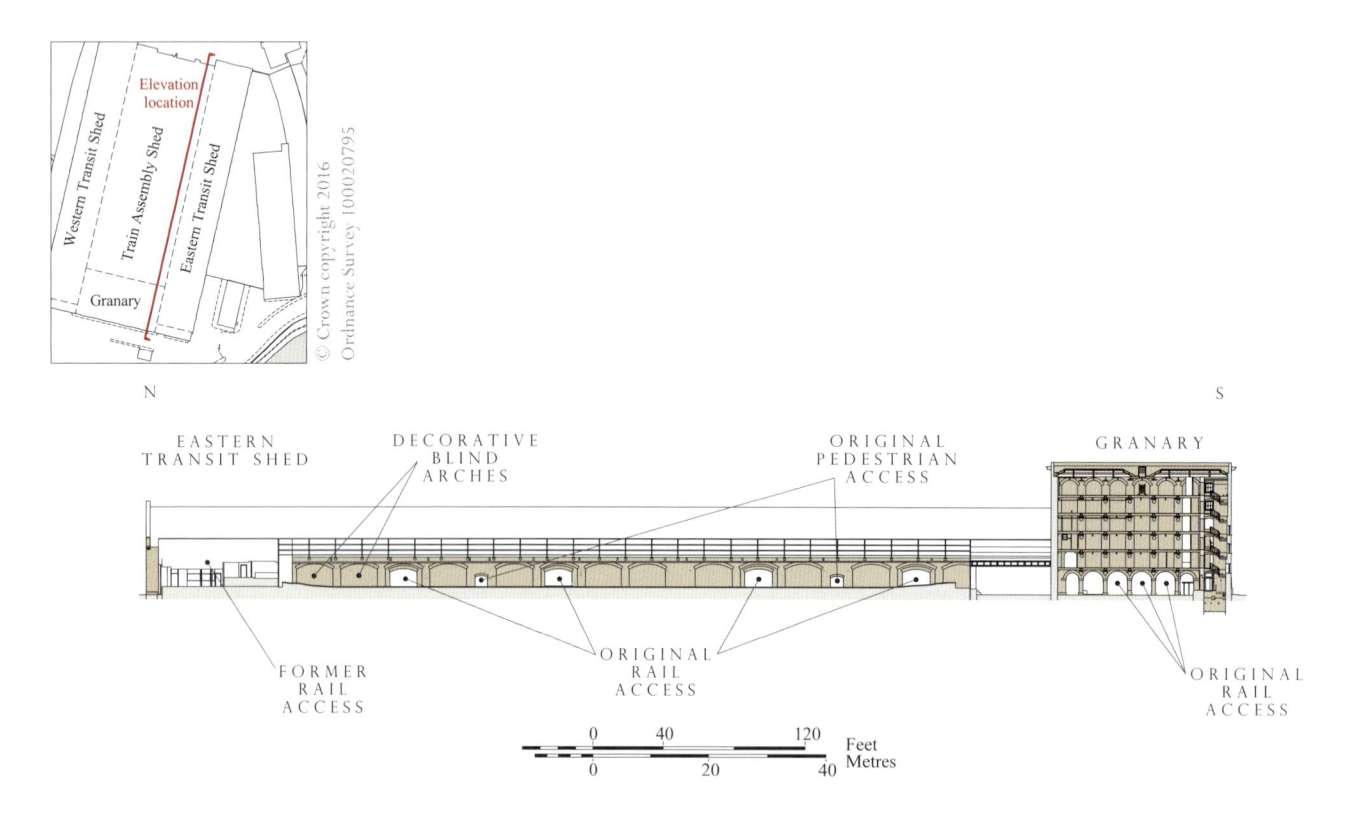

Fig. 3.22 Cross-sectional elevation through the Train Assembly Shed (Landscape Feature 11) showing the external face of the west wall of the Eastern Transit Shed (Landscape Feature 8), the internal face of the east wall of the Granary (Landscape Feature 12) and the 1850–1 masonry, scale 1:1,250

Fig. 3.23 Circular niche incorporating an 1850–1 turntable base in the Eastern Transit Shed (Landscape Feature 8), photograph faces northeast

indicating the former presence of a dock (Dock 4) and a subterranean stable block (Landscape Feature 10) was also identified. Like their western equivalents, the Eastern Stables (Landscape Feature 10), Platform 2 and Dock 4 were structurally integral to one another, demonstrating that they were erected in one continuous build. Consequently, five semi-circular structures were once again incorporated in the edge of the platform, each of which accommodated a well for a single turntable (Fig. 3.23) whilst a further three such wells were found in a semi-ovoid structure that was integral to the southern end of the structure (Fig. 3.24). The foundations of three of the five semi-circular examples continued downwards into the Eastern Stables (Landscape Feature 10), where they formed part of the western wall of that subterranean block (Fig. 3.25; Fig. 3.31). No rails or sleepers associated with the shed's original 'up' railway line were discovered, but the locations of the associated turntable wells belied its former presence. The location of the track as extrapolated from the turntable positions accorded perfectly with that shown on the Humber plan (Fig. 3.8).

Like its western counterpart, archaeological, documentary and cartographic analysis has enabled the workings of the Eastern Transit Shed to be better understood. Incoming goods would be brought into the building by train, before being temporarily offloaded onto Platform 2 through the use of pairs of hydraulic or manually operated cranes. These items would then be transferred to barges moored in Dock 1 or road vehicles stationed on the cart way for dispatch to London and the south.

Fig. 3.24 Semi-ovoid niche incorporating three 1850–1 turntable bases in the Eastern Transit Shed (Landscape Feature 8), photograph faces south

3.9 The Western and Eastern Transit Shed Stables by 1852

Landscape Features 9 & 10

Lewis Cubitt's earliest architectural plans indicate that he intended to integrate subterranean stabling in the foundations of the Transit Shed platforms from the outset of the design process (TNA RAIL 236/239: 80). Cubitt intended that the Western Transit Shed would contain stabling sufficient for 100 horses, access to which would be gained from the outside via a 'shallow incline' (TNA RAIL 236/239: 80). Access to the stables from inside the shed would take the form of a ramp leading from Platform 1, which can be seen on the Humber Plan (Fig. 3.8). The latter plan also suggests that the Eastern Transit Shed Stables were remarkably similar, albeit lacking any means of direct communication with the outside.

Construction of the Western Stables progressed according to schedule well and the structure was reported to have been ready for use in the early months of 1851. In contrast, Joseph Cubitt's regular reports on the construction process at King's Cross suggest that there was considerable uncertainty regarding the eventual function of the Eastern Stables during this period.

In mid January 1851, the Executive Committee decided to accept an offer made by Mr Sherman (the Great Northern's contractor for the cartage of goods in London) of £200 per annum for the sole use of the stables under the Western Transit Shed (TNA RAIL 236/71: 164, 198). The stables were not available for immediate occupation, as construction of the subterranean entrance was scheduled to commence the following month (TNA RAIL 236/273: 18/02/1851). Despite Joseph Cubitt's suggestion that Sherman transfer his horses to the incomplete Eastern Goods Shed Stables in order to make way for those of Edward Wiggins (the Great Northern's contractor for the delivery of coal in London), Sherman remained in possession of the Western Stables into the following September. The eastern stable block did not remain empty, but instead became a depository for goods placed in long term storage (*ibid.*: 07/02/1851; TNA RAIL 236/275/23: 27/09/1852). At the end of October of that year the General Manager, Seymour Clarke, was informed by the Directors that no more money was to be allocated for the construction of the underground stables in the Eastern Transit Shed until a decision had been reached regarding their future use either for stabling or for goods storage (TNA RAIL 236/71: 198). Later that month, the Board finally yielded to Clarke's request thus assuring the future of the Eastern Stables (TNA RAIL 236/18: 99). Sherman then moved his horses into the stables, freeing up the Western Stables for the use of Wiggins' animals (TNA RAIL 236/17: 228). Following the end of Sherman's contract that December, the Great Northern moved its own horses into the Eastern Stables (*ibid.*).

Built heritage and archaeological evidence relating to the construction, appearance, footprint and function of the Western Stables by 1852

Landscape Feature 9

The subterranean walls of Landscape Feature 9 formed a structure that was rectangular in plan with a floor that was laid in granite setts (Fig. 3.25). It was largely built from stock bricks, which were once again stamped with the letters 'JJ' (presumably indicating 'John Jay'). This landscape feature represents the underground Western Transit Shed Stables that were completed early in 1851, the walls of which continued above ground to form the sides of the central section of Platform 1.

A cobbled ramp, which could be accessed from the cart road to the west, was unearthed in the northern part of the Western Transit Shed (Fig. 3.25; Fig. 3.26). It led downwards from the west side of Platform 1 to the Western Stables (Landscape Feature 9). The ramp turned 90° to the north before turning 180° to the south and entering the subterranean space via a central doorway that was capped by a semi-circular arch. This feature represents the Western Transit Shed Stables access ramp. In his design brief, Lewis Cubitt described a 'shallow incline' that provided external access to the Western Transit Shed Stables from the western side of the Goods Yard, which records suggest was constructed in February 1851 (TNA RAIL 236/239: 80; TNA RAIL 236/273: 18/02/1851). The presence of this feature was confirmed archaeologically: as anticipated, it took the form of a tunnel that connected with the low ground on the western side of the western terrace (Fig. 3.25). The western end of the tunnel was integral to the western face of the Retaining Wall (Landscape Feature 4) whilst the eastern end opened directly into the stables (Fig. 3.25; Fig. 3.27). Two infilled voids were observed on the southern face of the latter doorway along with a cement filled feature on the northern side, which suggests that the tunnel was originally gated at the stables end (Fig. 3.27).

It is probable that the tunnel and the ramp were used flexibly, with horses moving in and out of the stables by way of either opening. A one-way system with dedicated access and egress points would have been unnecessary, since the animals most probably clocked on and off duty at specific times, which would have limited congestion at 'pinch' points within the stables. Indeed, the Eastern Transit Shed Stables (Landscape Feature 10) possessed one ramp for many decades, demonstrating that a second was not essential to the efficient working of the system. In all probability, the two access points each connected the Western Stables with a different part of the yard. The ramp provided direct communication with the Western Transit Shed,

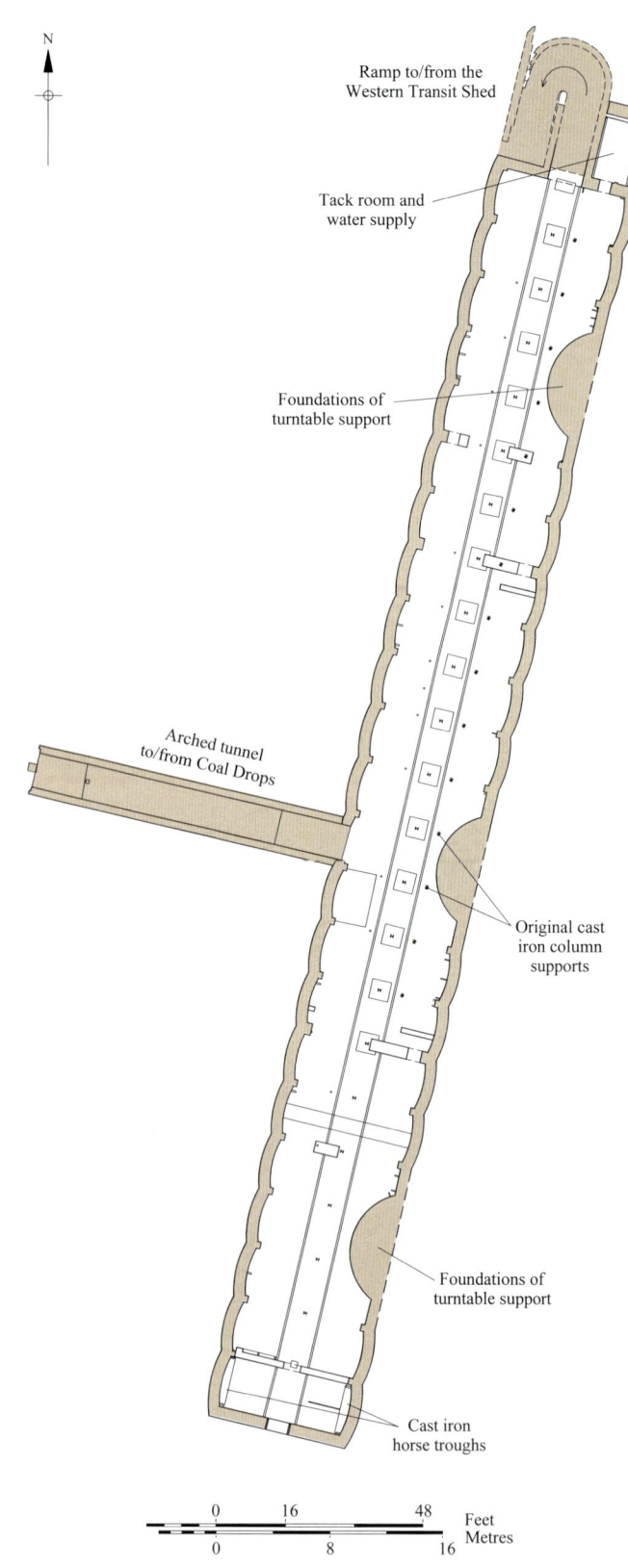

Fig. 3.25 Basement plan of the Western Stables (Landscape Feature 9) *c.*1851 reconstructed from archaeological remains, built heritage evidence and cartographic sources, scale 1:500

Fig. 3.26 Ramp, constructed in 1850–1, providing a link between the Western Transit Shed (Landscape Feature 7) and the underground stables (Landscape Feature 9), photograph faces south

Fig. 3.29 Original 1850–1 horse trough against the east wall at the south end of the Western Transit Shed Stables (Landscape Feature 9), photograph faces east

Fig. 3.27 Tunnel, constructed in 1850–1, connecting the Western Transit Shed Stables (Landscape Feature 9) with the low ground on the western side of the Retaining Wall (Landscape Feature 4), photograph faces west

Fig. 3.30 Tack room of 1850–1 at the north end of the Western Transit Shed Stables (Landscape Feature 9) showing tack posts (upper left), brick sink and drain (lower right), photograph faces north

Fig. 3.28 The Western Transit Shed Stables (Landscape Feature 9), photograph faces south

Fig. 3.31 The foundations of a turntable support of 1850–1 in the east wall of the Western Transit Shed Stables (Landscape Feature 9); note the timber posts in the wall upon which tack was hung and the roof supports (in the form of padstones and circular cast iron columns supporting cast iron beams) from which the brick jack arches of the roof were sprung, photograph faces east

whilst the tunnel enabled access to and from the Coal Drops (Landscape Feature 17), a facility that required a large number of horses for the onward transshipment of coal.

Inside the stables themselves, 22 equally spaced pilasters were observed in the west wall, either side of a series of curved alcoves (Fig. 3.28). A further sixteen pilasters interspersed with identical bays were also incorporated in the east wall, which differed from its western counterpart due to the existence of three large semi-circular protrusions that accommodated turntable wells at above ground (Fig. 3.25), an example of which can be seen in Fig. 3.31. All the pilasters were aligned perfectly with their above ground equivalents in the Western Transit Shed. They flanked 36 alcoves, 22 of which were positioned along the western wall with a further fourteen along the eastern wall (Fig. 3.25). Originally east–west divisions protruded from each pilaster in order to partition each bay into stalls, as demonstrated by the survival of a series of truncated wall stubs. Together these divisions and alcoves formed 36 individual stabling areas (Fig. 3.25). Cast iron horse troughs were originally positioned against the end wall of each alcove to feed and water the animals. With the exception of two troughs positioned at the far southern end of the stables (Fig. 3.25; Fig. 3.29) all of these had been removed at some point in the past, evidence for their existence being provided by scars in the form of lines of cast iron observed along the side walls of each bay. The surviving troughs had tethering rings attached to them to secure the stabled animals (Fig. 3.29). The presence of two rings per trough suggests that it was the intention to stable up to two horses in each alcove. If this was the case then a total of 72 animals could have been kept in the Western Transit Shed Stables at any one time, somewhat less than the figure of 100 cited by Lewis Cubitt in his design brief to the Stations Committee in 1850 (TNA RAIL 236/239: 80). This indicates that either fewer horses were kept in the stables than the architect first proposed, or that his figure of 100 was a rough approximation, or that the animals were circulated, occupying the facilities in shifts.

Timber posts for hanging tack were inset into the side walls of the stables. Originally there appears to have been one set into each pilaster with a further two set into each alcove. The majority of these had been broken at some point in the past leaving scars in the form of stumps of wood in the faces of the walls. In a few places, iron equivalents of these timber pegs were found, which were attached to the cast iron columns that once supported the roof. The upper part of the east wall was infilled in several places with late 20th-century brickwork suggesting that these areas were once open, providing ventilation.

A small room to the east of the access ramp at the north end of the stables appears to have been used as an additional storage space for tack as well as housing a much needed source of water for the animals (Fig.

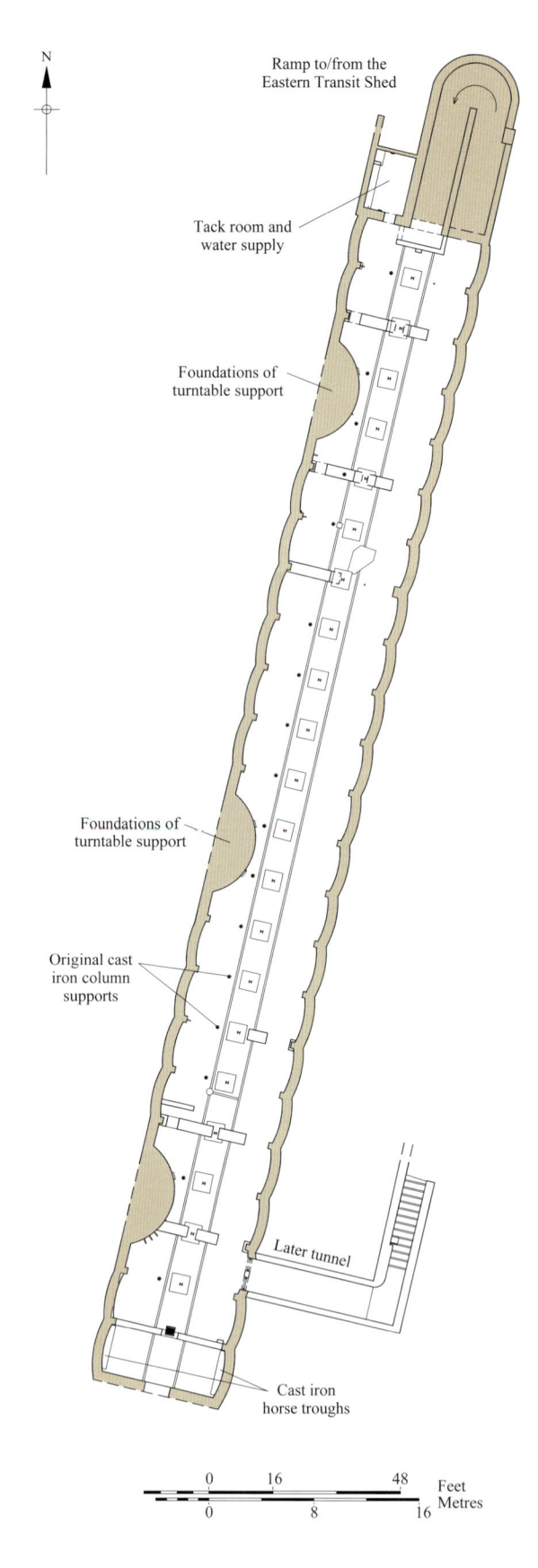

Fig. 3.32 Basement plan of the Eastern Transit Shed Stables (Landscape Feature 10) c.1851 reconstructed from archaeological remains, built heritage evidence and cartographic sources, scale 1:500

3.30). Several timber posts protruded from its walls from which spare harnesses and other items of horse furniture would have been hung. A sink was situated in the northeast corner, which consisted of a red brick support with a hollow centre into which a triangular cast iron basin pivoting on an iron rod was set. Water entered from a pipe positioned above and waste liquid was tipped into the hollow beneath the sink before passing into a red brick drain that ran off to the south.

The stables were originally spanned by north–south brick jack arches (Fig. 3.17), which were supported by east–west cast iron beams held up by padstones that either capped the pilasters or were built into the turntable surrounds in the side walls. Circular cast iron columns were also used which were either tied into each of the dividing walls between the stable bays or were positioned to the west of the turntable surrounds (Fig. 3.25). Above ground in the Western Transit Shed, this brick vaulting supported Platform 1 within the footprint of the subterranean stables. Presumably these structures represent the 'brick arches turned against Wrought Iron Girders' as described by Joseph Cubitt in one of his regular reports to the Board (*ibid*.: 80, 105).

Built heritage and archaeological evidence relating to the construction, appearance, footprint and function of the Eastern Stables by 1852

Landscape Feature 10

Landscape Feature 10, the Eastern Transit Shed Stables, was situated below the Eastern Transit Shed (Landscape Feature 8). The earliest incarnation of the structure mirrored the Western Transit Shed Stables (Landscape Feature 9), the only difference being the lack of a secondary exit communicating with the exterior of the Goods Station (Fig. 3.32; note that the tunnel shown on this drawing is a later addition). This meant that an extra bay could be incorporated in the eastern wall of the structure which in turn meant that the building was capable of stabling fractionally more animals than its western equivalent. Based on an assumption that each bay was wide enough to provide sufficient space for two beasts this would take the maximum number of animals that could be comfortably accommodated at any one time to a total of 74.

3.10 The Train Assembly Shed by 1852

Landscape Feature 11

The Train Assembly Shed was the third major element in Lewis Cubitt's original Goods Station design, the Granary representing something of an afterthought. Cubitt envisaged that the Train Assembly Shed would be accommodated in the open area between the Transit Sheds, and that it would be enclosed by a roof. This space would be crucial to the way that the station would function. Trains arriving from the north were to be shunted into the Train Assembly Shed for division into single wagons for unloading, whilst empty or newly loaded wagons would be returned to the area for reassembly into trains for northerly dispatch.

A letter written by Lewis Cubitt in January 1850 contained some interesting details regarding the intended function and appearance of the Train Assembly Shed (TNA RAIL 236/239: 78). Cubitt wrote that he intended to cover the tracks between the two Transit Sheds (i.e. the Train Assembly Shed area) 'in four spans with Iron Columns- Girder roofing, Slated and Glazed with Ventilating Louvers [*sic*.] to carry off steam and smoke' (*ibid*.). The reference to 'steam and smoke' suggests that Cubitt originally envisaged that locomotives would be admitted to the shed, a proposal which did not come to fruition as engines were never permitted in any of the Goods Station buildings, owing to the risk of fire that they posed to the goods stockpiled within. In place of locomotives, goods wagons were instead shunted along the sidings by horses, assisted by a number of manually operated capstans, whilst they were transferred from one siding to another by means of turntables (Fig. 3.58).

Lewis Cubitt's drawings of the iron roofing of the Train Assembly Shed of March 1850 represent the earliest illustration of the structure (ICE Cubitt 1850). Because the Stations Committee had yet to decide whether to include a granary, he depicted the iron roofs extending the full 600' length of the Transit Sheds (Fig. 3.5). Although the plan showed roofs covering only the two outermost bays of the four bay Train Assembly Shed, in his accompanying specification Cubitt anticipated that the two central bays would be covered 'within the next six months' (*ibid*.). Each of the four roofs was to be 45' 2¼" wide and span three tracks, with the outer edges of the two roofs supported by the brickwork of the Transit Sheds and the inner edges by alternating cast iron columns and brick piers (Fig. 3.5). A third row of cast iron columns and brick piers would provide longitudinal support where the two innermost roofs met. The roofs themselves were of a design strikingly similar to that which Cubitt used for the roof of the Temporary Passenger Station. It was intended that they would be constructed from wrought iron components including framed struts, suspension bars, braces, rivets, bolts, plates, washers and horizontal tie rods running the length of the roofing (*ibid.*). The canopy would be drained by longitudinal iron gutters, which would discharge rainwater into the hollow cast iron columnar supports (*ibid.*). Where brick piers alternated with the columns, the guttering on the roof was 'to be raised in continuation of the same incline and the water … in such cases to be brought by the next nearest column'. An arrangement of 4" rainwater pipes were to be fitted at the eaves of the roofs at intervals of 245' corresponding with the rows of columns (*ibid.*).

Although John Jay's contract for building the Goods Station included the glazing, slating and boarding of the completed iron roofs of the Train Assembly Shed, it did not include the erection of the iron frames of the roofs of the Train Assembly Shed and the Temporary Passenger Station. It was not until 20th March 1850 that the Stations Committee of the Great Northern invited tenders for their construction (TNA RAIL 236/273: 27/03/1850). The contract was awarded to the Glaswegian firm of Robertson & Lister on 17th April 1850 at a cost of £4 12s 9d per square foot (TNA RAIL 236/239: 113). It appears that the contractors started work shortly afterwards and progress was sufficiently advanced by late June for the Stations Committee to grant Cubitt permission to order roofing materials in order to enable Jay and his men to begin work on the canopy that would cover the Train Assembly Shed's two central bays (*ibid.*: 123).

The prolific architectural writer John Weale described arrangements in the Train Assembly Shed during his tour of the Goods Station construction site early in 1851. Although the building had not been completed by the time of his visit, enough had been built for him to gather a clear impression of how the station was supposed to work. Weale observed that the turntables in the Transit Sheds allowed wagons to be moved into the Train Assembly Shed 'through a corresponding number of sliding doors, exactly opposite to other doors in the outer walls, communicating with the access roads' (Weale 1851: 811). He also noted that the Train Assembly Shed was divided into four bays, each with three sets of tracks. This arrangement allowed as many as twelve trains to be made up or divided simultaneously, all under cover (*ibid.*). A reporter for The Observer newspaper simply described the Assembly Shed as a 'very commodious lay-by' for wagons (Anon 1851b).

Built heritage and archaeological evidence relating to the construction, appearance, footprint and function of the Train Assembly Shed by 1852

Landscape Feature 11

Landscape Feature 11 was a large single storey structure, some 155m in length (north–south) by 55m in width (east–west) that was divided into four equally sized north–south bays that were covered by symmetrically pitched roofs (Fig. 3.2; Fig. 3.33). Its above ground components were recorded in detail in a Historic Building Recording Survey (KXF07), whilst the below ground elements were investigated as part of an archaeological excavation and watching brief (KXI07). The feature represents the Train Assembly Shed, which formed the central section of the Lewis Cubitt's Goods Station.

The northern end of the structure was largely open to the elements, comprising three short lengths of brickwork that divided it into four equally sized, open bays (Fig. 3.33; Fig. 3.34). These structural elements were built with red bricks in Flemish bond. In combination with the sides of the Eastern and Western Transit Sheds and a series of internal supports, they propped up the three valley gutters of the four gables that ran the length of the Train Assembly Shed, each gable spanning one of the four bays (Fig. 3.34). Internally, each valley gutter was supported by six short lengths of brick wall and twelve columns, with the latter being set from north to south between the former in groups of two, two, three, two and three as shown on the Humber Plan of 1866 (Fig. 3.8). Most of these elements had been removed at a later date. However, a total of eight original columns survived in the southwestern part of the shed, which were exactly as Cubitt had specified, i.e. circular, hollow and made of cast iron (Fig. 3.35). Their locations confirmed the configuration indicated on the Humber Plan (Fig. 3.8). Each shaft bore a maker's name plaque close to the base, which stated that they had been cast by 'W. RICHARDS & SONS, MAKERS, LEICESTER' (Fig. 3.36).

Fig. 3.33 The Train Assembly Shed (Landscape Feature 11) looking towards the Granary (Landscape Feature 12); the original slated and glazed roof was replaced, largely with asbestos sheeting, in the 1950s, photograph faces south

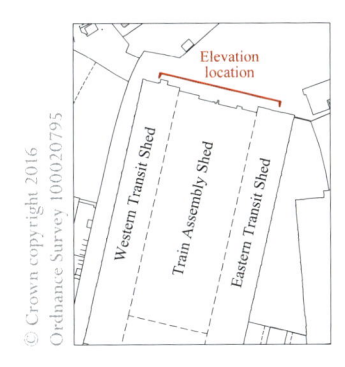

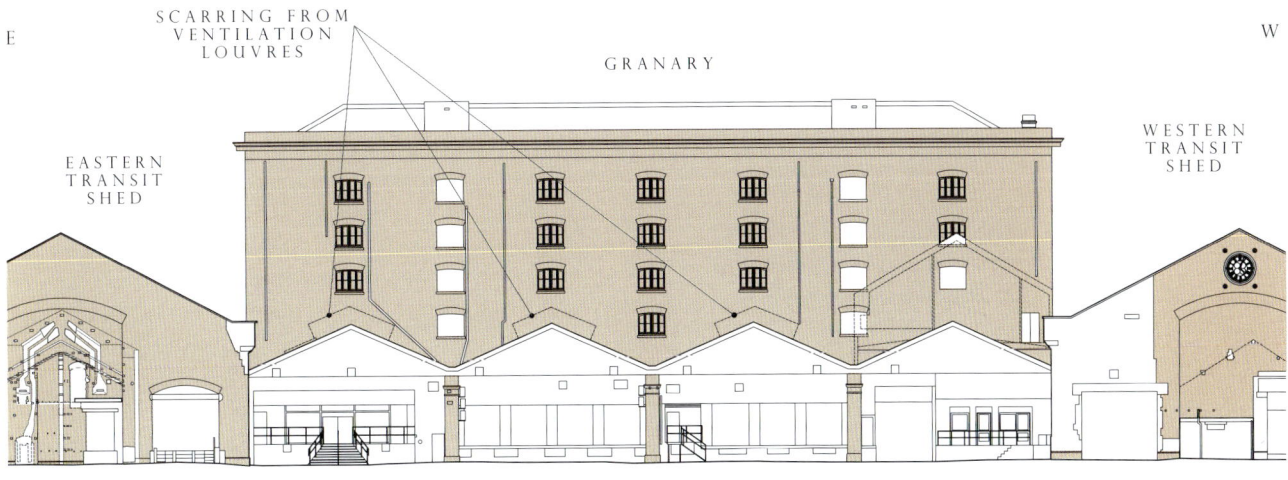

The Train Assembly Shed by 1852 53

Fig. 3.34 External elevation of the north sides of the Western Transit Shed (Landscape Feature 7), the Eastern Transit Shed (Landscape Feature 8), the Train Assembly Shed (Landscape Feature 11) and the Granary (Landscape Feature 12) showing the original 1850–1 masonry, scale 1:500

Fig. 3.35 The Train Assembly Shed (Landscape Feature 11) after the removal of the roof; note six of the eight 1850–1 hollow cast iron columnar roof supports in the right (west) side of the image and the three masonry turntable supports of the same date in the foreground, photograph faces south

Fig. 3.36 Maker's mark on an early 1850s column within the Train Assembly Shed (Landscape Feature 11)

Roof scars on the north elevation of the Granary (Fig. 3.34) indicate that the canopies of the adjoining Train Assembly Shed were originally fitted with ventilating louvres that were designed to vent smoke and steam as shown on Lewis Cubitt's drawings of 30th March 1850. This suggests either that the decision to prohibit locomotives from the Train Assembly area was made at quite a late stage, after the architect's plans for the roofs were executed in April 1850, or that steam engines were originally permitted into the northern end of the Train Assembly area.

Cast iron brackets projecting from the eastern wall of the Western Transit Shed are the probable means by which the original roof trusses of the Train Assembly Shed were secured to this wall. Similar fixtures were presumably used to fix the roof to the western side of the Eastern Transit Shed.

Historical maps (e.g. Fig. 3.8) and contemporary descriptions (e.g. Weale 1851: 811) indicate that a total of twelve railway tracks were situated in the Train Assembly Shed, with three railway sidings being situated in each of the four bays. This arrangement was confirmed when sections of these tracks were observed archaeologically. Historic building recording also substantiated the suggestion (based on cartographic evidence) that the north side of the building was largely open to the elements (e.g. Fig. 3.8). This would have enabled wagons to be easily shunted in and out of the structure via the twelve sidings that ran into the shed from the north (Fig. 3.8). The Humber Plan also indicated that when first built the Train Assembly Shed did not contain any platforms, presumably because it was dedicated to assembling and disassembling trains rather than goods handling (Fig. 3.8). No evidence to the contrary was unearthed during the archaeological monitoring exercise.

3.11 The Granary by 1852

Landscape Feature 12

It has previously been noted that the granary was not an original element of Lewis Cubitt's Goods Station design, and that its eventual inclusion appears to have been something of an afterthought. Seymour Clarke, the Great Northern's General Manager, was keen that a grain store be placed near the south end of the Goods Warehouses (i.e. the Transit Sheds), a location that Cubitt approved due to the efficient use of the available space and the ease with which the new building could be accommodated within his earlier draft design (TNA RAIL 236/239: 83–5). In a report to the Stations Committee written in December 1850, Cubitt noted that 'it [i.e. the Granary] will in some respects affect the general plan and will absorb some of the space of the 200' remaining from the allotment to the Goods Building …your Engineer will be pleased to consider how far the workings with the Rails and Turntables and the Platform will conform with the general system he has laid down … access will also be had from the dock' (*ibid.*: 84–5).

Cubitt refined his drawings in the days that followed, enabling him to submit a set of working plans and elevations of the Goods Offices, the Goods Warehouses and the Temporary Passenger Station to the Stations Committee two weeks later. At this point the Board had yet to decide whether to include a granary in the finished Goods Shed, so Cubitt indicated that the foundations of the other buildings were to be built 'so as to admit of a Granary at the end of the building being added hereafter if necessary' (*ibid.*: 105). Construction of the Goods Station therefore began before the question of providing a granary had been resolved. The matter was still under discussion on 27th March, when Joseph Cubitt reported that the contractor John Jay and his men had completed 'a considerable part of the foundations of the large Goods Buildings' (TNA RAIL 236/273: 27/03/1850). Three days later, Lewis Cubitt signed off plans of the iron roofing of the future Train Assembly Shed which showed the roofs of that structure extending the whole length of the goods building between the Eastern and Western Transit Sheds (Fig. 3.5). Despite the conspicuous omission of the Granary from these drawings, both the north and south walls of the future Granary were shown in the positions that they would eventually occupy suggesting that its foundations could have been laid before its construction was authorised by the Board of the Great Northern.

Although he had yet to receive formal permission to proceed with the works, on Tuesday 21st May 1850 Lewis Cubitt wrote to the Stations Committee to recommend that construction of the Granary be 'commenced and carried on with all possible despatch' (*ibid.*: 05/06/1850). Enclosed with Cubitt's letter was a set of three plans of the structure (since lost), which were reviewed and finally approved at a meeting held on Wednesday 5th June (*ibid.*: 295). If construction of the Granary foundations had not already commenced, work on them presumably began shortly thereafter, although the exact date was not documented.

Neither Joseph nor Lewis Cubitt recorded the date when the Granary was completed in their regular progress reports to the Board, however a newspaper report published on 5th January 1851 describing a fatal accident at the site suggested that the upper levels were approaching completion by then (Anon [The Times] 1851a). Towards the end of that month the Executive Committee of the Great Northern sanctioned a request from Lewis Cubitt that the company accept a tender from W.G. Armstrong & Co. to supply and install the hydraulic cranes and pumping equipment in the Granary 'at once', suggesting that the building may have been almost ready to fit out by then (TNA RAIL 236/71: 295). It was certainly complete by March 1851, when Seymour Clarke and Lewis Cubitt presented the Board with a proposal to convert it to temporary dormitories in order to accommodate visitors attending the forthcoming Great Exhibition, which was due to be held between May and October of that year (TNA RAIL 236/16: 130). Although the latter proposal was turned down by the Board, the fact that it was made at all suggests that the Granary was complete but empty during the spring of 1851, and that the company's senior officers entertained serious concerns regarding its readiness for use during the forthcoming harvest.

Clarke's and Cubitt's concerns were not without foundation. At the end of July, Clarke warned the Board that it was necessary to take immediate steps to install the hydraulic machinery if the Granary was to receive grain from that year's crop at all (*ibid.*: 273; Smith 2008: 4). In response to Clarke's warnings, the Board instructed Joseph Cubitt to impress upon William Armstrong the importance of completing the works without delay and to summon him to London to explain himself (TNA RAIL 236/16: 279). At a meeting of the Board that August, Armstrong assured the Directors that he would 'have [the machinery] ready for use with the least possible delay', but conceded that the sack machinery for the Granary would not be available until October (*ibid.*).

It is not entirely clear when the hydraulic machinery in the Granary became fully operational. The Great Northern did not pay Armstrong's bill until early in 1852, which may suggest that some components did not arrive as quickly as he promised (Smith 2008: 4). The twenty hydraulic cranes that Joseph Cubitt ordered at the end of May 1850 appear to have been initially operated by hand and it is possible that the machinery in the Granary was also worked manually until problems with the hydraulic power supply were finally resolved (TNA RAIL 236/70: 275). In fact, the supply pipe that provided the system with water was still to be laid by March 1852, indicating that hydraulic power cannot have been activated until after that date (TNA RAIL 860/44: 120, 133; see Section 3.14 for further details).

Despite its late inclusion in Cubitt's overall scheme, the Granary came to be considered by many visitors to King's Cross to have been the centrepiece of the completed Goods Station. Contemporary accounts of the Goods Yard often stressed the scale of the finished building and the ingenuity of the means by which goods were loaded or unloaded from wagons, carts and canal vessels. The earliest reports were written before the hydraulic system became operational; a report published in The Observer towards the end of April 1851 noted that the Granary could store as many as 20,000 quarters (60,000 sacks) of corn, whilst John Weale's guidebook of the same year commented approvingly on the 'intimate junction' between the building and the basin (Anon 1851b; Weale 1851: 811). The first account to describe the fully operational station was published in The Illustrated London News at the end of May 1853 (Anon 1853). The writer of the latter piece described the mechanisms by which grain sacks were hoisted by hydraulic crane to the upper levels and how they were returned to the bottom of the building by chutes for onward distribution by waiting carts (*ibid.*: 427). The report also included the earliest reference to the 150,000 gallon capacity water tanks which had been installed in the roof space, presumably in order to supply Armstrong's hydraulic system (*ibid.*; TNA RAIL 236/70: 295).

Once complete, the Granary would store grain that was brought in from the northern and eastern provinces. It was destined for the London flour mills and would help to feed the city's burgeoning population.

Fig. 3.37   The Granary (Landscape Feature 12), flanked by the Eastern and Western Goods Offices (both later additions), photograph faces northwest

Built heritage and archaeological evidence relating to the construction, appearance and footprint of the Granary by 1852

Landscape Feature 12

The Granary (Landscape Feature 12) is a six storey building that is approximately 21m tall, 55m long (east–west) and 30m wide (north–south). The structure is located between the southern ends of the Western and Eastern Transit Sheds (Fig. 3.2; Fig. 3.37). It represents Lewis Cubitt's Granary, a building that was studied in detail in a historic building recording survey (KXF07).

William Humber's Plan of 1866 provided an insight into the original layout of the Granary (Fig. 3.8). A brief summary of the content of the plan is therefore included here as it played a vital part in the interpretation that follows. The Humber Plan suggested that three transverse east–west railway lines ran between the Granary and the Transit Sheds that were connected to the Train Assembly Shed via six wagon turntables positioned at junctions with the two north–south tracks (Fig. 3.8). The same source also suggested that two platforms or banks existed in the Granary. The northern example was divided into three sections by the two north–south tracks (henceforth termed Banks 3a to 3c from west to east), whilst the southern example consisted of one long, continuous bank (henceforth titled Platform 4; Fig. 3.8; Fig. 3.3). Two subterranean canal docks were also present (Fig. 3.3), which enabled corn to be loaded onto barges for onward dispatch through large trapdoors that were built into the platforms. A report into the workings of the Goods Station written by Humber which accompanied the plan stated that the Granary received inwards goods by means of two railway tracks at the centre of the building, whilst empty wagons were removed via the two outer tracks (Humber 1866: 273).

The western and eastern walls of the Granary were built against the southern ends of the eastern and western walls of the Western and Eastern Transit Sheds. Since the adjoining sections of those walls were not bonded together, they must have been erected in separate phases of construction. What is more, an internal pedestrian doorway that linked the Granary and the Eastern Transit Shed at first floor level had been cut into the fabric of the latter structure but was integral to the former providing further evidence that the construction of the Eastern Transit Shed was well advanced by the time that work began on the Granary. Cubitt's plan of the iron roofing of the Train Assembly Shed dated 30th March 1850 indicated that the building of the Granary may have unofficially commenced before its creation was formerly approved by the Board, perhaps when construction work on the Transit Sheds began (see above and Fig. 3.5). The built heritage evidence strongly contradicted this interpretation since the foundations of the Transit Sheds and the Granary were entirely separate. This in turn suggests that the official story is correct: work on the Transit Sheds commenced in the early months of 1850, probably around the beginning of February, their foundations being largely complete by the end of the following month (TNA RAIL 236/273: 27/03/1850). Dithering over the inclusion of the Granary then continued, meaning that work on that structure did not begin until its inclusion was approved by the Stations Committee the following June (*ibid.*: 295).

Many original features have been retained in the exterior façades of the existing Granary building. The brick built structure was topped by a decorative Millstone Grit cornice with an undecorated frieze and a stone parapet wall and was underpinned by a small brick plinth with chamfered sandstone coping. The exterior was bonded in relatively decorative Flemish style brickwork whilst the interior was characterised by plainer English bonding. The exterior was further embellished in a modest fashion by alternating the colour of the pointing from white to greyish brown in order to create six horizontal decorative dark bands that were eight courses high and were aligned with each floor level.

The south (front) elevation of the extant Granary (Fig. 3.39) is nine bays wide, the bays at the far west and east being narrower than the others. They accommodated original cantilevered stone staircases

Fig. 3.38 The timber chutes of 1850–1 that were used to convey sacks of grain from the upper floors of the Granary (Landscape Feature 12) through the easternmost ground floor window in the south wall to waiting carts

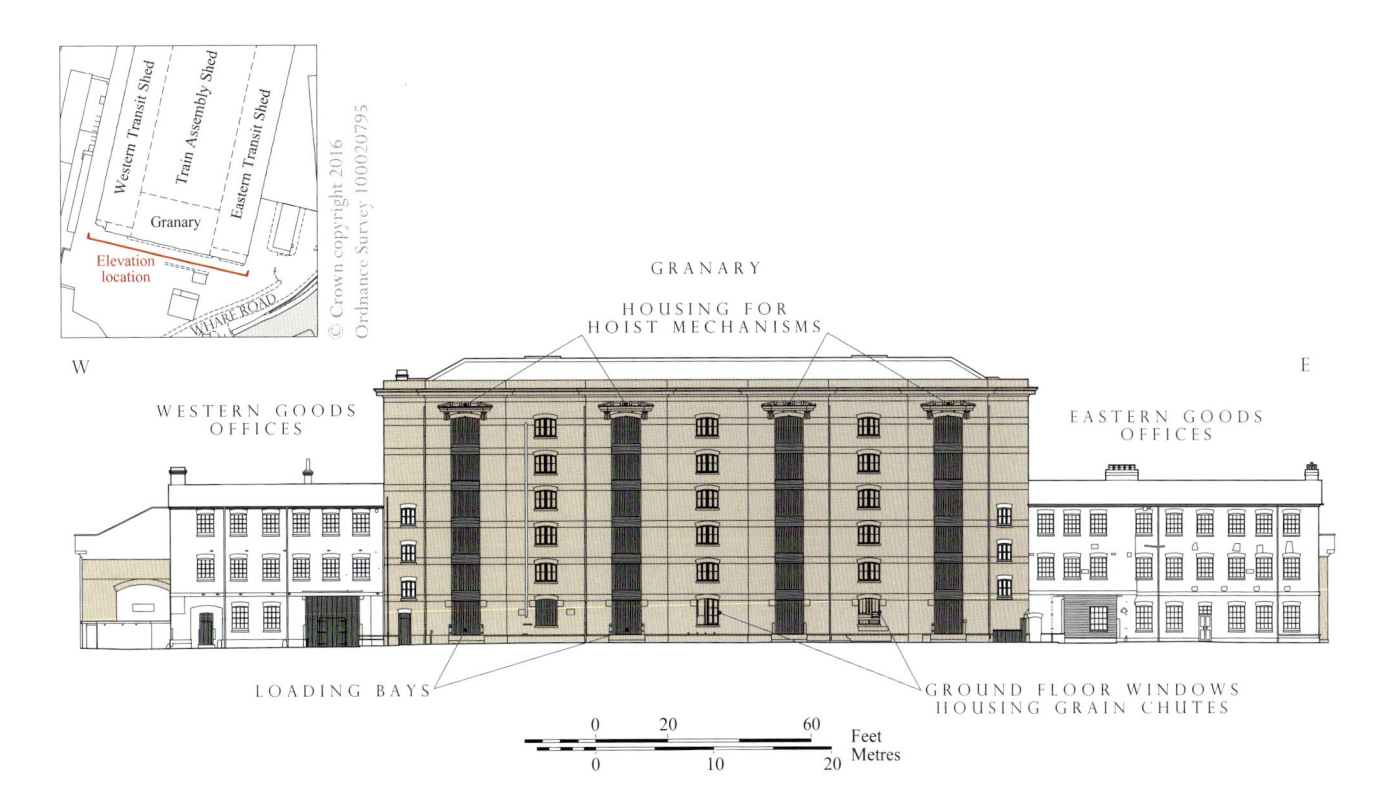

Fig. 3.39 External elevation of the south wall of the Granary (Landscape Feature 12) showing the original 1850–1 masonry, scale 1:625

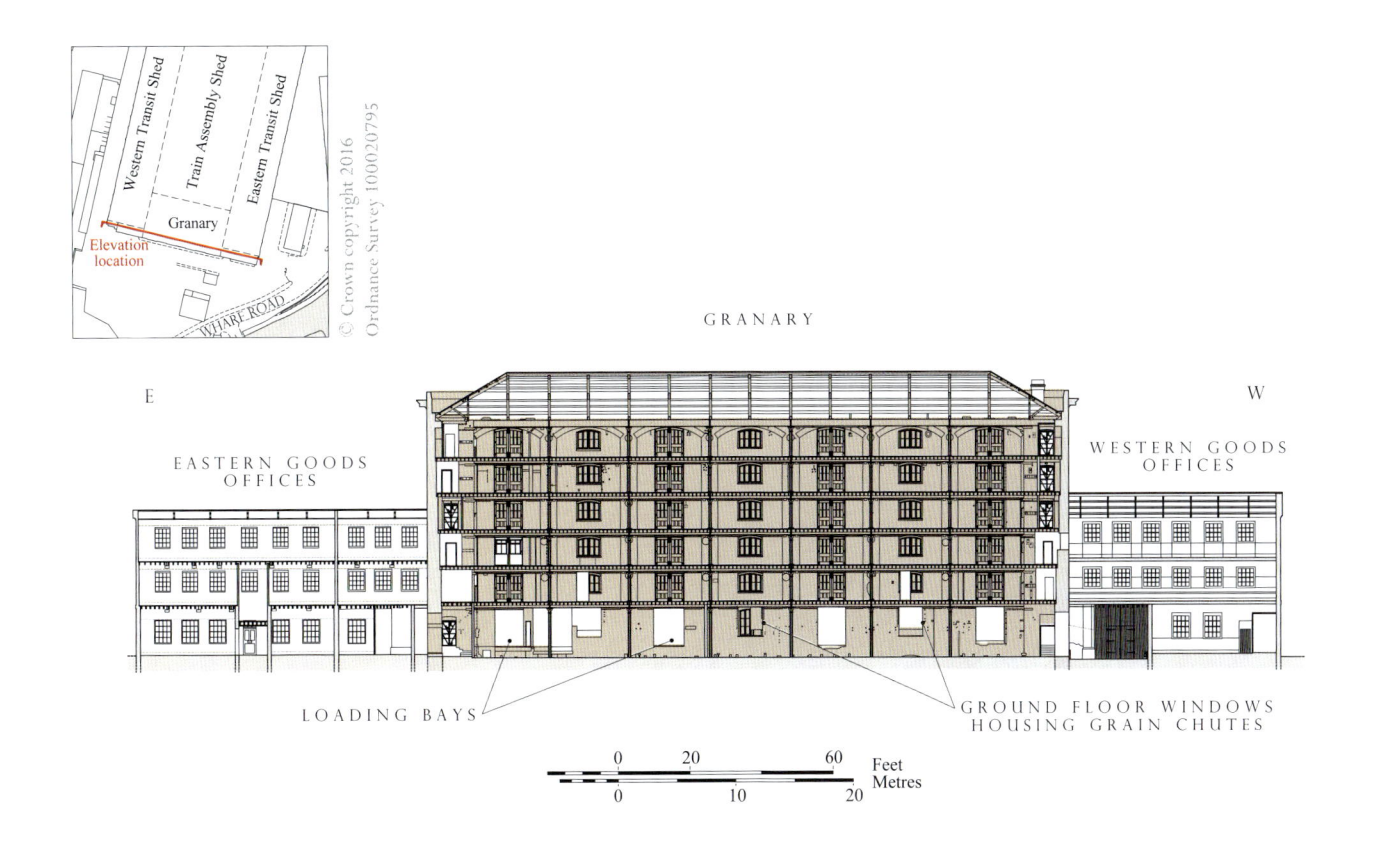

Fig. 3.40 Cross-sectional elevation through the Granary (Landscape Feature 12) showing the internal face of its south wall and the original 1850–1 masonry, scale 1:625

Fig. 3.41 An example of the 1850–1 hoist mechanisms and hipped canopies that surmounted the Granary's (Landscape Feature 12) fifth floor loading bays, photograph faces north

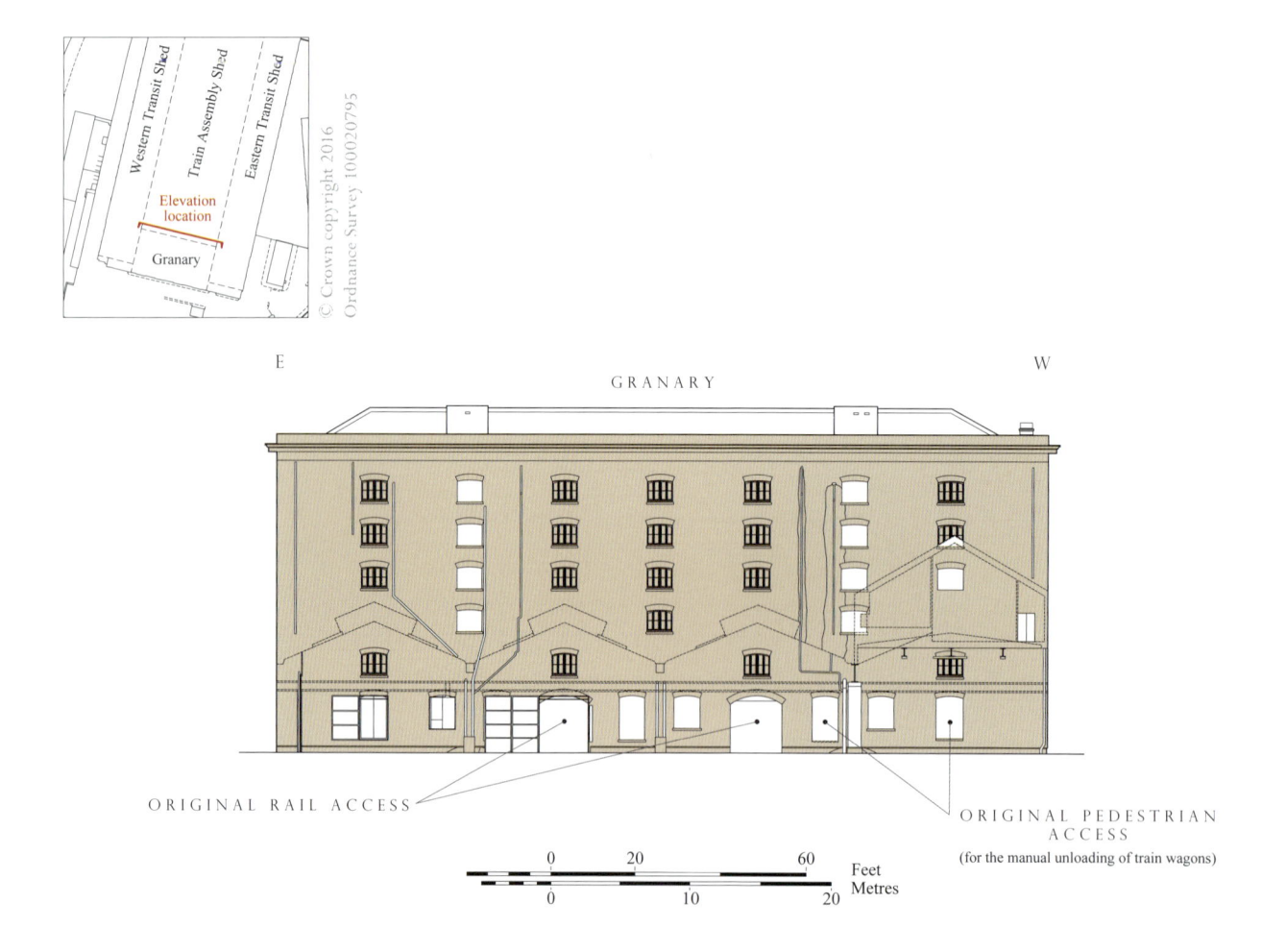

Fig. 3.42 External elevation of the north wall of the Granary (Landscape Feature 12) showing the original 1850–1 masonry, scale 1:500

that ran up the southwest and southeast corners of the structure. The original doorways and windows that opened onto the western and eastern staircases were housed in the western and eastern walls. The staircases respectively rose clockwise and anti-clockwise, mirroring each other.

Moving in from the staircase locations towards the centre of the building, 24 loading bays were arranged in four vertical strips of six (one per floor) in the southern elevation, alternating with the bays that contained the windows (Fig. 3.39). They were not visible on the exterior face of the south wall, where they were covered up by modern, black painted timber boards. However, an inspection of the interior face demonstrated that 20 of the 24 original timber loading bay doors survived, the four ground floor examples having been removed and infilled with bricks (Fig. 3.40). Nineteen of the remaining timber doors were identical, the one oddity being the most easterly example on the second floor that had no doubt been modified at some point after its installation. The doors opened inwards, were timber framed and the nineteen unaltered examples had six opaque glass panes in three vertical sections with a thin horizontal glazing bar in their upper halves. Without exception the ground floor doors had large external sills that were made from two pieces of granite. Either side of them, large vertical timbers had been fixed to the wall as a protective measure against vehicle strikes. Whilst only three of these remained *in situ* at the time of the historic building survey, scars and holes substantiated the former presence of others.

Externally, each loading bay had a shallow housing for a hoist mechanism that was secured above the highest example. Each of these hoists consisted of a pair of iron pulley wheels that were partially recessed into the outer face of the south wall and were protected by shallow, sloping hipped canopies. These decorative awnings were supported by two pairs of stone brackets that were capped by Roman style roof tiles and lead flashings (Fig. 3.41). Decorative embellishments on the canopies included classical stone mouldings and cast iron bars moulded in an identical style in the span between the two pairs of brackets. Internally, iron plates fitted with pairs of iron spindles for horizontal wheels would have guided the hoist cables through to the internal side where, offset to the west, there would have been an iron bracket that held another guide wheel that rotated parallel with the wall. Whilst none of the internal wheels survived, one of the brackets was observed, the positions of the others being indicated by a series of scars.

The remaining three bays were characterised by segmental arched windows on all floors, set into recessed brickwork in between the projecting bays (Fig. 3.39). These openings had stone sills and the windows themselves were composed of eight glass panes in four vertical panels, the central pair of which formed inward opening casements. With the exception of those on the lowest storey the windows were equally sized, those on

Fig. 3.43 Southeast stairwell in the Granary (Landscape Feature 12) taken from the fifth floor

Fig. 3.44 Iron framework around the base of a column in the Granary (Landscape Feature 12), which was used as a fairlead

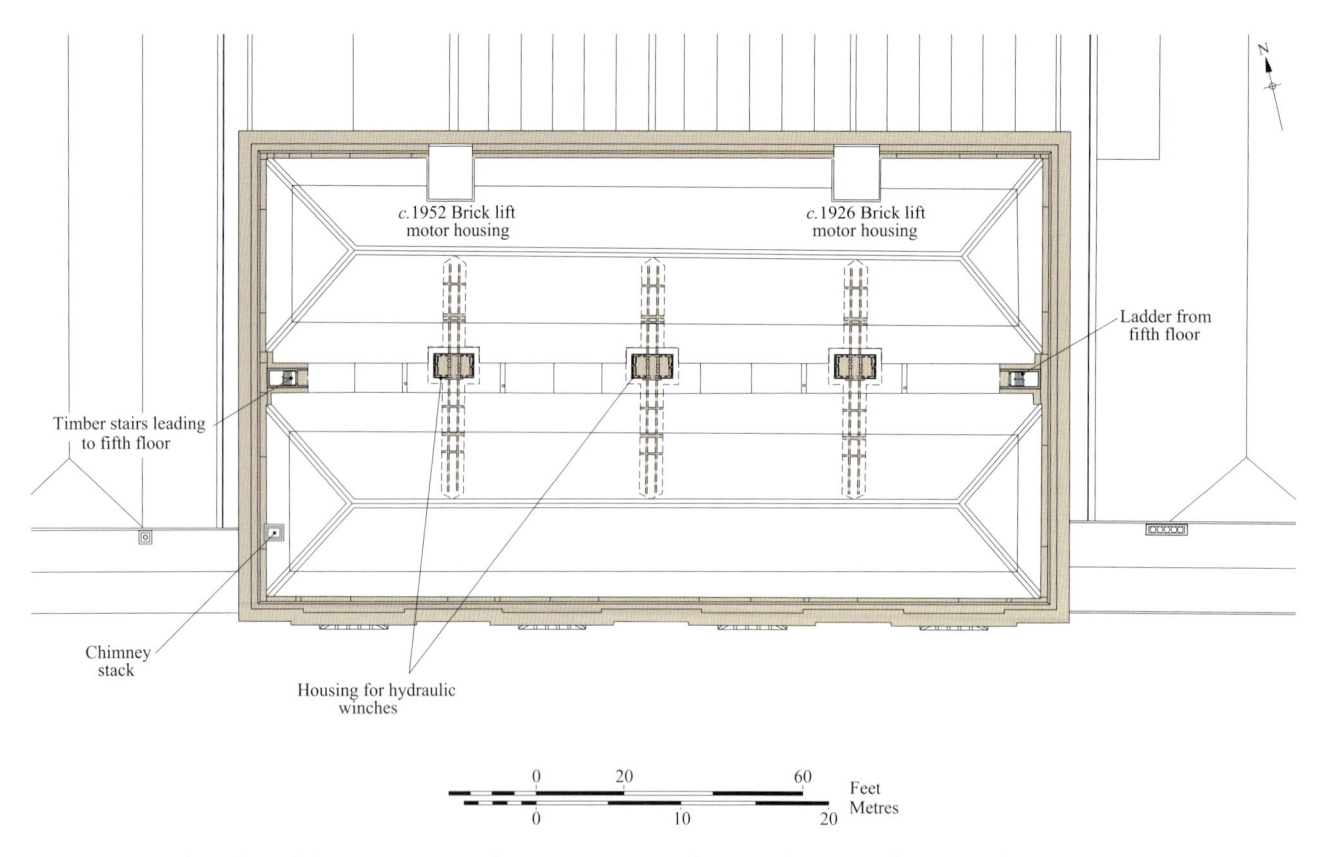

Fig. 3.45 Roofing plan of the Granary (Landscape Feature 12) showing the original 1850–1 elements, scale 1:500

Fig. 3.46 Central roof valley of the Granary (Landscape Feature 12) surmounted by infrastructure associated with the sack hoists' hydraulic winches of 1850–1, photograph faces east

Fig. 3.47 Housing for the sack hoists' hydraulic winches of 1850–1 on the roof of the Granary (Landscape Feature 12), photograph faces north

the ground floor being taller due to the fact that they were used as conduits for timber chutes that channelled grain from inside the Granary to the exterior of the building. These chutes were still *in situ* at the time of the building recording survey. Two exited through the easternmost window opening, one descending from the first floor and the other from the second floor, an arrangement that was also evident in the westernmost window (Fig. 3.38). The central example accommodated just one wooden chute on its western side that descended from the first floor. Repaired sections of flooring indicated that four direct connections existed between the chutes on the second floor and the overlying three storeys. Due to the extensive nature of the repairs it is hard to say whether the openings originally housed chutes that have since been removed or whether they took the form of small vertically aligned trapdoors that accommodated winches.

The north elevation of the Granary closely resembled its southern equivalent but was slightly different in certain respects as it was designed to be integrated with the Train Assembly Shed to the north (Fig. 3.42). Two large doors were present at ground floor level that originally each housed one of the two north–south railway sidings that ran into the building from the Train Assembly area to the north (Fig. 3.42). In addition to these, there were five tall and three

Fig. 3.48 Gate on the second floor of the southwest stairwell of the Granary (Landscape Feature 12)

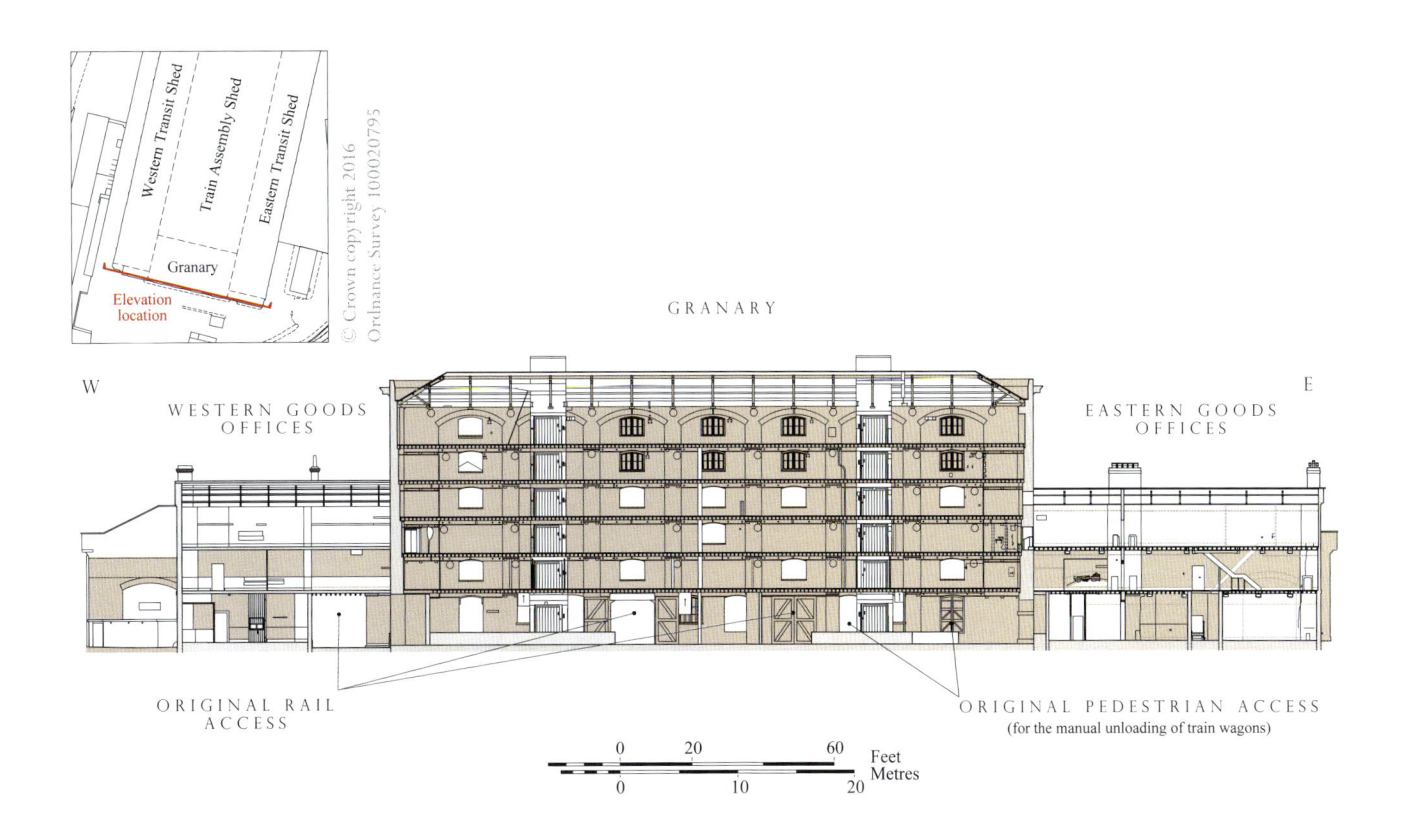

Fig. 3.49 Cross-sectional elevation through the Granary (Landscape Feature 12) showing the internal face of its northern wall and 1850–1 masonry, scale 1:625

short openings, all of which had either been infilled with brick or modified to varying degrees during later periods. The former were presumably used as doorways through which sacks of grain were manoeuvred by hand after being unloaded from wagons that were stationed on the adjacent Train Assembly Shed rail sidings, hence their sills being the same height as the Granary platforms. The short openings, which had higher sills, probably functioned as windows in the first instance, although at the time of the built heritage survey they accommodated timber chutes. The chutes probably represent additions that were installed after a cart road was inserted in the southern end of the Train Assembly Shed; as such they will not be considered further in this section (see Section 5.1 for further details).

Unlike the southern elevation, no loading bays were present in the above storeys. Instead, 28 identical windows were arranged in seven vertical rows of four. Whilst each of the top three storeys were illuminated by seven windows, the first and second floor window positions alternated with one another in order to avoid the Train Assembly Shed roofs, leading to a total of four windows at first floor level and three at second floor level (Fig. 3.42).

The ground floor section of the western (side) elevation was characterised by six open segmental archways on the ground floor, three of which originally admitted one of the three railway lines that communicated with the Western Transit Shed, the others being reserved for pedestrian access (Fig. 3.17; also Section 3.7 for further details). The original ground floor entranceway to the western stairwell could be found to the south of these arches, which has since been bricked up (Fig. 3.17). On the floors above this, five windows originally illuminated the stairwell. Only two of these have survived unaltered as those on the first, second and third floors were modified when the Western Goods Offices (Landscape Feature 20) were

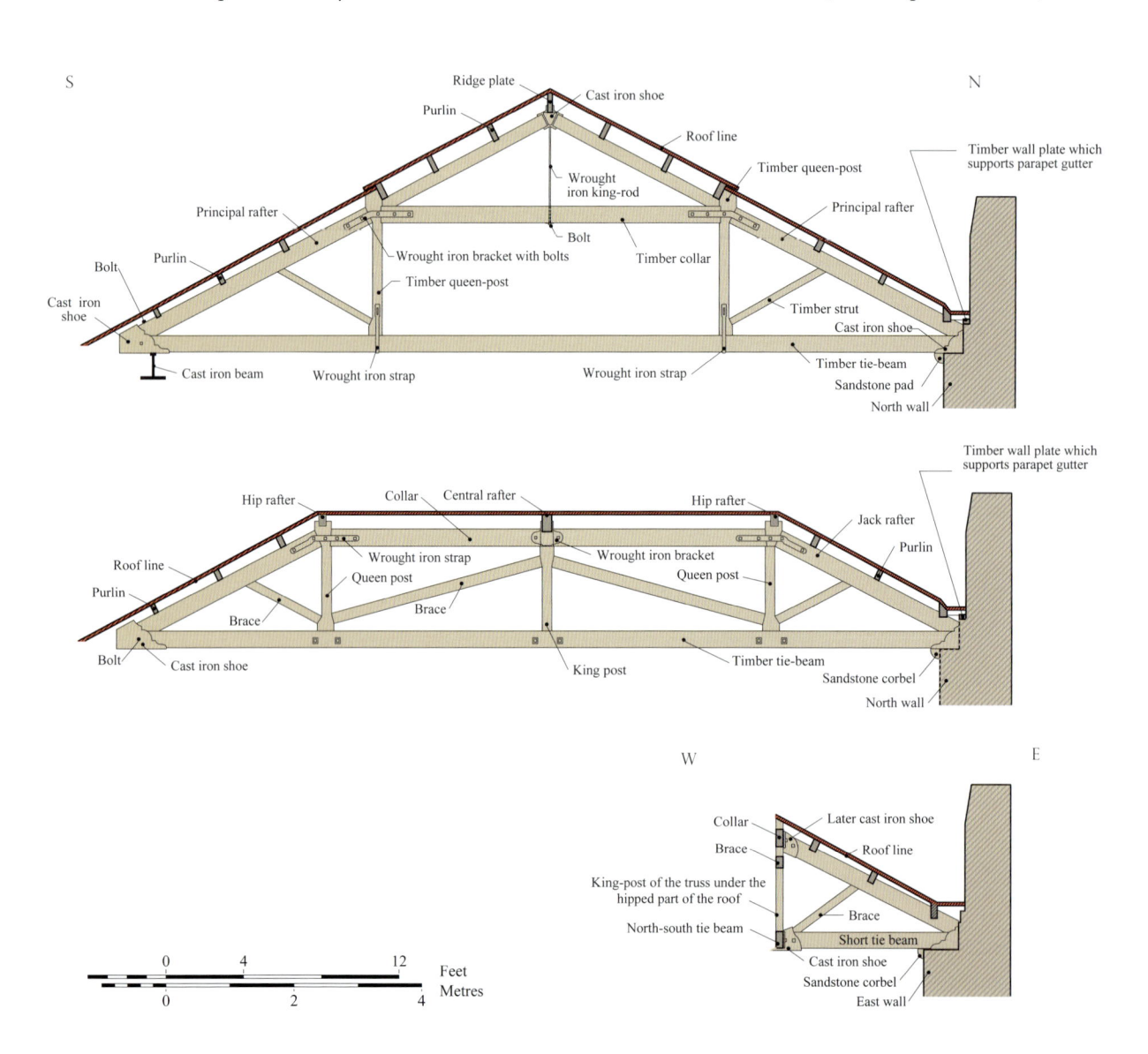

Fig. 3.50 Roof trusses of 1850–1 in the Granary (Landscape Feature 12), scale 1:100

constructed. With a few minor differences, the original form of the eastern (side) elevation of the Granary mirrored its western counterpart.

The only enclosed areas on the ground floor of the Granary were the stairwells that were situated in the southwestern and southeastern corners of the structure (Fig. 3.52). The remaining floor space constituted a single open area dominated by a grid of circular, hollow cast iron columns that supported the first floor and divided the area into nine east–west and seven north–south bays (Fig. 3.52). The columns were all made to the same design with cruciform ribbing, although those on the ground floor were taller (at 4.10m or 13'6") than those on the upper floor (which were 2.4m or 8') because greater floor to ceiling height at ground floor level was needed in order to accommodate the swing of the arms of the hydraulic cranes. Several columns had been fitted with vertical iron bars that were held in place by collars (Fig. 3.44). Grooved wear marks on these bars appear to have been caused by friction from ropes or cables, suggesting that they were used as fairleads. The columns supported east–west cast iron beams that carried the floor boards on all levels apart from the roof. The large east–west spacing of the columns (6.9m) was needed in order to span the turntables and the canal docks at ground floor level.

Cast iron columns were distributed across the upper storeys in an identical arrangement to those on the ground floor.

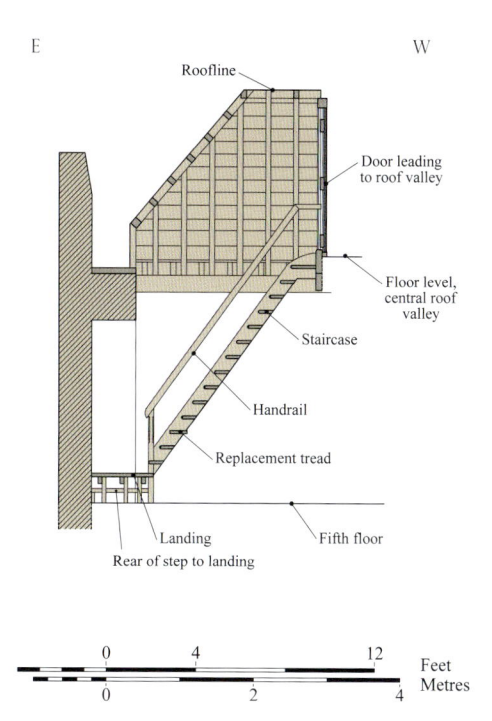

Fig. 3.51 Cross-sectional elevation through the eastern stairs of 1850–1 that led to the central roof valley of the Granary (Landscape Feature 12) showing their southern wall, scale 1:100

Access to the first floor was provided by two north doorways in the stairwells in the southeast and southwest corners (Fig. 3.40). The north end of the east wall contained a segmental arched doorway, infilled at a later date, which would have enabled the movement of goods between the Granary and the Eastern Transit Shed. The floorboards had been replaced in numerous locations, presumably to patch holes that were left when chutes, trapdoors and hoist cables were removed (Fig. 3.52). The second floor of the Granary was very similar to the first, although its southwest stairwell was gated at the base of the flight of stairs to the third floor (Fig. 3.48). As on the first floor, repairs to the floorboards marked the locations of former hatches and the cables that connected with the hoist mechanisms on the roof. The third floor was generally similar to those found above and below.

The staircases leading to the fourth floor were lit by original windows that were situated in the east wall of the southeast stairwell and the west wall of the southwest stairwell. Unlike their equivalents on the lower storeys these fourth floor windows had not been removed when the three storey Goods Offices were constructed.

The fifth floor was basically the same as the floors below bar a number of minor differences. The cast iron columnar roof supports were identical, apart from those on each side of the central east–west bay, each of which had a water inlet cast into their capitals that collected water from pipes that came down from the central roof valley above. Rainwater then passed down the inside of the interconnected columns to the ground floor, before being taken away by drains that emptied into the canal inlets below. The column lines supported single rather than paired cast iron beams since they carried the roof of the building rather than one of its heavier floor surfaces. A series of slightly shorter beams supported the inner ends of the roof trusses that themselves carried a timber beam that held up the ceiling joists. These in turn supported the timber construction of the central roof valley.

The roof of the Granary comprised two hipped roofs aligned east–west with a broad central flat valley, one bay wide (Fig. 3.45). Each had fifteen roof trusses that included two under the hips (Fig. 3.40; Fig. 3.49). Access to the roof top was gained from the east and west ends of the fifth floor of the Granary via two steep timber ladders with timber handrails that emerged at the east and west ends of the central roof valley from low timber cabins (Fig. 3.51). Between the access stairs at the east and west ends of the central roof valley were a further three timber cabins that formerly housed the hydraulic winches that operated the Granary's sack hoists. The mechanisms have since been removed leaving these winch houses empty. Each winch system was characterised by two timber gantries that extended to the north and south, either side of the winch house, straddling the expanse between the peaks of the north and south roofs. Suspended below these gantries on either side of the winch houses were pulley wheels, which guided the pulley cables that ran through one of the vertical shafts that lay below (these were formed when the correct combination of trapdoors were opened as shown in and described subsequently). This allowed the sack hoists to travel vertically through all floors of the building, depositing or more probably collecting goods from train wagons stationed at ground floor level.

Evidence in the form of scars, replaced flooring and patching arranged across the entire building (Fig. 3.52) suggested that two east–west rows of three trapdoors were once arranged in a grid pattern on every floor of the Granary, directly above the outermost east–west railway lines that ran through the structure at ground floor level. The trapdoors could be opened to create six vertical shafts that extended through the entire structure to the hydraulic winches on the roof. It is likely that the winches were used to lift sacks from train wagons on the ground floor through these hatches to the upper floors (Fig. 3.45; Fig. 3.46; Fig. 3.47). Cut-off vertical timber boxing was visible between some of the joists, which took the cable for the hoist system that operated across all floors.

Remains indicative of the former presence of five probably related chutes were found across the northern half of the Granary, one on each storey (Fig. 3.52). Their distribution strongly suggested that their functions were connected and it seems likely that they conveyed sacks of grain downwards under the influence of gravity. They were repeatedly offset to the west on each floor, effectively creating one large conduit that a sack could traverse in increments as it descended (Fig. 3.52; Fig. 3.53). The pathway that they made effectively cut across the Granary from the northeastern corner of the fifth floor to the northwestern side of the ground floor, the outgoing sacks landing either on Platform 3 (west) if the trapdoor to Dock 2 was closed or directly onto a waiting barge if it was open (Fig. 3.53). A virtually identical arrangement of former openings occupied the southern half of the Granary, although these cut across the structure in the opposite direction (i.e. west to east), thus linking the upper levels of the building with the eastern end of Platform 4 and 'Dock 3' (Fig. 3.53).

Functioning of the Granary c.1852

Landscape Feature 12

It is possible to determine how Landscape Feature 12 functioned when it first opened to traffic in the early 1850s. Figures 3.52 and 3.53 reconstruct the flow of goods through the building, revealing how and why the structure's railway lines, loading bays, grain chutes, winches and trapdoors interacted. The Granary was designed in order to receive, store and distribute grain grown in the provinces and transported to London by rail; therefore the following interpretation is predicated

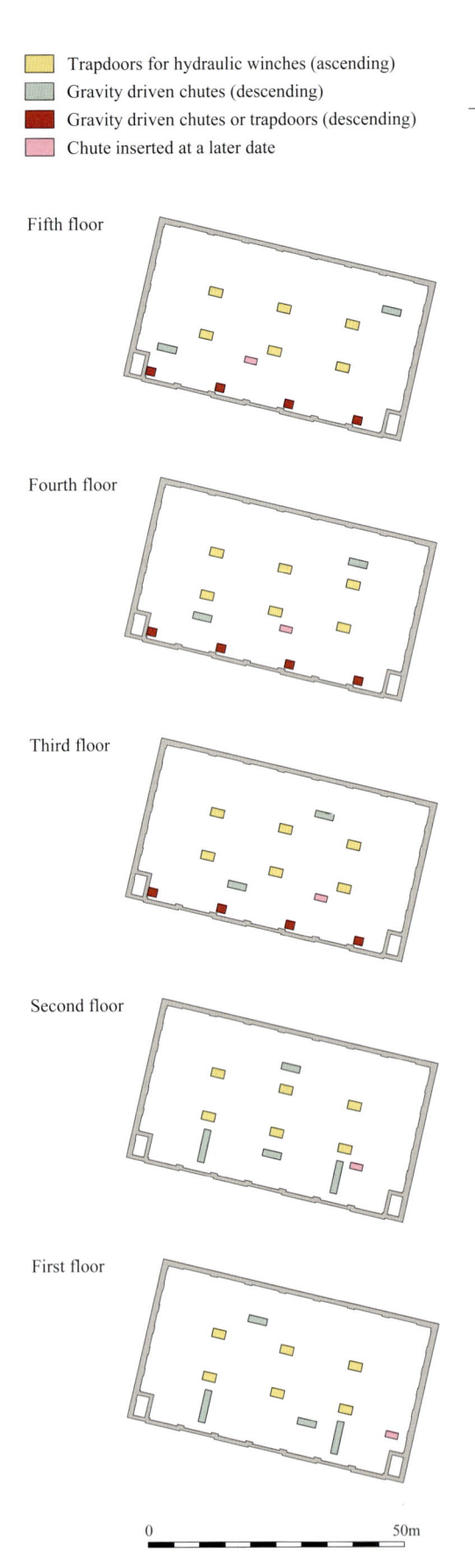

Fig. 3.52 Schematic diagram showing the distribution of trapdoors and chutes across all floors of the Granary (Landscape Feature 12) based on the scarring that was observed in the floors and ceilings of the building, scale 1:1,250

on the assumption that the movement of the commodity into and out of the Goods Station was heavily or entirely skewed in one direction.

Grain entered the building by a variety of routes. The track arrangement shown on Humber's plan suggested that railway wagons could be stationed at the southern termini of the Train Assembly Shed's sidings against the north wall of the Granary. Sacks could then be manually carried into the building via the loading bay doors in the northern edifice. However, the most efficient way of receiving incoming sacks was by bringing them directly into the building by railway wagon. Humber's 1866 report states that grain entered on the two central north–south railway tracks that connected with the Train Assembly area, from where it could be offloaded direct from the wagon and transported upwards by the hydraulic winches (Fig. 3.53). Humber noted that the empty trucks were then dispatched by way of the outer tracks (Fig. 3.8; Fig. 3.58).

After the repeal of the Corn Laws in 1849, grain imports from abroad increased rapidly and it is therefore reasonable to assume that substantial quantities of the commodity began to flow into the port of London at a time that coincided with the opening of King's Cross Goods Yard. Although the Granary was directly linked to the docks by the canal, the possibility that grain was transported to King's Cross by barge seems to be an unlikely one. The port of London possessed its own granaries so shipment to King's Cross would have been unnecessary unless it was intended to forward the commodity northwards by train, which seems somewhat implausible under

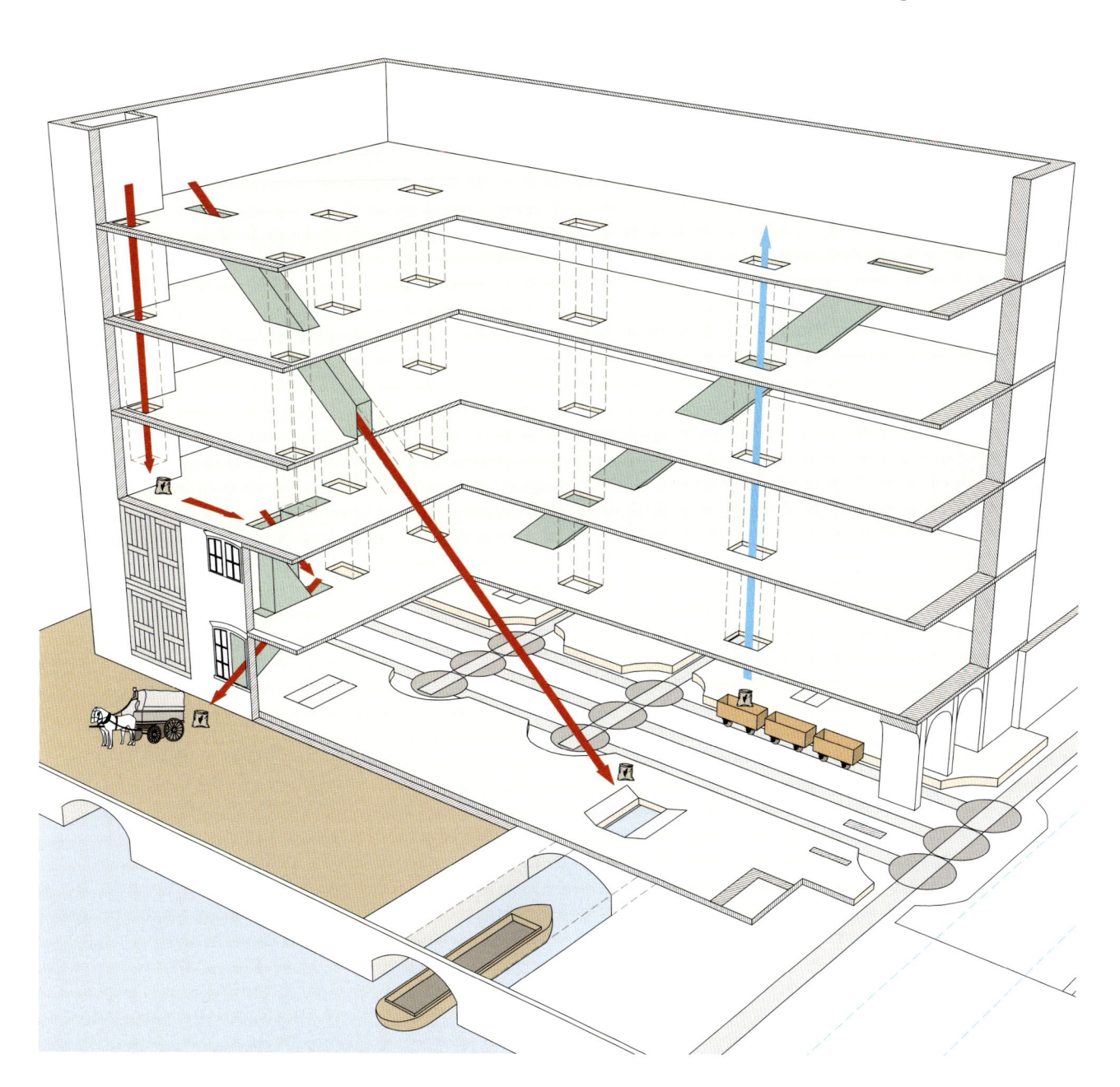

Fig. 3.53 Reconstruction of the rail and canal infrastructure and the trapdoor and chute network inside the Granary (Landscape Feature 12) c.1851 (Landscape Feature 12), not to scale

Fig. 3.54 An engraving from ***The Illustrated London News*** of 28th May 1853 (Anon 1853) showing horse-drawn vehicles lining up against the southern wall of the Granary (Landscape Feature 12) to receive sacks of grain; two barges sail into the building via Canal Tunnels 2 and 3

normal circumstances. Therefore it appears that the role of the waterway at King's Cross was to forward grain from the north to London and the south, a conclusion that is supported by the infrastructure that appears to have been present in the Granary. Grain sacks could be deposited in waiting barges with ease via the gravity driven chutes and conveyed to horse-drawn drays for onward transit to London by road via the loading bay doors and winches in the southern façade. In contrast the apparent lack of hydraulic apparatus beside the canal inlets would have made the task of removing heavy sacks from boats for storage within the building extremely arduous and rather impractical.

Most of the grain brought into King's Cross by train was presumably stored in the Granary for a period of time prior to onward shipment to London and its hinterland. It is unlikely that sacks were manually winched upwards via chutes or pulleys owing to the strenuous nature of the task, especially given the ease with which a mechanized system would have achieved the same objective. This strongly suggests that the three hydraulic winches that were housed in the roof took sacks upwards through the six vertically aligned trapdoors (Fig. 3.46; Fig. 3.47; Fig. 3.53). These winches could lift sacks directly from railway wagons parked on the two outer east-west sidings (Fig. 3.53).

Grain stored on the upper floors of the Granary was returned to ground floor level by a variety of means. Sacks stored on the first and second floors would have been dispatched via one of the five chutes that exited through the three ground floor windows in the southern elevation (three being linked to the first floor, the remaining two with the second floor). Horse-drawn carts would have lined up below these in order to receive the sacks directly (Fig. 3.53; Fig. 3.54). Grain from the upper floors could also have exited in this way after being conveyed to the second floor via the four sets of openings that were evenly distributed along the southern side of the building (Fig. 3.53). In a similar way, bags could have been lowered from the upper storeys for direct deposition in waiting carts through one of the twenty loading bay doors that were arranged across the first to the fifth floors in the southern elevation (Fig. 3.38). Sacks leaving in this way would have been secured to one of the four winches that were installed at eaves level (Fig. 3.41). They would then have travelled downwards to a waiting cart, the descent being manually or hydraulically controlled to prevent free-fall. Grain could also have been sent from any upper storey via the northern network of gravity driven east–west chutes for direct deposition in canal barges moored in Dock 2 if the trapdoor in Platform

3 (west) was open or upon the platform itself if it was closed (Fig. 3.52; Fig. 3.53). Similarly, sacks could have been sent downwards via either of the two southern sets of east–west openings (Fig. 3.52; Fig. 3.53). In the case of the most westerly set of chutes they would have been deposited in barges moored in Dock 3 if the trapdoor in Platform 4 was open or upon the platform itself if it was closed (Fig. 3.52; Fig. 3.53). Sacks dumped upon the platforms must either have been stockpiled there to speed up the loading of barges or manually carried through the loading bay doors in the southern elevation for onward shipment by road or barge.

Across the first to the fifth floors of the Granary a series of black circles with white numbers had been painted onto the interior of the building, generally on the piers or pilasters that strengthened the walls. For example, on the first floor the numbers '13' to '18' and the numbers '60' to '67' were respectively painted on the south and north walls, whilst those on the east and west internal edifices were not legible apart from '19' on the west wall of the southwest stairwell. Exactly when these numbers were painted remains uncertain, although they were evidently added after the building became operational. They probably formed part of a system that enabled the correct goods to be retrieved from storage for onward transfer to waiting carts or barges.

A similar system of notation was observed on the exterior of the building. Adjacent to all the ground floor loading bay openings in the southern elevation was a sequence of white numbers painted on blue squares. From east to west the sequence ran as follows: the first loading bay had a number '1' on its west side; an adjacent window accommodating chutes had a number '1' on its east side and '2' on its west side; the second loading bay had '2' on its east side and '3' on its west side; the next window had a number '3' on its west side only, probably because it only accommodated a single chute; the next bay had a number '4' on both sides; the penultimate position had a number '5' on both sides and the most westerly loading bay had a number '6' on both sides. This system might have been used to direct road vehicles to the correct position for loading and may also have enabled operatives to send outgoing goods through appropriate loading bay openings.

In summary, the evidence discussed above confirms that the Granary was designed to receive and accommodate grain imported from the north by rail, before onward shipment to the metropolitan and wider regional markets by canal and by road. This conclusion was strongly supported by the overall design of the Granary, which would only have functioned efficiently when goods were flowing in by rail and out via the canal and the road.

3.12 Railway Infrastructure in and around the Main Goods Shed and the Granary by 1852

Landscape Features 7–12

At the beginning of July 1850 Chief Engineer Joseph Cubitt informed the Board of the Great Northern that the foundations of the turntables in the Transit and Train Assembly Sheds were 'in a state of great forwardness, most of them …framed and ready to lay in; the greater part of the turntables, switches and crossings and the water cranes are on the ground' (TNA RAIL 236/273: 04/07/1850). Cubitt had ordered the turntables from the Ipswich based manufacturer Ransomes and May three months earlier, the majority of which were delivered during the summer and autumn of 1850 (TNA RAIL 236/70: 183). Four types of turntable were ordered for delivery at King's Cross Goods Station, all of which were manufactured to Wild's Patent (Smith 2009: 1):

16' diameter 12-ton turntables, each £115
13' diameter 8-ton turntables, each £75
12' diameter 7-ton turntables, each £65 10s
12' diameter 5¼-ton turntables, each £46 10s

Despite the contractors' best efforts to complete the Goods Station by the summer of 1850, Joseph Cubitt admitted in his half yearly report to the Board that goods traffic was not yet running, 'owing to the station arrangements not being sufficiently forward for the advantageous reception of Merchandize, Coals, Cattle &c…but the works connected with these arrangements are in a state of active progress' (TNA RAIL 236/273: 23/08/1850). The Western Transit Shed was ready for traffic by the end of October (*ibid*.: 24/10/1850), whilst the Eastern Transit Shed was not completed until the following March (*ibid*.: 18/02/1851).

The earliest detailed depiction of the railway infrastructure in and around the Goods Station can be found in the form of the Humber Plan and the reconstruction that follows is drawn heavily from this source. However, since other documents indicate that modifications were made to the Goods Station's railway infrastructure before it was published in 1866, it cannot be used uncritically. Several changes were made during the intervening period, including a scheme to extend 'the lines north of the Goods Shed to enable loaded wagons to be passed out of the loading shed and formed into trains' and the addition of an extra turntable on 'No. 4 Up Line' (TNA RAIL 236/20: 338; TNA RAIL 236/21: 286). The following account therefore takes into account these other documentary clues.

The Humber Plan indicated that the Goods Shed complex housed a total of fourteen north–south railway tracks. Each Transit Shed accommodated one of these, the remaining twelve being arranged in four groups of three in the Train Assembly Shed (Fig. 3.8). Standard railway protocol identified London-bound lines as 'up' lines, whilst outgoing tracks were referred to as 'down' lines; these were differentiated from one another by assigning a number to each (e.g. 'No. 1 Up Line'; 'No. 3 Down Line' and so on). Five 'up' and five 'down' lines in the Assembly Shed terminated to the immediate north of the structure's southern wall, whilst one 'up' and one 'down' line continued into the Granary before terminating at the junctions with the most southerly east–west siding found therein. Those in the Transit Sheds exited these buildings to the south, terminating near the Granary Basin (Landscape Feature 5). They were crossed at right angles by ten east–west lines (Fig. 3.8). Six of these have been termed 'long' lines in this document as they traversed the entire complex crossing all the 'up' and 'down' lines (three ran from the Eastern Transit Shed, through the Assembly Shed and into the Western Transit Shed, whilst the other three connected the Transit Sheds with the Granary). Two of the remaining four, henceforth termed 'short' lines, ran from the Western Transit Shed across the most westerly group of 'down' lines in the Train Assembly Shed, an arrangement that was mirrored on the opposing side by the other two 'short' lines that linked the Eastern Transit Shed with the most easterly group of three 'up' lines in the Train Assembly area (Fig. 3.8). In terms of track layout, the complex exhibited east–west symmetry at this time, although this did not apply to the arrangement of turntables.

The Humber Plan indicated that turntables were not placed at every junction within the Goods Shed complex. Whilst turntables could be found at 63 of the 70 junctions by 1866, only 62 of these were present in 1852 since an extra turntable was added to 'No, 4 Up Line' in 1855, which is assumed to be the fourth line from the right within the Train Assembly Shed (TNA RAIL 236/21: 286). This addition must be shown on Humber and it is likely that it is the solitary example that is depicted at the northern end of that siding (Fig. 3.8). Since it is probably a later addition it will not be discussed here.

It is also worth considering the arrangement of the sidings that immediately surrounded the station as these also affected the way in which the complex functioned. The Humber Plan of 1866 suggests that a network of tracks encircled the Main Goods Shed (Landscape Features 7–11) and the Granary (Landscape Feature 12) and extended down around the eastern and western sides of the Granary Basin (Landscape Feature 5) as shown in Fig. 3.8. Since the extension down the eastern side of the basin was added in 1856 it will be omitted from the reconstruction that follows (TNA RAIL 236/27: 264). The Humber Plan shows turntables at all the intersections between the north–south and east–west sidings that surrounded the Goods Shed and the Basin with one exception. No turntable appears to have existed at the junction between the Eastern Transit Shed's 'up' line and the siding that ran parallel with the northern edge of the Granary Basin (Fig. 3.8). However, as the following discussion demonstrates, it would have been necessary to fit a turntable in that location in the first instance, although it must have been removed before 1866.

Archaeological evidence relating to the construction, appearance, footprint and function of the railway infrastructure within the Main Goods Shed and the Granary by 1852

Landscape Features 7–12

In order to assess the accuracy of the historical maps and plans and to better understand the nature of the railway infrastructure in the Goods Shed complex (i.e. the Eastern Transit Shed, the Train Assembly Shed and the Western Transit Shed) five archaeological transects were excavated across the complex (Fig. 3.3). These sample areas targeted various intersections between the north–south lines and the six 'long' east–west lines. Whilst most of the railway tracks had been removed, numerous turntable supports were identified, the distribution of which strongly suggested that the arrangement of sidings corresponded to that shown by Humber (Fig. 3.8). Thirty circular brick structures were uncovered in the Goods Shed complex within these excavation areas, although the Humber Plan demonstrated that only 28 of them were original. Those at the northern end of No. 2 and No. 3 'Up' Line were added later as they are not shown on the 1866 map (Fig. 3.8); these are excluded from this discussion as shown (Fig. 3.3). In five instances, turntable supports were absent from their anticipated locations, but these anomalies were easily explained by the impact of later truncations.

The turntable supports were circular or sub-circular in shape with diameters that varied between 3.90m and 4.18m (they were presumably once uniform in shape and size, this variance being due to post-depositional subsidence resulting from cracking and tipping). Predominantly header bonded, machine pressed bricks were used to construct the bases, which were two courses wide and at least four courses deep. A deposit of concrete had been poured into and around them, cementing them in place. All those that were found were situated near the base of the archaeological sequence and no doubt represent early features. They were interpreted as masonry supports for 28 of the 62 turntables that were situated in the Goods Shed complex when it first opened to traffic in the early 1850s (Fig. 3.3).

In addition to the excavations that took place within the Main Goods Shed (Landscape Features 7–11) and the Granary (Landscape Feature 12), a further two external interventions were dug between that complex of buildings and the Granary Basin (Landscape Feature 5). These trenches each targeted one of two original turntables. For clarity, the eastern example has been called 'A' whilst the western has been identified as 'Turntable B' (Fig. 3.3). Both were extremely well preserved with virtually complete *in situ* mechanisms (Fig. 3.55; Fig. 3.56; Fig. 3.57).

Turntable A was situated to the immediate south of the southeast corner of the Eastern Transit Shed. Its recorded remnants demonstrated not only how it functioned but also how it had been constructed and installed.

Initially a large cut was excavated through the ground-levelling deposits. A rectangular timber frame was then erected within this, which was held in place with poured concrete. Eighteen timber chocks were then set in the top of the concrete in a circular formation along with an identical chock at the centre. These supported the outer metal drum of the turntable, which consisted of six curved metal plates that were riveted together to form a cylindrical structure that was 5.20m in diameter and 0.80m tall. An inner circular drum was then bolted to the sides of the outer drum at the base of the structure, which had a diameter of 4.85m and a height of 0.25m. Four metal arms radiated from its centre and these formed a cross shape that reinforced the outer drum enabling it to support a greater weight. A revolving mechanism was installed above this, which consisted of fifteen spokes radiating from a central column that sat directly above the central wooden chock. Metal wheels were found at the end of each spoke and these slotted into a groove that ran around the inner face of the turntable drum thus supporting the mechanism and allowing it to rotate. Two railway tracks crossing at right angles were mounted on a metal plate that sat above the revolving mechanism and it was these that directly supported the rolling stock that used the device. The revolving mechanism was 4.90m wide (approximately 16') making this turntable the largest example that was unearthed during the archaeological investigations in the Goods Yard.

Turntable B was situated to the south of the southeast corner of the Western Transit Shed. It was smaller than Turntable A, its central revolving mechanism being 3.66m in diameter. It was also well preserved and was constructed in a similar manner to Turntable A. Once again, a timber frame was positioned within a large cut before being cemented in place and infilled with concrete. As was the case with Turntable A, the stationary outer and cross-braced inner iron drums were set upon the concrete infill. The lining once again supported the revolving mechanism, which was similar to the one that was found in Turntable A except that it was smaller, having twelve wheeled spokes rather than fifteen. The turning mechanism upheld a cross-braced metal support upon which two sets of tracks, crossing at right angles, sat.

Similar or identical 12' turntables would have been housed inside the circular masonry bases that were unearthed inside the Goods Shed. The excavation of Turntable B was therefore important as it not only provided information concerning its mode of construction and operation but also illustrated the probable arrangement of the turntables inside the Goods Station. However it is necessary at this point to account for a number of discrepancies between historical maps and plans with regard to the latter turntable. If

Fig. 3.55 Turntable A of 1850–2, photograph faces west

Fig. 3.56 Turntable B of 1850–2, photograph faces south

Fig. 3.57 Internal view of the turning mechanism of Turntable B of 1850–2

the cartographic evidence is taken at face value, the excavated remains of Turntable B may not represent the actual device that was installed in 1850-2. The turntable appears on the Humber Plan of 1866 but is not shown on the Ordnance Survey Map of 1871; it then reappears two years later on Crockett's Plan of 1873 (COL MS 15627). Whilst this anomaly could well represent an error on the part of the Ordnance Survey, it is also possible that the turntable was temporarily removed, perhaps for maintenance purposes, before the mechanism was reinstated. Alternatively, a replacement could have been moved to the site from elsewhere or a new mechanism could have been acquired. No records suggestive of such a purchase have been found, indicating that the mechanism is most probably original.

Turntables A and B were presumably used to direct traffic around the Granary Basin, the latter being partnered by a heavy duty hydraulic crane that was used to transfer bulkier goods between the canal and the railway.

A reconstruction of the traffic management procedures that were in place during the formative years of the Goods Shed has been achieved by initially considering the function of each part of the complex. These roles would have dictated the order in which wagons visited each section giving a rough impression of traffic movement. Documentary evidence clearly indicates that the station was divided into defined working areas: it was split with 'up' (i.e. arrival) lines towards the east and 'down' (i.e. departure) lines towards the west. However, written information concerning the arrival of wagons in the Granary suggests that the split was probably uneven, When William Humber inspected King's Cross Goods Station in 1865 he observed that the Granary received inwards goods by way of two railway tracks at the centre, whilst empty wagons were removed via two outer tracks (Humber 1866: 273). Humber's account suggests that incoming trains were probably received on one of eight 'up' lines that ran into the eastern and central sections of the Train Assembly Shed, where they were split into single wagons for ease of movement through the station. After disassembly, individual wagons would have been manoeuvred into the Eastern Transit Shed or, if laden with grain, the Granary for unloading. Rolling stock would then either be returned to a 'down' line in the western half of the Train Assembly Shed for re-assembly into empty trains for immediate dispatch or sent to the Western Transit Shed for re-loading. Once fully laden, the wagons would join their empty counterparts on one of the 'down' lines in the western half of the Train Assembly Shed for re-assembly into north-bound goods trains. Although it was technically possible to directly receive a south-bound train on the Eastern Transit Shed's 'up' line and dispatch a train northwards directly from the Western Transit Shed's 'down' line the historical evidence suggests that this did not occur or was not common practice.

In order to operate the system described by Humber, a turntable would have been required at the junction of the Eastern Transit Shed's 'down' line and the east-west track that ran along the edge of the Granary Basin (Fig. 3.8). If a turntable was not situated in that location then empty wagons could not have been conveyed from the Granary's easterly outer track to the Western Transit Shed for reloading (providing a strict one-way system was in operation, a scenario that seems highly likely in order to avoid collisions). Although two sets of railway tracks crossing each other at right angles appear in that location on Humber's Plan of 1866 (Fig. 3.8), no turntable is shown either on that source or on any subsequent cartographic depictions. In fact, later maps imply that the north-south track (i.e. the southernmost section of the Eastern Transit Shed's railway line) was subsequently lifted, demonstrating that a turntable was never reinstated there (see Chapter 4). The most plausible explanation for the contradictions that exist between Humber's written account and the infrastructure that he mapped is that the traffic management system in the Goods Station was on the brink of change when his inspection began in 1865. By the time Humber began to map the complex, the turntable may already have been removed and the flow of traffic through the Granary altered as a result. For the purposes of this chapter, it shall therefore be assumed that a turntable was present at the junction of the Eastern Transit Shed's railway line and the east-west track that ran alongside the Granary Basin from at least 1852 until it was lifted in 1865 or 1866 (Fig. 3.58).

Consideration of the layout of the railway lines in light of the inferences that are detailed above has enabled the probable traffic management system that was in use during the Goods Station's earliest years (c.1852-65) to be reconstructed (Fig. 3.58). An assumption has been made that, with a small number of exceptions, a one-way system was indeed in place for train wagons throughout the entire station. This is presumed to have been necessary in order to maximise efficiency by minimising congestion and mitigating the probability of collisions. The following rules would ensure that rolling stock could be conveyed in a straightforward manner to the Goods Shed's various departments in the correct and necessary order.

The six rules are as follows:

1 The eight north-south 'up' lines that entered the eastern and central sections of the Train Assembly Shed and the one north-south 'up' line that entered the Eastern Transit Shed carried wagons in a southerly direction only (exceptions to this are outlined in the third rule).

2 The four north-south 'down' lines that ran through the western half of the Train Assembly Shed and the one north-south 'down' line in the Western Transit Shed carried wagons in a northerly direction only (exceptions to this are outlined in the third rule).

3 With the exception of the two north-south lines that continued into the Granary, the most southerly sections of ten Train Assembly Shed sidings south of the most southerly turntable row must have been two-way as they possessed 'dead ends' at their southern termini.

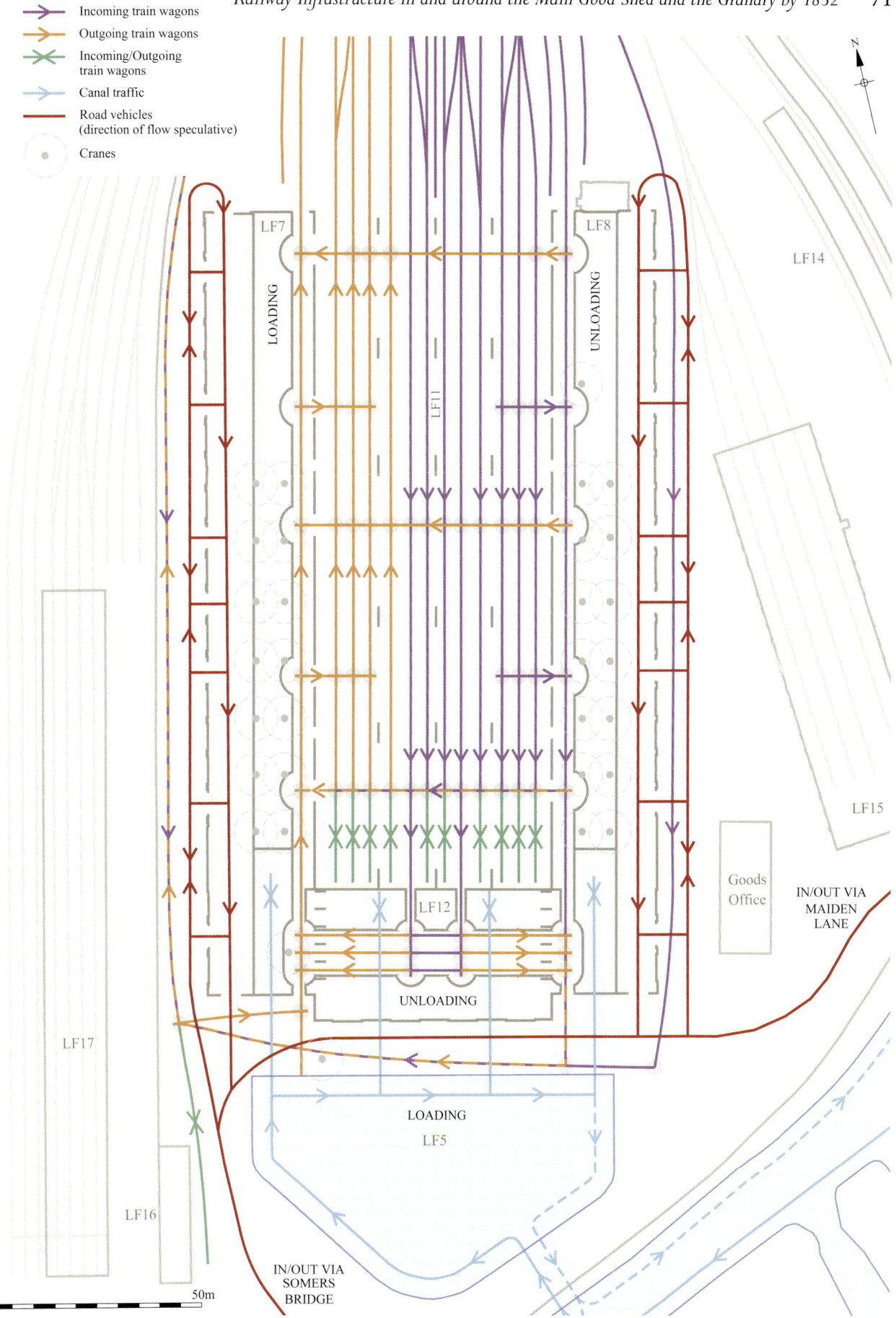

Fig. 3.58 The ground floor of the Main Goods Station (i.e. Landscape Features 7–11) and the Granary (Landscape Feature 12) reconstructed from archaeological remains, built heritage evidence, historic documentation and cartographic sources showing the probable traffic management procedures that were in place and the flow of goods through the complex *c.*1852, scale 1:1,250

72 The Construction of the King's Cross Goods Station: 1847–1852

4 The three 'long' east–west lines that traversed the entirety of the Train Assembly Shed carried wagons in a westerly direction only.

5 The remaining four 'short' east–west lines that ran into the Train Assembly Shed carried wagons in an easterly direction only.

6 In accordance with Humber's account, wagons entering the Granary by way of the east-central track would initially have been conveyed eastwards before being returned to the Western Transit Shed via external lines as shown (Fig. 3.58) whilst those entering via the west-central track would have been sent westwards for immediate return to the Western Transit Shed.

This system could sequentially marshal train wagons to the correct areas in the following order:

1 To the east and central sections of the Train Assembly Shed for reception and disassembly into individual wagons

2 To the Eastern Transit Shed or to the Granary for unloading

3 To the Western Transit Shed for reloading

4 To the west side of the Train Assembly Shed for reassembly and dispatch.

These fairly simple rules create a complex array of options that are best illustrated by way of a diagram (Fig. 3.58). This straightforward system of traffic management would have enabled each element of the Goods Station to fulfil its intended function.

The north–south track that ran down the eastern side of the Eastern Transit Shed (i.e. the arrivals shed) was almost certainly used by incoming wagons only. Many of the wagons that used this line probably carried heavier goods that were destined for onward transshipment to London and the south by barge rather than road vehicles, since they could have easily reached the 10-ton crane that was situated by the northwest corner of the Granary Basin (Fig. 3.58). A similar pathway would have facilitated the receipt of heavy goods from the canal for onward transshipment to the north (Fig. 3.58).

The external north–south line that ran along the western side of the Western Transit Shed continued to the south, beyond a turntable to the Hydraulic Station (Landscape Feature 16). Although the Humber Plan suggests that it then met another turntable before turning and running alongside an extension to that structure (i.e. the future Horse Provender Store), documentary evidence categorically states that this section was not constructed until 1856. Since it is not original it will not be considered in this chapter (see Section 4.17). In 1852, this section of track terminated in a 'dead end' beside the Hydraulic Station. South of the turntable it must therefore have been two-way. Together, these sidings were most probably used to deliver fuel to the Hydraulic Station's boiler house as shown in Fig. 3.58.

3.13 The Lamp Room and Coffee Shop by 1852

Landscape Feature 13

Two rectangular structures are shown abutting the northern façade of the Eastern Transit Shed on the earliest historical maps and plans of the Goods Station (Fig. 3.2; Fig. 3.7). As previously discussed, the western projection probably represents a loading platform associated with the Western Transit Shed (Landscape Feature 7), however the eastern example appears to have been used for a different purpose. The Humber Plan of 1866 was the first source to illustrate and annotate it in detail, indicating that it consisted of two north–south aligned bays, the most westerly of which was labelled 'V' or 'Y' and the most easterly of which was subdivided into two separate rooms both labelled 'G.R.' (Fig. 3.8). On later cartographic plans, it was labelled 'Refreshment Room', suggesting that it housed canteen facilities. It is likely that the structure was referred to by Seymour Clarke in November 1855 when he submitted a proposal to the Board requesting that 'three rooms be constructed above the Lamp Room and Coffee Shop in the Goods Shed to provide accommodation for [guards] when they come in cold and wet and for the inspectors to keep their memoranda etc.' (TNA RAIL 236/21: 126), hence the label 'G.R' (guard room?). This evidence demonstrates that the structure functioned as a welfare area during the early 1850s, whilst the term 'lamp room' suggests that sections of it were set aside to store guard's lamps.

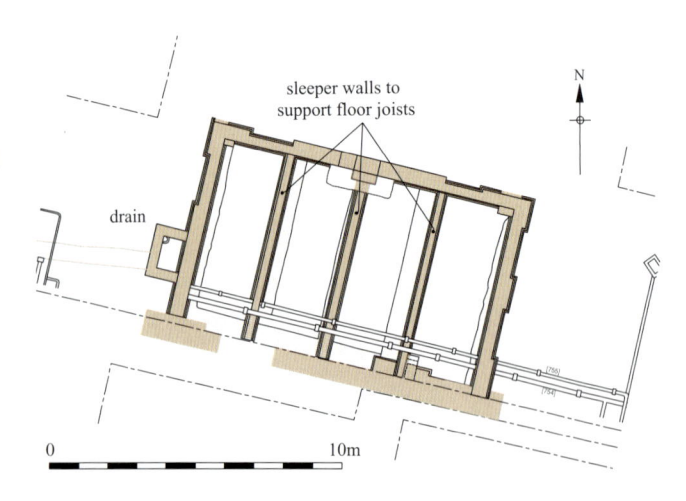

Fig. 3.59 Plan of the Lamp Room and Coffee Shop (Landscape Feature 13) *c.*1852 reconstructed from archaeological remains and cartographic sources, scale 1:250

Archaeological remains relating to the construction, appearance, footprint and function of the Lamp Room and Coffee Shop by 1852

Landscape Feature 13

A rectangular foundation constructed with red bricks was unearthed during an archaeological monitoring exercise (KXI07) to the immediate north of the Western Transit Shed in the anticipated location of Landscape Feature 13 (Fig. 3.59). Like the loading platform that abutted the north wall of the Western Transit Shed, the above ground portions of the building would have partially obscured the Eastern Transit Shed's decorative blind arch, suggesting that the building did not form part of Cubitt's original design. This notion was supported by the fact that the foundations abutted the Western Transit Shed's north wall, demonstrating that they were erected after the shed had been completed. However, the structure appears on the very earliest cartographic depictions of the complex, indicating that it must be an extremely early alteration that was in position from 1852 or fractionally earlier (Fig. 3.7). This in turn demonstrates that it must have been present around the time that the yard opened for business, hence its inclusion in this formative chapter. Therefore it is highly likely that these remains formed part of the welfare and storage facility that was known as the 'Lamp Room and Coffee Shop'.

3.14 The Temporary Passenger Station and the Great Northern Carriage Shed by 1852

Landscape Features 14 & 15

In September 1849, Chief Engineer Joseph Cubitt submitted plans of the Great Northern's future Temporary Passenger Station to the company's Board of Directors. In addition to the station's platforms and booking offices, Cubitt announced that he also intended to build a carriage shed capable of containing as many as eighty carriages (TNA RAIL 236/273: 27/09/1849). Owing to the temporary nature of the termini, he specified that this building should be constructed 'in such a way and of such permanent material as will admit of its removal and refixing elsewhere' (*ibid.*). This statement suggests that the Great Northern may have originally intended to re-erect the structure on the site of the permanent passenger termini once this was made ready. Such frugality would presumably have been music to the ears of anxious Directors and shareholders after the sudden decline in investment in the railways at the end of the 1840s.

Unlike the Temporary Passenger Station, the adjacent carriage shed did not feature amongst the works that were listed in the schedule of prices for the station buildings that was issued to the Stations Committee of the Great Northern in January 1850 by Joseph's uncle, the architect Lewis Cubitt (BL 8244f5:

Fig. 3.60 Queen Victoria and Prince Albert at the Temporary Passenger Station (Landscape Feature 14) in August 1851. This image, taken from **The Illustrated London News** of that year, shows them departing with their children from Maiden Lane for one of many trips to their beloved Scotland, many of which would be made in a purpose built royal saloon over Great Northern rails (Anon 1851c). Their decision to travel by train proved to be a huge public relations boon to the company, helping it to overcome the problems that had blighted its difficult formation (Vernon 2007: 66).

Cubitt 1850, 4). It was, however, depicted in outline form on a plan of the iron roofing of the Temporary Passenger Station and the Goods Depot that was submitted at the end of March of that year (ICE Cubitt 1850; Fig. 3.5). This suggests that the company had decided to proceed with the construction of the carriage shed by this date at the latest.

The Temporary Passenger Station was erected rapidly during the spring and early summer of 1850. By the beginning of August, sufficient progress had been made to enable the commencement of passenger services to and from the new station. Since Joseph Cubitt made no further mention of the adjacent carriage shed in his reports to the Board, it may be assumed that it too had been completed by the August deadline.

Cubitt's stated intention of making the carriage shed a temporary and moveable structure is somewhat contradicted by a report published in *The Observer* newspaper, which described the shed as a single storey structure, 300' (91m) long by 80' wide, which the Great Northern hoped to retain and reuse, presumably in its original location. It was proposed to reuse the shed in order to accommodate both workshops and additional goods traffic (Anon 1851b). This evidence suggests that by April 1851 Joseph Cubitt's original plan to move the structure to the site of the Permanent Passenger Terminus had been shelved.

Complications with the acquisition and clearance of the site of the new King's Cross Passenger Station, coupled with the engineering challenges presented by the series of steep gradients along its approaches meant that the Great Northern would not complete the structure quickly (Biddle 1990: 60; Thorne *et al.* 1990: 94). During this ferociously competitive period in railway history, the failure to establish an inner London passenger terminus would have prevented the Great Northern from competing effectively with the services offered by the company's rivals running out of Euston. It is therefore reasonable to assume that the Great Northern would have gone to great lengths to ensure that it was able to run passenger services in and out of the metropolis as quickly as possible in order to placate the company's shareholders and generate a much needed revenue stream at a time when finances were squeezed. In light of this evidence, it is unsurprising that company pulled out all the stops to ensure that the temporary passenger terminus entered service according to schedule.

Archaeological evidence of the construction, appearance, footprint and function of the Temporary Passenger Station by 1852

Landscape Feature 14

Landscape Feature 14 is illustrated on a series of historic plans and maps dating from 1849 onwards, in the location of the Temporary Passenger Station (Fig. 3.2). The earliest incarnation of the structure represents the Temporary Passenger Station at Maiden Lane completed in August 1850.

With the exception of a set of spandrels that were subsequently incorporated into a later structure known as the East Handyside Canopy, no remnants of the Temporary Passenger Station survived above ground (Christopher 2012: 62). However, some very small sections of the structure's foundations were identified below the modern tarmac ground surface during an archaeological watching brief that was carried out by PCA in 2007, which was enough to demonstrate their survival (Mazurkiewicz 2008: 18). These were subsequently investigated in greater detail during an archaeological recording exercise that was undertaken by MoLA (see Braybrooke 2012 for further details).

Built heritage evidence of the construction, appearance, footprint and function of the Carriage Shed by 1852

Landscape Feature 15

In contrast to fragmentary remains of the temporary passenger station, the above ground portions of Landscape Feature 15 were well preserved. This feature was situated to the immediate west of the former location of the Temporary Passenger Station in the anticipated location of the Carriage Shed that originally accompanied the temporary passenger terminus. A Historic Building Recording Survey undertaken as part of this project (KXM08) demonstrated that, whilst the structure had been extensively modified during later phases of construction, elements of the original single storey shed did survive. These remains were extremely limited, however, suggesting that the above ground portions of the building had been largely rebuilt at a later date. Since only a few courses of original brickwork survived on each face of the structure, any conclusions concerning its initial appearance are limited in scope. Despite this, several useful inferences could be drawn.

The earliest phase of the Carriage Shed was constructed with purple stock moulded bricks in Flemish bond, the fabrics of which suggested a mid-19th-century date. This date range is in keeping with the documentary evidence, which states that the building was operational by August 1850.

The distribution of these building materials in the fabric of Landscape Feature 15 suggested the following. The western and eastern elevations originally appear to have been characterised by 24 bays with recessed panels of brickwork that alternated with brick pilasters (Fig. 3.61). With the exception of the base of the northwest corner pier, the entire northern and southern elevations have been rebuilt during later periods.

Despite the fact that so little of the original fabric of the Carriage Shed survived, it is possible to reconstruct certain aspects of its layout by considering the surviving remains against the cartographic evidence. The Humber Plan of 1866 suggests that two railway tracks ran into the structure from the north, whilst another two terminated against its northern wall to the east. Space for an additional two sidings is shown to the west, suggesting that six may originally have run up to the shed (Fig. 3.8). Humber indicates that the northern end of the Carriage Shed must have contained *at least* two openings for the accommodation of the railway sidings detailed above. Given that Humber's plan was surveyed in the mid-1860s, several years after Landscape Feature 15 ceased to be used as a carriage shed, it is unlikely that it depicted the original track layout. It is probable that the northern edifice consisted of six bays (as it does now), the positions of which have been preserved by later rebuilds. Each bay most likely contained an opening that accommodated one of six sidings that ran into the building from the north. Owing to the fragmentary nature of the extant remains and the fact that no contemporary facsimile of the earliest track layout in the shed survives, the flow of traffic between the Carriage Shed and the Temporary Passenger Terminus remains uncertain.

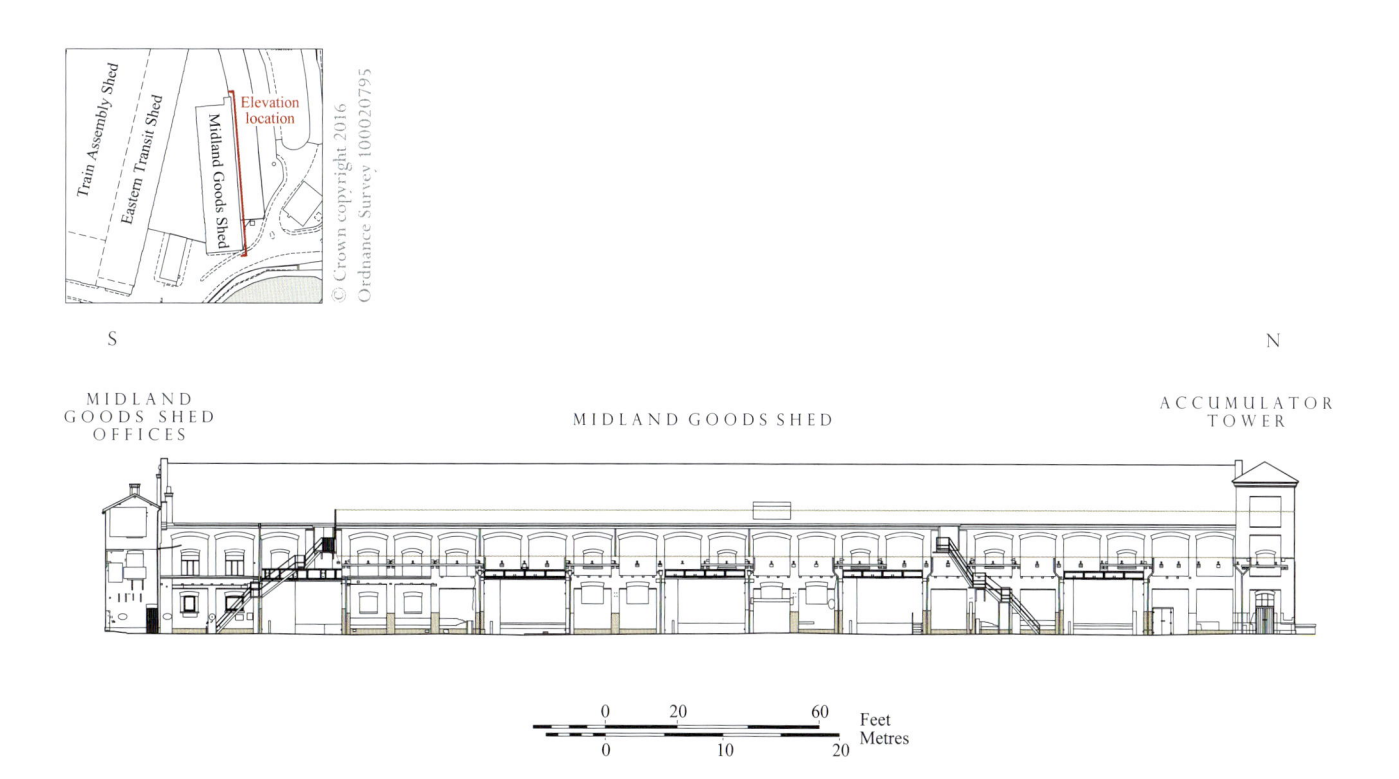

Fig. 3.61 External elevation of the east wall of the Carriage Shed (Landscape Feature 15) showing the original 1850 masonry, scale 1:625

3.15 The Hydraulic Station: Hydraulic Machinery and the Hydraulic Network by 1852

Landscape Feature 16

A detailed account of the evolution of the machinery that made up the hydraulic network at King's Cross Goods Station and the intricacies of the engineering that was involved can be found in Chapter 13 of this volume.

The earliest surviving documentary reference to the planned hydraulic power system at King's Cross Goods Station was made by Chief Engineer Joseph Cubitt on 28th May 1850, when he recommended that the company adopt William Armstrong's hydraulic cranes for use at the new station (TNA RAIL 236/70: 275). Cubitt sought permission from the Executive Committee of the Great Northern Railway Company to place an order for 40 such cranes at a cost of £30 each, which he proposed would be 'worked in the first instance by hand, but ultimately by hydraulic power', the latter 'to be applied by means of a steam engine' (*ibid.*). In response the Committee approved the purchase of twenty of Armstrong's cranes, and it is likely that these accounted for the twenty 2-ton cranes that appeared on Armstrong's order books the following year (TWCA MS 1975/1: 1847–52). It has been observed that whilst eight of the original twenty 2-ton cranes had a lift of 14', the remaining twelve had a greater lift of 24' and is likely that some of these larger examples were positioned around the canal arms in the Transit Sheds where a greater reach would have been needed (Smith 2008: 4).

Eight months later in January 1851, Cubitt recommended that the company make arrangements 'at once' to acquire the hydraulic machinery necessary to work the cranes already ordered, all of which were scheduled to be installed 'in the sheds and granary and on the wharf at King's Cross' (TNA RAIL 236/71: 295). The Executive Committee approved a tender worth £5,477 submitted by Armstrong for the supply of the materials, their carriage, delivery and erection, but which excluded the cost of building the Hydraulic Station itself (*ibid.*). In addition to the twenty 2-ton hydraulic cranes already ordered, Armstrong agreed to supply sixteen 1-ton cranes for the Goods Shed, one 10-ton crane, ten grain hoists, one steam pumping engine, two boilers, one accumulator with an 18" diameter ram with a 15' stroke, two accumulators with 8½" diameter rams with a 15' stroke, 933 yards of pressure pipe, stop valves and momentum valves and 400 yards of light cast iron pipe 'for conveying the water out of the sheds' (TWCA MS 1975/1: 1847–52).

Most of the hydraulic machinery that was ordered from Armstrong was destined for the Transit Sheds and the Granary and this order brought the number of cranes that were to be installed therein to a total of 36, plus one heavy crane that, owing to its size, must have been external to the complex. Although these buildings had been completed by March 1851, Armstrong's order was far from ready by this date and both Lewis Cubitt (Chief Architect) and Seymour Clark (GNR General Manager) became increasingly concerned that the machinery would not be operational in time for that autumn's harvest (see Section 3.11).

Little written evidence concerning the precise arrangements of the earliest hydraulic machinery in the Goods Yard survives and much of that is contradictory. The writer John Weale commented on the proposed arrangement of cranes in the Transit Sheds after visiting the Goods Station construction site in 1851. He was informed that when complete, each of the Transit Sheds was to be fitted out with eighteen goods cranes (Weale 1851: 812) a number that tallies with Armstrong's order books but is at odds with the earliest detailed depiction of the interior of the complex, William Humber's plan of 1866 (Fig. 3.8). According to this source, there were fifteen cranes in the Eastern Transit Shed and seventeen in the Western Transit Shed, bringing the total number of cranes in the Goods Station to 32 rather than 36 (Fig. 3.8). It is unlikely that four cranes were removed between 1851 and 1866 when the volume of goods passing through the station was rapidly increasing. A more likely explanation for the discrepancy is that fewer cranes were initially installed in the Goods Shed than were ordered from Armstrong. Indeed, Weale himself includes the following caveat in his account: 'Railway boards are notoriously given to change their plans, nor are Railway engineers stable in their intentions. What we may say would very possibly [be built], therefore, [could] be found totally at variance with the works as finished' (*ibid.*).

Given Lewis and Joseph Cubitt's predilection for mirror imaging and inverted symmetry within their earliest plans for the Goods Station, coupled with the information that was provided by Humber, it is probable that each Transit Shed originally housed just fifteen cranes. It appears that the number in the Western Transit Shed was increased to seventeen following Joseph Cubitt's request to the Board for an additional two cranes for the 'departure Goods Shed' in November 1852 (TNA RAIL 236/74). This would mean that the original number of cranes in the Goods Station amounted to 30 rather than 36. It is therefore probable that the Great Northern chose to redeploy the six spare cranes that were originally ordered from Armstrong elsewhere.

Humber's inspection report of 1865 noted that the cranes within the Transit Sheds were arranged along the platforms in alternating 1 and 2-ton pairs (Smith 2008: 4, 6), an observation that has enabled the distribution of the machines to be approximated (Fig. 3.58). Given the vagaries of the evidence, the reconstruction as illustrated is hypothetical but is thought to be likely for the following reasons. Firstly, the arrangement of the 30 original cranes in the Eastern and Western Transit Sheds as shown in the reconstruction exhibit inverted symmetry relative to each other thus respecting that wider trend. Secondly, larger, 2-ton cranes were most probably situated beside the canal docks where it is assumed their longer reaches were required. That provided a starting point from which the alternating pairs of 1 and 2-ton cranes could be extrapolated northwards along the platforms. Whether the solitary crane at the northern end of the Eastern Transit Shed represents a 1 or 2-ton machine remains uncertain.

Soon after operations began, the inverted symmetry described above was broken at the request of Joseph Cubitt, presumably in order to account for actual working practices (Fig. 3.8; Fig. 3.58). Since the new cranes represent later additions that probably were not available until early in 1853, they will not be discussed any further within this chapter.

Information from several documentary sources has enabled the whereabouts of the 10-ton crane that was ordered from Armstrong in 1851 to be deduced. The crane would have been large and would therefore have been situated outside the Goods Shed. Humber's plan illustrates four external candidates. Two 10-ton cranes (identifiable through labelling) are shown along with an 8-ton example to the immediate north of the Goods Shed whilst a further two external but unlabelled heavy cranes are shown to the northwest and to the east of the Granary Basin. The vital clue appeared in a request to made by Seymour Clarke in 1853, in which he sought permission to build a new heavy crane on the east side of the Basin, similar to the one already in place on the northwest side (TNA RAIL 236/21: 334). This indicates that the heavy crane shown on Humber's plan on the northwest side of the Granary Basin almost certainly dates to the Goods Station's formative phase, suggesting in turn that those to the north of the Goods Shed were installed later. This proves beyond doubt that the example located on the eastern side of the Basin was added as a result of Clarke's 1853 request. The location of the original heavy crane is illustrated in Fig. 3.8.

Work began on the erection of the Hydraulic Station itself in the early months of 1851. In contrast to the issues that preoccupied Joseph Cubitt at this time, such as the completion of the Eastern Coal Drops (Landscape Feature 17) and competing demands for stabling in the Goods Yard, the Chief Engineer made no direct reference to the Hydraulic Station in his regular reports to the Board, suggesting that its construction passed without incident (TNA RAIL 236/273). The earliest depiction of the building can be found in Cubitt's watercolour painting of September 1852 (Fig. 3.6). This image shows a two storey structure in the background to the far left of the picture, which must represent the Hydraulic Station's accumulator tower. This housed the largest of the three accumulators that were supplied by Armstrong, the other two being located remotely at the northern end of the Goods Shed (Smith 2008: 15). With the exception of this painting and Captain Galton's sketch plan of 1852 (Fig. 3.7), no historical sources detailing the internal or external layout of the Hydraulic Station in the early 1850s have been discovered. Consequently, the following account relies on later sources.

It was not until December 1851 that Joseph Cubitt met with the Regent's Canal Company to discuss arrangements for the installation of a supply pipe from the adjacent canal 'by which the Great Northern [wished] to obtain waters for working the hydraulic cranes at their station' (TNA RAIL 860/44: 103). At the end of January 1852, the canal company granted the Great Northern permission to lay the pipe on payment of an annual rent of £30, although this had yet to be done by March (*ibid.*: 120, 133). It is therefore not entirely clear when the hydraulic machinery in the Granary and the Transit Sheds was switched on, but a date after March 1852 seems likely. This suggests that the cranes were operated by hand for longer than Cubitt had originally envisaged.

No contemporary documents describing the water supply to the Hydraulic Station in the early 1850s have been found. However, documents dating to the early 20th century identify the Granary Basin as the source for the water, which was drawn into a well located close to the Hydraulic Station via an intake main in the west wall of the basin (TNA RAIL 1189/1426: 'Agreement for the Supply of Water at King's Cross' 02/10/1916). A 1916 agreement between the canal company and the Great Northern Railway confirmed that this arrangement had been in force since at least 1869 when it was first formalised, so it is not unreasonable to assume that this configuration had actually been in place since 1852 when the supply pipe was originally laid (*ibid.*; Smith 2008: 11).

Archaeological evidence relating to the construction, appearance and function of the hydraulic machinery by 1852

The heavy crane that was ordered from Armstrong in 1851 was made the target of an archaeological excavation (KXP08). Identified as 'Crane Base 1' here, it represents the 10-ton crane that is shown on the Humber Plan on the northwest side of the Granary Basin to the immediate southeast of Turntable B (Fig. 3.62; see Fig. 3.3). This crane was used for moving heavy goods between train wagons on the adjacent railway siding and barges moored in the Granary Basin. The crane base consisted of a rectangular masonry body, orientated east–west, with a circular head at the western end. The east, north and south walls of its main body formed a rectangular brick lined chamber that was 9.14m long and 2.60m wide, the walls of the structure being 0.38m thick and over 2.38m deep. The rectangular chamber was connected to a circular recess at the 'head' end of the structure, which had an internal diameter of 2.32m.

Red, yellow and purple bricks in English bond were used to construct the walls and floor of the crane base.

The east, north and south walls of the rectangular chamber were capped by one course of stretcher bonded granite blocks that were clipped together with metal butterfly clamps. This was absent from the circular chamber, perhaps because it had been robbed after the structure fell out of use. Three large sandstone blocks were found inside the circular chamber, which had been cut to fit next to each other within the curved structure. They must have supported the turning mechanism of the crane as four evenly spaced bolts were observed, protruding through specially cut holes in the sandstone, which would have secured this in place. The floor of the rectangular chamber sloped from a height of 22.74m OD at the western end to 22.02m OD at the eastern (head) end where it ran below the wooden brace. It then levelled out to a height of 21.81m OD, abutting the sandstone blocks in the circular chamber at the southern end. When the crane was operational the only part of the base that would have been visible would have been the granite cap, the rest of the structure being subterranean.

A second, smaller rectangular structure butted the circular head of the crane base. This was constructed from irregularly bonded red, yellow and purple bricks and was also capped by one course of granite. Excavation demonstrated that the chamber was 1.35m deep, its base being formed by concrete that had been poured around a hydraulic pipe that itself was 140mm in diameter. The chamber must therefore have been associated with the crane's hydraulic power supply.

The crane remained operational throughout much of the Goods Yard's life and it therefore had to be maintained in order to keep it in working order. As such, the crane base's brickwork had been partially replaced with concrete on at least two occasions.

Archaeological evidence relating to the hydraulic pipe network

Large stretches of hydraulic pipe work were unearthed during the excavations at King's Cross Goods Yard but few sections were securely phased. The precise layout of the earliest incarnation of the hydraulic network therefore remains uncertain. In contrast, archaeological excavation revealed a great deal about the layout of the Hydraulic Station itself.

Archaeological evidence relating to the construction, appearance, footprint and function of the Hydraulic Station by 1852

Landscape Feature 16

An archaeological excavation (KXO08) revealed the foundations of a rectangular building in the anticipated location of the early Hydraulic Station, stratified below a later incarnation of the structure (Fig. 3.2; Fig. 3.63). The following reconstruction of the appearance and layout of this building relies on archaeological evidence, supplemented by cartography where appropriate. Other documentary sources are also referred to where appropriate, in order to establish a comprehensive picture of the nature and function of the earliest version of the building.

The Hydraulic Station was built against the southern side of the Retaining Wall (Landscape Feature 4) before the high ground that abutted its eastern face was raised. After the ground was elevated this meant that only one storey of the Hydraulic Station was visible above ground level at the top of the terrace to the east and north, whilst two storeys were exposed at the bottom of the terrace to the west and south. The upper levels of most 19th-century engine houses could be accessed via internal metal gantries that were installed so as to enable the upper reaches of the large machines that they housed to be accessed for maintenance and cleaning purposes. This topographical position would have meant that it would

Fig. 3.62 Crane Base 1 of 1851 exposed by excavation, photograph faces west

The Hydraulic Station by 1852

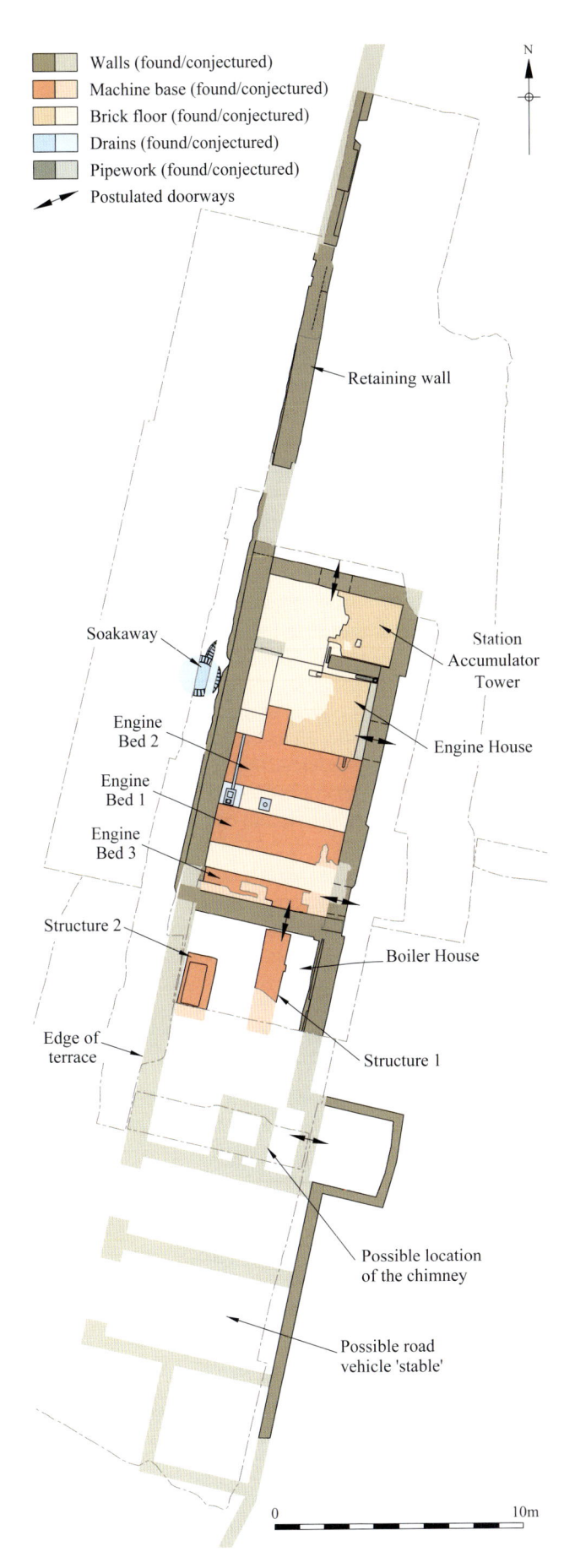

Legend:
- Walls (found/conjectured)
- Machine base (found/conjectured)
- Brick floor (found/conjectured)
- Drains (found/conjectured)
- Pipework (found/conjectured)
- Postulated doorways

Labels on plan: Retaining wall; Soakaway; Engine Bed 2; Engine Bed 1; Engine Bed 3; Structure 2; Edge of terrace; Station Accumulator Tower; Engine House; Boiler House; Structure 1; Possible location of the chimney; Possible road vehicle 'stable'.

Fig. 3.63: Plan of the Hydraulic Station (Landscape Feature 16) c.1852 reconstructed from archaeological remains and cartographic sources, scale 1:250

have been possible to walk directly onto such a raised area inside the Hydraulic Station through a doorway that connected with the high ground on the eastern side of the terrace.

The northern and eastern external walls of the Hydraulic Station (Landscape Feature 16) were built in red bricks in English bond, held together with light grey, indurated lime mortar. The brickwork was tied in on either side of the returns of these walls, indicating that they were constructed in one continuous build. This contrasted with the western external wall (formed by the Retaining Wall i.e. Landscape Feature 4, described in Section 3.4), which the rest of the Hydraulic Station abutted, indicating that the former was built before the latter was constructed against it. After the foundations of the Hydraulic Station had been completed, made ground was piled against the eastern and northern sides of the structure, forming the high ground on the western side of the terrace. This ground raising episode formed part of the landscaping work that was begun by John Jay and his men from 1849 onwards, as detailed in Section 3.1.

After the internal and external walls were constructed, a probable pitched roof was added. When the Hydraulic Station was rebuilt in 1898, the contractor for the works was instructed to take down almost the entire canopy, leaving only that of the accumulator tower intact (TNA RAIL 236/449: 27). Though the original roof was not described, an instruction to reuse old slates 'insofar as they are sound and good' gives an indication of the probable fabric (*ibid.*: 28).

Stretcher bonded red brick floors were found inside the northern half of the building which, at a level of 21.41m OD to 21.25m OD, were virtually flush with the 19th-century ground surface that formed the low road at the bottom of the terrace to the west. The surviving sections of the floor were uneven and had clearly suffered subsidence.

The surviving walls formed a rectangular structure that was over 33.80m long, the southern end having been destroyed by later building activity, and up to 6.95m wide (22' 9⅔"). The locations of the two internal divisions that were found archaeologically (Fig. 3.63) correlated perfectly with the two most northerly dividing walls that are shown on the Humber Plan of 1866 (Fig. 3.8) confirming the reliability of this source. Archaeological evidence demonstrated that the building was divided into a minimum of three rooms (Fig. 3.63). As the physical remains of the southern end of the Hydraulic Station did not survive, historic map detail in the form of Captain Galton's plan (Fig. 3.7) and the more detailed 1866 Humber Plan (Fig. 3.8) have been used to reconstruct its footprint. Whilst Galton's plan (Fig. 3.7) is imprecise and highly schematic, it is sufficient to demonstrate that the size and outline of the Hydraulic Station was not significantly altered between 1852 and 1866 when the Humber Plan was drawn up, the only exceptions being some unrelated structures of an entirely different function that had been tacked onto the southern end of the building (Fig. 3.7; Fig. 3.8).

The most northerly room was rectangular in plan and smaller than the other three, with internal dimensions of 2.52m north–south by 5.50m east–west. Other than a brick floor, no internal features were found. At 13.86m², it would have been too small to contain the two locomotive type boilers or the 24hp Armstrong engine but could easily have housed the station accumulator. This interpretation also tallies with Cubitt's 1851 watercolour, which depicts the Accumulator Tower towards the northern end of the Hydraulic Station (Fig. 3.6). The housing for the device presumably sat directly on the brick floor, offset to the east to avoid the doorway that linked the tower with the Engine House to its immediate south (Fig. 3.63).

The specification for the rebuilding of the Hydraulic Station in 1898 indicated that the engines were located to the immediate south of the accumulator (Smith 2008: 11), a situation that may also have been the case before the station was enlarged to the west in the early 1880s. The larger, 9.61m north–south by 5.16m east–west, room to the south of the Accumulator Tower was therefore interpreted as an Engine House. Within this room a red brick base aligned north–south (Engine Bed 1) spanned the entire width of the room from east to west, butting the eastern and western external walls of the building (Fig. 3.63). It was 1.52m wide and survived to a maximum height of 0.79m. Unfortunately, no bolt holes or other means of securing machinery onto the structure survived, presumably because of the high degree of horizontal truncation that it had been subjected to. The masonry that was observed therefore represented the foundations of an engine bed rather than the bed proper.

A second, similar feature, Engine Bed 2, was unearthed to the immediate north of Engine Bed 1. It had an asymmetrical, inverted 'T' shape in plan, the main body of the structure being 2.00m wide. It had also suffered severe horizontal truncation and was in a similar state to Engine Bed 1.

The order for hydraulic machinery placed with Armstrong at the end of January 1851 specified a single steam pumping engine, which it is assumed was delivered no later than the autumn of 1851 (Smith 2008: 4) and installed in the station soon afterwards; presumably on one of these beds. Engine Bed 1 may be the most likely candidate as it was approximately the correct size and shape to support such a device and was located in a logical position for the Hydraulic Station's very first pumping engine due to its proximity to the boilers. If so, then Engine Bed 2 must date to a fractionally later time, having been erected in 1856 in order to support a supplementary engine that was ordered during that year (Tim Smith pers. comm.). Unfortunately, the nature of the archaeology precluded relative dating of these beds so it remains possible that their roles as described here were in reality reversed.

A small rectangular drain was incorporated in the brick floor of the Engine House to the north of Structure 1, which presumably removed waste water that would otherwise have puddled on the floor. This suggests that the space could become wet, presumably due to condensation caused by steam escaping from the engine. A brick lined soak away with a diameter of 1.58m was identified on the external side of the west wall. Water from the drain discharged into this feature via a pipe that ran through the foundations of the building.

A third brick feature, Bed 3, also spanned the entire width of the Engine House. It abutted the entire length of the southern wall of the building so that the east-central portion of the structure effectively formed a small step that led up to a doorway that provided internal access to the Boiler Room. It was 0.88m wide and 0.17m deep, the top of it being slightly higher than floor level at a height of 21.45m OD. Whilst the exact function of the structure remains unknown it is reasonable to assume that it supported supplementary apparatus of an uncertain function that must have been associated with the engines. It has been speculated that the bed could have supported a very small engine, perhaps used to pump water out of the canal or into the Granary roof tanks. Whatever the nature of the apparatus, it must have been set away from the doorway that was present in the southern wall of the Engine House so as to avoid impeding access to the Boiler House (Fig. 3.63).

A third room to the south of the Engine House must represent the Boiler House. The two locomotive type boilers that were ordered from Armstrong in January 1851 were probably accommodated there. Two rectangular structures were found within the confines of the room.

The first, Structure 1, was orientated north–south and was 2.74m long, 1.10m wide and 0.70m tall. It consisted of a concrete foundation that supported red, frogged bricks that were capped by a single stone slab. At 21.31m OD, the top of the structure correlated well with the level of the floor surfaces that were unearthed in the Engine House (varying between 21.41m OD and 21.25m OD due to differential degrees of subsidence). Since Structure 1 was perfectly aligned with the doorway that connected the Engine House with the Boiler House, it must have been associated with this threshold in some way. Given the similarity between the height of the structure and the floor surfaces of the Engine House, it is possible that it formed a small step that led down into the Boiler House. A second, very similar rectangular base, Structure 2, was identified to the southwest.

One or both of these bases could have supported supplementary machinery that was associated with the boilers, such as a small boiler feed pump for example (*ibid.*). In the case of Structure 1, the location of this machinery must have been offset towards the south otherwise internal access between the Engine and Boiler Houses would have been impeded. Locomotive type boilers were, for the most part, mounted above ground so it is not surprising that limited traces of them were definitively identified during the archaeological excavation (*ibid.*).

No contemporary descriptions of the original 1850s chimney have been found, although an invitation for tenders to alter the Hydraulic Station in June 1898 included a specification for the complete demolition of the old stack. This was said to be brick built, square and 80' (24.38m) tall (TNA RAIL 236/449), although it is not clear whether the shaft thus described was erected in 1850–2 or whether it represents a later replacement that was constructed when the boilers were subsequently changed. Although the original chimney was not identified archaeologically having been comprehensively demolished, its likely position can be extrapolated from the Humber Plan of 1866 (Fig. 3.8). An internal square structure situated in the southeast corner of the Boiler House is shown on the plan, which may represent the chimney. Placing it close to the boilers would make logical sense from an engineering perspective (Tim Smith pers. comm.).

The Humber Plan of 1866 indicated that a fourth room abutted the southern end of the Boiler House (Fig. 3.8; Fig. 3.63). Perhaps not part of the Hydraulic Station, this could have formed a storage area, perhaps for stowing horse-drawn carts (*ibid.*). Several strands of evidence give credence to this interpretation, not least the proximity of the structure to known stables. The Humber Plan suggests that carts could gain access to the structure via the low road that ran along the western boundary of the Goods Yard, entering it with ease due to the fact that it was three sided and entirely open to the west. Internally, it was divided in two by an east–west wall and it is probable that one or more carts were stowed in each of these bays. Unfortunately, this interpretation could not be tested archaeologically as the structure here had been obliterated by later building activity.

3.16 The King's Cross Coal Depot, the Coal Drops, Coal Offices and Wharf Road Viaduct by 1852

Landscape Features 17–19

From the outset, the Great Northern set out to corner a significant portion of London's large and lucrative coal trade (Thorne 1990: 112). In a move calculated to maximise profits, the company decided to sell coal directly, thereby bypassing the capital's established coal merchants. Whilst independent traders were not prohibited from obtaining coals imported by the Great Northern, the high rates charged by the company deterred any serious competition (*ibid.*: 117). Plans to develop coal handling facilities at the King's Cross Goods Station were therefore well advanced by September 1850, when the Great Northern's newly appointed Coal Manager, Coles Child, and the buildings contractor John Jay presented the Board with plans to build drops or 'staiths' for the reception of the commodity (TNA RAIL 236/71: 48). Construction of the Coal Drops (Landscape Feature 17) began that autumn, and they were almost complete by the following January, when Joseph Cubitt's announced that the importation of coal would commence the following month (TNA RAIL 236/273: 08/01/1851). The Coal Drops were erected to the east of the Coal and

Fig. 3.64 The south elevation of the Coal Offices (Landscape Feature 18) as seen from the Regent's Canal, photograph faces north

Stone Basin (Landscape Feature 6), which was itself largely dedicated to the onward shipment by barge of coal that had been imported from the north by train (see Section 3.5 for further details concerning the Coal and Stone Basin).

In order to administer the company's coal trade at King's Cross, it was necessary to provide office space at the Goods Station for coal managers and clerks. A range of purpose built Coal Offices (Landscape Feature 18) were erected close to the southwest corner of the Goods Yard against the north bank of the Regent's Canal. Both Edward Wiggins, the Great Northern's contractor for the delivery of coal in London, and the Coal Manager, Coles Child (and the latter's successor, Herbert Clarke), occupied rooms in the complex.

The Coal Offices are mentioned in a report prepared by Joseph Cubitt in February 1851, which suggests that they had been completed by that date. In the report in question, Cubitt presented a proposal to the Board for additional stabling at the coal yard (*ibid.*: 07/02/1851). He criticised the existing plans to accommodate horses belonging to Edward Wiggins, the Great Northern's contractor for the delivery of coal in London, 'at the end of the Coal Offices', pointing out that this area 'was soon to be occupied by the road leading from Somers Bridge' (*ibid.*). Furthermore, Cubitt argued that the space available in that location was not large enough to house the 50 or more horses that Wiggins intended to stable there. Instead, he advocated the construction of an entirely new range of stables beneath the raised roadway that subsequently came to be known as the Wharf Road Viaduct (*ibid.*; Fig. 3.2). Whilst Cubitt's suggestion concerning alternative stabling was realised through the incorporation of a stable block in the Wharf Road Viaduct (constructed later in 1851), a rent return of September 1852 suggests that a small number of Wiggins' horses were additionally stabled at a location 'near the Canal', separate from the majority of his animals (TNA RAIL 236/275/23). This description presumably refers to stabling at the Coal Offices, although it remains unclear as to whether this was situated at the eastern or western end of the building.

Archaeological evidence relating to the construction, appearance, footprint and function of the Coal Drops by 1852

Landscape Feature 17

The Coal Drops (Landscape Feature 17), consisted of a long, linear structure orientated north-northeast–south-southwest that was constructed on the low side of the terrace to the immediate west of the Hydraulic Station (Landscape Feature 16) and the Retaining Wall (Landscape Feature 4) as shown in Fig. 3.2. As the drops were situated outside the western boundary of the Area of Investigation, an archaeological assessment of the structure was impossible. The Coal Drops are of relevance here however, as they were a vital component of the Great Northern Coal Department's infrastructure at King's Cross. Consequently, the following description comprises a brief synopsis of earlier field observations made by Robert Thorne and English Heritage in the late 1980s, in conjunction with an analysis of relevant historic maps and plans (Thorne 1990: 114–7).

The viaduct that effectively formed the Coal Drops consisted of 48 arched 'cells' arranged against a central 'spine' wall that supported the railway sidings (*ibid.*: 114). These were capped by a second tier of identical arches that supported a pitched roof, which sheltered the otherwise largely open structure from the worst of the elements. Train wagons accessed the Coal Drops on one of four lines that ran the length of the viaduct. Each 'cell' within the structure housed a coal drop mechanism that included a hopper that was located between the tracks and two ground level cart loading bays (*ibid.*). The Great Northern's 7½ton coal wagons were bottom opening, so they could be manoeuvred over a corresponding hopper in order that their cargo could be discharged directly into it. The coal was then channelled straight into sacks that were situated in waiting carts. These horse driven vehicles were able to draw up inside the cells by way of the cart road that ran along the low ground on the western side of the terrace to the immediate east of the Coal Drops (this was known as the Laser Road at the time of writing). Along with the canal barges that were moored in the Coal and Stone Basin to the east, these carts were responsible for the onward distribution of coal to the metropolis. The horses were specially trained to cope with the noise of the coal dropping into the waiting cart.

For further information concerning the architectural details of the Coal Drops, the reader should refer to the third monograph in this series (forthcoming) and Thorne's text. As Thorne points out, the Coal Drops' precise mode of operation remains uncertain as virtually all traces of the railway tracks and the coal drop mechanisms have been removed whilst the surviving remains were also badly damaged by fire in the 1980s.

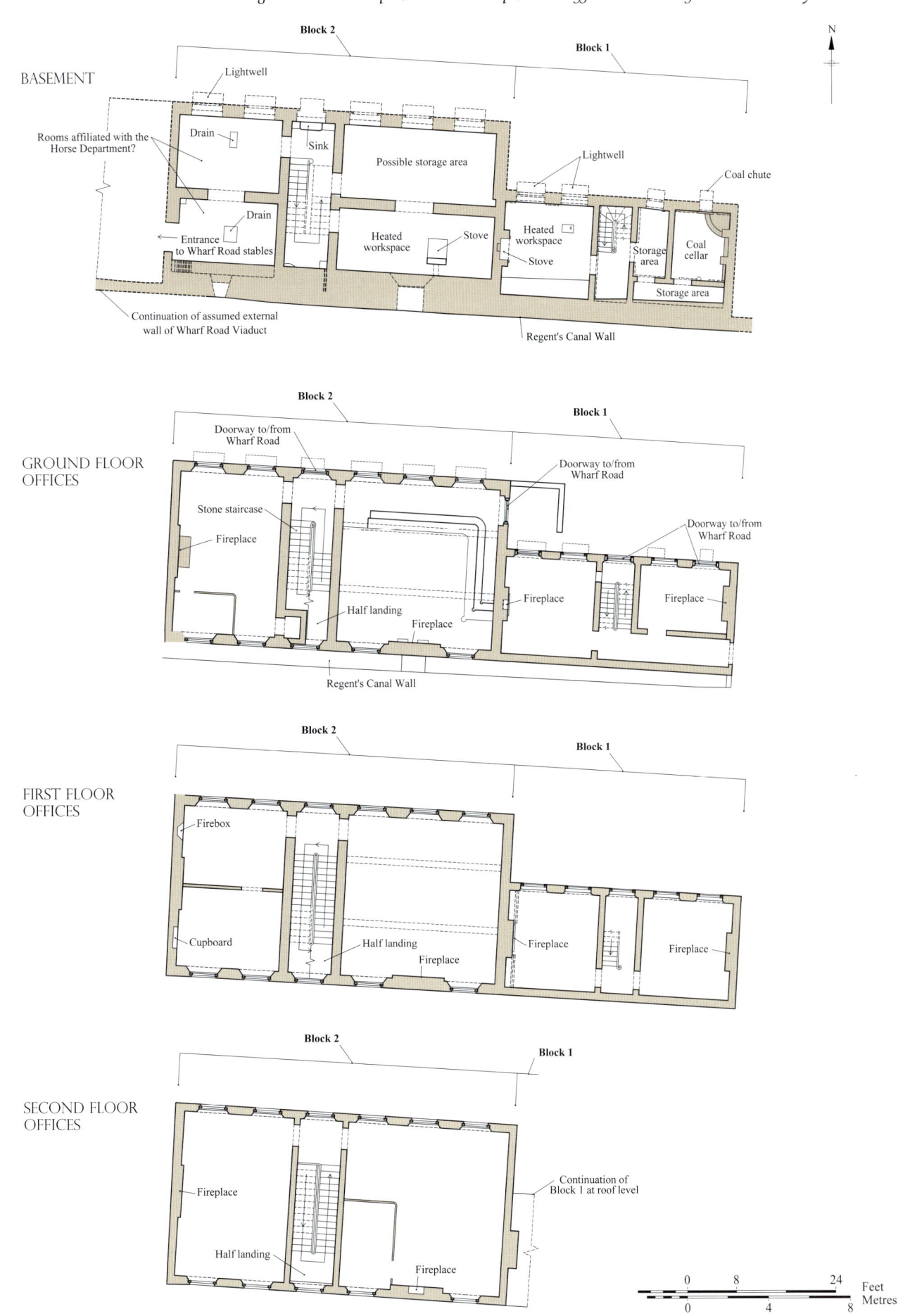

The King's Cross Coal Depot, the Coal Drops, Coal Offices and Wharf Road Viaduct by 1852 83

Fig. 3.65: Floor plans of the Coal Offices (Landscape Feature 18) *c*.1851 reconstructed from built heritage evidence and cartographic sources, scale 1:250

Archaeological evidence relating to the construction, appearance, footprint and function of the Coal Offices by 1852

Landscape Feature 18

The Coal Offices (Landscape Feature 18) (Fig. 3.64; Fig. 3.65) comprise a graceful, curvilinear complex of buildings that is situated between Wharf Road and the north bank of the Regent's Canal (Fig. 3.2). A sketch plan drawn in April 1853 indicated that only the easternmost two blocks (numbered Blocks 1 and 2 herein, Fig. 3.65) are original, the other sections being later additions (TNA RAIL 236/276/6: 20/04/1853). Consequently, only these early elements will be considered in this chapter.

The earliest incarnation of the Coal Offices was built using decorative Flemish bonding, their windows and doorways being framed by segmental arches with York stone sills and their doorsteps being composed of an identical material. Although a straight joint was observed between the northern façades of Blocks 1 and 2, (Fig. 3.66), the southern faces of the two structures were integral to one another demonstrating that they were built together (Fig. 3.67). Both blocks possessed basements, but from the northern vantage point of the Wharf Road Viaduct these subterranean structures were externally obscured by the made ground that formed the elevated terrace upon which the bulk of the Goods Station sits (Fig. 3.66). Their presence could be more easily discerned when viewing the offices from the low ground in the vicinity of the Regent's Canal, from where a sliver of the Basement of Block 2 could be seen (Fig. 3.67). The basement was visible in this location since the upper reaches of the canal-side wall had been deliberately cut away and capped when the offices were built and a vertical slice removed to create a lightwell. Although the Coal Offices and the Towpath Wall were erected within a few years of each other, this demonstrates that the latter was complete before work began on the former.

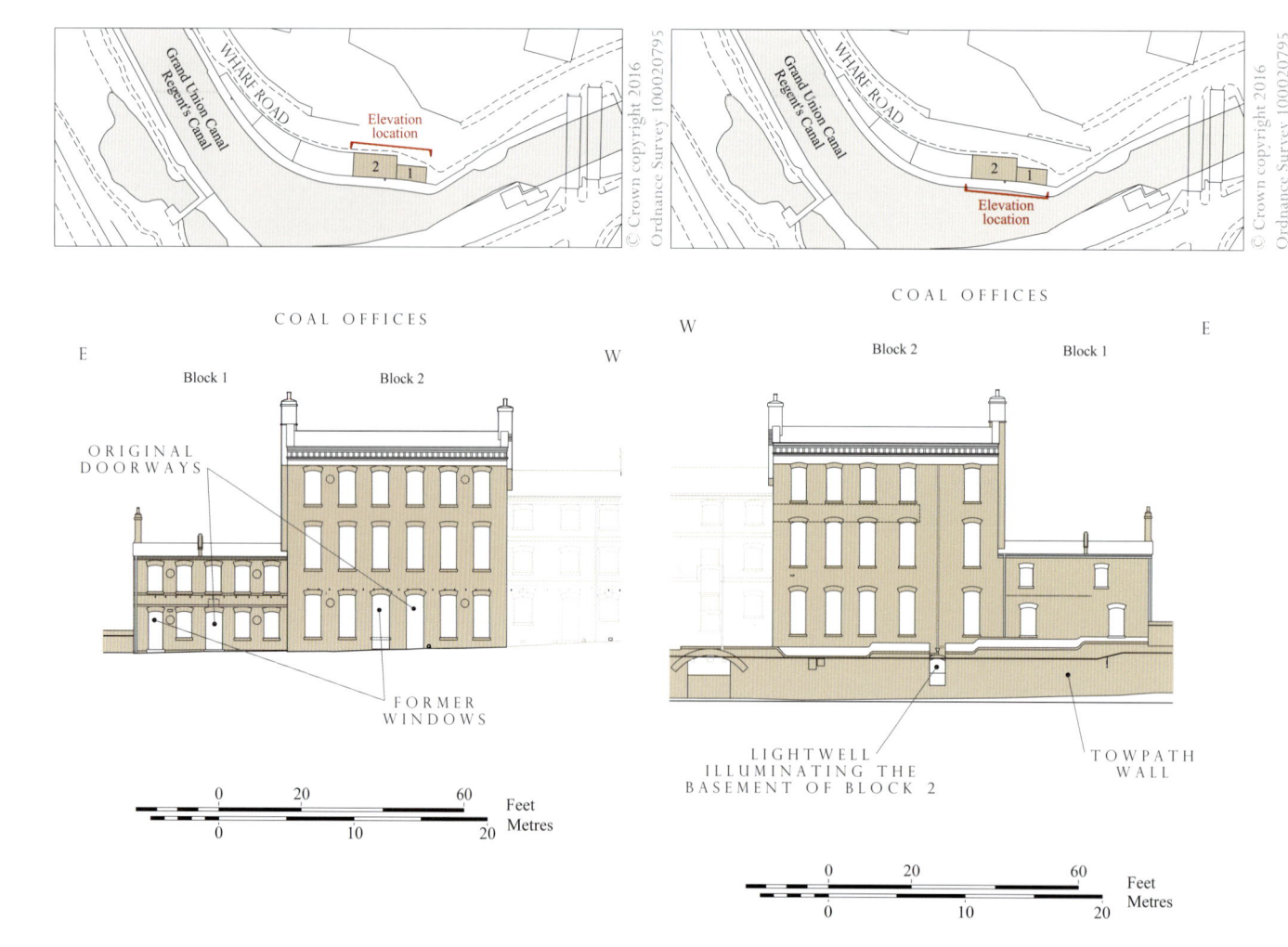

Fig. 3.66 External elevation of the north wall of the Coal Offices (Landscape Feature 18) showing original 1850–1 masonry, scale 1:500

Fig. 3.67 External elevation of the south wall of the Coal Offices (Landscape Feature 18) showing original 1850–1 masonry, scale 1:500

At three storeys high including the basement, Block 1 appears small and squat when compared with its partner to the west (Fig. 3.66; Fig. 3.67). Its original front doorway, which led from Wharf Road, was centrally located, opening onto a hallway that connected with a central staircase that in turn communicated with the basement and the first floor. Access to the ground floor offices was permitted via doorways that led off the aforementioned hallway (Fig. 3.65).

Block 2 was taller, wider and more imposing than its relatively modest companion owing to the presence of an extra storey and greater floor-to-ceiling height at every level bar the basement (Fig. 3.66; Fig. 3.67). Each of the block's six bays originally contained a window position on every floor with the exception of the west central bay of the northern elevation at ground level, where the threshold was situated (Fig. 3.65; Fig. 3.67). The windows and doorways were similar to those of Block 1, the main difference being that they were visually grander owing to their greater length, thus imbuing the building with a dignified appearance.

The basement of Block 1 was divided into four rooms of varying sizes that were presumably used for storage and other purposes. These dark and dingy spaces would have been utterly useless as office accommodation and early records relating to the leasing of the structure in 1860s demonstrate that they were never used as such (TNA RAIL 236/284: 22/11/1860). A coal chute was present in the north wall of the most easterly room, which suggests that it was used as a coal cellar, whilst the two narrow rooms that bordered it would have been impractical spaces for anything other than general storage. In contrast, the square room to the west of the stairwell was the largest open space within the basement of Block 1 and was also the only section that was heated thanks to the presence of a cast iron stove. The room was therefore presumably occupied either permanently or on a part time basis by a small contingent of staff that were probably tasked with the upkeep of the block.

Although the precise function of each of the four subterranean rooms in Block 2 is less obvious, it is possible that one or more of the smaller examples was again used for storing maintenance equipment, cleaning paraphernalia, clerical supplies or old records. Again, only one room appears to have been heated, in this instance the southeastern example, so it is possible that the cleaners and maintenance men who serviced the offices congregated there during periods of down time. Since a doorway connected the basement with the underground stables below the Wharf Road Viaduct, it remains possible that the western rooms were used by the horse department or an affiliated subcontractor rather than staff working out of the Coal Offices.

The remaining rooms at ground and upper floor levels in both blocks were the preserve of the Great Northern Railway's clerks, who administered the lucrative coal trade from the building under the auspices of their head of department, Coles Child. Although the office workers were kept warm in winter by a series of fireplaces in each room, those based in Block 1 may have found their accommodation wanting. Natural daylight would have been limited during the darker months as windows were not originally instated in the southern wall of that section, a deficiency that would have made life difficult for those working in the poorly illuminated rear of the building (Fig. 3.67). It is possible that these facilities were allocated to the junior clerks, while their superiors worked in the better appointed facilities of Block 2 alongside the Coal Manager, Coles Child.

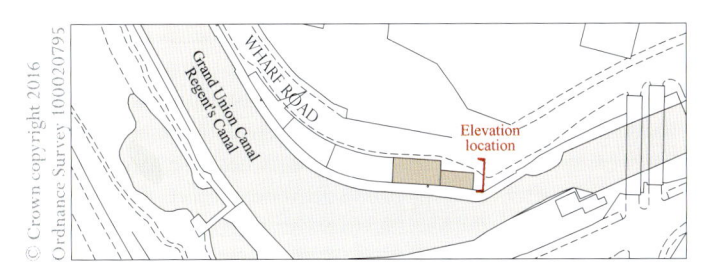

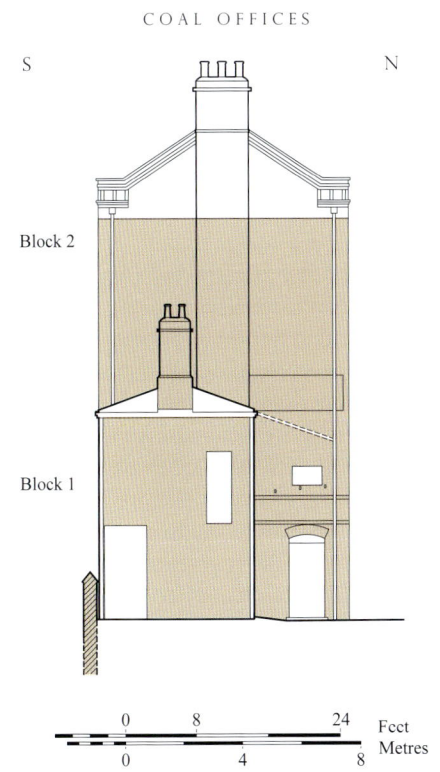

Fig. 3.68 External elevation of the east wall of the Coal Offices (Landscape Feature 18) showing original 1850–1 masonry, scale 1:250

Archaeological evidence relating to the construction, appearance, footprint and function of the Wharf Road Viaduct by 1852

Landscape Feature 19

Wharf Road runs along the southern and western perimeter of the Goods Yard, the viaduct being situated towards the western side near the Coal Offices (Fig. 3.2). The thoroughfare provided road vehicle access to the southern side of the yard from York Way. It was not recorded in detail as part of this project although it was subsequently investigated during a separate historic building recording survey, the results of which form part of the third monograph in this series. The southern face of the structure was, however, integral to the southern boundary wall of the Regent's Canal (Landscape Feature 1) which was recorded as part of this project under the site code KXD07. This section is therefore described here.

Twenty-six arches with later brick infilling were integral to the south wall of the Regent's Canal (Fig. 3.69), directly below Blocks 3 to 5 of the Coal Offices (Landscape Feature 18). These openings lead to the vaults that are situated below the Wharf Road Viaduct, demonstrating that the viaduct and the canal wall were erected at the same time (presumably in 1851 as demonstrated by historical sources). A contemporary rent return suggests that some horses belonging to the Great Northern's coal delivery contractor, Edward Wiggins, may have been stabled within the arches that were integral to the viaduct and the canal wall (TNA RAIL 236/275/23). Horses presumably entered the stables by way of the canal towpath, whilst pedestrian access was permitted through the Coal Offices by way of the basement of Block 2 (Landscape Feature 18).

Fig. 3.69 Arched entranceway of 1851 (since infilled) integral to the north wall of the Regent's Canal (Landscape Feature 1) leading to the stables in the vaults below the Wharf Road Viaduct (Landscape Feature 19), photograph faces north

3.17 Hard Landscaping and Weighbridges by 1852

Very few areas of the Goods Station were paved during the early years of its existence, one rare exception being the tramway or stoneway[2] that enabled coal to be transported between the Coal Drops (Landscape Feature 17), the Hydraulic Station (Landscape Feature 16) and Somers Bridge (TNA RAIL 236/273: 08/01/1851; Fig. 3.8). This paved track crossed one of the yard's earliest weighbridges, at the southern end of the Coal Drops as shown on the Humber Plan of 1866, where it is depicted as a rectangle crossed by a saltire (Fig. 3.8; TNA RAIL 236/273: 18/02/1851).

In the mid-19th century, weighbridges consisted of a metal plate flush with the surrounding ground surface that connected with a set of scales. The carter would drive his dray onto the plate and the weight of his vehicle would be formerly recorded by an attendant, thus preventing any embezzlement on the part of the cartage team. The system also enabled the railway to keep a close track of the overall quantities of goods that were flowing in and out of their yard by road and operate a reward system for the speedy completion of rounds (Holden 1985: 102). The allotting of bonuses was often complicated, taking into account a range of factors that included the distance that had to be travelled, the number of deliveries that had to be made and the overall weight of the dray (ibid.). Deadlines were hard to meet, since carters not only had to deliver and collect frequently cumbersome loads but also had to collate the associated paperwork (ibid.).

All road vehicles entering or exiting any goods yard had to pass over such a weighbridge and King's Cross was no exception. A reporter who visited the yard in 1859 described the process as follows: 'In the morning, before the numerous teams leave the station, the contents of each wagon are weighed. The wagon is driven upon an iron platform, which communicates with scales inside an office, when deducting so much for the weight of the vehicle, the number of tonnage brought up and transmitted by the company can be daily ascertained' (Anon 1859).

The Humber Plan of 1866 suggests that at least seven weighbridges existed in the vicinity of the Coal Drops, however the records of the Great Northern demonstrate that no more than six of these were original, the example to the immediate north of the Provender Store (Landscape Feature 25) having been instated in 1854 (TNA RAIL 236/20: 112; Fig. 3.8). The same source also suggests that further weighbridges could be found elsewhere in the yard, including another 1854 addition that was situated beside 'the engineers office' in an unknown location towards the station's northern boundary (TNA RAIL 236/20: 112).

---

2   Humber's use of the term 'tramway' suggests that it probably comprised iron or timber rails inset in a surround of stone setts, which would have improved the traction of heavily laden wagons pulled by horses. This, together with its location, suggests that this roadway was used by wagons carrying coal unloaded from railway wagons in the Coal Drops. Alternatively, the roadway shown on Humber's plan may have been a 'stoneway', a paved road with longitudinal granite blocks set at the spacing of the cart.

3.18 The Building Materials and the Evolution of the Great Northern's Track Network by 1852

The original Goods Yard buildings were, for the most part, made from materials that were produced in London or the surrounding area. By far the most common brick type that was used was a purple moulded frogged stock example made from a local London or Medway fabric (type 3032), many of which were stamped 'JJ'. Unless reused, the stamped examples were only identified in original sections of the structures erected between 1850 and 1852, such as the Main Goods Shed and Granary Complex (Landscape Features 7–12), the foundations of the earliest platforms, turntable bases, crane bases and docks and within the walls of the Granary Basin (Landscape Feature 5). Although no historical documentation has been found to link the initials 'JJ' with any particular brick maker (Kevin Hayward pers. comm.) they almost certainly relate to the prolific builder and railway contractor John Jay, who constructed a significant section of the early Goods Yard and its approaches. This interpretation is supported by the fact that the occurrence of the 'JJ' stamp was restricted to the earliest phases of the structures that John Jay built. Moreover, the records of the Great Northern Railway show that Jay had been paying brick tax on the materials that he was using to construct the goods termini at Maiden Lane and since that duty was only levied on brick producers it is clear that Jay was manufacturing bricks himself (TNA RAIL 236/15: 420, 509). He had been passing the cost on to the Great Northern, so after the levy was abolished in 1850 the company went to great lengths to ensure that they would not pay the monies that were payable under the former tax unnecessarily. On 20th June 1850 the Board issued a response to a letter from Jay, who had requested that the Great Northern should not deduct any sum from his contract prices in response to the repeal of the brick tax (*ibid.*: 509). Unfortunately for the builder, the Board 'resolved that this request cannot be complied with' (*ibid.*).

It is likely that Jay's bricks were manufactured fairly locally, given that the King's Cross area was a tile and brick manufacturing centre during the period that the Goods Yard was built. Jay may even have erected a temporary brick making facility on or near the site itself, employing either clamps or kilns. Either of these options would enabled Jay to keep the cost of transporting these bulky materials to a minimum. Alternatively, the bricks could have been transported to the site with reasonable ease from another area of London or the Medway Valley that had access to a waterway that connected with the Regent's Canal.

Numerous other kinds of bricks were used in the earliest phase of the Main Goods Shed and Granary Complex (Landscape Feature 7–12) and its associated infrastructure, all of which were made of local fabrics. These included (fabric 3035) frogged and unfrogged yellow stock bricks, observed in the earliest railway platform supports, whilst purple, red and maroon unfrogged bricks (3032), yellow machine frogged bricks (3032) and transitional yellow stock bricks (3032/3035) were noted in the various turntable supports (see Appendix 3). Amongst the stone setts that formed the floors of the Western and Eastern Stables (Landscape Features 7 and 8), kiln bricks were observed, whilst plinth bricks were incorporated within the ramps that led into the stables from the platforms above. All of these materials were produced in the London region, their fabrics being compatible with the construction of the Goods Station in 1850–2.

The Hydraulic Station (Landscape Feature 16) also appeared to have been constructed of locally made bricks although none of these bore the 'JJ' stamp. Instead, the internal and external walls and the engine beds were made from a mixture of frogged and unfrogged (fabric 3032) purple and red stock bricks with occasional (fabric 3261) kiln bricks. The fragments of floor that survived were made from purple stock bricks (type 3032).

Various building materials that were imported from further afield were also used to construct the earliest phase of the Goods Station at King's Cross. These included the Aberdeen granite and basalt setts that formed the floors of the Western and Eastern Stables (Landscape Features 7 and 8) and the granite and York stone cobbles that were observed in the ramps that lead to those underground structures. The glazed ceramic drain pipes that were lain down during the yard's formative stages were of a fabric type that suggests a Yorkshire origin.

The first goods services to enter the station did so in the last quarter of 1850 whilst the Goods Yard was still under construction (Thorne *et al.* 1990: 96). Although the Great Northern's railway network was still incomplete, the company had established excellent rail connections with Yorkshire and the industrial centres of the West Riding by that date. The Great Northern would therefore have been able to transport building materials to King's Cross at highly competitive rates from an early stage in the company's history. For that reason it is unsurprising that building materials such as York stone and drain pipes heralding from Yorkshire formed part of the assemblage that was retrieved from the station's formative phase.

Good connections with the north of Scotland would take longer to establish. Although Great Northern rails did not enter the country, the company collaborated with other northern carriers including the North Eastern and the North British, which enabled it to run express trains between King's Cross and Scotland as early as 1852 (Simmons and Biddle *et al.* 1997: 192). However, the latter connection most probably arrived too late to be used to enable it to be used to carry the Aberdeen granite that was used in the depot's earliest phases of construction. What is more, although some Aberdeenshire quarries would become connected to the national network via the Alford Valley Railway, this did not take place until 1859 (*ibid.*: 137). Given that the Great Northern did not establish a rail connection to the north of Scotland until after the Goods Yard was completed in 1852, it seems likely that the granite that was used in the earliest phases of the construction of the Goods Station at King's Cross was brought to London by sea from the port of Aberdeen. It was then presumably shipped onwards to the yard by way of the Regent's Canal or via a cartage contractor either directly from the docks or from a London based stone merchant.

3.19 Synopsis

The location chosen by the Great Northern to build its Permanent Goods Station was practical and well considered. Lying to the immediate north of the New Road, the site was as close to the centre of the city as the company was permitted by the Government to construct such a facility, in an area that also had excellent established transport links for the onward shipment of goods. The proximity of the Regent's Canal was of crucial importance, enabling the company to establish a modern and highly sought after canal and rail interchange. This setting enabled the Great Northern's London Goods Station to compete with its main rival in the capital, the London & North Western's Goods Station at Camden, which incorporated a similar, though smaller, interchange facility.

In order to get ahead of the competition, the company made a series of astute decisions: its London goods head would be bigger than those of its competitors, its route to the north would be faster, its yard would be partially mechanised through the use of a pioneering hydraulic system and its decision to take the coal trade in house enabled it to corner the lucrative coal market. Not everything went according to plan; the Granary was not completed as soon as was expected, the hydraulic system did not enter service as quickly as was originally anticipated and the company failed to secure a complete monopoly over the Sheffield coal fields. Despite causing anxiety at the time, these short term concerns did not result in significant longer term problems. The relatively direct nature of the Great Northern's line to London had the effect of greatly reducing the journey times and lowering carriage costs for goods. This was attractive to potential customers and reduced the importance of achieving a complete monopoly. The company's competence soon created an influx of goods to the yard from the north, pouring money back into the depleted coffers of the Great Northern.

Unlike the site of the Permanent Passenger Terminus to the south, the location of the Goods Station was largely undeveloped prior to the arrival of the Great Northern. Commencing in April 1849, construction therefore proceeded rapidly and, whilst it was not entirely without incident, few serious delays were incurred. The Goods Station gradually became operational between late 1850 and 1852, becoming fully functional just three years after the construction process began. The archaeology encountered suggested that a logical system of goods handling and traffic management was in place that was largely but not entirely one directional. This is reconstructed here in Figure 3.58, which shows rail, canal and road traffic movements throughout the Main Goods Sheds and their immediate periphery. Whilst the likely flow of train wagons and barges is reasonably well understood, the direction that road vehicles took through the Transit Sheds remains speculative. The route as shown in the reconstruction was decided upon as it would have made logistical sense for the carters to form orderly queues parallel with the external sides of the Transit Sheds before entering at the northern end, thus limiting congestion around the Granary Basin and inside the station itself. The road vehicular routes as illustrated would also have permitted easy access to the southern façade of the Granary where the cartage contractors received outgoing corn. On paper, the system looks as though it would have worked well, but at the time of its inception it was entirely untested. Its success or otherwise and the response of the Great Northern will be explored in the next chapter, which will consider what happened in the decades immediately after the Goods Station become operational.

The relative ease with which the Goods Station was built clearly illustrates why the company decided to build a Temporary Passenger Station next door to its goods railhead. Legal wrangling over land purchases meant that it would take time to gain possession of the site of the Permanent Passenger Station, part of which was already occupied by buildings before the arrival of the Great Northern. Once these obstacles had been overcome, it was necessary for the company's engineers to master the challenges presented by the steep northern approaches to the passenger station to the locomotives of the day, an impressive feat of engineering that would take time to resolve fully. Without the provision of a temporary substitute, the lack of a passenger terminus would have prevented the Great Northern from fully graduating from a local passenger carrier to a major player in the railway industry as quickly as the company's directors and shareholders would have liked. The Great Northern therefore required a temporary passenger terminus that could be built and brought into service as quickly as possible. That this goal was achieved as early as August 1850 owed much to the persuasive powers of William Cubitt and the architectural skills of his brother, Lewis, who devised a 'temporary' structure that could be erected swiftly at an affordable cost.

In summary, the construction of the Great Northern's Permanent Goods and Temporary Passenger Stations proceeded efficiently and without major incident. Whilst challenges were encountered that caused much anxiety at the time, these were quickly overcome and any delays that did ensue were generally insignificant. The survival of the fledgling company during a difficult time for the railway industry in general owes much to the rapid construction of these London termini and the capable management of the men that oversaw the process.

Building the Goods Yard: The Human Cost

By Jonathan Butler and Rebecca Haslam

'Yesterday afternoon, shortly after 3 o'clock, an accident of a very alarming, as well as fatal, character took place at the terminus of the Great Northern Railway, at King's Cross'

From the London Evening Standard 17th December 1850 (Anon 1850)

'Yesterday afternoon, shortly after three o'clock, an accident of an alarming and lamentable character took place at the present temporary terminus of the Great Northern Railway, by which one unfortunate young man has lost his life, and three other persons have been seriously injured...The young man Kendal is unmarried, but it is stated that the other unfortunate men...have large families dependent entirely upon their labour'

From Lloyd's Weekly London Newspaper 12th January 1851 (Anon 1851a)

The construction site at the Goods Yard was the scene of two fatal accidents within three weeks, either side of Christmas 1850. The first occurred on Monday 16th December 1850 during the excavation of a tunnel under the Regent's Canal. It was widely reported in the newspapers of the day:

'It appears that a large body of miners and other workmen having been for some time past engaged in excavating a tunnel which is to pass from the eastern side of the Maiden-lane-bridge, under the Regent's Canal, into the permanent station on the sites of the late London Fever and Smallpox Hospitals, at King's-cross. These men were in the employ of Mr. Hodge, the sub-contractor to the Messrs. Jay, the contractors for the terminus, and at the time mentioned, the miners had returned from their dinner about an hour, and were at work in groups in various parts of the tunnel. One group of miners, headed by a man named Abel Wynne, were at work in the tunnel about 40 yards from its mouth, and were engaged in making ready for what is technically termed the "cill" – a portion of timber used in supporting tunnels before the brickwork is introduced. He had just exclaimed, "Come, my men, we must pitch into this work, and get the 'cill' in," and had made one or two strokes with his pick, in conjunction with a man named Samuel Edwards, when a large amount of earth gave way and fell upon them and other workmen. An alarm was instantly raised, and after some delay the men were dug out, when it was found that the poor man Wynne had his head crushed in a frightful manner. He was conveyed to Randall's Cottages, near Randall's Tile-kilns, in Maiden-lane, but life was quite extinct. The other men were but slightly injured. No cause can be assigned for the falling of the earth in the tunnel, which is of a clayey character, and very dry.' (Anon 1850)

Abel Wynne was living locally in Maiden Lane and was buried at St Mary Islington on 22nd December 1850; he was 40 years of age (http://interactive.ancestry.co.uk/1559/91280_197215-00038).

On Saturday 4th January 1851 the second accident took place during the construction of the Granary (Landscape Feature 12), shrouded in scaffolding and complete up to fourth floor level, and Dock 3, the eastern canal inlet that ran into the building:

'The Great Northern Company...are having constructed an immense granary. This building is situate on the southern side of the station, abutting on a spacious dock [i.e. the Granary Basin]. From this dock there exist two cuttings or creeks, running into the granary, constructed for the purpose of enabling barges to run in and receive the grain under cover of the building itself, and to facilitate the loading and unloading of goods. In this building a large number of workmen in the employ of Mr Jay, the contractor, were at work in various departments, and in the creek on the eastern side of the granary...carpenters, sawyers &c....[were] working on a kind of floating saw pit.

At the time above-mentioned, some workmen were employed in raising an iron girder to one of the upper floorings [of the Granary], by means of 'sheers'. This girder, it is stated, is in weight from five to six tons, and the men had raised it to the fourth story, when suddenly one of the 'sheer legs' broke, and the ponderous mass fell with a frightful crash, crushing everything in its progress downward to the creek described...

...The frightful crash which the falling mass created attracted the attention of every one within, and many without, the building, and on reaching the creek several poor fellows were seen floating about in the muddy water below, clinging to the masses of floating timber which had been carried into it. From the difficulty to get at them, upwards of quarter of an hour elapsed before any one could be rescued, and when they were got out, it was found by Mr Thompson, surgeon, of York-row, Maiden-lane, who was at once sent for, that two of the men named Green and Rolfe were so seriously injured that immediate removal to University College Hospital was necessary. They were without delay conveyed to the hospital on shutters, where it was ascertained that the poor fellow Green had both his ankles broken, while his companion had sustained very dreadful injuries. A third was conveyed to his own residence.

Some time having elapsed, it was deemed advisable by the foreman of the works to call a muster of the men, and a young man named James Kendal, a carpenter, was found to be missing. Drags were produced, and after dragging for about an hour the body of the young man in question was found embedded in the mud; and on being got out, it was discovered that he had a frightful lacerated wound over the left eye and temple; and as the poor fellow Kendal was a carpenter, and was known to have been working on one of the upper floors of the granary, it is believed that he must either have been standing on some planks which the girder struck at the time it fell, or that the girder itself must have struck him during its descent. The body of the young man...was at once removed to the City of York Tavern.

The young man Kendal is unmarried, but it is stated that the other unfortunate men, Green and Rolfe, have large families entirely dependent upon their labour.

The unfortunate man Green expired on Tuesday morning. From the first there had been no hope of his recovering from the severe injuries he had received. The upper dorsal vertebrae were fractured, as well as several of the ribs on both sides, the small bone of the leg was broken, and the foot dislocated. Rolfe's injuries are also very serious, but they are internal, and cannot yet be ascertained with accuracy' (Anon 1851a)

An inquest was subsequently held by Mr Wakley M.P., the coroner for west Middlesex, at the City of York Tavern in Maiden Lane on Tuesday 7th January 1851:

'The coroner remarked that in this case it did not appear that blame attached to any person. It was very easy to make suggestions after an accident, and it seemed that "luft" tackle ought to be used in these cases, which would be a very useful precaution, and he had no doubt that in future it would be carefully adopted. He was often astonished to see the carelessness of the men themselves employed in these undertakings, and really, considering the magnitude of the works, that about 600 men were employed on them, and that they had been in progress upwards of two years, it was extraordinary that more lives had not been lost. This showed that great care and caution had been used.

Mr. Vallance wished to observe that Mr. Jay had given directions for the interment of the unfortunate men, and everything should be done for their widows and families they could possibly require.' (Anon 1851a)

A summary of the incident in Jackson's Oxford Journal of 11th January 1851 gave the names of the other men as William Rolf and James Green (Anon 1851b) although some reports stated the other injured man to be called Tyler rather than Rolf (Anon 1851c). The Kentish Independent reporting on the incident and subsequent coroner's inquest stated that Kendal's forename was in fact John rather than James (Anon 1851c) and a John William Kendall of William Street is recorded in the parish registers of St Mary Islington as being buried on January 10th 1851, aged 21 (http://interactive.ancestry.co.uk/1559/91280_197215-00040). It was reported that James Green subsequently died of his injuries and an inquest was held on Thursday 9th January where a similar verdict of 'Accidental Death' was recorded (Anon 1851c).

From the burial records of Abel Wynne and John Kendall it was apparent that both men lived very close to the construction site. Abel Wynne lived in Maiden Lane which formed the eastern boundary of the site, whilst John Kendall was from William Street which was a road off the eastern side of Maiden Lane immediately to the south of the Tile Kilns on the 1846 Parliamentary Plan of the proposed Great Northern terminus (Fig. 3.1). On the 1846 plan terraced housing is shown to occupy the southern side of William Street and it is possible that John Kendall lived as a lodger in one of these houses. The 1851 Census records a number of railway labourers and carpenters living in Little and Great William Street (TNA HO107/1500/395–402; HO/107/1500/340–348). However, it was reported at the inquest that 600 men were employed on the construction site (Anon 1851a) and it is highly likely that many of the men may have been living in tents or shacks as part of an itinerant navvy workforce on undeveloped land surrounding the site as the living conditions during the early days of railway building were often notoriously bad.

Construction sites are notoriously dangerous places to work with fatal accidents a common occurrence well into the late 20th century and still occurring today despite the raft of Health and Safety legislation. However, in the mid 19th century during the boom in railway building across the country and the other major construction projects which were linked to the growing industrialisation of Britain, accidents must have had happened with such frequency that only the very worse were reported on. Saying that, the coroner Thomas Wakley M.P. stated that it was surprising that more accidents had not occurred at such a large construction project as King's Cross, which was employing in the region of 600 men and this could only be due to the due care and attention shown on the site presumably by Mr Jay as well as his employees. One of the more surprising features of the coroner's summing up is that he makes no mention of the fact that there had been another fatal accident on the site only a matter of three weeks before. As coroner he could hardly have been unaware of the earlier incident, especially as it was likely that the inquest took place in the same tavern on Maiden Lane. So this might raise the prospect that a fatal accident every three weeks was not something of note to the coroner, however it is strange that no mention was made of it and rather that the accident record of the site was in fact commended.

No other reports of accidents during the construction of the Good Terminus at King's Cross have been discovered so it is possible that these were the only two fatal accidents and that Mr Jay was to be commended for looking after his employees. He is reported as seeing to the funeral arrangements and doing everything that could be done for the widows and families of the dead men (Anon 1851a). This would either suggest that he was indeed an enlightened employer who looked after his men or was perhaps more cynically trying to keep the coroner appeased to avoid an unfavourable verdict. Mr Jay's credentials as a good enlightened employer may perhaps be questioned by another occurrence later the same year. In October 300–400 carpenters, bricklayers and labourers went on strike following a meeting in the same Duke of York Tavern in Maiden Lane over the non-payment of the hour and a half when they were accustomed to leaving at 4pm on a Saturday rather than 5.30pm (Anon 1851d). The dispute was quickly settled as all the men were offered jobs on other sites by other firms forcing Messrs Jay to accede to their demands (Anon 1851e).

CHAPTER 4

Competition and Congestion at the King's Cross Goods Station: c.1853–1879

'The prodigious extent to which the railway system has extended within the last dozen years is totally unparalleled in the annals of commercial enterprise in all past ages…A trip to Greenwich would at one time satisfy our longings for locomotion in one day…but such petty distances are now treated with comparative contempt. A hundred miles a day is quite a common affair, and is looked upon as such. If, however, the locomotion of passengers has so greatly increased, the freight of goods and merchandise has almost exceeded it in the production of revenue. The facilities afforded by railways for the transmission of merchandise has resulted in an amount of business of which none but those who have visited some of the head warehouses of the goods stations can form any conception…We were lately conducted over the stables and warehouses … at King's-Cross, and what we saw was quite sufficient to enlighten us as to the prodigious amount of merchandise that is transmitted through their hands…In addition to general merchandise, a vast trade is carried on in coals, corn, potatoes and the transport of woollen and cotton stuffs from the manufacturing districts'

A journalist describes his visit to the Great Northern Railway's Goods Station at King's Cross (Anon 1859)

The first section of the Great Northern mainline opened in 1848, with the London end becoming operational between 1850 and 1852. The company's trunk route ran from London and York via Hitchin, Peterborough and Grantham with a loop line to Bawtry near Doncaster via Boston and Lincoln whilst branch lines extended to Sheffield and Wakefield. This network gave the Great Northern access to the London markets and a host of economically productive areas such as the agricultural heartlands of East Anglia and Lincolnshire, the northern coal fields that bordered Doncaster and the steel and textile producing areas around Sheffield and Wakefield.

From the outset, the Great Northern faced a series of difficult challenges. The company had spent vast sums on the construction of its new venture at a time when the British economy was still mired in a recession (Nock 1974: 24). Meanwhile the Great Northern's Euston-based rival, the London & North Western Railway, had under the capable supervision of its General Manager, Captain Mark Huish, revived old alliances with the Midland and other railway companies in an attempt to undermine it. The London & North Western's position was strong since the Great Northern required running rights over sections of its tracks, including the entry into York Station, which led to the establishment of a number of agreements regarding the pooling of traffic that were highly unfavourable to the new company (*ibid.*: 24–9). A series of potentially damaging price wars then erupted over passenger fares, one of which ended with the Great Northern offering to undercut any price offered by the Midland and the London & North Western by 6d (*ibid.*: 29, 34). On the other hand, passenger numbers were significantly boosted by the opening of the Great Exhibition in London's Hyde Park in 1851. The event attracted about 6.2 million people, approximately one third of the entire population of England and Wales, the majority of whom would travel into the city from the provinces by train (Wolmar 2007: 112). Although the exhibition would provide an important revenue stream exactly when the Great Northern needed it the most, the event could scarcely sustain a railway over the long term.

From the outset however, the promoters of the Great Northern had envisaged that the company's fortunes would lie in the carriage of goods, agricultural produce, raw materials and minerals, specifically coal. The company's Euston rivals were powerless to prevent it from quickly gaining a huge market share of the traffic that originated from the Sheffield coalfields which, thanks to an alliance with the South Yorkshire Railway, gave the company a virtual monopoly (Nock 1974: 29–30). From the mid-1850s onwards in partnership with the North Eastern, the Great Northern enticed coal shipments away from London-bound sea colliers (Thorne 1990: 112). Then, thanks to a series of confrontations with the Midland Railway, the company consolidated its access to the Nottinghamshire coalfields thanks to a working relationship with the 'Ambergate' Railway (Nock 1974: 30–1; Williams

1988: 66; Simmons and Biddle *et al.* 1997: 191, 354–5). London's appetite for coal was such that the resulting traffic provided the new company with a huge source of revenue at the expense of the London & North Western and the Midland in particular. The Great Northern also made the radical decision to sell the coal direct to the consumer by cutting out London based middlemen, thus boosting their income further. As the first railway company to pioneer the carriage of the commodity to the capital in vast quantities, the venture was extremely successful (Brandon 2010: 83).

Important gains were also made in other areas. In 1853, a direct line linking Peterborough, Grantham and Retford opened, meaning that yet more agricultural areas could be reached, whilst the coming of the railway to Peterborough caused that city to rapidly grow into another important manufacturing centre (Simmons and Biddle *et al.* 1997: 191). By the end of the 1850s, agreements with other railway companies also meant that the Great Northern had gained access to most of the important West Yorkshire mining and manufacturing towns including Bradford, Halifax and Leeds.

The extent and connectivity of the Great Northern's rail network ensured that the company had built up sufficiently robust finances to withstand the financial crisis of 1866, triggered by the collapse of the banking house Overend & Gurney, which drove several other railway companies to bankruptcy (Wolmar 2007: 144).

London was both a huge consumer of goods and a major centre of production and redistribution so it is not surprising that both railway imports and exports increased in tandem with the growth of the city, which boomed throughout the Victorian period. The Goods Yard at King's Cross therefore witnessed a more or less constant rise in goods handling, with 'natural' growth being further accelerated by the efforts of senior officers of the company to attract new business from the very beginning. The company was quick to spot that money could be made from the rapid transport of fresh fish from North Sea ports to London markets. As early as the summer of 1852, Seymour Clarke made overtures to the capital's leading fish salesmen by offering them favourable rates for the carriage of fish from the northern ports (TNA RAIL 236/275/17: 'Fish Traffic from the North into King's Cross' 1852).

The late 1850s saw the volume of traffic entering the yard increase significantly, in part thanks to the contrasting fortunes of the Great Northern's rivals. The London & North Western had long desired to merge with the Midland, despite the fact that the latter company was inclined to retain its independence (Williams 1988: 67; Wolmar 2007: 114). However, the Midland company's outlook changed dramatically after it failed to profit from the Great Exhibition of 1851 owing to its lack of a London railhead (Wolmar 2007: 114–5). When a Bill proposing a merger with the London & North Western was placed before Parliament, the Government took a dim view of the monopolies that amalgamations of that kind would generate, and the Bill was defeated in 1854 (Nock 1974: 32; Williams 1988: 67). Following the defeat of the Bill, Captain Mark Huish of the London & North Western set about making a series of covert agreements with the Midland regarding a 'common purse' arrangement that effectively amounted to a merger in all but name. These illegal actions were later exposed whilst an unrelated railway matter was being contested in Parliament (Nock 1974: 33; Wolmar 2007: 115). The consequences of exposure were potentially devastating for the Midland, which saw its carefully crafted alliances with the Euston Confederacy torn apart (Nock 1974: 33, 35). Although the Midland had started making its own way to London in 1853, the company remained a predominantly provincial concern and had yet to reach the capital when the Confederacy finally disintegrated in 1857 (Williams 1988: 70). Following the demise of the Confederacy, the Midland was left with little choice other than to seek an arrangement with its former arch rival if it wished to regain access to the metropolis (Nock 1974: 36). The collapse of the Confederacy also meant that another former Euston ally, the Manchester Sheffield & Lincolnshire, could join the Great Northern fold, thereby granting access to the industrial heartlands in the western side of the country. A working arrangement with that company also meant that from 1857 onwards the Great Northern could run through goods and passenger services between London and Manchester, whilst in 1866 the two companies would jointly purchase the West Riding & Grimsby Line (Wakefield to Doncaster), thus further improving their already excellent links with Yorkshire (Simmons and Biddle *et al.* 1997: 191).

These developments soon impacted upon King's Cross goods and passenger stations. In June 1858 the Great Northern formally granted the Midland Railway to right to run trains on Great Northern tracks from Hitchin into the latter's termini at King's Cross (Biddle, 1990: 62, 65). Whilst this agreement gave the company's rivals renewed access to the metropolis, the Midland was charged a substantial fee for the privilege (Nock 1974: 36). Underused facilities at the goods station were converted for the reception of the Midland Company's goods services, whilst part of the Coal Depot to the north-west of the goods sheds was earmarked as the site for a new shed (the Midland Roundhouse) for the new arrival's locomotives. In order to ensure that its services did not suffer undue disruption as a result of the additional traffic, the Great Northern naturally prioritised its services over those of the Midland, which increasingly led to lengthy delays to the latter company's traffic (Williams 1988: 75). Despite the Midland's dissatisfaction at this state of affairs, the arrangement provided an invaluable opportunity for the company to consolidate its share of London-bound traffic before it sought Parliamentary approval for the construction of its own stations in the capital.

When the Midland Railway finally submitted its Extension to London Bill to Parliament in 1863, the Great Northern was quick to object. Ironically, however, the company had ensured its swift passage by granting the Midland running powers into King's Cross in the first place. The severe congestion that ensued demonstrated that another main line was needed, meaning that no serious arguments could be raised against the Midland Bill. Having obtained Parliamentary sanction for the construction of its line, the Midland would plough through Agar Town and St Pancras, destroying the homes of approximately 20,000 people as well as disinterring the remains of thousands more laid to rest in the overcrowded burial grounds of St Pancras Old Church (Wolmar 2007: 117).

Having acquired land in Agar Town upon which to build its own goods depot as early as 1859, the Midland was able to vacate the King's Cross Goods Yard in July 1862, five years before the company's mainline into London was completed (Townend 1975: 16). Despite the cessation of Midland goods traffic, the latter company continued to use the locomotive roundhouse in the Great Northern Coal Depot until St Pancras Station was opened to passenger traffic in October 1868 (*ibid*.: 20–22).

The insatiable demands of the metropolis were such that the Goods Yard at King's Cross continued to suffer from the effects of congestion even after the departure of the Midland. When the civil engineer William Humber inspected the facilities in 1865 he was informed by Seymour Clarke that the complex had already reached its maximum operating capacity. In order to reduce delays to traffic, the Great Northern had already placed a Bill before Parliament which sought authorization to build additional accommodation for goods wagons at Holloway Road (Thorne *et al.* 1990: 101). This was a prudent move, as the late 1860s to the 1870s would see the company extend its reach even further. During this period the creation of new branch lines and the formation of new alliances would ensure that the amount of territory across which the Great Northern could competitively carry freight would grow considerably (Nock 1974: 54–62). Most crucially, the company would cement its links with Scotland through an affirmation of its working relationship with the North British and the North Eastern Railways. In 1860 the three companies established a shared pool of locomotives and passenger rolling stock known as the East Coast Joint Stock, thus consolidating the primacy of the East Coast Mainline in railway history (Simmons and Biddle *et al.* 1997: 191). The company also went on to gain a foothold in the lucrative Derbyshire coalfields, which had previously been the preserve of the Midland (Nock 1974: 58–9). Through the establishment of a joint line with the Great Eastern, the company also increased its dominance over the agricultural heartlands of Lincolnshire (Simmons and Biddle *et al.* 1997: 191).

In summary, the entire period from the early 1850s to the end of the 1870s was characterised by a more or less constant increase in the volume of goods that flowed into and out of King's Cross as the population of London and the reach of the Great Northern Railway grew.

The number of canal barges on Britain's waterways had been falling since the 1840s as customers slowly began to turn to the faster and more efficient railways

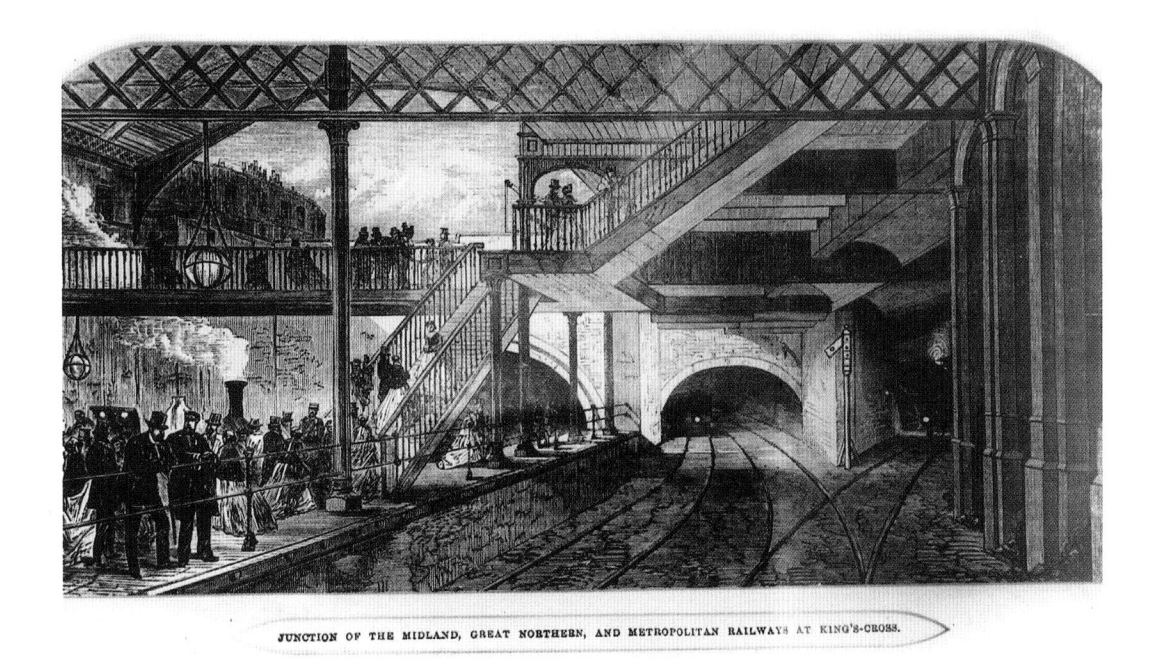

Fig. 4.1 Taken from ***The Illustrated London News*** of 1868, this image shows the recently constructed junction between the Great Northern and Metropolitan railways at King's Cross, which enabled the former company to carry their goods to and from the heart of the City (Anon 1868)

(Denny 1977: 55). Acutely aware of the declining market share of the canals, many canal owners attempted to establish their own railway concerns during this period. These included the Regent's Canal Company, which applied to establish a railway along the course of the canal on several occasions between the 1840s and the early 1890s. These efforts came to nothing, owing to Government objections, financial problems and logistical issues (*ibid.*: 79–81; Faulkner 1990: 53). Despite the gradual decline of the canal sector, the Regent's Canal still handled a great deal of traffic from the 1850s to the 1870s, which ensured that the canal facilities at King's Cross Goods Yard were still very much in use at the end of that period.

In the London area the woes of the waterway companies were compounded by the continuing improvement of the capital's railway network. These included the creation of the first cross London railway lines, such as the Metropolitan in 1862, the London Chatham & Dover Railway's London extension in 1864 (linked with the Metropolitan in 1866) and the West London Extension of 1863 (Harrison 1962: 196; Simmons and Biddle *et al.* 1997: 279). These developments enabled goods to be transported through the heart of London entirely by rail without the need to call upon the services of road vehicles or waterborne transport. The Great Northern made the most of these developments by establishing alliances with other London carriers, a decision that enabled them to decentralise their London goods handling operation. Although King's Cross did not lose its status as the company's principal goods depot, by the end of the

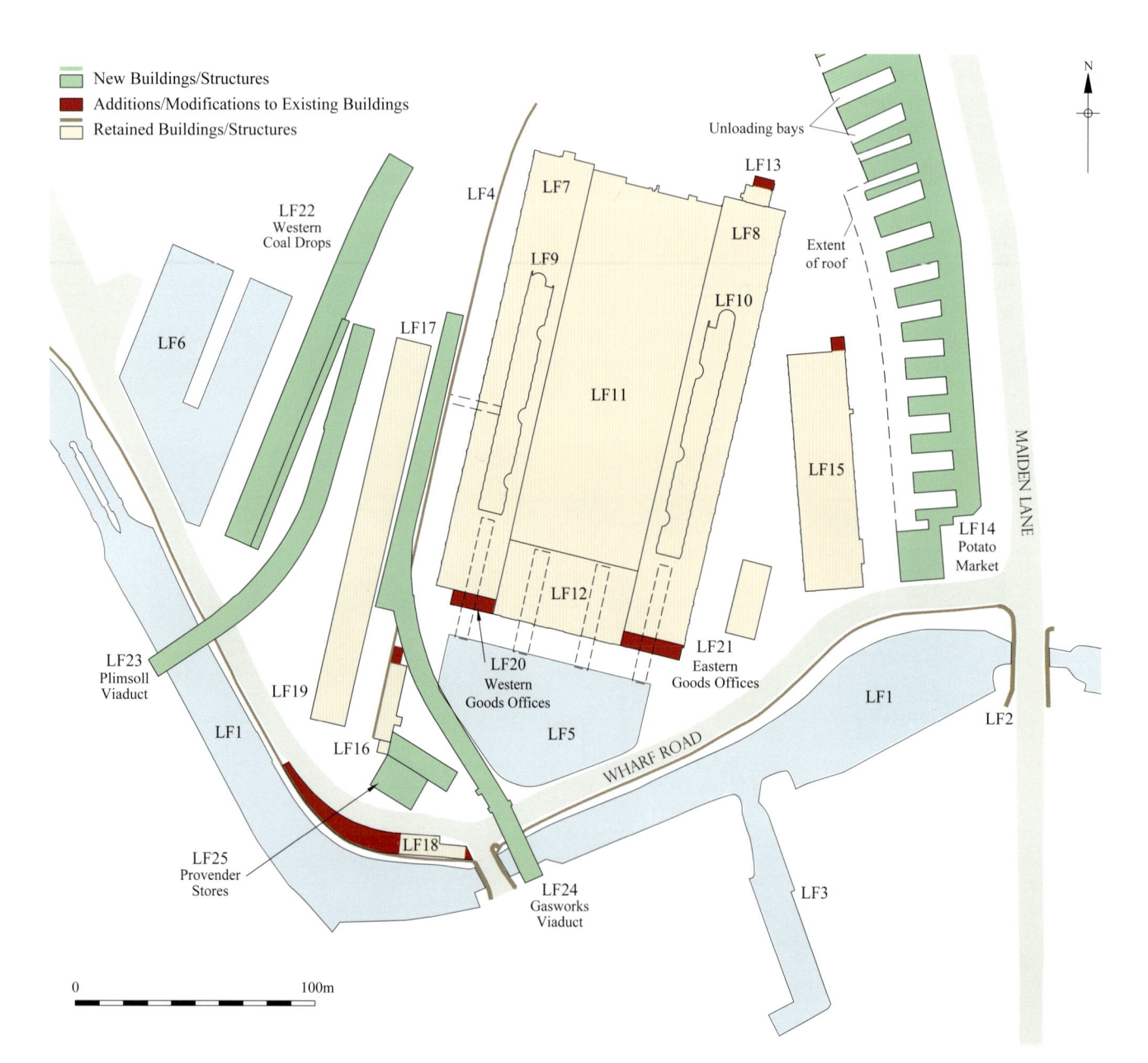

Fig. 4.2 The locations of the Landscape Features that were present in the Goods Yard from the mid 1850s to the 1870s (previously existing buildings indicated by LF number, new features are by both name and number), scale 1:2,500

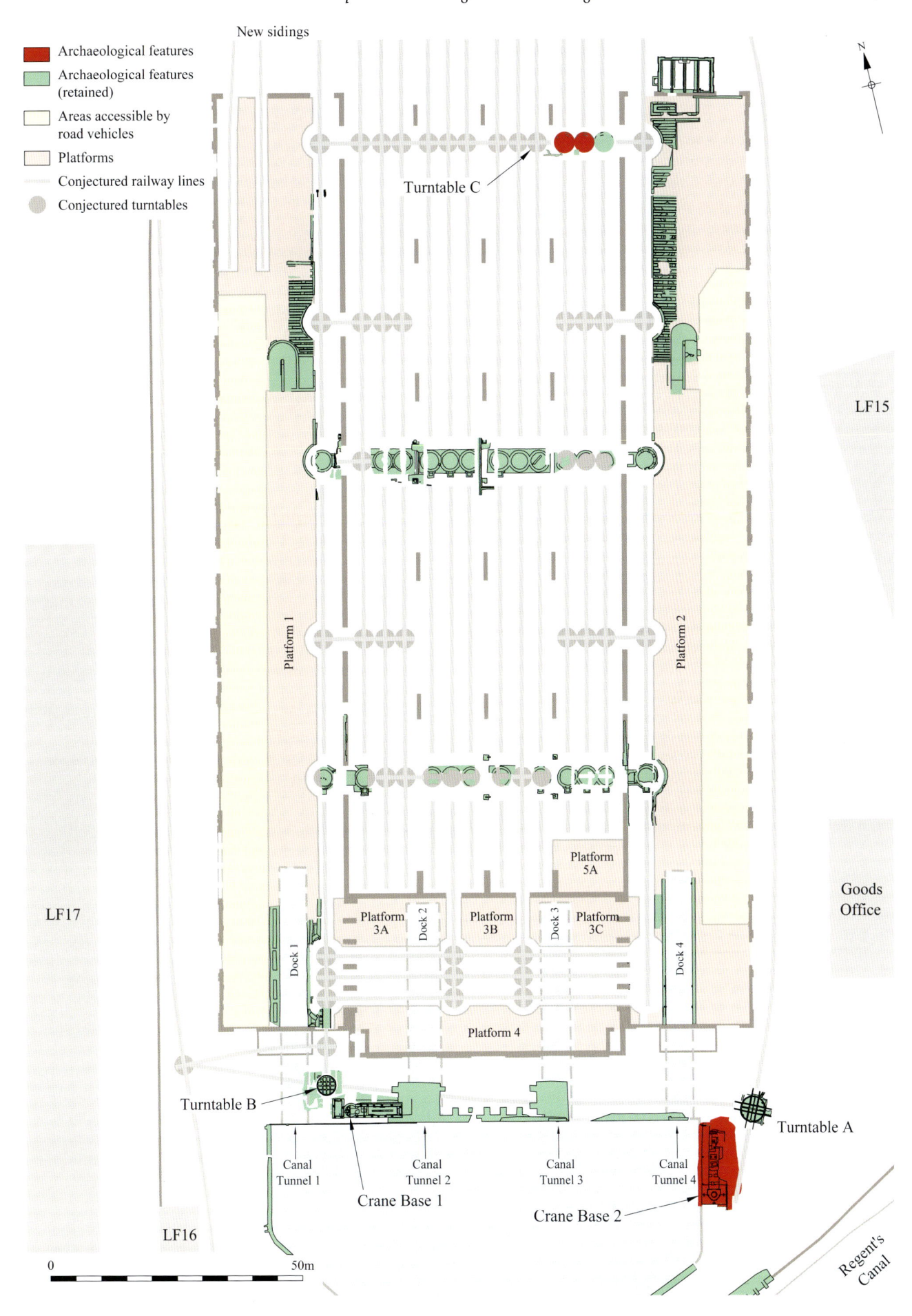

Competition and Congestion at the King's Cross Goods Station: c.1853–1879   97

Fig. 4.3   The locations of the significant archaeological features dated from the mid 1850s to the 1870s, scale 1:1,000

1880s it was one of many Great Northern goods depots in the metropolis.

The first of the Great Northern's new goods facilities in the Greater London area to enter service was the yard at Holloway Road, which commenced operations in the early 1860s. It was joined soon afterwards by the Farringdon yard, after Great Northern trains were given running powers over the Metropolitan Line, an event that enabled the company to penetrate the Square Mile (Nock 1974: 64). Meanwhile, the development of cross-London rail arteries enabled Great Northern coal shipments to be carried south of the Thames by rail for the first time. The venture was undertaken in conjunction with the London Chatham & Dover Railway (LCDR), via the Snow Hill tunnel and over Blackfriars Bridge (*ibid*.: 72). From the late 1860s the Great Northern continued to improve its distribution networks south of the river, eventually constructing a large coal depot of its own between Brockley Cross and Nunhead, as well as another shared with the Midland at Hither Green (Goslin 2002: 20; Brandon 2010: 84). These developments necessitated the creation in 1866 of a marshalling yard at Ferme Park between Hornsey and Haringey (Nock 1974: 69; Brandon 2010: 84). This facility further streamlined the passage of goods through the city (Brandon 2010: 84). Many London-bound goods trains would terminate at Ferme Park before being divided, sorted and reassembled so that a percentage of their wagons could be sent to King's Cross, Holloway Road or Farringdon or be dispatched across the Thames to the south (Nock 1974: 69). Together, these new goods reception and processing facilities must have eased the burden at King's Cross.

Owing to the limitations of the twin track Copenhagen and Gasworks Tunnels, congestion on the main line into London persisted throughout most of this period. Both goods and passenger trains destined for King's Cross and Farringdon were forced to use the former, whilst goods services for Farringdon and passenger services for King's Cross had to pass through the latter thus creating two serious bottlenecks in the north of the city. Pressure was eased when the Great Northern built the Canonbury diversion, which allowed through goods traffic to and from East London and the docks to be siphoned off at Finsbury Park and sent on via the North London Railway yet congestion problems persisted. It was not until the Board sanctioned the doubling of the Copenhagen Tunnel in 1875 that the matter was more effectively addressed (*ibid*.: 64).

The means by which the King's Cross Goods Station met the challenges of the 1850s, 1860s and 1870s are discussed in greater depth in the section that follows. In order to assist the reader, the landscape features that were in use during this period are illustrated in Fig. 4.2.

4.1 Hard Landscaping throughout the Yard: the mid-1850s to the 1870s

As the volume of rail traffic grew, so the wear caused to the unmetalled rail and cart roads of the Goods Yard by horse shunting became a growing inconvenience. Consequently, in November 1853 Seymour Clarke recommended to the Board that certain (rail) roads in the Goods Yard be paved for a total cost of £2,550 (TNA RAIL 236/19: 89). Although the extent of the scheme is not known, it is clear however that it did not extend to the south side of the Goods Shed complex, where the rail and cart roads in front of the Granary lacked permanent paving for several years to come.

The Board's willingness to pay for the improvement of rail and cart roads in the Goods Yard appears to have waned as the 1850s progressed and it was not until March 1856 that Seymour Clarke eventually persuaded it to approve the paving of the Potato Market sidings (Landscape Feature 14), an undertaking that was only sanctioned once he halved the estimated cost of the works (TNA RAIL 236/21: 319). By August of that year the condition of the unmetalled railroads in front of the Granary and those leading to the manure sidings had deteriorated to such an extent that Clarke recommended to the Board that both be paved with granite owing to the growing volume of traffic (TNA RAIL 236/22: 215). Clarke estimated that the Granary road could be paved for about £650, while the road by the manure sidings could be improved for an additional £670. The Directors demurred and resolved instead that Clarke merely present them with an account of the cost of repairing the roads on an annual basis. Clearly dissatisfied with the Board's response, Clarke subsequently reported that the road leading to the manure sidings was 'so rotten as to render some permanent repair really essential' (*ibid*.: 264). In support of his case Clarke recruited Walter Marr Brydone, the newly appointed Chief Engineer of the Great Northern, who argued that the approved method of repair (scattering broken Guernsey granite on the damaged surface) constituted a straightforward waste of money (*ibid*.). Despite Clarke's entreaties, the Board would only consent to paving half of the road. Having agreed to at least part of the necessary works, the Board authorised the purchase of 720 yards of 4" x 4" Mountsorrel granite pitching (TNA RAIL 236/227: 05/11/1856). In January 1858, Brydone instructed that this material was the stone of choice for paving all of the Goods Station's roads in preference to the inferior Guernsey variety (*ibid*.: 13/01/1858).

In February 1858 Clarke again sought authorisation to pave the Granary (rail) road; however his request was deferred by the Directors, who requested that an annual account of the cost of repairs be prepared and submitted for consideration at a later date (TNA RAIL 236/26: 281). Thirteen months later Brydone reported that the cost of maintaining the Granary (rail) road amounted to £90.2.9, which contrasted favourably with the sum of £172.3.10 spent the preceding year. The reduction in cost appears to have been a consequence of an undocumented decision to coat the ground surface around the track with pitch Macadam[3] in order to protect it from the ravages of horse shunting. In response to Brydone's report, the Board 'resolved that the present system of macadamizing the road be continued' (*ibid*.).

Despite the partial success of Brydone's experiment, Clarke continued to push for a lasting solution to the damage to surfaces caused by horse shunting in the Goods Yard. In January 1860 the Board approved his recommendation that 'certain portions of the King's Cross Goods Yard where there is much horse shunting should be paved with common stone', admitting that in wet weather the ground in these areas was 'almost impassable' (*ibid*.: 202). It was not until November 1861 that the Board finally relented to the permanent paving of the Granary rail road with Mountsorrel granite (TNA RAIL 236/29: 199). Although a poor harvest brought about a reduction in the Great Northern's receipts from grain traffic in 1860, trade picked up somewhat the following year, which may have prompted them to acquiesce and improve the road for the relatively modest sum of £191 (TNA RAIL 236/283: Clarke to the Board 19/12/1860).

An indication that the Great Northern eventually came to pave station cart roads as a matter of course is given by an engineer's report from May 1870, in which reference was made to the considerable expense incurred at the company's stations and yards in London, Peterborough, Grantham, Boston, Lincoln and Doncaster as a result of purchasing large quantities of new granite to replace existing cart roads damaged during the preceding harsh winter (TNA RAIL 236/303: 18/05/1870).

Archaeological evidence relating to the installation of hard landscaping: the mid-1850s to the 1870s

Removal of the modern tarmac in front of the Granary (Landscape Feature 12) revealed evidence of an earlier surface composed of square and rectangular granite setts. The patchwork nature of the bonding materials that cemented them in place revealed that the setts had clearly been relaid on numerous occasions as a result of routine maintenance on the surface itself and during the installation and repair of service runs. Despite this, it remains possible that some of the stones in the location of the siding that ran in front of the Granary Basin could have originally been instated as early as 1861 when

---

3   Pitch Macadam was a rudimentary tarmac-like substance formed of broken stone, sand and tar (the latter material would have been easily obtained from the adjacent gasworks on the south side of the Regent's Canal).

that siding was paved. Others may have been added later, perhaps during the improvement programme of 1870. In order to improve traction, the long sides of the rectangular setts had been positioned at right angles to the direction of traffic flow, whilst drainage guttering and paving associated with services were aligned with the road.

Similar paved areas were encountered throughout the yard. This demonstrates that almost all of the station was eventually paved with granite, despite the Great Northern's initial resistance to paving even their most intensively used cart roads.

4.2 The Granary Basin: the mid-1850s to the 1870s

Landscape Feature 5

By March 1853 Seymour Clarke had become concerned about the impact upon business caused by the slow rate at which bulky materials such as stone and timber were transferred from wagons in the Arrivals (Eastern Transit) Shed into barges moored in the Granary Basin (TNA RAIL 236/21: 334). To alleviate the problem he requested permission to build a new heavy crane and track on the east side of the Basin, similar to the one already in place on the northwest side, which was used to unload goods from barges prior to onward distribution by rail from the Western Transit Shed (*ibid*.). It was intended that

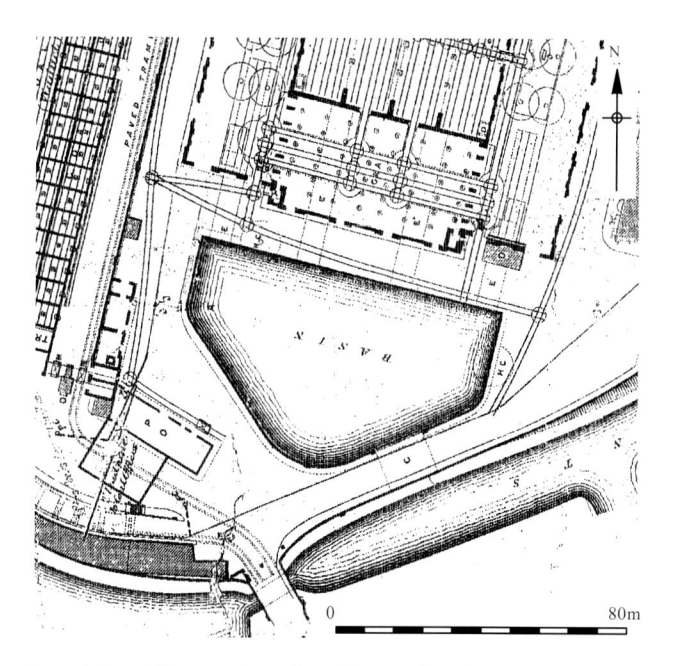

Fig. 4.4   The Humber Plan of 1866 detailing the southern end of the Main Goods Shed (Landscape Features 7–11), the Granary (Landscape Feature 12) and the Granary Basin (Landscape Feature 5), scale 1:2,000 (NNRG DMFP 00026266)

the new crane would accelerate the rate at which bulky goods were loaded into barges, thereby eliminating delays and congestion amongst the vessels using the Basin. The Board approved Clarke's request, authorising the purchase of a new crane for a cost of £615 and granting permission to extend the necessary track for an outlay of £648 (*ibid.*).

One other alteration appears to have been made to the railway infrastructure that surrounded the Granary Basin during this period. Cartographic evidence suggests that the railway wagon turntable identified here as Turntable B (located to the immediate north of the Granary Basin) was temporarily removed between 1866 and 1871 before being reinstated in 1873 (Fig. 4.3; Fig. 4.4; Fig. 4.5; Fig. 4.6). Its removal would have impacted upon the flow of traffic through the Goods Station since the turntable linked several important elements of the complex, including the Western Transit Shed's 'down' line and the railway infrastructure that surrounded the Granary Basin. Given the importance of the structure, its omission from the 1871 Ordnance Survey Map might be a cartographic error. If it was genuinely removed for repair or reuse elsewhere it appears to have been quickly reinstated.

Archaeological evidence relating to the modifications made to the infrastructure surrounding the Granary Basin in 1853

Landscape Feature 5

The northern end of the new track that was commissioned by Seymour Clarke in 1853 was unearthed in its anticipated position during an archaeological excavation (KXP08). It connected with the southern side of Turntable A, (Fig. 4.3) enabling railway goods wagons to traverse the eastern side of the Granary Basin as far as the new heavy crane (Fig. 4.4). The support for that device, Crane Base 2, was also identified archaeologically (Fig. 4.3). It was remarkably similar in nature to Crane Base 1. The new section of track terminated in a 'dead end' to the south of the crane indicating that multi-directional traffic flow was operational along its length.

The physical remains of Turntable B shed no further light on whether or not the structure was removed and reinstated between 1866 and 1873.

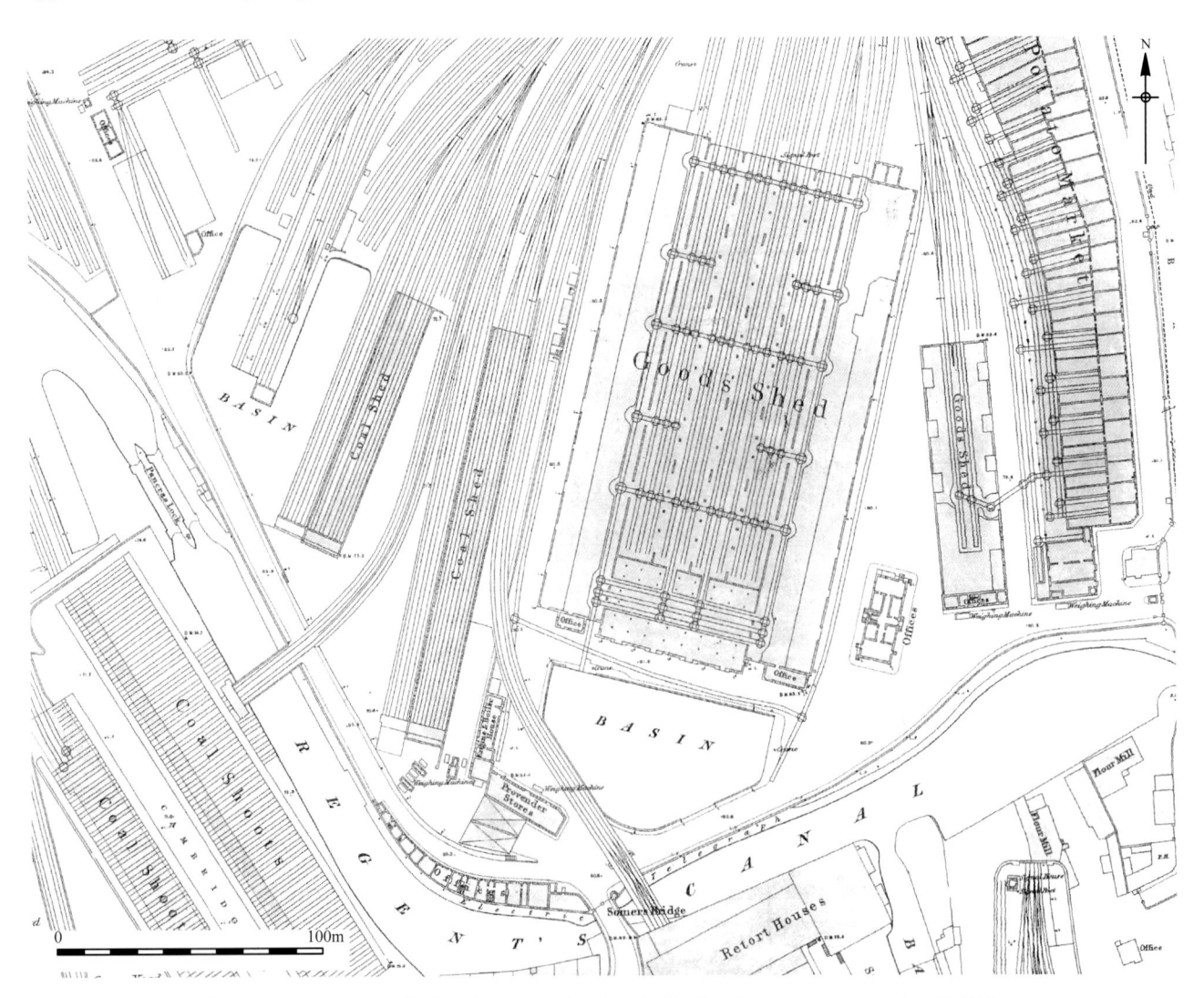

Fig. 4.5 The Ordnance Survey Map of 1871 depicting the Goods Yard and its environs, scale 1:2,500

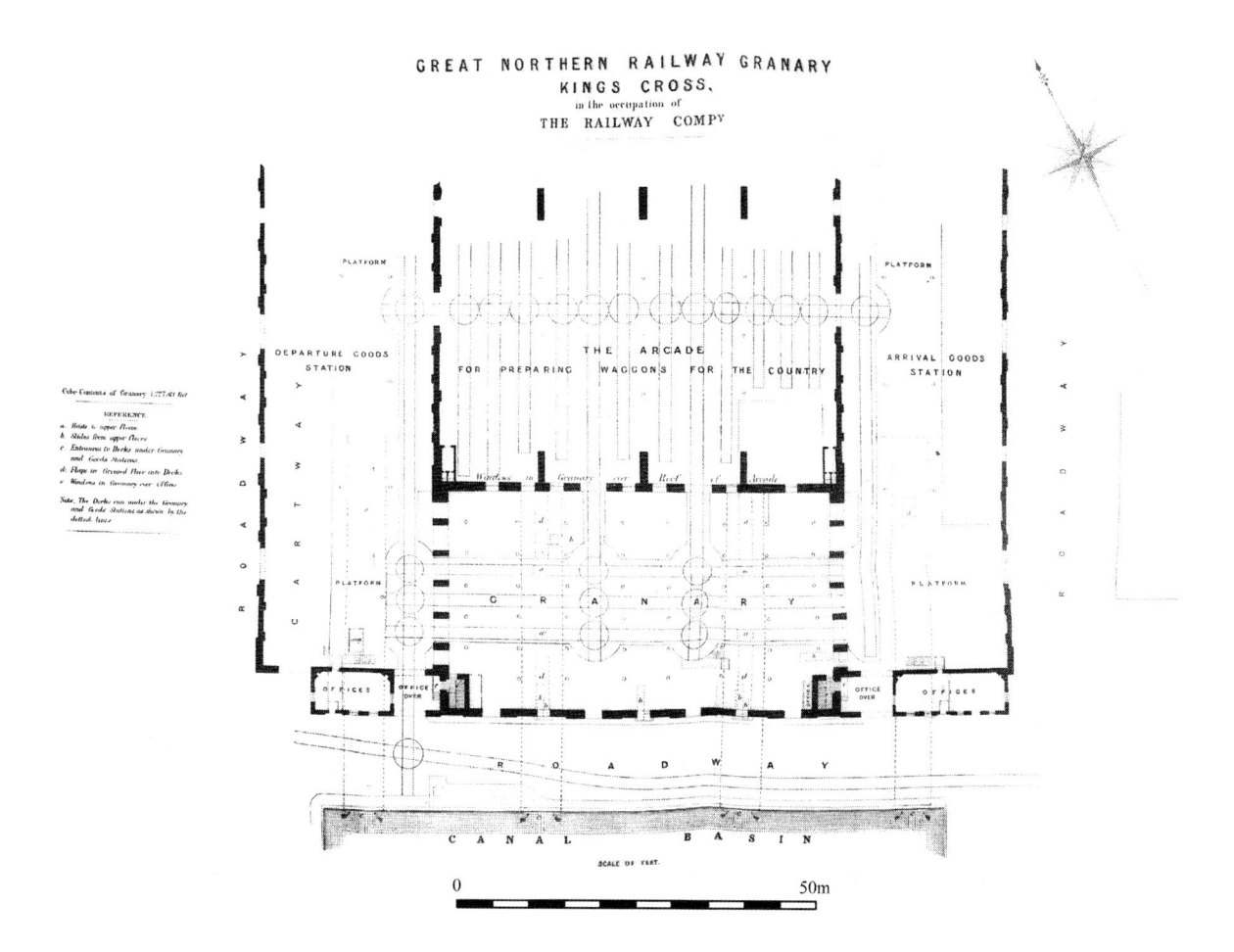

Fig. 4.6 The Crockett Plan of 1873 detailing the southern end of the Main Goods Station (Landscape Features 7–11) and the Granary (Landscape Feature 12), scale 1:1,000 (COL MS 15627)

4.3 The Inward (Eastern) and Outward (Western) Goods Offices: the mid-1850s to the 1870s

Landscape Features 20 and 21

The Eastern and Western Goods Offices, which became the administrative hubs of the Goods Department, did not form part of Lewis Cubitt's original Goods Station design. Company records suggest that much of this work was initially undertaken in the Transit Sheds themselves, rather than in purpose-built offices. For example, in October 1854 Seymour Clarke informed the Board that it was 'desirable [that] the invoicing of Goods at King's Cross should be done on the shed' to which end he proposed that the Inwards Goods Office should be altered at a cost of £123 (TNA RAIL 236/20: 77). Both the language used in the Board minute and the relatively insubstantial sum spent suggested that this office was located within the Eastern Transit Shed itself. Such an arrangement can hardly have been ideal, especially given the growing volume of traffic and therefore administration at this time. In October 1856 Clarke claimed that the health of the clerks working in the 'offices of the Goods Shed King's Cross Station' was being harmed 'by the want of adequate ventilation' (TNA RAIL 236/22: 240). Clarke suggested that their discomfort could be alleviated by means of 'a shaft… carried through the roof', the construction of which could be achieved for an estimated cost of less than £8.

Evidently this measure did little to resolve the problems caused by the lack of dedicated administrative facilities in the Goods Station. As the volume of goods traffic grew, the company was obliged to recruit additional clerical staff to process the accompanying paperwork. It is likely that their accommodation in the Transit Sheds was becoming ever more cramped as the decade progressed, prompting Clarke to press for the provision of new office facilities. In late July 1858, Clarke presented the case for additional office accommodation to the Board (TNA RAIL 236/25: 356). The following month Clarke sought authorisation to build a pair of 'additional buildings for offices', which he estimated could be constructed for a total cost of £1,130 (*ibid.*). The proposal was referred to the Executive and Traffic Committee, which authorised the construction of the Eastern Goods Offices and most probably the Western Goods Offices on the 10th of that month. The contract for their construction was promptly awarded to Messrs W. Rummens, a company that was already undertaking unspecified works within the confines of the Goods Yard (*ibid.*).

Historically, the Eastern and Western Goods Offices were respectively referred to as the 'Inwards' and 'Outwards' Goods Offices, in keeping with the direction of traffic that each was responsible for administrating. Descriptions of the subsequent enlargement of the Eastern Goods Offices suggest that it was originally two storeys high, and it is highly likely that this was also the case for its western counterpart (TNA RAIL 236/170: 78).

The earliest cartographic depictions of the new offices can be found on Edward Stanford's 'Library Map of London and its Suburbs', which was published in 1862 (not illustrated). The map suggests that the Inwards and Outwards offices were erected against the southern faces of the Eastern and Western Transit Sheds in the spaces between the railway and cart way exits of those buildings. Interestingly, whilst the Outwards (Western) Goods Office is shown on the Stanford map, it does not appear on William Humber's highly detailed plan of 1866 (Fig. 4.4), although it did appear on the First Edition Ordnance Survey map of 1871 (Fig. 4.5). The building was certainly in existence by early January 1866, when Seymour Clarke proposed adding an extra storey to 'the King's Cross Outward Goods Office' (TNA RAIL 236/34: 34). In all probability, the Outwards Goods Offices were established in 1858 around the same time as the Inwards Goods Offices, the anomaly on the Humber Plan of 1866 representing a rare error.

As the quantity of goods handled by the yard grew throughout the 1860s and 1870s, so did the volume of accompanying paperwork. In response the company was periodically obliged to recruit additional clerical staff to handle the expanding workload (TNA RAIL 236/90: 88; TNA RAIL 236/91: 124). In 1865 Clarke recommended that the Inwards (Eastern) Goods Offices be enlarged, owing to the fact that 'the existing building [was] much too small for the clerks engaged within' (TNA RAIL 236/33: 113). He estimated that the necessary works could be undertaken for a cost of £231 and his scheme was authorised by the Board on the 11th of April of that year. Whether these measures improved the lot of the clerks is not known, however the records of the Great Northern Board suggest that the bureaucratic burden grew remorselessly as the company's goods business continued to expand.

In early January 1866 Clarke turned his attentions to the Western Goods Offices, informing the Board that it was 'desirable for the sake of the health of the clerks and the dispatch of business to afford more room in the King's Cross Outward Goods Office' (TNA RAIL 236/34: 34). He went on to recommend 'that another storey be added to the existing office and that it be extended' for an estimated cost of £526. The contract for the works was subsequently awarded to William

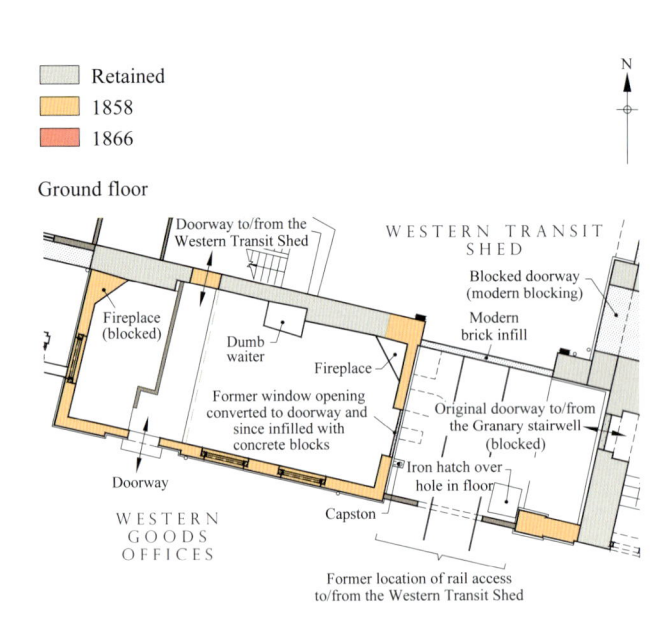

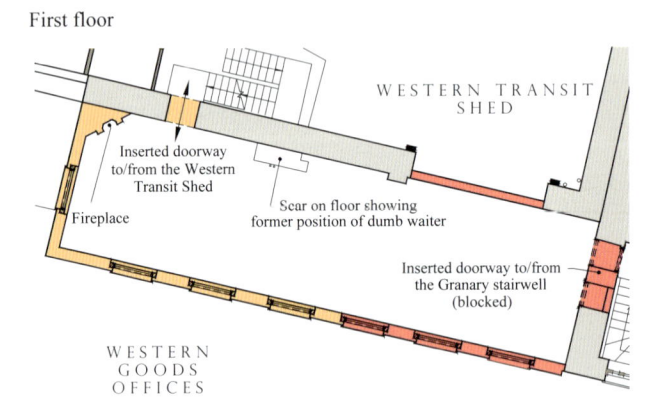

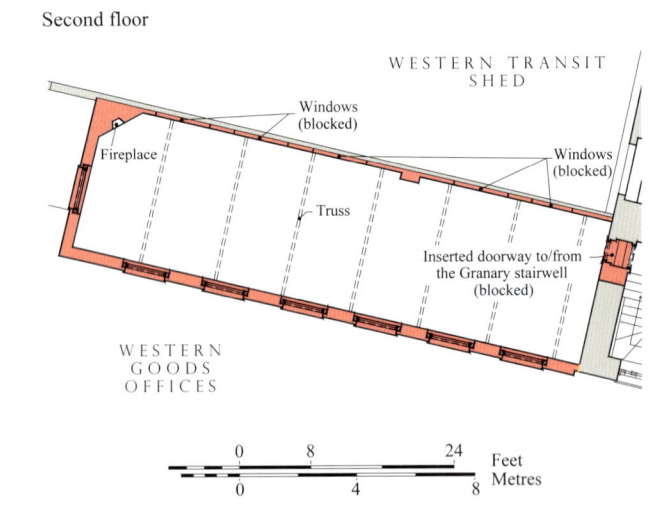

Fig. 4.7 Plan of the Western Goods Offices (Landscape Feature 20) by 1866 reconstructed from built heritage evidence and cartographic sources, scale 1:250

Prince (TNA RAIL 236/92). The footprint of the enlarged building is shown in considerable detail on a plan surveyed by Edwin Crockett for the Wharf and Warehouse Committee of the London Fire Offices in 1873. Crockett's plan suggested that the extension built by Prince incorporated an upper level of offices that sat over the arch that provided access to the Eastern Transit's Shed's rail siding (Fig. 4.6).

The minutes associated with this episode form the earliest documented references to the Western Goods Offices. This suggests that the facilities necessary to administer the work of the Outwards Goods Department were relatively satisfactory up to that point in time, in contrast to those provided used by the Inwards Goods Department, which had been enlarged once already and were rapidly becoming overcrowded again.

Soon after the Inwards (Eastern) Goods Offices were extended in 1865–6, it became apparent that the enlarged facilities continued to provide insufficient space for the growing contingent of the Inwards Goods Department's administrative staff. In October 1867, Seymour Clarke submitted 'several schemes' for further enlargement to the Way and Works Committee. He once again argued that the works were necessary because 'the health of the Clerks suffers in the existing confined space' (TNA RAIL 236/170: 55). The Committee chose to recommend a proposal for the further enlargement of the offices worth £327.10.0 to the Board (*ibid.*). The preferred scheme comprised the addition of another storey (i.e. the third) to the Inwards Goods Offices (*ibid.*: 78). Contractors were invited to tender for the work at some point between October and the beginning of December 1867. Once again, the successful applicant was William Prince, who submitted a tender amounting to £335 (*ibid.*).

Despite the addition of an extra storey, the office space was soon found wanting and yet another expansion scheme was proposed in February 1873, when plans were submitted for the addition of 'two large rooms' to the offices of the Inwards Shed, 'so as to provide additional accommodation for the clerks employed there' (TNA RAIL 236/145: 25). The rooms were to be 'built over an archway' with construction running at an estimated cost of £390 (*ibid.*; TNA RAIL 236/39: 198–9). The works comprised the westwards extension of the offices on the first and second floors over an arch that adjoined the original train wagon exit that linked the Eastern Transit Shed and the Granary Basin. Edwin Crockett's survey of the same year indicated that pedestrians could enter the ground floor of the offices via a doorway in the east wall of that arch whilst the upper levels could be accessed by way of the eastern Granary staircase.

The Eastern Goods Offices were enlarged on three occasions during this period, whereas their counterparts to the west were extended only once, with the result that the former became approximately one third larger than the latter. This suggests that the clerical department that dealt with incoming commodities grew more quickly than the outbound division, which in turn indicates that the quantity of imports that were handled by the Goods Station grew at a faster rate than the volume of exports.

No other notable events associated with the offices took place during this phase with the exception of one incident that occurred in March 1873. A cash box containing £21 was stolen from an unlocked drawer in the desk of the Outward Goods Counter Clerk, who was obliged to resign from the company's service after the theft was discovered (TNA RAIL 236/145: 36).

Built heritage and archaeological evidence of the construction and extension of the Western Goods Offices in 1858–66

Landscape Feature 20

A straight construction joint observed between the Western Goods Offices and the Western Transit Shed indicates that the office block was a later addition, confirming the documentary evidence from the records of the Great Northern (see above).

The extant building is long and narrow in shape, being six bays wide from east to west and one bay deep from north to south. Exterior brickwork consisted of a mixture of yellow London stock (3035 fabric) and purple multicoloured bricks, usually with yellow surfaces, laid in Flemish bond (i.e. 3032 and 3034 fabrics; see Appendix 3 for further details on brick coding). Ornamental flourishes were sparse and plain. They included three courses of brick corbelling forming a cornice at eaves height, a brick stringcourse located between the ground and the first floor levels that was four courses high and a brick plinth at base level. Recessed segmental arches were found over the doors and windows, the latter being characterised by sandstone sills with horned sashes, each of which had eight panes. The building was capped by a gabled and slated roof.

Documentary evidence indicates that the earliest incarnation of the Western Goods Offices consisted of a structure that was a mere three bays wide, one bay deep and two storeys high. The First Edition Ordnance Survey map of 1871 (Fig. 4.5) suggests that the original entrance was located towards the western side of the southern elevation as it is today (Fig. 4.8) and that the northern side of the structure was formed by the pre-existing southern elevation of the Western Transit Shed, which it abutted. Few modifications were made to the building, with the exception of two new entranceways that were inserted at ground and first floor levels (Fig. 4.7). These provided pedestrian access between the Transit Shed and both storeys of its offices by way of a staircase that was attached to the northern face of the wall on the Transit Shed side. This was the only means of access to the first floor of the Western Goods Offices as no internal stairwell was constructed.

The built heritage survey also identified a previously undocumented component of the 1858 scheme. A brick wall was constructed against the western wall of the Granary, which was seamlessly joined to the Western Goods Offices through the use of a decorative string course (Fig. 4.8). The two structures were also built with identical materials, suggesting that these two structural elements were erected at the same time. Together with the eastern side of the Western Goods Offices, the wall that abutted the Granary may have supported a canopy that covered part of the adjacent rail siding that communicated with the Western Transit Shed. Such a covering would have protected open wagons as they queued up to gain access to the congested facility. Alternatively this area could henceforth have been converted into a small supplementary loading space through the provision of shelter.

The eastern extension to the building was no doubt added along with an additional storey in 1866 at the request of Seymour Clarke. The newly constructed second floor contained one large room that was equal in size to the newly enlarged chamber that was situated immediately below it. In contrast, the amount of office space at ground floor level remained unchanged in order to accommodate the active rail siding that the upper storeys now spanned. A new means of access to the first and second floors was provided through the insertion of new doors that connected with the Granary's southwestern stairwell. This meant that the ground and first floors could be accessed either from the north by way of the original staircase that linked the building with the Western Transit Shed or from the southeast via the southwestern stairwell of the Granary. Admittance to the newly constructed second floor could only be obtained by way of the Granary stairwell (Fig. 4.7).

Built heritage and archaeological evidence of the construction and extension of the Eastern Goods Offices in 1858–73

Landscape Feature 21

The extensive historic building recording survey of the Eastern Goods Offices (KXF07) has enabled the evolution of the structure from its inception in 1858 through to its extension in 1873 to be reconstructed. Certain elements of this have been based upon cartographic and historical evidence only, as much of the original fabric of the building's southern and eastern walls did not survive a devastating air raid that occurred on 9th November 1940 (Fig. 4.10; Fig. 4.11). Like the Western Goods Offices (Landscape Feature 20), the earliest sections of the structure were built in Flemish bond with a mixture of London stock bricks (3035 fabric) and purple multicoloured bricks with yellow surfaces (3032 and 3034 fabric).

The north wall of the Eastern Goods Offices was formed by the south wall of the Eastern Transit Shed, all elements of the former having been built against the latter as indicated by a straight construction joint between the two. The only significant changes that were initially made to the shared structure were identical to those that were made in the Western Transit Shed at this time. They involved the insertion of doorways at ground and first floor levels on the shed side, along with an accompanying staircase. Documentary and cartographic sources indicate that the offices initially consisted of a two storey building that was practically identical to its western counterpart (Fig. 4.9). The Ordnance Survey map of 1871 suggests that the original doorway was situated in the westernmost bay of the southern elevation

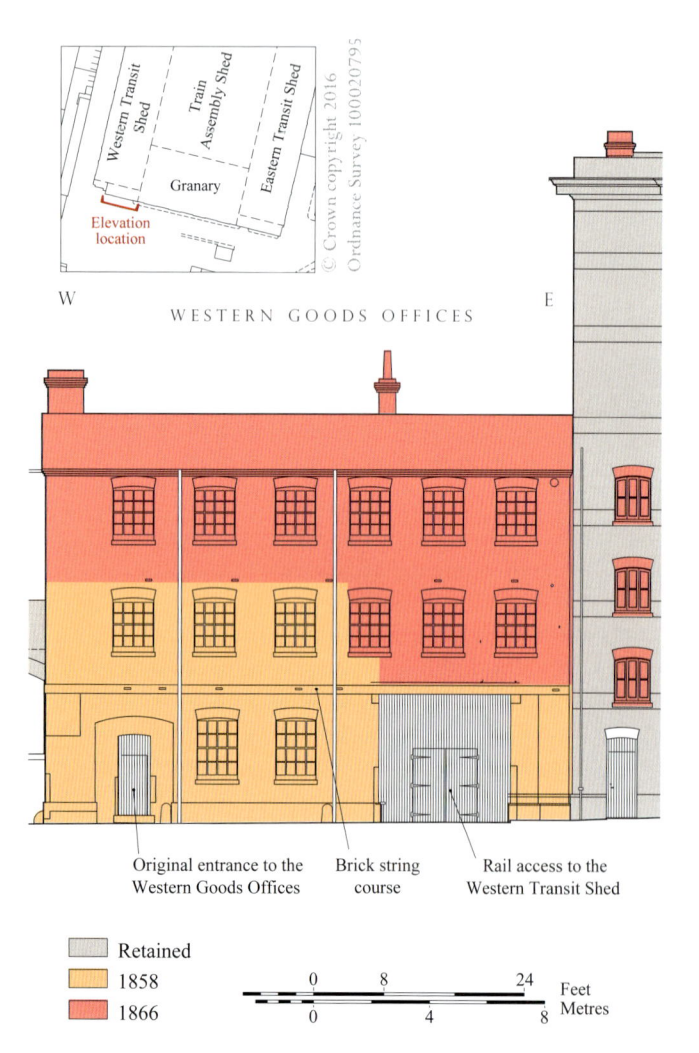

Fig. 4.8 External elevation of the south wall of the Western Goods Offices (Landscape Feature 20) showing the changes that took place from 1858–66, scale 1:250

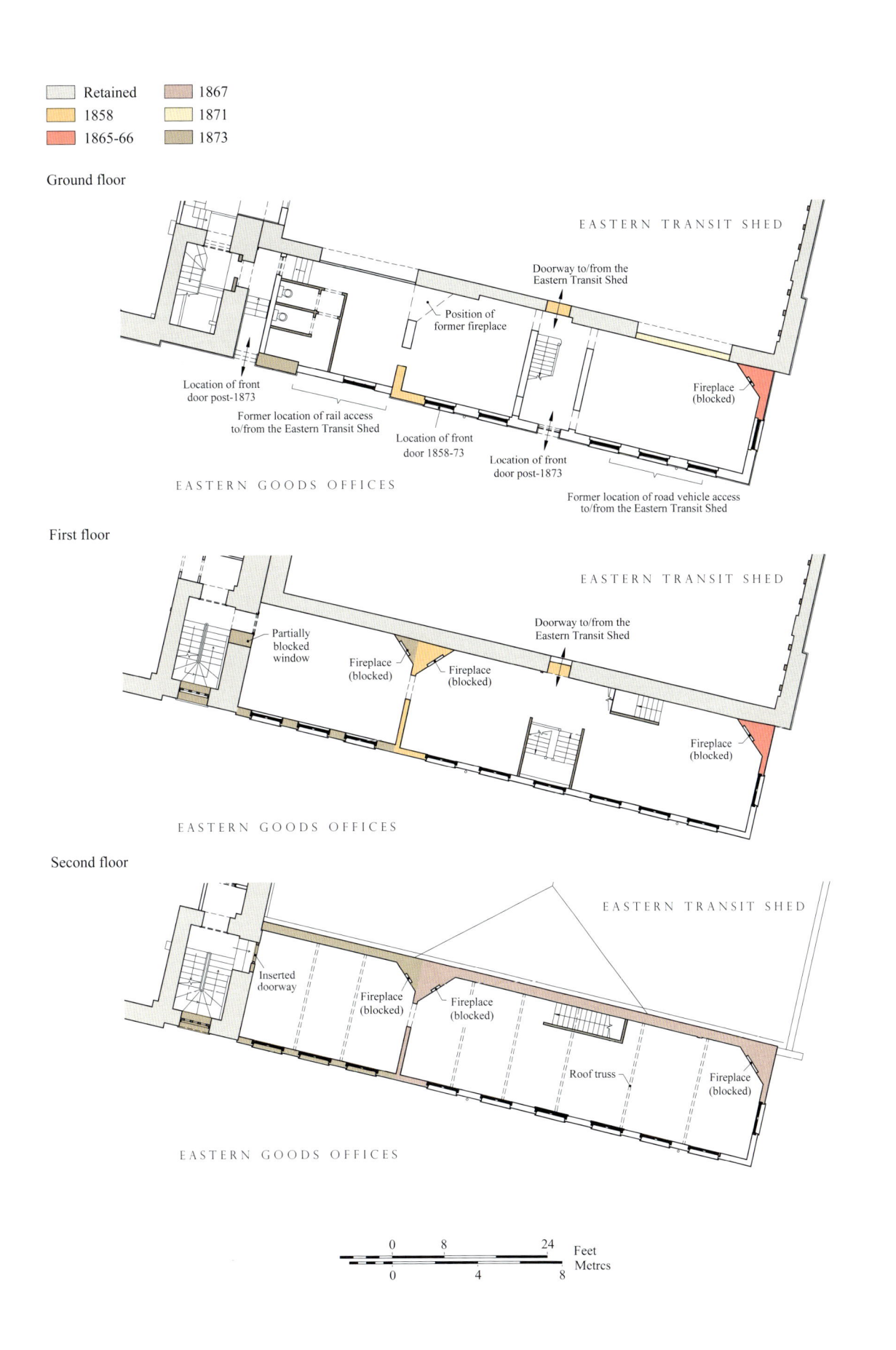

Fig. 4.9   Plan of the Eastern Goods Offices (Landscape Feature 21) by 1873 reconstructed from built heritage evidence and cartographic sources, scale 1:250

but has since been relocated (Fig. 4.5; Fig. 4.9). The same sources suggest that an additional three bays were added to the eastern side of the structure, most probably during the 1865–6 office extension programme (Fig. 4.5). These were positioned in the path of the cart way that ran along the eastern side of the Eastern Transit Shed. Recording of the dividing wall that separated the Eastern Transit Shed from the Eastern Goods Offices indicated that the cart way was most probably retained in the first instance, which in turn suggested that a cart way arch must have been incorporated in the southern façade of the office extension so that road vehicles could continue to be admitted to the shed (see Section 4.4). No physical evidence relating to such an opening survives in the southern façade of the Eastern Goods Offices since the latter was largely destroyed during an air raid in November 1940 (Fig. 4.10). An extra storey was added to the Eastern Goods Offices in 1867, most of which was destroyed in 1940. Slivers of masonry dating to the period did survive in the eastern and southern elevations (Fig. 4.10; Fig. 4.11).

During the summer of 1871, the southern end of Platform 2 in the Eastern Transit Shed was extended across the southern end of the Shed's cart way. It is therefore likely that the cart way exits in the southern façades of the Eastern Transit Shed and the Eastern Goods Offices fell out of use and were infilled at this time (Fig. 4.9). This meant that the space that had been taken up by the defunct section of the cart way that was situated below the offices could henceforth be used for additional office accommodation (see also Section 4.4).

In 1873, all three storeys of the Eastern Goods Offices were extended to the west, the location of the former railway siding being accommodated below an arch that was integral to the newly built section of the building (Fig. 4.9; Fig. 4.10; see also Section 4.4). Although significantly altered in modern times, this opening was still extant at the time of the building recording survey (Fig. 4.10).

The original doorway to the Eastern Goods Offices seems to have been replaced at this time by two new entrances (Fig. 4.9; Fig. 4.10). One of these was situated

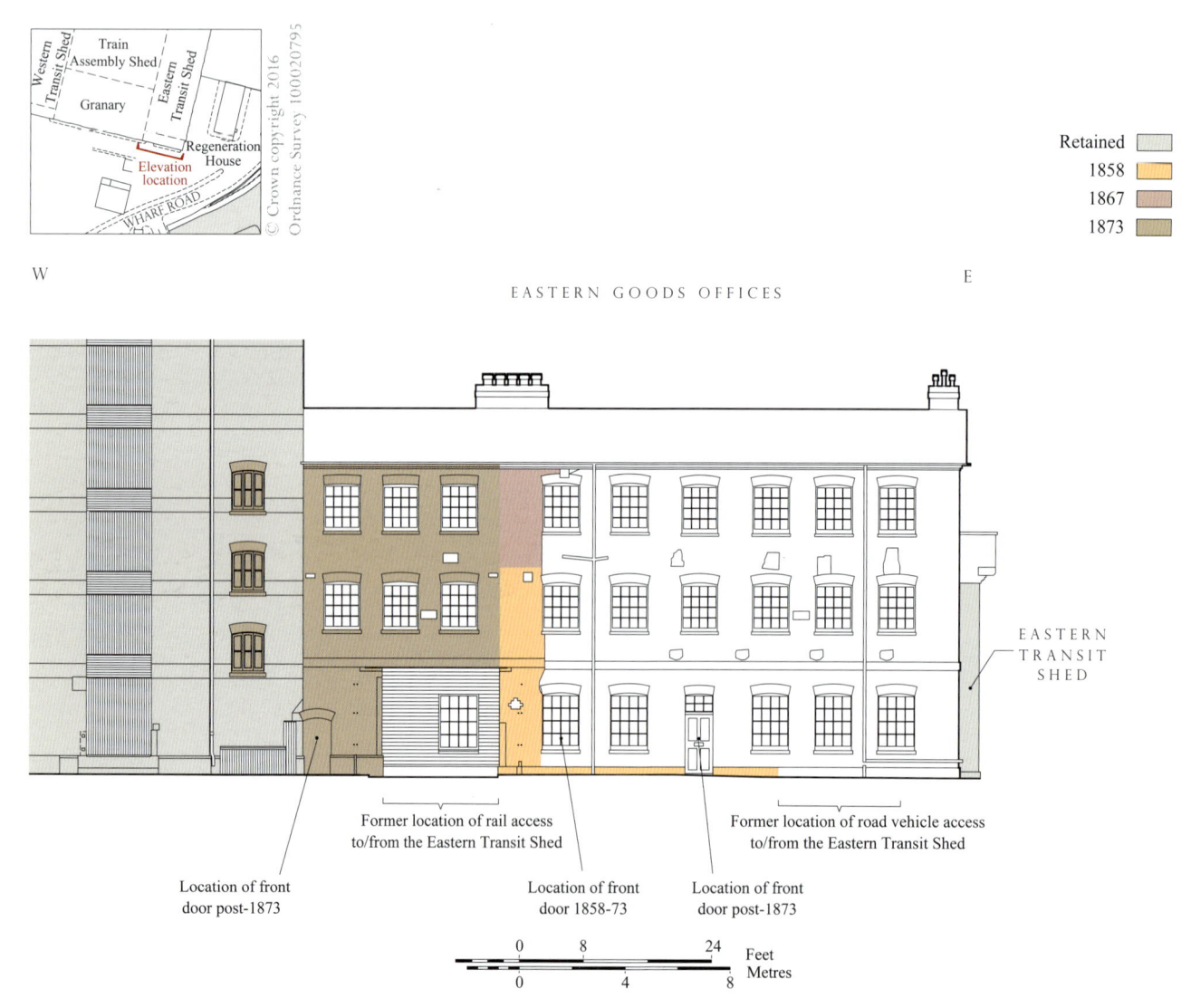

Fig. 4.10 External elevation of the south wall of the Eastern Goods Offices (Landscape Feature 21) showing the changes that took place from 1858–73, scale 1:250

to the centre-east of the southern elevation of the building, leading into the Goods Offices themselves. The other was situated to the far west of the same elevation, being partially cut into the earlier fabric of the Granary (Fig. 4.10). This enabled pedestrian access to and from the southeast stairwell of the Granary, which in turn provided access both to it and all three storeys of the Eastern Goods Offices.

The windows in the southern elevation of the Eastern Goods Offices were not original, having been replaced following the air raid of November 1940 (Fig. 4.10). Six were present on the ground floor of the southern elevation, which were taller than the eighteen that characterised the upper levels (Fig. 4.10).

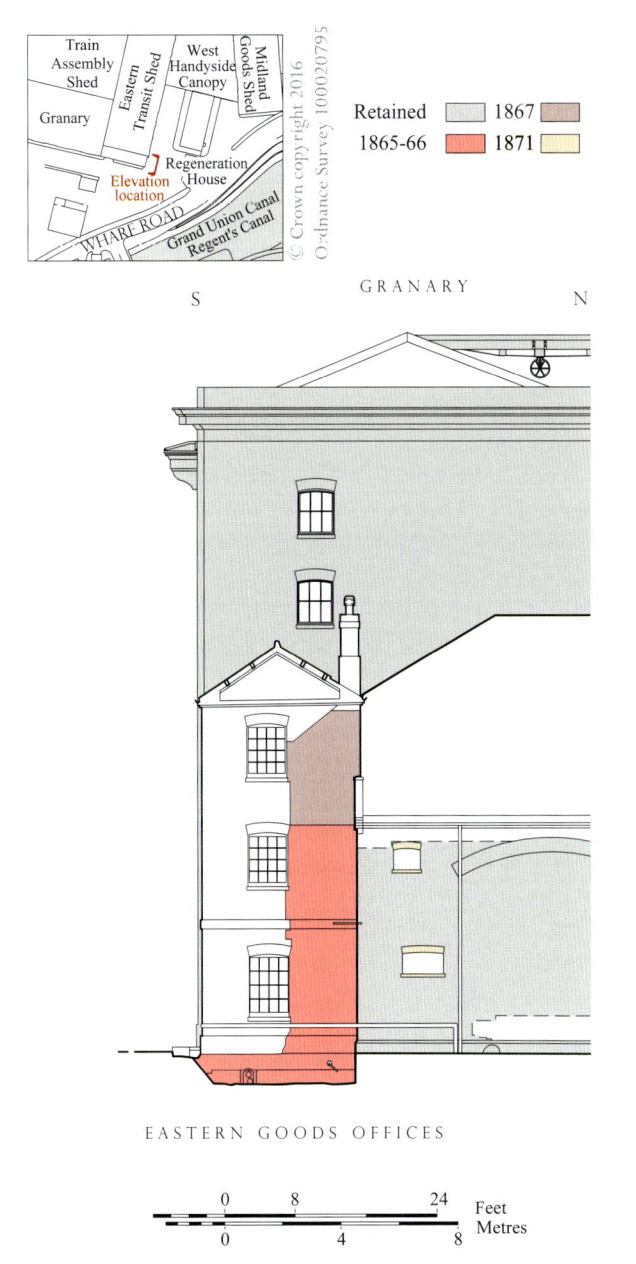

Fig. 4.11 External elevation of the east wall of the Eastern Goods Offices (Landscape Feature 21) and southern end of Eastern Transsit Shed showing the changes that took place from 1865–71, scale 1:250

In the eastern elevation, a further three windows were observed, one per floor. Comparison with cartographic sources suggests that they were reinserted in their post-1873 locations in the wake of the bombing.

Much of the masonry inside the Eastern Goods Offices had been rebuilt but nevertheless the internal layout of the structure as it was by 1873 has been reconstructed from the surviving remnants (Fig. 4.9). Each floor appears to have housed three separate offices by this time, which were connected by two staircases (including the Granary stairwell) and associated corridors.

4.4 The Western and Eastern Transit Sheds: the mid-1850s to the 1870s

Landscape Features 7 & 8

By the middle of the 1850s it had become apparent that additional loading and storage facilities for outward and inward-bound goods were required in the Transit Sheds. At the end of March 1855, Clarke recommended that additional accommodation be provided for 'Bale Goods from Scotland' in the Eastern Transit Shed by building a new platform at the north end of the Cart Road that would be fenced off from the rest of the goods handling facilities (TNA RAIL 236/20: 228; Fig. 4.4). Also in the spring of that year, Joseph Cubitt drew up a plan for 'extending the lines north of the Goods Shed to enable loaded wagons to be passed out of the loading shed [i.e. the Western Transit Shed] and formed into trains' (TNA RAIL 236/24: 78). Just over eighteen months later the Board approved Clarke's proposal to 'floor over' the southern end of the 'Up Goods Warehouse [i.e. the Eastern Transit Shed] next the office at King's Cross' for a cost of £69 (TNA RAIL 236/22: 332). This presumably created a modest amount of additional floor space for the handling of incoming commodities, however it cannot have taken the form of an elevated bank since this would have obstructed the southern end of the shed's cart way, which appears to have remained in use until 1871 (see also Section 4.3).

In August 1857, Seymour Clarke submitted a proposal to the Board concerning alterations to be made to the Eastern Transit Shed 'platform at the entrance to the stables beneath it…so as to increase the area of platform near two of the cranes' (TNA RAIL 236/24: 78). The stable entrance would also be modified at the same time. Clarke estimated that the work could be completed for an outlay of £42 for which approval was granted on the 18th of that month. Minutes taken during a Board meeting early the following January indicated that the necessary alterations were made, although their precise nature was not described. The same minutes also detail Clarke's subsequent application to make similar changes to 'the platform and approach to the stables in the Down Goods Shed [i.e. the Western Transit Shed]'.

Towards the end of October 1870 Seymour Clarke recommended that the north end of the Western Transit Shed be altered 'in order to facilitate the transfer of Traffic for the north coming from the City' for an estimated cost of £800 (TNA RAIL 236/171: 134). The proposal was approved by both the Traffic and Way and Works Committees and was recommended to the Board for approval. A further sum of £41.0.9 was spent upon unspecified 'Alterations in [the] Down Goods Shed [i.e. the Western Transit Shed]' at some point before 20th May 1871 (TNA RAIL 236/306/19: 18/07/1871). The Ordnance Survey map of 1871 and a plan of the Great Northern Goods Station of 1882 suggest that two tracks and additional platform space were installed in the north end of the shed at some point between these two dates, so it is possible that these minutes refer to their creation (Fig. 4.13). The provision of the extra tracks would have accelerated the speed at which outward-bound trains could be dispatched and thus the rate at which the platforms could be cleared.

By the turn of the 1870s, complaints about delays caused by congestion in the vicinity of the Eastern Transit Shed were becoming increasingly common, prompting the Board to invite contractors to tender for the construction of an 'additional storey' at the southern end of the structure in the summer of 1871 (TNA RAIL 236/171: 212). Although contemporary Board and Traffic Committee meeting minutes shed little light upon the nature of the proposal, a contract for the works worth nearly £6,000 was awarded to Mr. J. Kirk at the end of September (*ibid.*). Comparison between the 1871 Ordnance Survey map (Fig. 4.5) and Edwin Crockett's plan of 1873 (Fig. 4.6) suggests that Kirk's contract was associated with the creation of additional platform space at the southern end of the Eastern Transit Shed, where the existing platform was extended over the southern end of the cart way. This presumably commenced either at the end of 1871 or early in 1872 at the latest. Crocket's plan also suggested that the enlarged platform area necessitated the loss of three turntables that had previously facilitated the transfer of wagons from the platform to the transverse tracks on the ground floor of the Granary (Fig. 4.6).

The importance of maintaining a constant flow of goods from cart to train was clearly illustrated by an incident that took place at the beginning of April 1879, when a fire broke out on the roof of the Western Transit Shed as a result of the inattentiveness of a plumber who had been working there (TNA RAIL 236/146: 365). Though it was quickly extinguished by the clerks of the Goods Department, the outcome could easily have been much worse. A large quantity of gunpowder had been stored in the shed at the time, a practice that Henry Oakley, Seymour Clarke's successor as General Manager of the Great Northern, instructed should be halted immediately (*ibid.*).

Archaeological evidence relating to the modification of the Western Transit Shed: 1858–*c.*1882

Landscape Feature 7

The Western Goods Offices were erected against the northern end of the Western Transit Shed at the instigation of Seymour Clarke in 1858. It was at this time that two doorways were cut into the southern face of the Western Transit Shed at ground and first floor levels, thus providing direct internal access between the shed and the newly constructed offices to the east (see Section 4.3).

At some point after the Western Goods Offices were erected in 1858, an outbuilding or a canopy with a sloping roof was built against the southern face of the Western Transit Shed and the western side of the aforementioned offices. The former presence of the structure was discerned from a roof scar that ran horizontally across the cart way entrance to the Western Transit Shed before sloping diagonally down the side of the offices (Fig. 4.12). It may have provided shelter to road vehicles waiting to access the outbound shed from the south; they would have been fully loaded with goods intended for onward transport by rail, which may explain why they were provided with this protection.

When the Western Goods Offices were enlarged eastwards in 1866 it was decided that the extension should not compromise the Western Transit Shed's rail siding, which was to be retained. In order to achieve this, the upper storeys of the extension were built over the top of the railway, whilst a rectangular opening in the southern wall of the offices accommodated the line at ground floor level.

The undertaking impacted upon the fabric of the Western Transit Shed. For structural reasons, it necessitated that the original arched opening in the southern elevation of the building be narrowed and reinforced with a riveted compound steel girder so that it could support the weight of the upper storeys that had been built against it. This resulted in a straight joint in the brickwork towards the western side of the original opening and the insertion of a smaller arch above. The original timber sliding door frame then appears to have been reinstated on the internal side of the reconstructed opening.

Cartographic evidence demonstrated that two new railway sidings were inserted in the northern section of the Western Transit Shed between 1870 and 1882, most probably in October 1870 (Fig. 4.5; Fig. 4.13). This led to the truncation of the northern end of the shed's cart way and the removal of the loading platform that abutted its northern face. It also rendered the northernmost arched cart way opening in the western façade and the western cart way opening in the northern façade inaccessible to road vehicles, which must have subsequently been diverted through more southerly access points. A narrow,

Fig. 4.12   A diagonal scar was just visible running down the west face of the Western Goods Offices (Landscape Feature 20) caused by a structure that was built against it and the southern face of the Western Transit Shed (Landscape Feature 7) at some point after 1858, photograph faces northeast

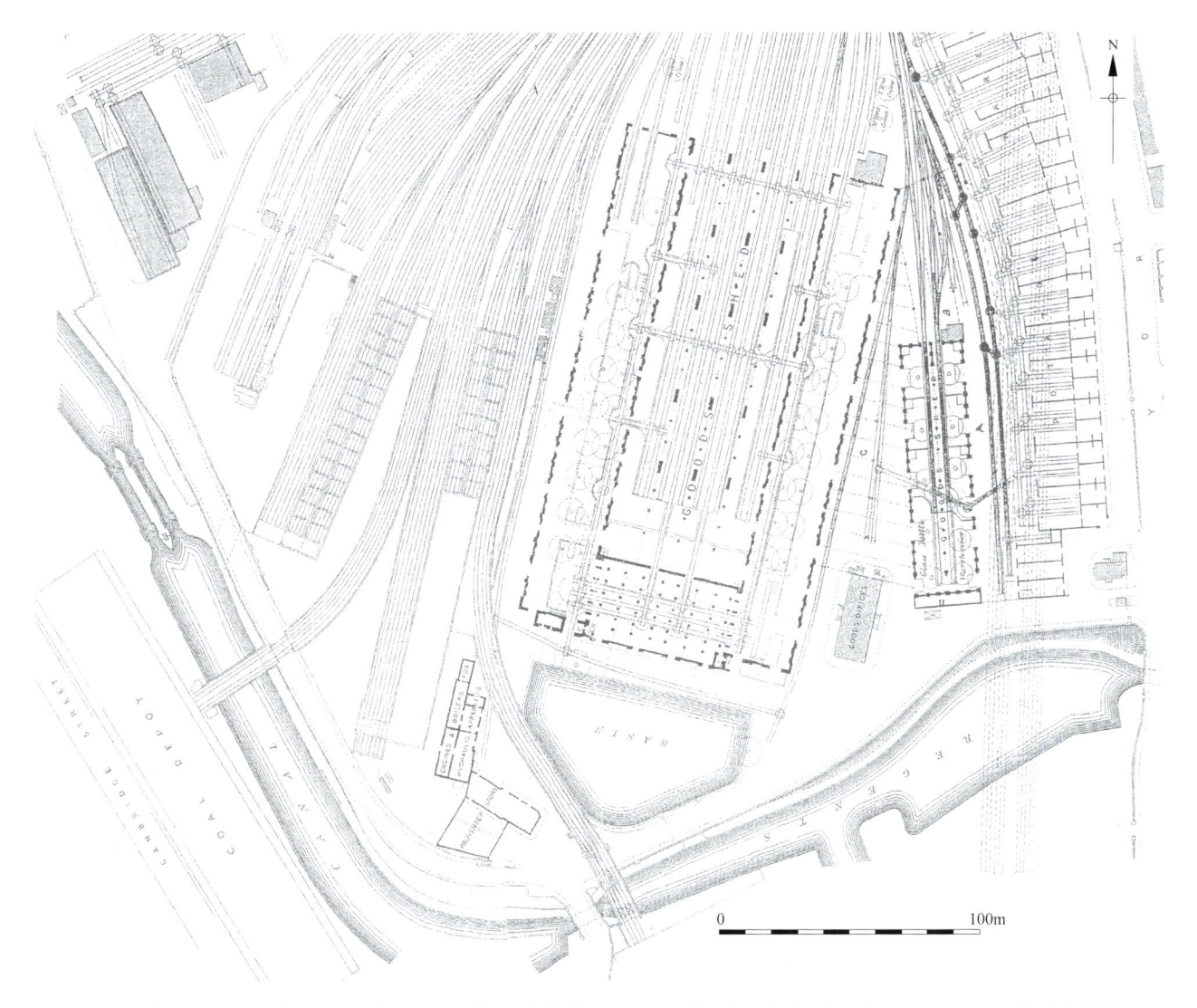

Fig. 4.13   The Great Northern Goods Station Plan of 1882 depicting the Goods Yard and its environs, scale 1:2,500

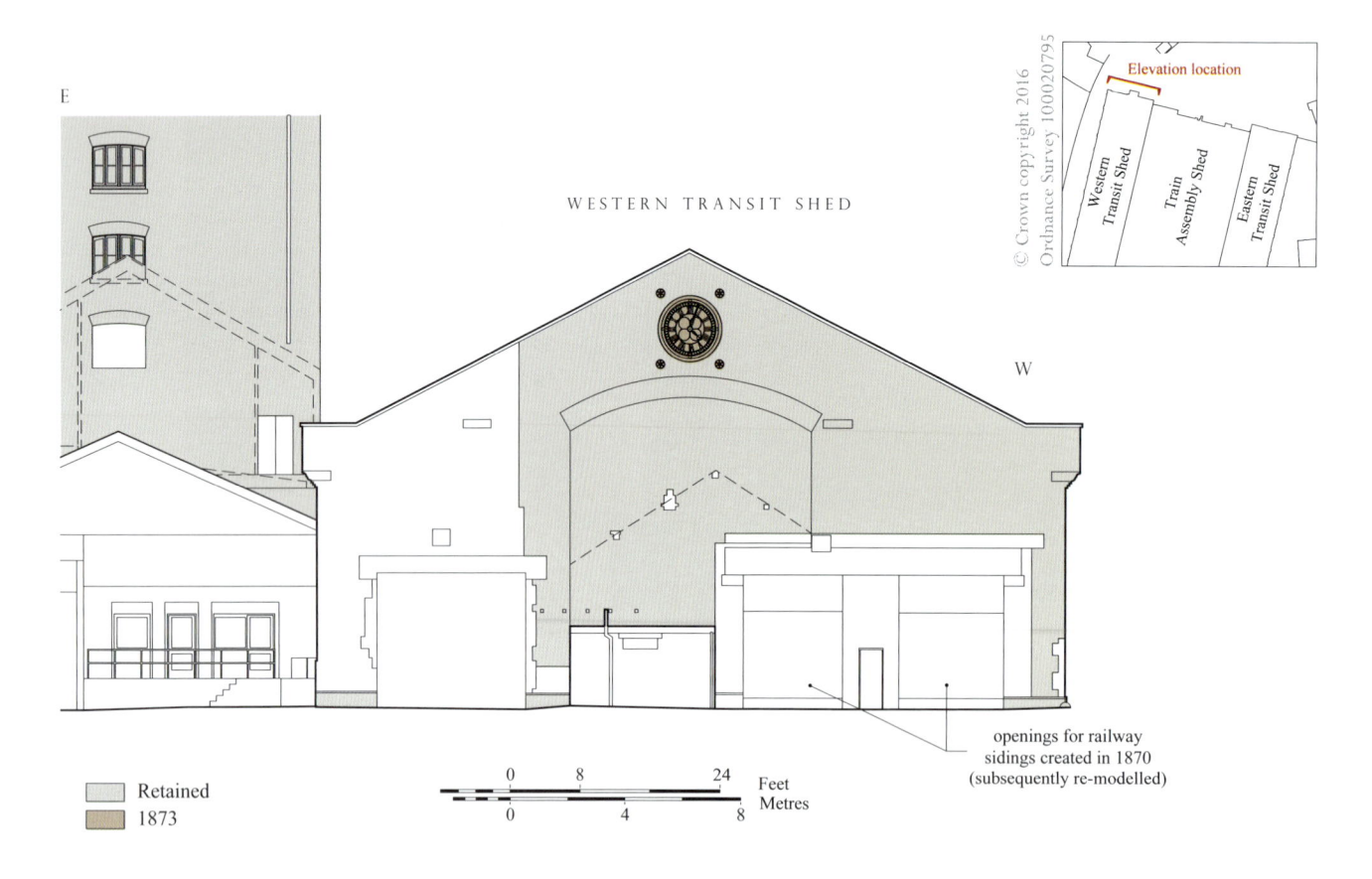

Fig. 4.14 External elevation of the north wall of the Western Transit Shed (Landscape Feature 20) showing the entrances that accommodated the two new sidings that were probably originally created in October 1870 (later remodelled) and the clock that was inserted in 1873, scale 1:250

Fig. 4.15 The clock mechanism inside its housing in the north wall of the Western Transit Shed (Landscape Feature 7), photograph faces north

north–south aligned platform extension was constructed between the two sidings, which would have allowed wagons that were stationed on them to be loaded. The 1882 map also suggests that an entirely new opening was inserted in the north wall of the Western Transit Shed to accommodate one of the new tracks, whilst the pre-existing cart way arch was used to admit the other (Fig. 4.13). Field observations undertaken as part of an historic building recording survey (KXF07) tell a slightly different story, suggesting that the original cart way arch was in fact widened and perhaps partitioned into two smaller entrances, each of which housed one of the two 'new' sidings (these were remodelled at a much later time and little of the original fabric survives externally; Fig. 4.14). Whilst no trace of the westernmost of these sidings survived below ground level, part of the eastern example was exposed during an archaeological watching brief.

It is probable that these modifications resulted from the General Manager's request of October 1870 involving the construction of new sidings in the northern end of the shed (TNA RAIL 236/171: 134). It is therefore probable that building work began soon after his proposal was approved by the Board, late in 1870 or early in 1871. Documentary evidence indicates that the works were carried out in order to 'facilitate the transfer of Traffic for the north coming from the City' or, in other words, to relieve congestion by creating additional room for the loading of wagons in the outbound goods warehouse.

An iron clock face, complete with its mechanism (Fig. 4.15) was observed in the outer face of the north wall of the Western Transit Shed, centrally placed above the arched recess (Fig. 4.14). The surrounding brickwork strongly suggested that it was not integral to the original fabric of the structure, a fact that was confirmed by the 1873 date of manufacture that was stamped on the back. Great Northern Railway company records indicate that the timepiece was purchased at some point after April 1872 from a Mr John Walker of Cornhill for £68 (TNA RAIL 236/144: 303). The clock bore the inscription: 'JOHN WALKER 68 Cornhill & 230 Regent St. London'. It is probable that Walker built the clock to order, hence the disparity of one year between the date of purchase and the date that was stamped on the instrument itself.

Cartographic sources suggested that two additional pedestrian access points were cut into the eastern wall of the Western Transit Shed at some point between the publication of the Humber Plan of 1866 and a plan prepared by Great Northern Railway in 1882 (Fig. 3.8; Fig. 4.13). Although the insertion of the doorways was not documented by the GNR, their presence was confirmed by a built heritage survey (KXF07; Fig. 4.16). The nature of the bricks that had been used to construct the openings strongly suggested a 19th-century insertion date.

Archaeological evidence relating to the modification of the Eastern Transit Shed: the mid-1850s to 1872

Landscape Feature 8

Like its western counterpart, the creation of the Eastern Goods Offices (Landscape Feature 21) in 1858 did not significantly affect the Eastern Transit Shed in the first instance. The shed's rail siding remained unaltered as indicated by the Humber Plan of 1866, as did its cart way (Fig. 4.4). The Eastern Goods Offices were neatly positioned between these thoroughfares against the south wall of the Eastern Transit Shed. Only one significant alteration is worthy of note, which involved the insertion of two doorways and a staircase through and against the internal face of the shed's southern wall (see also Section 4.3).

It was not until the Eastern Goods Offices were extended to the east in 1866 that more significant changes were made to the Eastern Transit Shed's southern wall. Cartographic evidence suggested that the cart road exit in the eastern side of that elevation was blocked and rendered redundant by the construction of the new office extension (Fig. 4.5) however the built heritage survey suggested a more complex version of events. It should be noted that physical evidence for this alternative story has only survived into modern times in the southern wall of the Eastern Transit Shed (i.e. the north wall of the Eastern Goods Offices) as the original southern façade of the offices had to be rebuilt after bombing during the Second World War.

The way that the cart way arch in the southern elevation of the Eastern Transit Shed was infilled contradicted the cartographic evidence. It strongly suggested that the opening was retained initially, which in turn suggested that vehicles continued to use the Eastern Transit Shed's cart way after the Eastern Goods Offices were extended over it. The logic for this is as follows.

A straight joint observed just below the springing of the original arch strongly suggests that a girder (since removed) was inserted so that the cart way entrance could support the weight of the storeys that were constructed over it (Fig. 4.17). Similar engineering solutions were used elsewhere in the yard, including the arch that accommodated the Eastern Transit Shed's railway siding in the same elevation (this had to be reinforced when the Eastern Goods Offices were extended again in 1873 as shown in Fig. 4.17). After the cart way fell out of use, a few bricks were removed from above the girder so that the latter could be prised out for reuse or recycling. The rest of the arch was then infilled with brickwork of a different colour (Fig. 4.17). Exactly when this occurred is uncertain, but given the nature of the mid to late-19th century bricks that were used to infill the bulk of the former opening a likely scenario is that it took place when the Eastern Transit Shed's platform was extended in 1871. These works, which were approved by the Board in the summer of that year, involved the construction of additional platform space at ground floor level and the construction of an 'additional storey' at the southern end of the Eastern Transit Shed (TNA RAIL 236/171: 212). The platform extension in question is shown for the first time on Crockett's plan of 1873 (Fig. 4.6); it clearly obstructs the southern end of the cart way, thus rendering the associated southern road exits redundant.

The works that accompanied the platform extension of 1871 resulted in a number of other alterations to the Eastern Transit Shed. Two new windows (since blocked) were inserted in the southern end of the eastern elevation of the building. Field observations demonstrated that they were more recent than the rest of the edifice as they were surrounded by lighter brickwork that was bonded with different mortar (Fig. 4.11). One window was found at ground level and the second was built at what would have been first floor level. Consequently, it seems likely that they accompanied the construction of the 'additional storey' at the southern end of the Eastern Transit Shed (ibid.). They cannot pre-date that new floor, otherwise the upper window would have had nothing to illuminate. A row of joist holes arranged at first floor level was noted between the two windows on the internal side of the wall, which almost certainly housed the supports for the new floor. The upper storey may have functioned as an office block or as an additional storage area for incoming goods.

Cartographic sources suggest that the Eastern Transit Shed's railway siding had been cut short by 1871

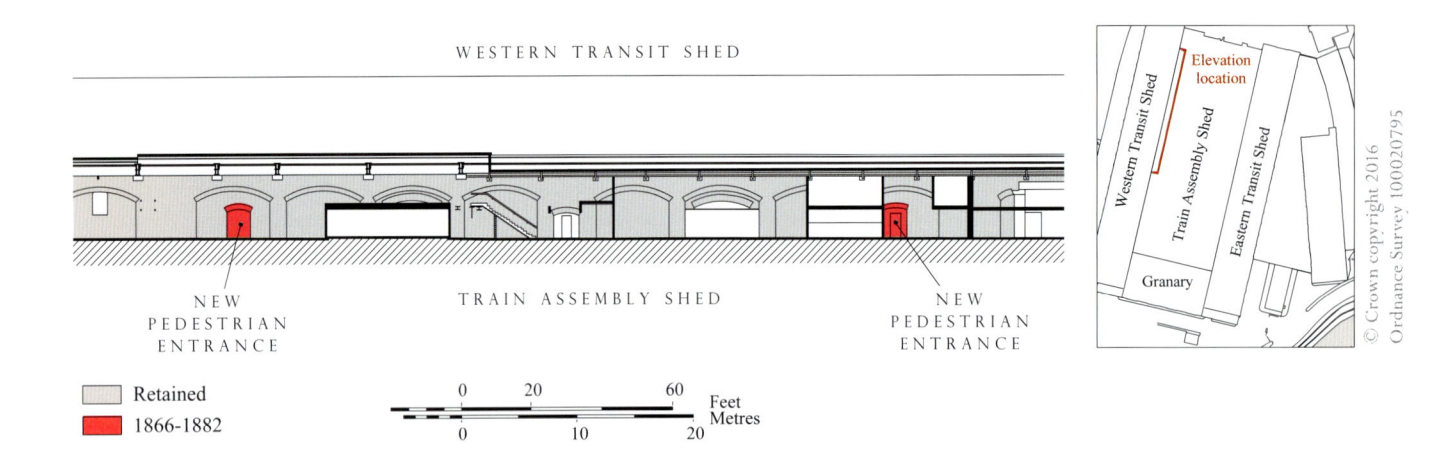

112 *Competition and Congestion at the King's Cross Goods Station: c.1853–1879*

Fig. 4.16   Cross-sectional elevation through the Train Assembly Shed (Landscape Feature 11) showing the external face of the east wall of the Western Transit Shed (Landscape Feature 7) and the doorways that were inserted in 1866–82, scale 1:625

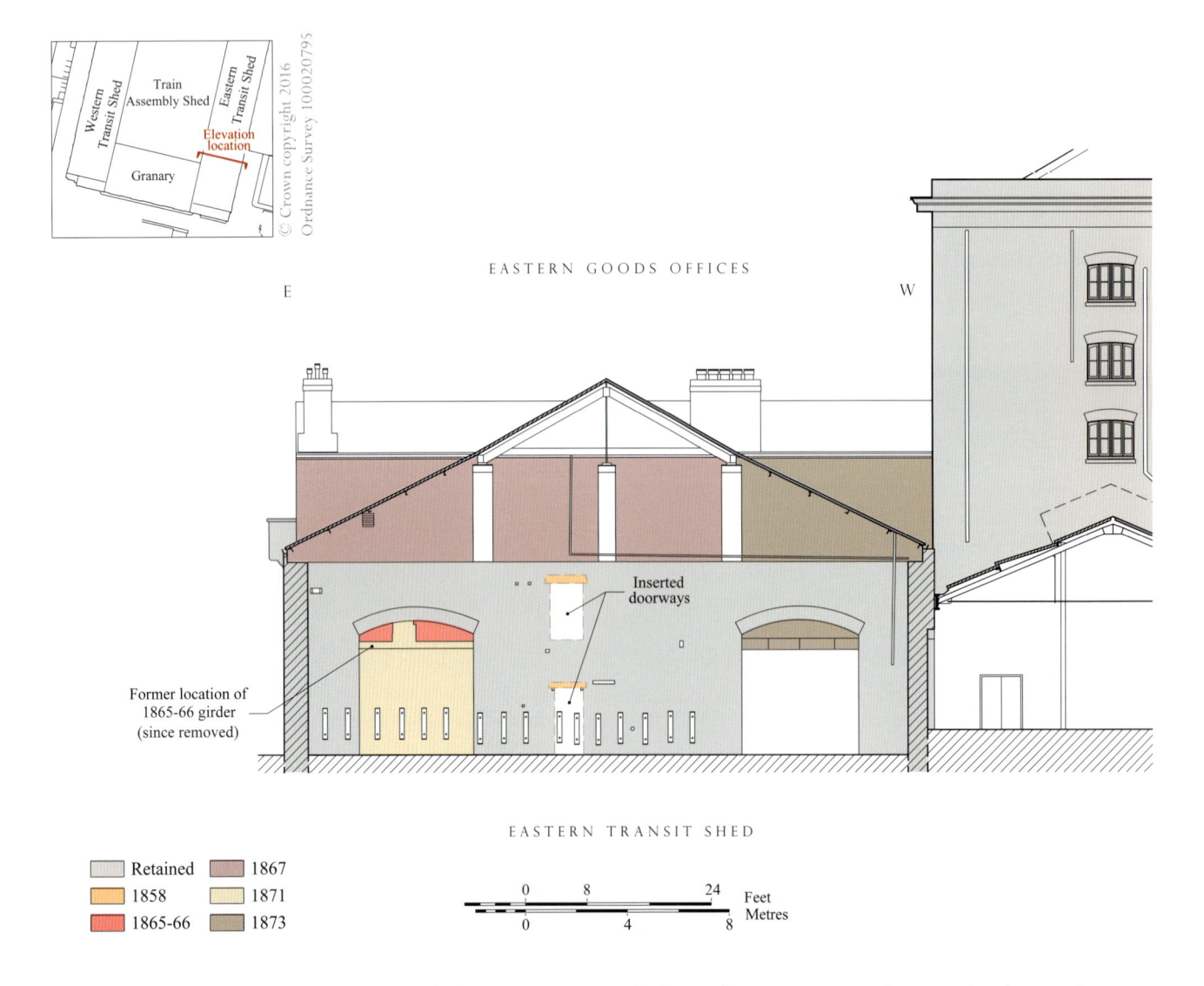

Fig. 4.17   Cross-sectional elevation through the Eastern Transit Shed (Landscape Feature 8) showing the changes that took place from 1858–73; note the modified exit for the railway siding and the infilled former cart way exit that were altered when the Eastern Goods Offices (Landscape Feature 21) were extended, scale 1:250

(Fig. 4.5). By then, it no longer ran up to the edge of the Granary Basin (Landscape Feature 5) terminating instead inside the shed at the most southerly wagon turntable. This would have had significant implications for the flow of traffic through the complex, prohibiting the transit of empty wagons from the Granary via the two outer tracks and eliminating the direct connection that previously existed between the Eastern Transit Shed and the Granary Basin's railway infrastructure.

A metal framed, multi-paned window was inserted in the east side of the northern elevation of the Eastern Transit Shed some time after the structure was completed (Fig. 4.18). The brickwork surrounding it had clearly been removed and rebuilt at a later date, demonstrating that it was a later addition. It is reasonable to assume that the window was inserted when the former 'Lamp Room and Coffee Shop' (Landscape Feature 13) was extended in 1872 (this once abutted the north side of the shed as discussed in Section 4.5). This seems likely as a number of other alterations were made to the northern section of the Eastern Transit Shed at this time. Joist holes on the internal side of the shed's north wall demonstrated that an earlier gantry that ran across that section of the structure had been removed and replaced with a new floor at a higher level. Goods were presumably winched to and from this raised area by way of a system of pulleys. The extant remains of two such mechanisms were observed on the internal side of the north wall (Fig. 4.17), whilst a pulley wheel survived on external side of the same facade (Fig. 4.18). The doorway that once granted access to the original gantry from the staircase that ran up the north wall of the shed fell out of use when the floor was replaced. This was made obvious by the fact that a joist hole for the replacement floor clearly penetrated the brick infill that was found within the former threshold. Furthermore, the height of the newly raised floor relative to the position of the door meant that retaining the latter feature would not have been viable. The new floor was also inserted below the level of the original timber beam supports for the sliding doors that once closed off the northern road and rail entrances to the Eastern Transit Shed. These shutters must therefore have fallen out of use beforehand, although the entrances that they once enclosed remained active. A new chimney seems to have been constructed at the same time as the replacement floor and it was the presence of this feature that conclusively dated this episode of building. It was shared with four fireplaces on the opposite side of the

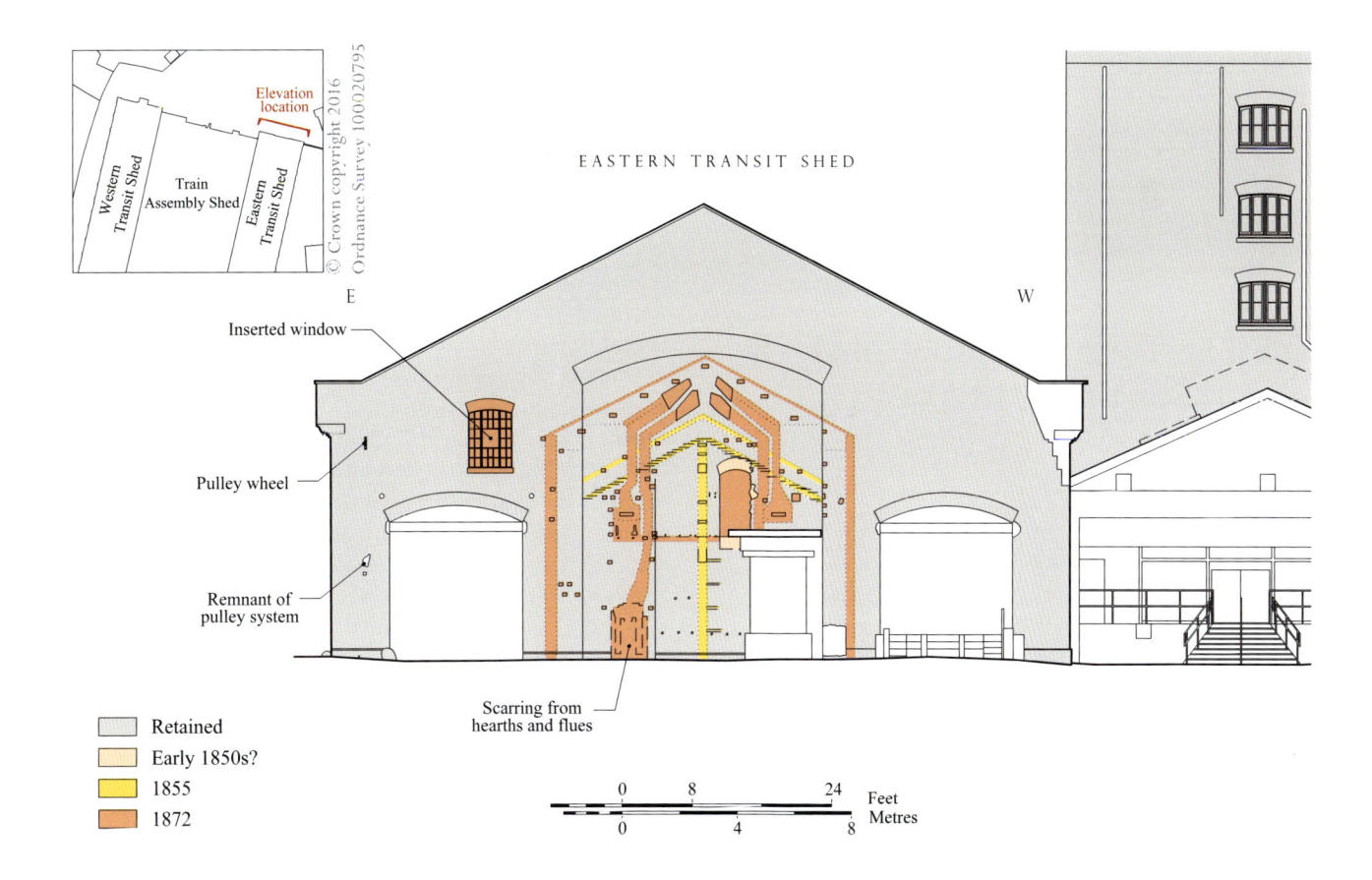

Fig. 4.18 External elevation of the north wall of the Eastern Transit Shed (Landscape Feature 8) showing the changes that took place from c.1853–72; note the scars of the former buildings that once abutted it, scale 1:250

wall that formed part of the extension to the former 'Lamp Room and Coffee Shop', the scars of which were clearly visible externally (Fig. 4.18). Documentary sources make clear that this extension was erected in 1872, strongly suggesting that the aforementioned modifications to the Eastern Transit Shed were also made at that time.

Another modification was made to the southern wall of the Eastern Transit Shed in 1873, when the Eastern Goods Offices were once again enlarged, this time to the west over the area that had until recently been occupied by the Shed's rail siding (Fig. 4.6). This meant that the arch in the south wall of the Eastern Transit Shed that formerly admitted the siding had to be altered so that it was capable of supporting the weight of the overhead office extension (Fig. 4.17). Like its counterpart in the Western Transit Shed, a compound, riveted iron girder was inserted above the springing of its arch and the space above was infilled with brickwork (Fig. 4.17). Exactly why this area needed to be spanned in this way is somewhat baffling as cartographic and documentary sources suggest that the portion of the railway siding that it formerly accommodated had actually fallen out of use and been removed at least two years before the Western Goods Offices were extended over it (most probably in 1871). One possibility is that the arch enabled road vehicle access to the southern tip of the Eastern Transit Shed's platform extension of 1871 after the southern cart way access points were bricked up, although constant use of such an arrangement would surely have generated a bottleneck. Perhaps the most plausible explanation is that the Goods Station's administrators wanted to retain the opening so that they could reinstate the Shed's railway siding as far as the Granary Basin, should circumstances require this in the future.

4.5 The Train Assembly Shed: the mid-1850s to the 1870s

Landscape Feature 11

The Great Northern took a number of steps to resolve rail traffic congestion in the Goods Yard between the mid-1850s and the late 1870s. In addition to the measures discussed above, the company also made a number of alterations to the layout of tracks and turntables in the Train Assembly Shed. In the spring of 1855 Joseph Cubitt devised a plan for 'extending the lines north of the Goods Shed to enable loaded wagons to be passed out of the loading shed and formed into trains' (TNA RAIL 236/20: 228). The following February the Board of the Great Northern approved a recommendation made by Seymour Clarke that an additional turntable be placed 'in the Middle Shed [i.e. Train Assembly Shed] on No. 4 Up Line', at a cost of £140 (TNA RAIL 236/21: 286). Given that the Eastern Transit Shed served the 'up' traffic, it is probable that 'No. 4 Up Line' was the fourth track to the west of the eastern wall of the Train Assembly Shed. The Humber Plan of 1866 showed an isolated turntable on this track at the western extent of the northernmost battery of these features (Fig. 3.8), and it seems probable that this represents the turntable that was added in 1856 (Turntable C, Figure 4.3).

The Ordnance Survey map of 1871 suggests that another seven turntables had been added to the northernmost east–west battery by that date (Fig. 4.5). This briefly brought the total number of turntables inside the Goods Shed and Granary to 70, however that number was soon reduced to 67 (see page 110).

Despite the improvements that were made to the Goods Station between the 1850s and the early 1870s, continuing congestion prompted Henry Oakley to propose additional measures to tackle the 'want of further accommodation for the Goods Traffic' at King's Cross to the Traffic Committee in September 1872 (TNA RAIL 236/144: 357). Although Oakley's report was subsequently passed to the Board for further consideration, neither the minutes of meetings held by that body nor those of the Traffic Committee give much indication of its content. However, it is likely that Oakley sought to secure further space to accommodate the increasing volume of goods that was arriving from the north by train. Since the southern end of the Eastern Transit Shed's platform had already been extended, there were few opportunities to create additional space within the arrivals shed itself. One possible solution might have been to create a new platform in the Train Assembly Shed by truncating one or more of the railway tracks nearest the Arrivals Goods Platform in the Eastern Transit Shed. Edwin Crockett's plan of 1873 showed that by the following April an open area had indeed been created in precisely that location (Fig. 4.6; termed 'Platform 5A' on Fig. 4.3). Three lines of track in the easternmost bay of the shed had been truncated to make way for the new storage space. This meant that whilst the Assembly Shed still primarily dealt with the construction and deconstruction of train wagons, the southeastern section of it was now devoted to the unloading of incoming goods. This significant structural and operational modification to the core of the yard is perhaps the clearest indication that Cubitt's original Goods Station design had become obsolescent less than twenty years after it was built.

4.6 The Granary: the mid-1850s to the 1870s

Landscape Feature 12

In the two decades after the King's Cross Goods Station opened for business, grain harvested in the wheat fields of Lincolnshire represented a significant proportion of the agricultural produce brought by rail into the metropolis. Having cornered the lion's share of this lucrative trade, the Great Northern came to realise that Lewis Cubitt's original design of the Granary required further adaptation if it was to fulfill its primary function efficiently during the decades that followed (TNA RAIL 236/30: 209; TNA RAIL 236/31: 263). As occurred elsewhere in the yard, the growth in traffic also necessitated the recruitment of additional clerical personnel, who in turn required new offices in which to work (TNA RAIL 236/20: 77).

In October 1854 Seymour Clarke recommended that a new 'Grain Office' be built adjoining the Granary (*ibid.*). The location and extent of this structure are unknown, although the amount budgeted for its construction (£116) suggests that it is unlikely to have been substantial, being housed inside an existing Goods Station building or perhaps in the Granary itself.

At the end of November 1862 the Board of the Great Northern authorised expenditure of £268 in order that 'certain alterations… for improving the means of delivering grain' from the Granary might be made (TNA RAIL 236/30: 209). The necessary improvements were outlined in reports submitted to the Board by Messrs Agard (Goods Manager) and Johnson (Great Northern Engineer since June 1861). Although Board meeting minutes did not elaborate upon the nature of these improvements, it is likely that they involved the enlargement or improvement of the existing system of timber sack chutes and trapdoors that directed the grain sacks through the building for loading into waiting vehicles.

Towards the end of December 1863 Seymour Clarke warned the Board that the Goods Yard was becoming increasingly overcrowded with unprocessed goods traffic, with the result that large quantities of grain and potatoes were left standing in wagons throughout the yard (TNA RAIL 236/31: 263). The backlog was so severe that Clarke was forced to inform the merchants based at the yard that no more grain or potatoes were to be brought to London until the issue had been resolved. Noting also that the Granary was 'quite full', Clarke requested permission to build additional goods sidings for wagons carrying both types of produce along with the creation of further accommodation at the Potato Market (*ibid.*: 264). In the meantime, Clarke advised grain merchants that the length of time they were allowed to store grain free of charge at the Granary was to be halved from fourteen to seven days (*ibid.*: 280–1).

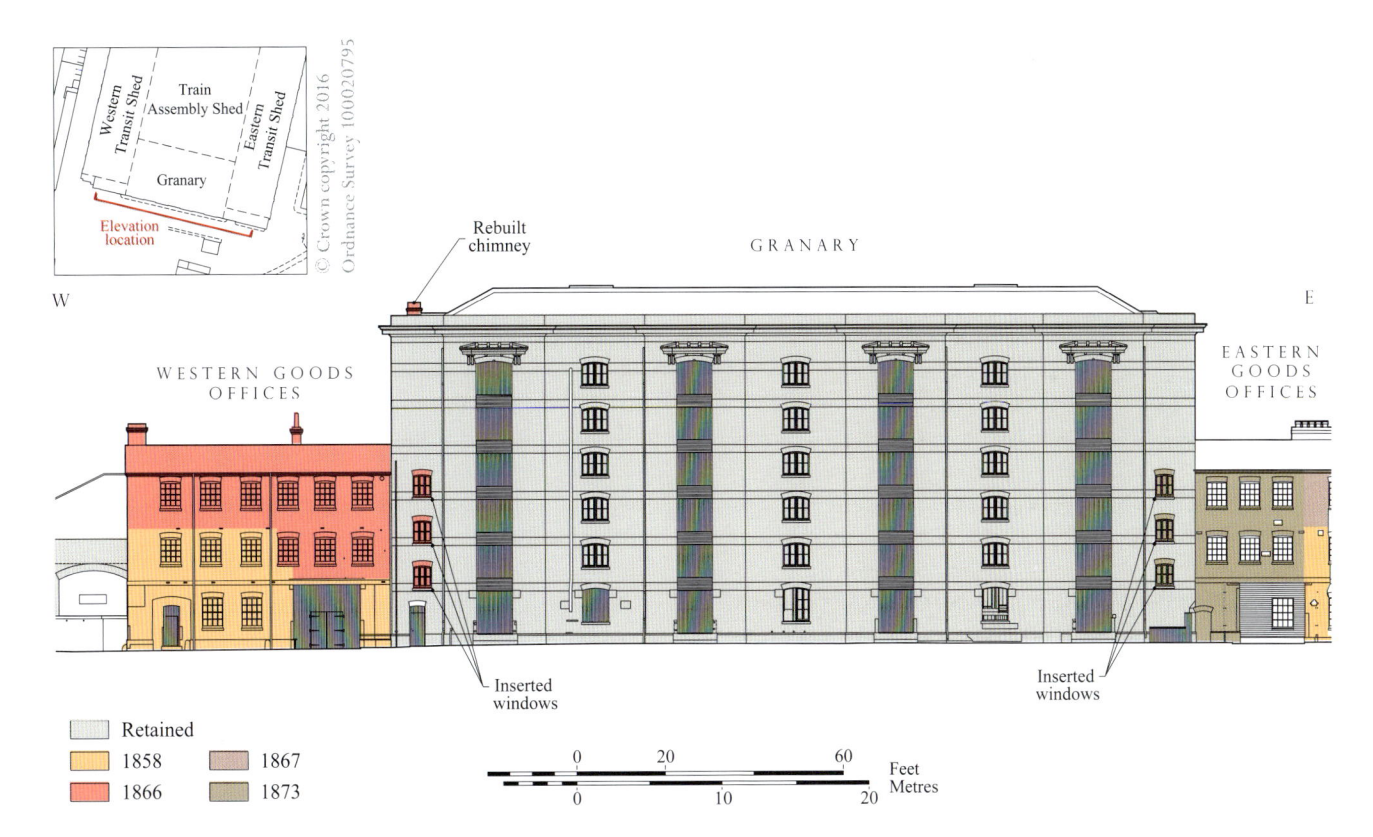

Fig. 4.19 External elevation of the south wall of the Granary (Landscape Feature 12) and the Goods Offices (Landscape Features 20 and 21) showing the changes that took place from 1858–73, scale 1:500

Built heritage evidence relating to the modification of the Granary: the mid-1850s to the 1870s

Landscape Feature 12

A network of four adjoining chutes and one trapdoor were arranged across the upper floors of the Granary so that they could deposit a descending sack at the far eastern end of Platform 4 (Fig. 3.52). Since their inclusion broke the regular pattern that the other chutes created they were probably inserted after the Granary became operational in order to account for actual working practices. The same logic suggests that a trapdoor, which provided a vertical link between the fourth floor and the top of one of the original chutes that descended from the third floor was also inserted at some point after operations commenced (Fig. 3.52). These additions may represent the modifications that were approved by the Board in November 1862.

The Western Goods Offices (Landscape Feature 20) were extended for the final time in 1866. Whilst the Granary's pedestrian entranceway in the building remained unimpeded (since it could still be accessed via the 'tunnel' that accommodated the Western Transit Shed's rail siding) the same could not be said for the windows that formerly illuminated the western stairwell. The windows at first and second floor level were converted into doorways that provided access between the Granary and the Western Goods Offices whilst the third floor window, which had been covered by the roof space of the extension, was bricked up entirely. To compensate for their loss, three new windows were inserted in the southern façade of the Granary (Fig. 4.19). The west wall was also partially rebuilt at this time in order to accommodate a new flue and stack. With the exception of that chimney, a similar set of alterations was made to the eastern side of the structure when the Eastern Goods Offices (Landscape Feature 21) were extended in 1873 (Fig. 4.19).

4.7 The Lamp Room and Coffee Shop (later the Refreshment Room): the mid-1850s to the 1870s

Landscape Feature 13

The Humber Plan of 1866 suggests that Landscape Feature 13 consisted of two north–south aligned bays by that time, the most westerly of which was labelled 'V' or 'Y' and the most easterly of which was sub-divided into two separate rooms both labelled 'G.R.' (Fig. 3.8). It is likely that reference was made to these rooms by Seymour Clarke in November 1855, when he submitted a proposal to the Board requesting that 'three rooms be constructed above the Lamp Room and Coffee Shop in the Goods Shed to provide accommodation for [guards] when they come in cold and wet and for the inspectors to keep their memoranda etc.' (TNA RAIL 236/21: 126). Clarke's request followed a spate of minor accidents in the Yard involving guards who travelled in to work at King's Cross from Peterborough. Although the matter does not seem to have been discussed further, it is likely that Clarke's request was granted, suggesting that the annotation 'G.R.' stands for 'Guard Room'. As indicated by cartographic sources (Fig. 3.8), these works do not appear to have altered the horizontal footprint of the structure.

At the beginning of March 1872, Henry Oakley recommended that 'a small room at the end of the Inwards Shed' should be converted 'for the accommodation of the Guards, Porters and others who will be provided with tea, coffee and be enabled to have their meals without leaving the premises' (TNA RAIL 236/171: 254). The proposal was approved by the Way and Works Committee on the grounds that the provision of an on site canteen would be likely to reduce the number of staff who went home to dine or, even worse, took their meals in the public houses of York Road. The resulting Refreshment Room was ready for use by 22nd April 1872, with victuals being provided by a Mr J.R. Cox, who paid £30 per annum to the Great Northern for the concession (TNA RAIL 236/144: 309). The Goods Station plan of 1882 indicates that the Lamp Room and Coffee Shop had been enlarged to the north by that time (Fig. 4.13) and it is probable that this extension housed the aforementioned canteen.

Archaeological evidence relating to the extension of the Lamp Room and Coffee Shop, later the Refreshment Club between 1855 and 1871

Landscape Feature 13

The north elevation of the Eastern Transit Shed (Landscape Feature 8) was dominated by a series of scars caused by successive incarnations of the welfare facilities (Landscape Feature 13) that once abutted it (Fig. 4.18). At least two, possibly three phases of scarring were visible in the wall, the earliest of which was characterised by a doorway (now blocked) at first floor level. The doorway may have been an original element of the facility, perhaps leading from the ground floor via a partially or fully external staircase to a small platform or gantry inside the northern end of the Eastern Transit Shed (Landscape Feature 8), the former presence of which was discerned from joist holes in the internal face of the wall. Sometime after the doorway was constructed, a pitched roof appears to have been added. The roof most probably post-dates the door as the marks that it left partially eclipsed the west side of the ornamental arch that surmounted the threshold (Fig. 4.18). This roof scar most probably relates to the vertical extension of the Lamp Room and Coffee Shop that was proposed by Seymour Clarke in 1855.

By the early 1880s the former Lamp Room and Coffee Shop (Landscape Feature 13) had undergone a further extension, this time to the north (Fig. 4.13). This was verified by an archaeological watching brief (KXI07), which unearthed the foundations of the extension. They were composed of four steps, constructed with red, English bonded bricks held together with indurated lime mortar, which sat on a poured concrete slab. The surviving remains of the extended structure demonstrated that the new footprint was roughly 'L' shaped in plan, in keeping with that shown on the Great Northern Railway Plan of 1882 (Fig. 4.13; Fig. 4.20). In all probability, the modifications relate to the 1872 extension that was carried out at the request of Henry Oakley.

The 1872 renovation appears to have been heated by hearths that left behind four sooty stains and scars indicating the courses of former flues and fireplaces that once abutted the north wall of the Eastern Transit Shed (Fig. 4.18). Two hearths abutted the elevation at ground floor level, along with another two at first floor level. Their flues rose vertically before converging in a chimney breast in the centre of the building, as observed on the internal side of the north wall of the Eastern Transit Shed. *In situ* traces of the original fabric of the eastern ground floor fireplace was identified in the form of two terracotta tiles stamped with the maker's marks 'O H C' in small decorative roundels. In all probability, similar fireplaces were situated within each room for the comfort of the staff. In the words of Seymour Clarke, guards frequently came in from work 'cold and wet' so this heating would have been important for their welfare, especially in bad weather.

The roof of the entire building appears to have been replaced and raised at this time, hence the presence of a second roof scar on the external north face of the Eastern Transit Shed (Landscape Feature 8). This scar indicates that the replacement roof was pitched at a similar angle to the original although at a higher level (Fig. 4.18).

Later sources suggest that a probable 'lean-to' incorporating a staircase with two lavatories below abutted the northeastern side of the building at ground floor level (see Section 6.5, Fig. 6.7). No trace of this was found during the archaeological excavation, suggesting that its insubstantial foundations were entirely destroyed. Nevertheless, its former existence was substantiated by the presence of two drains and some associated pipework that would have received the waste that came from the water closets (Fig. 4.20).

The extension to the former Lamp Room and Coffee Shop was most probably added soon after March 1872 at the request of Henry Oakley. It was almost certainly constructed to convert the building into the

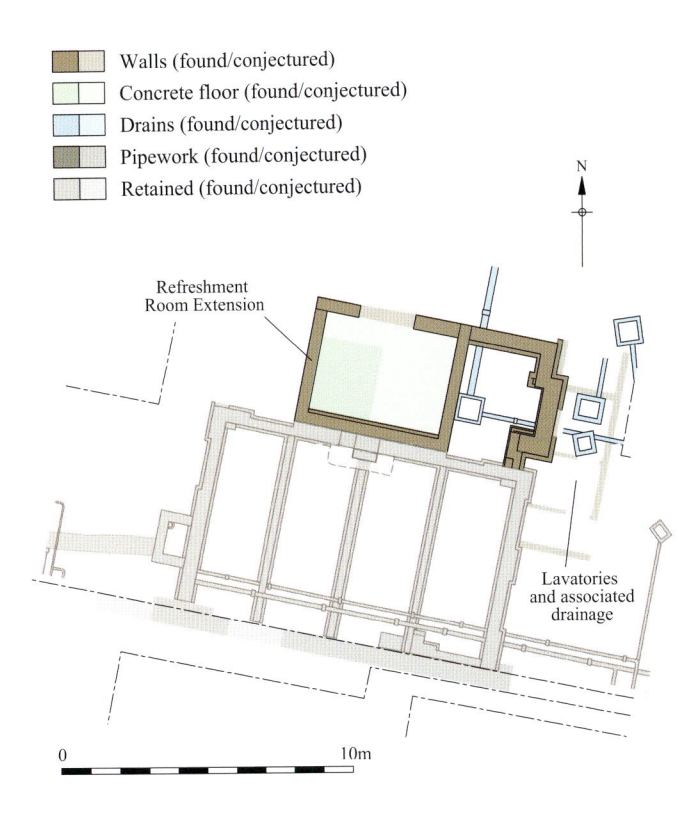

Fig. 4.20 Plan of the Refreshment Room by 1872 (Landscape Feature 13) reconstructed from archaeological remains and cartographic sources, scale 1:250

Refreshment Room, a welfare and canteen facility that was designed to increase the yard's efficiency by encouraging railway workers to take their meals on the premises thus deterring absenteeism and detrimental levels of alcohol consumption. The facilities appear to have been segregated according to an individual employee's rank within the company's workforce. While clerical staff took their meals in a relatively spacious 'dining' room, lower ranking staff ate in the 'mess rooms'. Despite the suggestion of hierarchical dining arrangements, the toilet facilities appear to have been shared by all parties, irrespective of status.

4.8 The Potato Market: the mid-1850s to the 1870s

Landscape Feature 14

From the outset the Great Northern Railway Company appreciated the commercial importance of securing the lucrative trade in potatoes in the capital. After the Goods Yard opened in 1850 it quickly began to attract a great deal of potato traffic at the expense of London's traditional potato trading centre at Tooley Street (Thorne *et al.* 1990: 101). Since King's Cross was far more conveniently sited for the onward transport of the commodity to the main vegetable market at Covent Garden, the amount of potato traffic entering the complex grew quickly until the volumes that were being handled

Fig. 4.21 Porters moving sacks of potatoes in the Potato Market (Landscape Feature 14) in 1951 (SSPL 1995-7233_LIVST_DF_175 © National Railway Museum and SSPL)

were second only to coal. Despite this, no dedicated facilities for the reception of that bulky commodity were included in Cubitt's original designs (*ibid.*).

The completion of the Permanent Passenger Station in 1852 provided the company with a priceless opportunity to lure London's potato traders to King's Cross, where they could do business in facilities provided the company, in return for a fee.

It was therefore decided to adapt and extend the buildings and roofs of the temporary passenger terminus in the Goods Yard for use by the potato trade, the two trainshed roofs providing partial cover for merchants to unload consignments of potatoes from specially designed wagons (Kay 2000: 364; Thorne 1990: 101). The former ticket offices and waiting rooms of the old station were used by the potato traders as offices, from which they sold their goods to the wholesalers who flocked from across the capital to King's Cross. In 1864, 39 small brick 'warehouses' were built for the salesmen and potato merchants between the 1850 station roofs and Maiden Lane, each provided with a short siding and storage facilities (Thorne 1990: 103; TNA RAIL 236/31: 264). However by the late the 1860s the potato merchants had grown increasingly dissatisfied both with the facilities at King's Cross, and with the terms of trade that were being offered by the company (TNA RAIL 236/90: 155). At the end of December 1869 the leading salesmen therefore requested that the Great Northern provide free rail passes, a reduction in their rents, security of tenure and longer leases (TNA RAIL 236/144: 1). Although the Great Northern did take steps to improve matters, their response was somewhat hesitant, and it was several years before the merchants' principal demands were met.

During the summer and autumn of 1877 Henry Oakley and Richard Johnson attempted to modernise and speed up goods handling in the Potato Market. Their plan involved the introduction of hydraulic capstan shunting to the facility, necessitating the acquisition of six hydraulic shunting capstans, a new accumulator and associated pipework from Tannett Walker and Co. of Leeds. The machinery was installed soon after in order 'to work the potato traffic satisfactorily' (TNA RAIL 236/43: 22, 68; TNA RAIL 236/146: 56, 74; TNA RAIL 236/173: 170). In 1878, the one of the merchants was provided with a new office, whilst the roof of the premises that were used by the firm of Thompson and Fulcher was extended at the beginning of 1882 (TNA RAIL 236/173: 330; TNA RAIL 236/146: 94).

Unfortunately for the Great Northern, these improvements failed to address the cause of the potato traders' principal grievance: the fact that the area where the majority of potato wagons were unloaded, between the Potato Market and the Midland Goods Shed, remained open to the elements. This resulted in inconvenient working conditions during inclement weather, a problem which persisted throughout the 1870s and the early 1880s.

4.9 The Construction of New Facilities between the Eastern Transit Shed and Midland Goods Shed: the mid-1850s to the 1870s

The continued growth of inward-bound goods traffic during the 1850s and 1860s led to increasing congestion in and around the Eastern Transit Shed (Landscape Feature 8). Despite the construction of additional platform accommodation at the southern end of the shed in 1871, congestion persisted and intensified in the years that followed (TNA RAIL 236/171: 212). The main problem lay with the dimensions of the Eastern Transit Shed, which only allowed for the creation of a limited area of additional platform space. Henry Oakley therefore proposed to provide 'a short additional siding…adjoining the Goods Office for loading certain descriptions of Goods into vans direct from trucks' in the triangular open area that was formed by the Eastern Transit Shed (Landscape Feature 8), the main Goods Office and the Midland Goods Shed (Landscape Feature 15), for an estimated cost of £500 (TNA RAIL 236/172: 339). The Way and Works Committee approved the application and a new inward goods siding, paved

Fig. 4.22 An engraving of the Potato Market taken from the Illustrated Times of October 1st 1864, showing the Potato Merchants trading from their 'offices' at King's Cross, which at that time were nothing more than a collection of wooden huts (Anon 1864)

unloading area and roadway were laid down shortly afterwards. The new facilities were left open to the elements, which led to unpleasant working conditions and operational inefficiencies in bad weather.

The construction of the siding, roadway and paved unloading area in 1871 represents the earliest formative phase of the West Handyside Canopy (Landscape Feature 26). Had this road and railway siding not been built in that location, its roof covering would not have been erected later.

4.10 The Temporary Smiths' Shed: c.1852–8

Landscape Feature 15

Following the completion of the Permanent Passenger Station in 1852, it was intended that the single storey Temporary Carriage Shed (Landscape Feature 15) of 1850 was to be converted for use as workshops and for the accommodation of goods traffic (Anon 1851b). The former carriage shed was subsequently handed over to the Great Northern Chief Engineer's Department following the cessation of passenger services from Maiden Lane (TNA RAIL 236/280: 01/06/1858). A document dating to the late 1850s stated that the department's facilities at the shed comprised both offices and smithies (*ibid.*). The smithies had been erected upon an existing tarmac roadway inside the shed, and were each fitted with chimneys, hearths and benches (*ibid.*: 5). The retention of the original bufferstops demonstrates that the existing tracks were also retained (*ibid.*: 3), presumably in order to allow rolling stock to be wheeled in for repair. When the shed was converted for the use of a new occupier in the late 1850s (see below), contractors were instructed to create 'two Door Openings... thru' wall at end of shed to communicate with present Engineer's Offices' (*ibid.*: 1), suggesting that at least part of the range of offices at the southern end of the structure had already been built by this date, although they were later enlarged.

4.11 The Midland Railway Goods Shed and Midland Offices: the late 1850s to the 1870s

Landscape Feature 15

Unlike the Great Northern, which had been established in order to provide a railway connection linking the capital and the provinces, the Midland Railway's origins were entirely provincial. In order to gain a foothold in the capital the company made arrangements first with the London & North Western Railway (LNWR), and subsequently with the Great Northern, to allow its passengers to travel to the London termini at Euston and King's Cross respectively. Though the Midland company was initially prohibited by the Great Northern from running its own trains into King's Cross, by the second half of 1857 the directors of the two companies had reached a provisional agreement that would permit them to do exactly that (Biddle 1990: 62–5). In mid December of that year the directors visited the Goods Yard to inspect Cubitt's old carriage shed, which the Great Northern proposed to let to the Midland for the accommodation of the latter's general goods traffic (TNA RAIL 236/280: 10/12/1857). The Midland's representative was happy with the proposal and Walter Brydone was instructed to draw up plans for the modifications that were necessary to improve both the shed and the lines that would carry the resulting goods traffic.

Preparations for the remodelling of those areas of the Goods Yard allocated to Midland traffic began in earnest in January 1858. That month the Resident Engineer at King's Cross submitted a request to the Company's Stores Department for 1,620 tons of Mount Sorrell or Markfield Granite for the 'new works ordered at King's Cross for Midland Traffic', in addition to 2,000 middle chains and 3,000 pairs of fish plates for 'the sidings of the new works for the Midland company's accommodation' (TNA RAIL 236/227: 13/01/1858). Notices to quit were prepared for those businesses occupying premises required for the Midland traffic, and in February Beart's Patent Brick Company, which occupied a storage shed in the northwest corner of the Yard, agreed to make way for the newcomers (TNA RAIL 236/25: 123–4).

A formal agreement for 'the interchange of Traffic and for the use of the Great Northern line between London and Hitchin and of the London Terminus by the Midland Railway' was signed by officials of the two companies at the beginning of June 1858 (TNA RAIL 236/280: 01/06/1858). The articles of agreement were accompanied by the specification for the reconstruction of the old carriage shed, which was presumably drawn up by Brydone following the directors' visit the previous December. The specification described the necessary works under six headings: 'Excavation; Brickwork, Stone Work &c; Wood Work; Removal of Engineers Smithys [sic] and Pitching up on Site of Smithys [sic]'. Improvements to the structure of the building included the insertion of new doorways and iron girders, the removal and refixing of downpipes, and the provision of fire hydrants and new gas fittings, while new facilities for the Midland company included the rebuilding of platform floors and the installation of ten new hand operated cranes (*ibid.*).

Within weeks of the original agreement between the two companies, the Midland submitted a request for additional office space for their clerks at the shed (TNA RAIL 236/25: 290). Walter Brydone calculated that the necessary facilities could be built for a cost of £900, a figure that was accepted by both companies (TNA RAIL 236/280: Clarke to Board 28/06/1858). Later, Brydone advised that a modest saving could be made if the proposed 'upper floor' was omitted, with additional space being provided on the 'second floor' in lieu (*ibid.*: Brydone and Oakley to Clarke 29/06/1858).

The Midland proposed further modifications that September, when it requested that the Great Northern supply a traversing crane capable of lifting 10 tons. This was to run along a new track beside the company's 'Goods Warehouse at King's Cross' together with new weighing machines, that were to be installed in the same location (*ibid.*: MR Engineer to MR Board 02/09/1858; TNA RAIL 236/26: 26). Though the Board of the Great Northern declined to supply either the crane or the weighing machines, it did agree to lay down the necessary length of rail for the crane, which the Midland must have supplied themselves.

The volume of goods and mineral traffic carried by Midland trains grew rapidly at the beginning of the 1860s. As early as November 1859 the company took steps to acquire possession 'of a piece of land [at St Pancras]... adjoining the North London Railway on one side and the land and premises of the Great Northern Railway on another' in order to accommodate the anticipated overflow from the company's facilities at King's Cross (TNA RAIL 236/280/310, 15/11/1859). Throughout the course of 1861 the Midland complained repeatedly about the delays that its goods and passenger trains were subject to on the line between Hitchin and King's Cross, which the company attributed to the Great Northern prioritising its own traffic at the expense of Midland trains. The problem became even worse the following year, when it was reported that 3,400 trains operated by the Midland were subject to delays on the Great Northern network (Stretton 1901: 175). The Midland also expressed dissatisfaction at what it perceived to be the tendency of senior officers of the Great Northern to ignore requests for additional goods and coal accommodation at the Goods Yard (TNA RAIL 236/28: 336, 344, 357; TNA RAIL 236/285 MR to Clarke 12/02/1861; MR to Clarke 18/12/1861). As the relationship between the two parties cooled, in 1861 the Midland announced its intention to build its own goods-handling facilities on the site of the former slums of Agar Town (Townend 1989: 16). The new Midland Goods Station was completed in 1862, and the company vacated its premises at King's Cross Goods Yard that July (*ibid.*). Despite the cessation of Midland goods traffic, the latter company continued to use the Great Northern's Engine Shed until St Pancras Station was opened to passenger traffic in October 1868 (*ibid.*: 20–2; TNA RAIL 236/29: 222).

It is not clear to what use the Great Northern put the building following the departure of the Midland, though a minute of the Executive and Traffic Committee noting that a police guard was appointed to watch over the 'recently vacated' building at the beginning of 1865 suggests that the company may have experienced difficulties finding a new tenant (TNA RAIL 236/91: 167). New sidings were laid to accommodate hay and straw traffic to the north of the shed in late 1863 (Fig. 4.4), though it is not known whether the building had any association with that trade at that time (TNA RAIL 236/294/6: 11/01/1864). It appears that several of the facilities previously used by the Midland company were somewhat under-used during the 1860s, and in 1870 the Traffic Committee

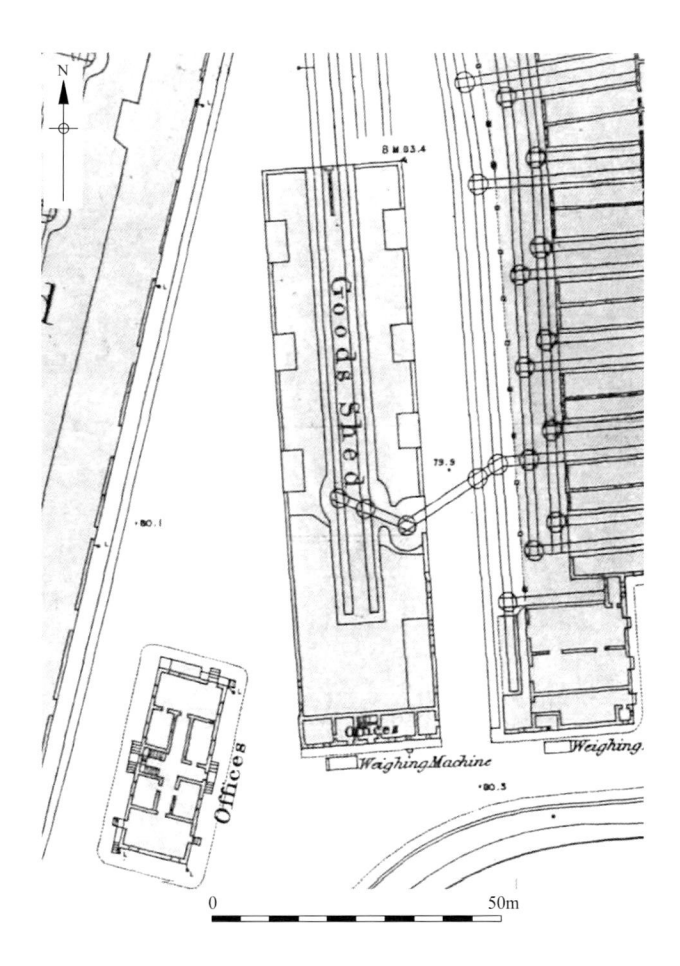

Fig. 4.23 The Ordnance Survey Map of 1871 detailing the Midland Goods Shed (Landscape Feature 15), scale 1:1,250

of the Great Northern considered a proposal that the former Midland Engine Shed be converted into a vast beer store (TNA RAIL 236/144: 95). Although the latter proposal came to nothing, by the late 1860s the Great Northern faced a growing demand from manufacturers and wholesalers of bottled goods, particularly beer, to provide warehouse facilities for the storage of their products as they passed through King's Cross Goods Station.

In 1869 the Great Northern granted the Yorkshire bottle manufacturers Kilner Brothers a lease for 21 years on the southern part of the Midland Goods Shed for use as a bottle warehouse (*ibid.*: 376). Within two years of taking possession of the shed, representatives of Kilner Brothers submitted a proposal to the General Manager of the railway company '...for additional accommodation to enable them to carry on their business as Bottle Merchants at the King's Cross Goods Depot' (*ibid.*: 169–70). The company offered to pay the railway for the construction of 'half of an additional floor over the Goods Shed' plus £75 in rent per annum. The proposal was reviewed by a committee chaired by Lord Culross (Chairman of the Great Northern), which recommended in March 1871 that an even larger scheme '...for putting an additional floor over the whole of the Goods Shed... be adopted' (*ibid.*: 181). The contract for the proposed works was awarded to John Kirk, of Woolwich, and construction was well under way by the end of November (TNA RAIL 236/306/19: 29/11/1871). Kilner Brothers had taken possession of the new storey by the end of October 1872, when the parties agreed upon a new lease for the extended premises (TNA RAIL 236/144: 376). Following the conversion, Kilner Brothers continued to commission further modifications to their premises in the Midland Goods Shed, including the insertion of 'a small counterbalance lift for £40' in June 1873 (TNA RAIL 236/145: 69).

In addition to their premises in the Midland Goods Shed itself, Kilner Brothers also leased space in the range of offices that stood at the southern end of the shed. As the firm continued to expand its operations at King's Cross the existing office accommodation appears to have been found wanting, and in March 1875 the company requested that the Great Northern provide 'additional office room' for £100, for which Kilner Brothers were prepared to pay an additional £30 in rent per annum (*ibid.*: 253). In June 1882 the Traffic Committee recommended that a further request from Kilner Brothers 'to make certain alterations in their offices' for an outlay of £159 also be approved (TNA RAIL 236/148: 185).

At least one room on the ground floor of the office block was used as the Goods Yard telegraph office from the mid-1850s (TNA RAIL 236/316: 1874). By the early 1870s the original telegraphic apparatus was becoming increasingly unreliable, prompting the company to purchase a new Wheatstones ABC Telegraphic Instrument from the General Post Office (GPO) in September 1873 (*ibid.*).

During the 1870s and 1880s, changes that occurred elsewhere in King's Cross Goods Yard impacted upon the fabric of the Midland Goods Shed. In March 1877 Henry Oakley suggested that the company install additional hydraulic apparatus in the Potato Market (TNA RAIL 236/146: 56). Though a considerable quantity was instated in the Goods Yard in the years between 1877 and 1882, it is likely that the erection of the subsidiary accumulator tower at the north end of the Midland Goods Shed was an element of the 1877 works owing to its proximity to the Potato Market. Within four years of installation, it was necessary to acquire another accumulator (with 18" diameter ram and 20' stroke) from Armstrong & Co. in order to replace one that had already worn out (TNA RAIL 236/175: 85).

Built heritage evidence of the conversion of the former Carriage Shed and Engineering Works to the Midland Goods Shed in 1858

Landscape Feature 15

When the original single storey temporary carriage shed and engineering works was converted into the Midland Goods Shed in 1858, the 1851 building was largely demolished, leaving only a few courses of original brickwork at the base of each elevation. It was then rebuilt using fabric 3035 bricks (see Appendix 3), although the footprint of the shed did not change (with the exception of the offices that would soon be added to the northern façade). It is therefore suggested that the original structure, described by Cubitt as being both 'temporary' and 'moveable', was composed of materials such as wood or iron that were replaced with brickwork in 1858.

The original eaves line of the resulting single storey Midland Goods Shed is marked by the top of a horizontal brick frieze that runs around the building. The brickwork below was laid in Flemish bond whereas English bond was used above, suggesting more than one phase of construction (Fig. 4.25; Fig. 4.26). Humber's plan of 1866 reflected the changes that were made to the building following its conversion from a carriage shed to a goods shed (Fig. 3.8). It indicated that by the mid-1860s the six railway tracks that entered the structure from the north had been reduced to two central ones that were linked via turntables to a new transverse track that communicated with the Potato Market sidings. The latter track exited through a new opening that had been inserted into the eastern wall of the shed along with ten road vehicular equivalents in the eastern and western sides of the structure (i.e. five in each; Fig. 3.8). These features would have been unnecessary in a carriage shed but were essential to the design of a goods shed so as to enable the easy transfer of commodities between road and rail vehicles (Bussell and Tucker 2004c: 7).

The two large openings in the north wall that brought the railway tracks into the building were still extant at the time of recording (Fig. 4.25) and were spanned by their original 1858 cast iron I-beams, the lower flanges of which are visible. Extant cast iron corbel brackets suggested that they could be shuttered off through the use of horizontally sliding two leaf doors. Although modified, the original locations of the ten vehicular loading bays and the one railway opening that are shown in the east and west walls on the Humber Plan of 1866 were also identified, although some access points have since been bricked up (Fig. 4.24; Fig. 4.26). Like the northern openings, they could also be closed off when required through the use of sliding doors.

Two wooden mountings for internal slewing cranes survived *in situ* to the immediate west of two of the road vehicular loading bays in the eastern edifice. In order to facilitate the transfer of goods between road vehicles and train wagons, the Goods Station plan of 1882 suggested that such cranes were originally stationed between each loading bay and the north–south railway lines that entered the building from the north (Fig. 4.13).

It is likely that two long, raised platforms originally ran the length of the shed, either side of the north–south tracks. Their surfaces would probably have been at the same approximate height as the wagon and van floors, thus enabling incoming and outgoing goods to be loaded and unloaded with relative ease. The entire ground floor of the shed has been raised to the level of the old platforms and capped with concrete during modern times, however a section of one of the original platforms was observed where a late 20th-century lift shaft had been removed. It demonstrated that the structure consisted of timber boards overlying joists that rested upon low brick walls.

The Goods Station plan of 1882 also indicated that rectangular indents were present where the platforms met the loading bays so that road vehicles could back into the shed (Fig. 4.13). Before the southern section of the building was converted into a bottle warehouse in 1869 it is likely that such indents existed beside every road vehicular access point. The built heritage survey indicated that the niches were edged by dwarf brick walls that had been clad with stone, their perimeters having been further reinforced with a variety of *ad hoc* materials that included large timbers, reused rails and in one instance a curved iron plate. These items may have been added to help protect the platform edges from vehicle strikes.

The January 1858 specification for the conversion of the building into the Midland Goods Shed alluded to the presence of 'Engineer's Offices'. It seems likely that these were situated somewhere within the shed itself since the office block that abuts the southern side does not appear on plans predating 1858. This in turn suggests that the two lowest storeys of the building were erected when the Midland Railway Company asked for additional space to accommodate its clerical staff (TNA RAIL 236/280: 01/06/1858). Since the Midland made the request in June 1858 and the conversion of the shed itself commenced the previous January, work on the main structure must have been well advanced or entirely complete by the time that work on the offices began. This explains why the two

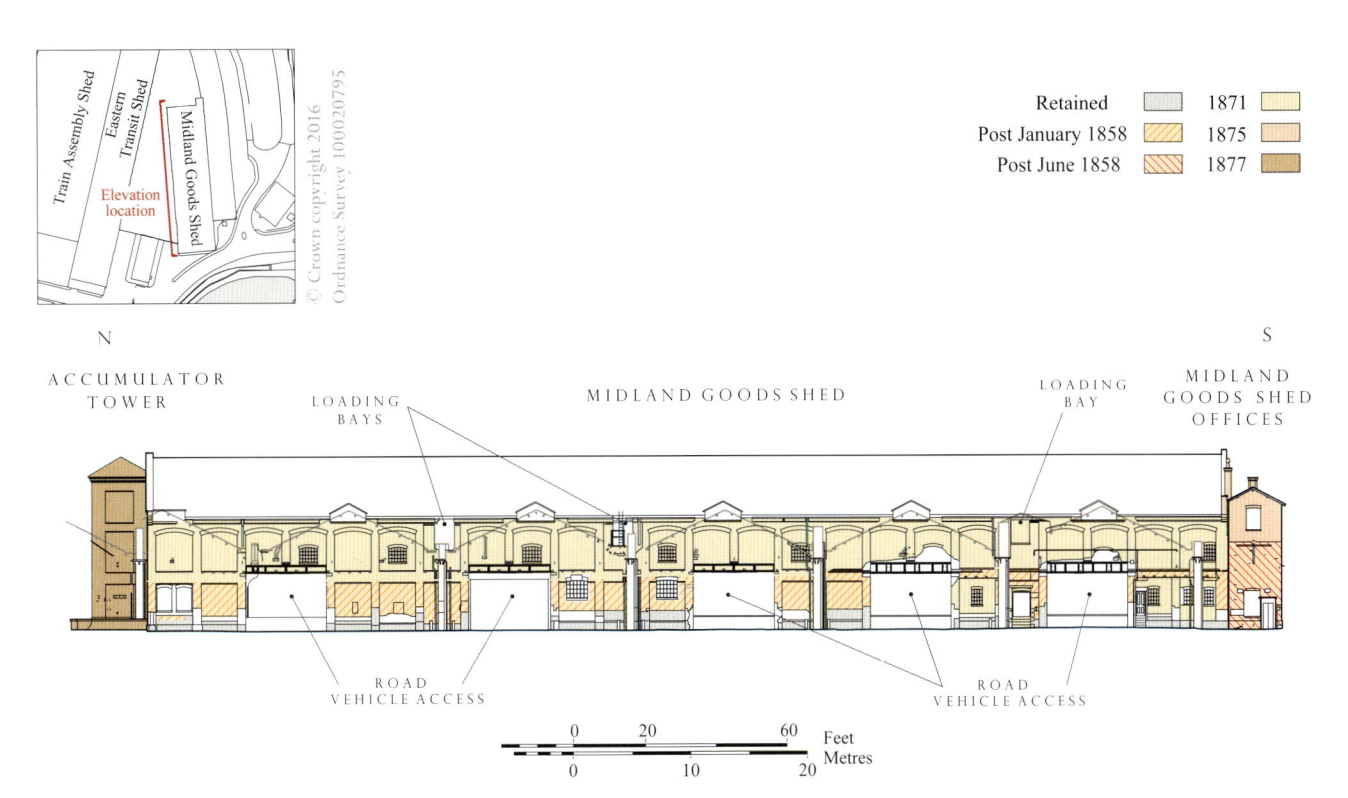

Fig. 4.24 External elevation of the west wall of the Midland Goods Shed (Landscape Feature 15) showing the changes of 1858–77, scale 1:625

structures are close together in date but are not bonded into one another thus forming two distinct phases of construction (Fig. 4.24; Fig. 4.26).

The offices appear on the Humber Plan of 1866 although the footprint that is shown is slightly different to the extant arrangement, which is illustrated for the first time on the Ordnance Survey map of 1871. However, no physical evidence was observed that suggested a phase of extension between 1866 and 1872, the entire northern edifice up to the top of the first floor instead consisting of one continuous build.

Built heritage evidence of the conversion of the southern end of the former Midland Shed to a bottle warehouse in 1869–75

Landscape Feature 15

Ordnance Survey maps surveyed between 1871 and 1915 identify the south end of the building as a 'Glass Bottle Warehouse' and the northern end as a 'Goods Shed' (Fig. 4.5). This suggests that the Great Northern retained a substantial section of the building for its own purposes after letting the southern section to Kilner Brothers in 1869.

In order to accommodate the bottle warehouse, a number of changes to the fabric of the building were

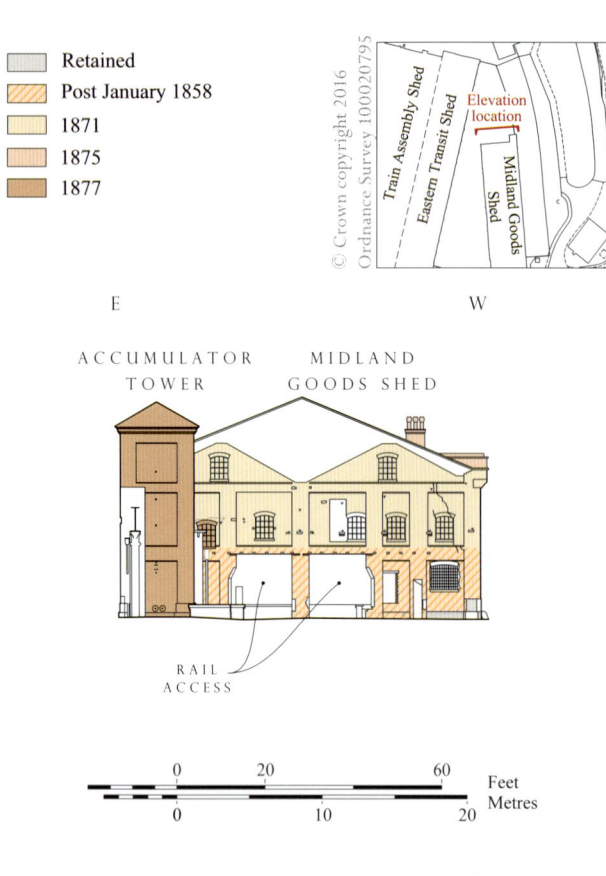

Fig. 4.25 External elevation of the north wall of the Midland Goods Shed (Landscape Feature 15) showing the changes of 1858–77, scale 1:500

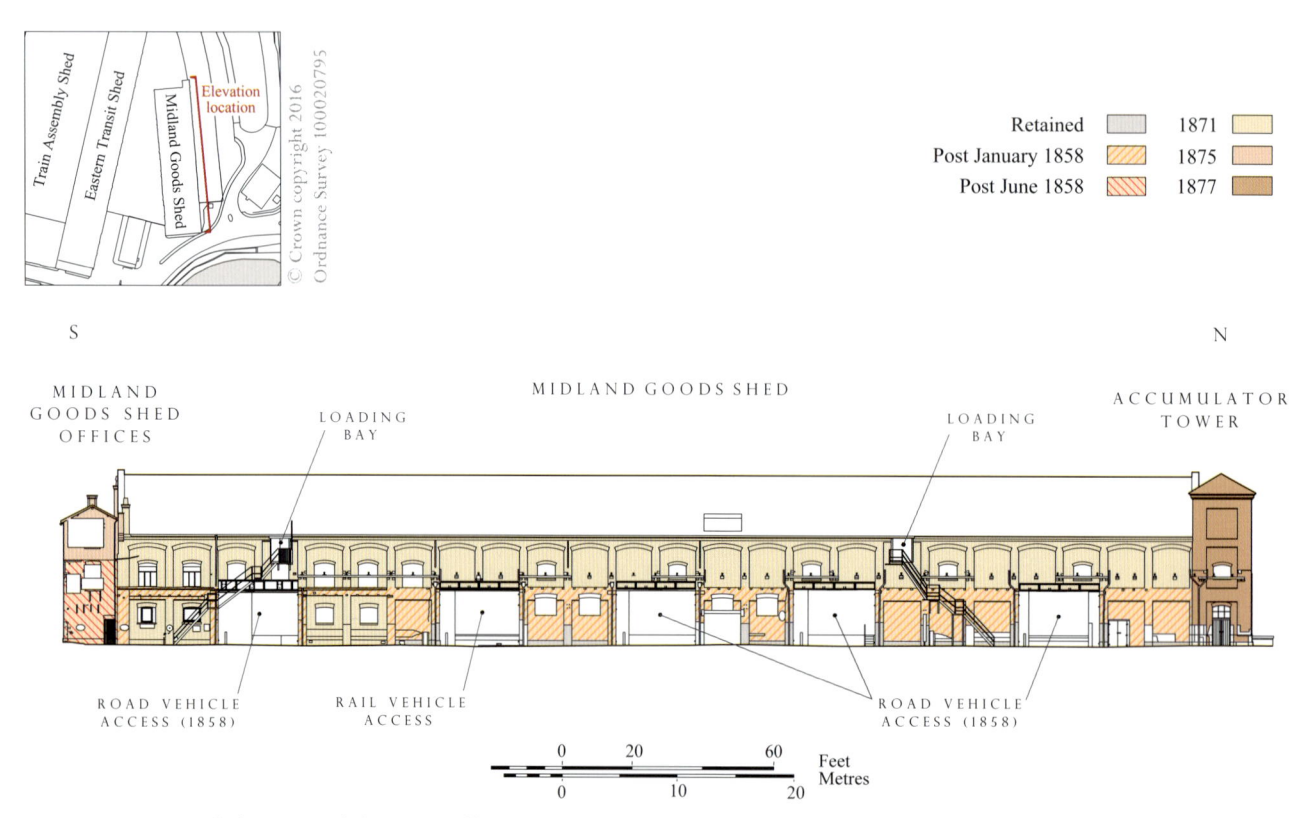

Fig. 4.26 External elevation of the east wall of the Midland Goods Shed (Landscape Feature 15) showing the changes of 1858–77, scale 1:625

made. Two former loading bay openings at the southern end of the western elevation were infilled (these were later reinstated; see Section 8.3) and a number of new cast iron framed windows were introduced so that the warehouse could be better illuminated (Fig. 4.24). Similar alterations were made to the opposing wall for the same reasons, the most notable being the infilling of the two most southerly road vehicle access points (one of these would be reinstated in a slightly different location at a later time; Fig. 4.26; see also Section 8.3).

The surviving loading bays in the section of the building retained by the Great Northern were enlarged, perhaps to compensate for the loss of their northern equivalents (Fig. 4.24; Fig. 4.25). This necessitated that the cast iron beams that formerly surmounted the originals be replaced with plate girders and that part of the decorative stepped cornice that once ran all around the shed be partially destroyed in order to accommodate new sliding doors.

Att the request of Kilner Brothers an upper storey was added to the shed, above the aforementioned brick frieze in 1871 (Fig. 4.24; Fig. 4.25). The double gable roof of the shed was not replaced, the original instead being reinstated at a higher level once work on the lower elements had finished.

The new section of the building was characterised by a succession of decorative upper panels that were constructed immediately above each of the ground floor examples in English rather than Flemish bond. In the western elevation, 21 of the 24 panels were adorned with decorative segmental arches, whilst the remaining three were characterised by horizontal heads owing to the fact that they accommodated loading bay doors (Fig. 4.24). The upper section of the eastern wall was much the same as its western counterpart, apart from the fact that it possessed two rather than three loading bays. In the western elevation, a particularly well preserved loading bay was identified. It contained a pair of timber doors that were surmounted by a concrete lintel with a quarter height wooden platform extending out below. The platform was connected to a set of iron chains that allowed it to be lowered and raised as required.

The upper storey of the Midland shed was constructed at a surprisingly high level, being positioned above most of the new window positions. Only two examples in the northern elevation illuminated the first floor (Fig. 4.25), which was mainly top lit by glass panels in the roof. The insertion of the new floor meant that a number of new columns needed to be added at ground floor level in order to support the weight of the additional storey.

An internal timber staircase in the southwest corner linked the ground and first floors of the shed. Although the first floor was entirely open at the time of recording, it was originally partitioned into several discrete sections. Documentary sources suggested that the southern end of the first floor was occupied by Kilner Brothers whereas the area to the north was used by the Great Northern. Since the areas were occupied by different companies it is reasonable to assume that a partition formerly separated them. The southernmost section of Kilner Brothers' bottle warehouse also appears to have been shuttered off from the rest of their working area at first floor level. A pair of fireplaces in the south wall demonstrated that it was the only section of the warehouse that was heated so it is assumed that it was used as an office space. Four circular glass discs with short, silvered tubes beneath were situated in the southeast corner (although the presence of a further four timber inserts indicates that eight were originally present). They probably functioned as lightwells that illuminated the space below.

Two hatches communicating with the railway lines below were integral to the floor in the northern section of the building. Winches were presumably used to transfer goods through the hatches thus enabling the loading or unloading of the goods wagons that were manouevered into position below. Another two hatches were present in the southern half of the structure, which were presumably used by Kilner Brothers to transfer bottles between floors.

The Ordnance Survey map of 1871 (Fig. 4.5) suggests that a rectangular structure, labelled 'offices' on later maps and plans (Fig. 4.27), had been inserted in the southeast corner of the former Midland Shed by that time. It was probably erected in 1869 for the use of Kilner Brothers. Physical evidence of the existence of the offices took the form of internal wall plaster that had been applied to the southern end of the east wall of the Midland Shed. A number of joist holes had also been inserted that presumably supported the north wall of the office, which may also have had an upper storey.

In addition to the office described above, Kilner Brothers also seem to have occupied one of the rooms in the main Midland Goods Shed Office block (Landscape Feature 14). However this does not seem to have been sufficient to meet their needs, leading the company to request additional office accommodation as early as 1875. This may account for the addition of a third storey to the office block, which presumably took place during that year, as well as the insertion of an intermediate floor or mezzanine (since removed) in the southern end of the main shed (Fig. 4.24; Fig. 4.26). The latter feature is shown on an east–west cross-sectional drawing of the Midland Shed dating to 1915, between the ground and first floors spanning the south-central and southeast sections of the building. According to the drawing, the mezzanine was supported by columns sitting on the raised platform surface. It may also have been inserted in 1875 in order to address Kilner Brothers' demands concerning extra office space. The fact that the southeast corner of the shed was used differently to the rest of the building is evident in the extant fabric. The soffits of the first floor joists in that section were stained, suggesting that they had once been covered with lath and plaster. They also incorporated narrow timbers with slots in them, orientated north–south, which would once have received narrow vertical timbers for the partition walls that divided the offices. The south wall of the ground floor of the shed also bore telltale scars that were associated with the former presence of the mezzanine. Within the footprint of the former structure,

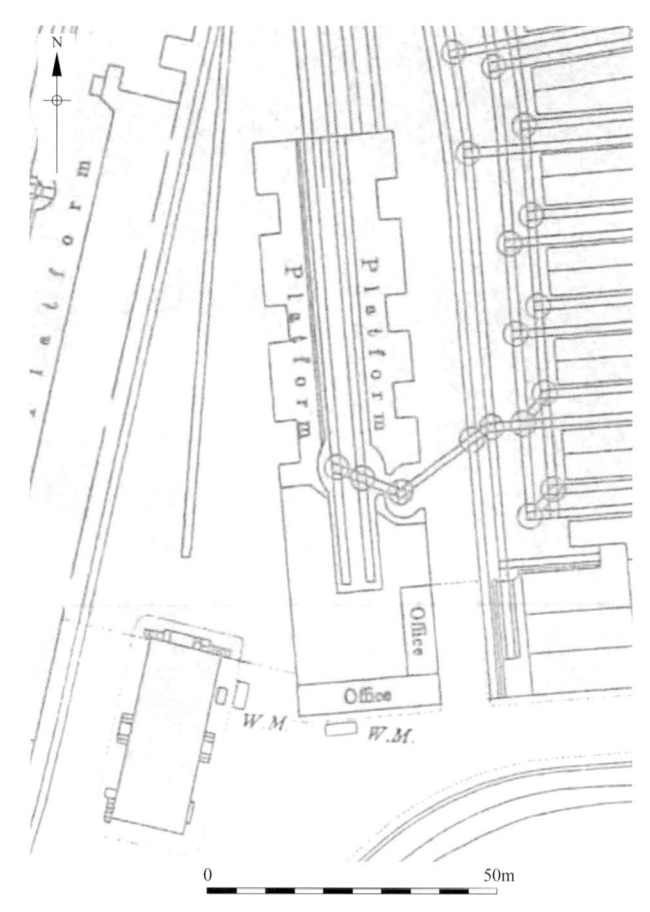

Fig. 4.27 The Ordnance Survey Map of 1894–6 detailing the Midland Goods Shed (Landscape Feature 15), scale 1:1,250

the wall of the shed had been plastered and still possessed a skirting board, suggesting that it formed part of an office rather than the bottle warehouse proper.

Built heritage evidence of the construction of the Accumulator Tower against the north wall of the Midland Shed c.1877

Landscape Feature 15

A square, brick accumulator tower [GLOSSARY] abutted the northeast corner of the Midland Shed. Executed in English bond, the decorative elements of the exterior included a cornice and recessed brickwork panels that were in keeping with the rest of the shed, although three rather than two panels were present owing to the fact that the tower was taller. Unlike the rest of the modern extant roofing materials that cover the bulk of the shed, at the time of the building recording survey the roof of the accumulator tower was still clad with slate.

The tower could be accessed by way of a pair of ground floor doors in the east elevation. It was illuminated from above via two arched windows, one of which immediately surmounted the door, the other being situated at a higher level within the central decorative panel (Fig. 4.26).

Internally, the tower housed the remains of the remote hydraulic accumulator mechanism. This tall circular structure practically filled the interior space with very little room between it and the walls. A metal ladder, fixed to the south wall, enabled the top of the mechanism to be accessed thus enabling inspections and repairs to be undertaken.

4.12 The Expansion of the Coal Trade and the Evolution of the Coal Offices: the mid-1850s to the 1870s

Landscape Feature 18

The Great Northern successfully maintained an effective monopoly over the lucrative coal trade at King's Cross throughout the 1850s. However, by the end of that decade a number of independent coal merchants had begun to trade from the yard, although they were compelled to work from temporary accommodation at the coal drops. In contrast, Herbert Clarke, the Great Northern's coal agent, continued his trade from the comparative luxury of the purpose-built Coal Office building (TNA RAIL 236/283/4: 18/08/1860). Clarke's monopoly began to break down in 1860, when a number of established independent merchants expressed an interest in acquiring permanent accommodation in the Coal Offices. This process was greatly accelerated after July of that year, when the Court of Chancery issued a judgement which condemned the 'very crafty and tricky contrivance' that the Great Northern had employed to conceal its practice of buying and selling coal at a profit from the judicial authorities (*ibid.*: 12/07/1860). The Court's judgement not only exposed these practices but also prohibited the Company from selling coal on its own account.

The decision left the Great Northern with no option other than to encourage independent merchants to establish permanent premises in the coal yard in order to maintain the coal trade at King's Cross. To accommodate the newcomers it was decided to refurbish and extend the existing Coal Offices, which were to be divided into individual units that would be let out at rates ranging from £10 to £100 per annum, depending upon size and location (*ibid.*: 08/12/1860). Following the cessation of the Great Northern's monopoly, Herbert Clarke began trading as an independent coal merchant, retaining No. 3 Office (from which he had overseen the trade during the 1850s) for the next four and a half decades (Thorne 1990: 118; Kelly & Co. 1861: 722).

Table 1: An inventory of occupants within the Coal Offices (Landscape Feature 18) as of October 1860 (based upon TNA RAIL 236/283/4: 22/11/1860)

Office No.	Floor	Name	Rent p.a.	Date Tenancy Started
1	Ground	Not finished	-	-
2	Ground	Not finished	-	-
3	Ground	Herbert Clarke	£100	29/09/1860
4	Ground	Silkstone Colliery Ltd	£80	29/09/1860
5	-	Not finished	-	-
5a	-	Rickett Smith & Co.	£70	-
6	-	Finney & Seal (not finished)	£70	-
7	-	Samuel Plimsoll (not finished)	£60	-
8	-	Not occupied (not finished)	£60	-
9	Ground	Goodwill & Co.	£40	25/10/1860
10	Ground	Edward Wiggins	£40	01/10/1860
11	Ground	Finney & Seal (empty)	£40	25/10/1860
12	Ground	Rickett Smith & Co.	£40	06/08/1860
13	Ground	Samuel Plimsoll	£15	09/07/1860
14	Ground	Day & Co. (out of business)	£15	09/07/1860
15	First Floor	To be let with No. 1		
16	First Floor	To be let with No. 2		
17	First Floor	To be let with No. 3		
18	First Floor	To be let with No. 4		
19	First Floor	To be let with No. 5		
20	First Floor	To be let with No. 6		
21	First Floor	To be let with No. 7		
22	First Floor	To be let with No. 8		
23	First Floor	To be let with No. 12		
24	First Floor	Cockburn & Jordan	£10	09/07/1860
25	First Floor	Thomas Thornicroft	£10	09/07/1860
26	Second Floor	Rickett Smith & Co.	£40	-
27	Second Floor	Rickett Smith & Co.	£40	-

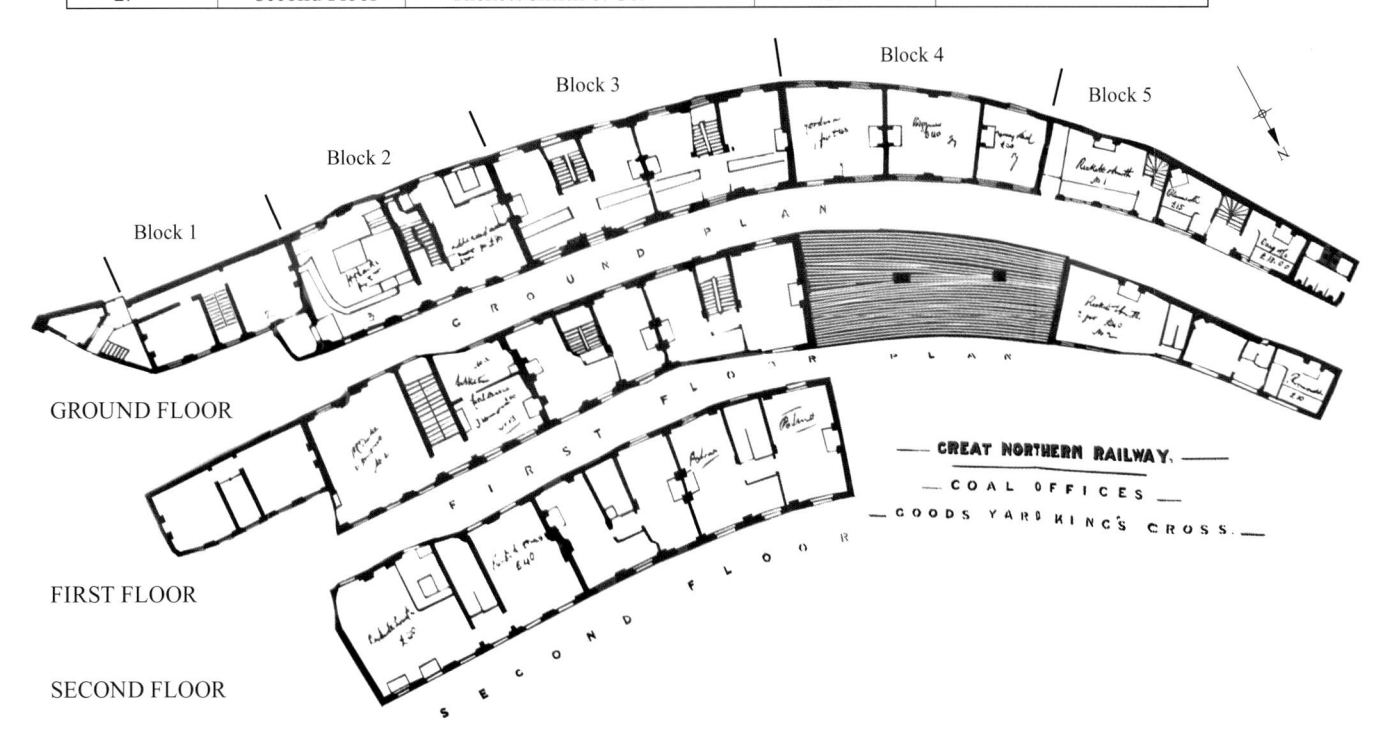

Fig. 4.28: Ground, first and second floor Great Northern Railway plan of the Coal Offices (Landscape Feature 18) printed in 1860, scale *c*.1:500 (TNA RAIL 236/283/4)

Demand for accommodation seems to have been high, with applications for office space being received from agents and colliery owners alike before work had even begun on the new buildings (TNA RAIL 236/283/4: 18/08/1860). Plans of the proposed new range were submitted to the Board by Seymour Clarke as early as January 1860 (TNA RAIL 236/27: 208). Eager to secure the best accommodation in the block, independent coal merchants such as Messrs Rickett Smith & Co. submitted drawings showing which offices they required. Though the Great Northern Board noted that that the latter company was interested in acquiring 'a larger space than that previously intended to be assigned to any one firm', the proposal was accepted on the condition that the privilege be granted by the payment of a greater proportionate rent (*ibid.*). Rickett Smith & Co. accepted the railway company's terms, submitting a formal application for accommodation in early February 1860. They were joined soon afterwards by Thomas Thornicroft, J. Oakes & Co., Cockburn & Jordan and Beale Walker & Co., all of whom had applied for office space by the beginning of May (TNA RAIL 236/283/4: 21/08/1860).

Both the refurbishment of the existing offices and the construction of the adjoining range appear to have been well advanced by the summer of 1860. Whilst no surviving records document exactly when the works were fully complete, the Stanford map of 1862 (not illustrated) indicated that all five blocks (henceforth termed Blocks 1 to 5 from east to west) of the enlarged Coal Offices were extant by that time. Independent merchants had begun to occupy the complex some time before that date. For example, documents confirm that Rickett Smith & Co. had moved into offices on the upper floor of the 1850s building at the beginning of June 1860, whilst at least part of the new extension was complete and available for use by the middle of July (*ibid.*: 18/08/1860, 22/11/1860). A list of occupants compiled by the end of October of that year demonstrates that the merchants that are listed in Table 1 had acquired premises in the partially finished Coal Offices by the latter date.

Fig. 4.29 The extension to the Coal Offices (Landscape Feature 18) seen from the south bank of the Regent's Canal, photograph faces northwest

An inventory dating to November 1860 indicated that Herbert Clarke & Co. had by then moved into Nos. 1 and 2 Offices, by which date Rickett Smith & Co. had added No. 4 Office to its complement of rooms in the building (TNA RAIL 236/147; TNA RAIL 236/172). A trade directory of 1861 listed fifteen merchants residing in the offices, plus the cartage contractor Edward Wiggins (Kelly and Co. 1861).

Charrington Sells & Co. moved into No. 1 Office during the early 1860s but the accommodation that was provided by the Great Northern soon proved to be too small for their purposes (Kelly & Co, 1871). In 1864 the coal merchant submitted a request for additional space (TNA RAIL 236/90: 93–4). The railway company approved the request, following which it was agreed that the Coal Offices should be further enlarged at a cost of £200 (later revised upward to £220). In return, Charrington Sells & Co. would pay an annual rent of £40, a sum that was subsequently adjusted to a more reasonable £25 (*ibid.*: 106–7). The 1871 Ordnance Survey map (Fig. 4.5) indicates that an eastern triangular abutment and various other small extensions had been added to the Coal Offices by that date, but the footprint of the complex otherwise remained unchanged. It is therefore likely that the enlargement works that were instigated by Charrington Sells & Co. in 1864 were largely or exclusively limited to the vertical addition of floors.

Demand from coal merchants for accommodation at the offices seems to have remained high throughout the early to mid-1860s, a period that saw the continued growth of the coal trade at King's Cross. As coal traffic continued to increase the Directors of the Great Northern authorised the outlay of nearly £25,000 on a new Coal Stacking Ground close to the northern boundary of the Goods Yard in the first half of that decade (TNA RAIL 236/33: 314, 356; Fig. 3.8). In the spring of 1865 the Company sanctioned the subdivision of No. 12 Office (on the ground floor in Block 5) for an outlay of £15 in return for a combined annual rental income of £20 (TNA RAIL 236/91: 195). Two years later the Company agreed to pay a nominal rent on the property to the Regent's Canal Company, following a representation from the latter that demanded compensation because 'certain windows on the buildings belonging to the Company [the GNR] overlooked the Canal Company's property' (TNA RAIL 236/170: 10).

Most coal merchants and colliery representatives also established premises at the Great Northern Coal Drops (Landscape Feature 17), where day to day trade was conducted. The Coal Offices appear to have been reserved mainly for administrative purposes, which process could be carried out remotely (TNA RAIL 236/35: 15; Thorne 1990: 117).

In 1866 the coal trader Samuel Plimsoll established his own coal drops at Cambridge Street (Landscape Feature 23), on the other side of the Regent's Canal from the Great Northern Goods and Coal Station. Plimsoll had previously rented space in the Goods Station, where he perfected a new coal dropping mechanism which resulted in fewer breakages and therefore less waste (see 4.14 below). Plimsoll's new drops at Cambridge Street began to draw trade away from the Great Northern Coal Depot, the fortunes of which subsequently fell into decline (Thorne, 1990: 120). Owing to the resulting loss of business, the Great Northern decided to convert part of the Eastern Coal Drops (Landscape Feature 17) into warehouse facilities in May 1875 (TNA RAIL 236/172: 234). The latter premises were acquired by Messrs Bagley and Wold for use as a bottle warehouse at some point before 1879 (TNA RAIL 236/146: 380). The Railway Company's Coal Offices (Landscape Feature 18) also declined in importance, since they were geographically distant from the rival depot at Cambridge Street.

Built heritage evidence of the extension of the Coal Offices in 1860–4

Landscape Feature 18

A built heritage survey recorded the Coal Offices in detail. Unfortunately the complex had been severely damaged by a fire in the early 1980s, necessitating the rebuilding of the roof and certain sections of the upper levels. Unless evidence exists to the contrary, it has been assumed that the repaired sections were reconstructed in a style that closely resembled the originals.

The extension that was made to the Coal Offices (Landscape Feature 18) in 1860 consisted of three additional blocks (Blocks 3 to 5), the easternmost of which abutted the west end of the 1853 complex (Fig. 4.2; Fig. 4.29).

The footprint of the new premises followed the line of the Regent's Canal, thus creating an attractive set of curved buildings that also tapered and narrowed towards the western end (Fig. 4.2; Fig. 4.29). Blocks 3 and 4 were built in English Bond, whilst Block 5 was constructed in Flemish Bond. Like Blocks 1 and 2, the windows and doors were all surmounted by segmental arches that were stylistically identical to their earlier counterparts, although the openings themselves were less visually dramatic, being somewhat shorter in length (Fig. 4.30). Like the original buildings, the window sills and all of the external door steps were made of York stone but in contrast, the extension was not basemented since it was built above a series of vaults that formerly accommodated the stables that were located below the Wharf Road Viaduct (Landscape Feature 18). The stables were earlier, being contemporary with Blocks 1 and 2. Whilst it had been possible to access the stables from the basement of Block 2 and vice versa since they were built, no access was created between them and the new offices that now directly surmounted them.

A straight joint in the brickwork was observed at the eastern end where Block 3 butted Block 2, demonstrating that they were not contemporary, as suggested by the documentary evidence (Fig. 4.31). In contrast, the lack of a straight joint between Blocks 3, 4 and 5 suggested that those elements of the offices were created during one continuous phase of construction (Fig. 4.31).

Originally, Block 3 was accessed from Wharf Road via doorways in the second and fifth bays from the east, the doorway in the fourth bay having been converted from a window at a later date (Fig. 4.31). Internally, each floor within Block 3 had the same layout of two

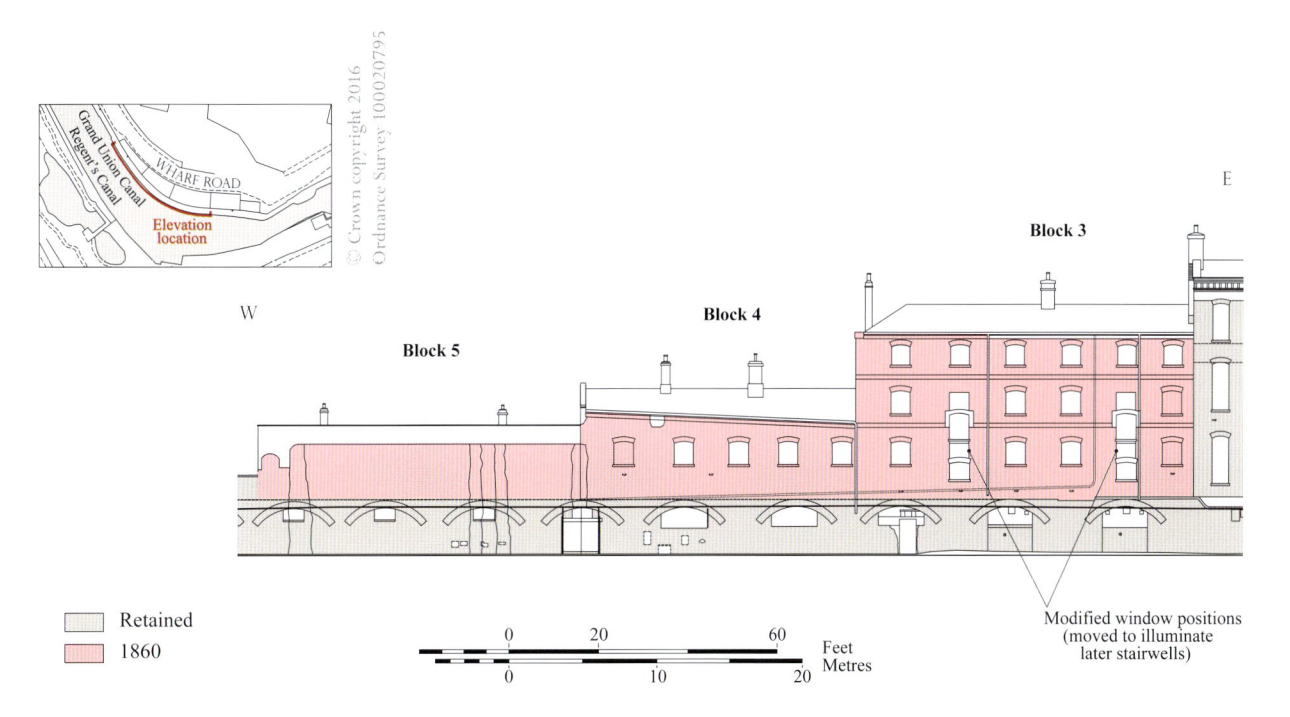

Fig. 4.30 External elevation of the south wall of the Coal Offices (Landscape Feature 18) showing Blocks 3 to 5 and the changes of 1860, scale 1:500

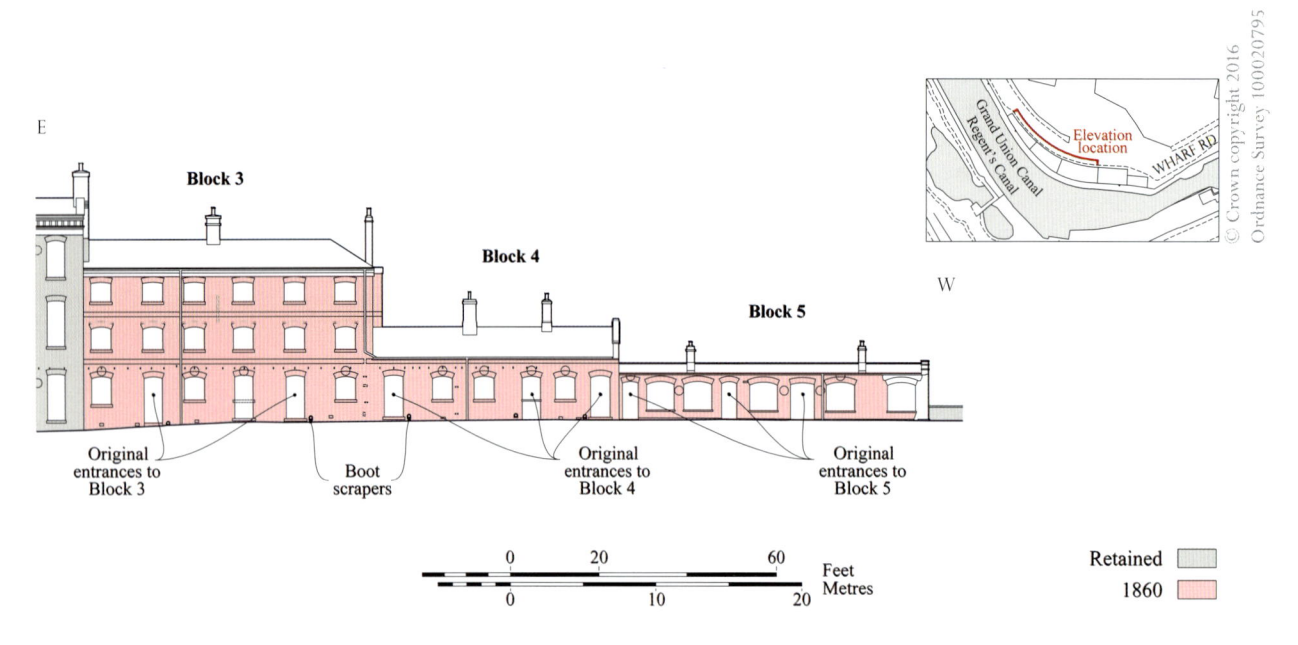

Fig. 4.31 External elevation of the north wall of the Coal Offices (Landscape Feature 18) showing Blocks 3 to 5 and the changes of 1860, scale 1:500

large offices, each of which was accessed via a central stairwell (Fig. 4.28). When at some point, the original staircases were replaced, the lower window openings in the staircase bays were moved downwards to illuminate the new landings (Fig. 4.30). While the stairs had not survived the early 1980s fire, their former presence could be discerned from scars that were visible in the plaster. These suggested that the eastern stairs ascended clockwise, whilst the western stairs wound in the opposite direction.

The front (north) elevation of Block 4 (Fig. 4.31) originally possessed three window positions and three doorways, one of which would later be converted to a window position. When Block 4 was constructed in 1860, each of the three doorways led to separate office units (Fig. 4.28).

The front (north) elevation of Block 5 originally contained three doorways in the first, fourth, and sixth bays from the east, the latter being double width (Fig. 4.31). Originally, the single width doorways led to one office unit at ground level that connected with another, equally sized unit at first floor level via a staircase that was positioned opposite the eastern street entrance. This access arrangement suggested that it would have been necessary to let these ground and first floor units together, a fact that was confirmed by a plan of the building surveyed in 1860, which indicated that both had been leased to Rickett Smith. The same source suggests that the double width doorway at the eastern end of the northern wall was a shared entrance that led to two separate ground floor units. These were arranged either side of a central corridor, thus affording sufficient privacy for the merchants who worked there. An identical arrangement was in place at first floor level, access being granted via a staircase that was positioned at the end of the corridor opposite the communal entrance (Fig. 4.28).

The 1860 plan of the offices indicates that a second storey originally surmounted Block 5. This was confirmed by the fact that several features that had been painted onto the southern wall had clearly been cut through, most probably in the wake of the fire that ravaged the building in the 1980s.

The blind bay at the eastern end of Block 5 appears to have been the only section of the structure that lacked an upper storey. The 1860 plan of the block indicated that it originally functioned as a toilet block, which would explain the absence of any windows. Historical maps indicate that the latter facilities were accessible from Wharf Road via a doorway in the western wall although no traces of this opening have survived owing to subsequent rebuilding.

Two significant alterations were made to the footprint of Blocks 1 and 2 during this episode of extension. Scoring caused by a sloping roof line (since removed) was observed on the eastern side of Block 2, which must relate to a lean-to that abutted the northern side of Block 1. Such a structure is shown on the 1860 plan (Fig. 4.28), taking the form of a rhomboid shaped addition. The plan also depicts a triangular extension against the eastern side of Block 1, which may explain why the bottom section of that elevation had been whitewashed (suggesting that it was once internal, the top of the paint marking the former position of the ground floor ceiling). This plan indicates that the triangular extension was originally one storey high. However, faint traces of whitewash on the brickwork above the ceiling line were observed, which suggests that an extra storey was added later, access to which was presumably granted via a new doorway (now infilled) that was inserted at first floor level in the eastern elevation of Block 1. One possibility is that the extra storey was created in 1864 at the request of Charrington

Sells & Co., who had moved into Block 1 by that date, but were unhappy with its size.

Only one internal change appears to have been made to Block 2 by 1860. The building plan of that year suggested that by then the western room at first floor level had been partitioned into two separate offices (termed '18' and '18a'). This was probably done in order to create a greater number of rooms that could be rented out to independent merchants. No trace of this partition was evident in the extant fabric of the building, perhaps because it abutted the original walls and was easy to remove without leaving an obvious scar.

It was probably around 1860 that numerous white circles were painted along the length of the northern façade of the Coal Offices overlooking Wharf Road. Although many were faded and were hard to discern, each originally contained a numeral that ascended chronologically, starting at '1' in the east at ground floor level. Directly above on the upper storeys, further circles containing numbers were observed. It seems likely that they identified specific offices within the complex, which from 1860 onwards were assigned to the various independent coal merchants that moved into the building. Indeed, the surviving numerals correspond to those that are shown on a plan dating to that year, which was drawn up to facilitate the allocation process (Fig. 4.28).

4.13 The Western Coal Drops: the mid-1850s to the 1870s

Landscape Feature 22

The growth of the Great Northern's coal trade during the 1850s necessitated the construction of an additional set of coal drops in the yard by the end of the decade (Thorne 1990: 113–4). Known as the Western Coal Drops (Landscape Feature 22), the new facility opened in 1860. A few years later the Great Northern added a huge coal-stacking ground to the northwest of the Goods Station.

The Western Coal Drops were similar in function and appearance to their eastern counterparts (Landscape Feature 17), but the rapid pace of technological change meant that the new facility was somewhat more advanced. These improvements included the provision of a wagon traverser in the southernmost bay and the use of simpler method of carrying the high level railway line into the depot (Duckworth and Jones 1990: 143).

Fig. 4.32 The Coal Offices (Landscape Feature 18), with the west (side) elevation of Block 5 in the foreground, photograph faces east

4.14 The Cambridge Street Coal Drops and the Plimsoll Viaduct: the mid-1850s to the 1870s

Landscape Feature 23

In 1863 the London coal trader Samuel Plimsoll acquired a number of leases for a portion of land south of the Regent's Canal in Cambridge Street. Plimsoll intended to construct a new coal drop facility that would be modelled on a patent of his own invention (TNA RAIL 236/30: 293; Thorne 1990: 120). The Great Northern was interested in the project and initially offered to buy the leases from Plimsoll, who agreed subject to being granted a tenancy on the land (TNA RAIL 236/30: 306–7). Although the railway company did indeed acquire the right to purchase Plimsoll's leases, the company neglected to take up the option (TNA RAIL 236/32: 278). This inaction may have encouraged Plimsoll to approach the rival Midland Railway, which was at that time in the early stages of developing its own passenger and goods stations next door to the Great Northern's own depot (*ibid.*).

When Plimsoll sold a number of these leases to the Midland company in 1864, Seymour Clarke of the Great Northern finally acted, although the Midland successfully retained a small interest upon which Plimsoll proposed to build part of his coal-handling facility. Having resolved the matter of ownership for the time being, the Great Northern Chief Engineer, Richard Johnson, prepared drawings of a railway viaduct and bridge to transport coal trains across the Goods Yard and the canal to Plimsoll's coal drops (*ibid.*). In order to accommodate the bridge and the northern end of the coal drops, it was necessary for the Great Northern to buy up the leases on the remaining houses, stables and wharves on the east side of Cambridge Street, the majority of which stood on ground that was owned by the Ecclesiastical Commissioners (TNA RAIL 1189/1423: 14). To add to the complex web of ownership in Cambridge Street, the west wall of the Regent's Canal and a large plot on the east side close to the junction with Wharf Road still belonged to the Regent's Canal Company, whilst land to the immediate south and west of the thoroughfare remained the property of the Imperial Gas Light and Coke Company (*ibid.*; TNA RAIL 783/110).

Plimsoll submitted formal proposals for the construction of his new coal depot to Seymour Clarke in January 1865, following which a contract for the site was drawn up. The Great Northern then set about constructing the bridge to carry their wagons to the new facility. It soon became obvious that the easiest way to achieve this would be to construct a viaduct across the Goods Yard, which would be carried over the canal by a new bridge. Although Richard Johnson finalised his design for both in 1865, construction was delayed over the months that followed, perhaps owing to technical complexities (TNA RAIL 236/33: 3). Consequently, it was not until early November of that year that the contracts for the new bridge and viaduct were finally put out to tender (TNA RAIL 236/92: 13). The erection of the wrought iron box lattice girder bridge across the canal was awarded to the London Engineering Company for £1,995, whilst the construction of the bridge's abutments and the timber and wrought iron viaduct were given to the Great Northern's preferred contractor, John Jay, who agreed to undertake the works for costs of £373 and £3,863 respectively (*ibid.*). The Cambridge Street Coal Drops opened in 1866, when they were depicted on the Humber Plan. They remained the property of Mr Plimsoll until they were eventually sold to the Great Northern soon after February 1891 (TNA RAIL 1189/1424).

4.15 The Gasworks Viaduct: the mid-1850s to the 1870s

Landscape Feature 24

The Great Northern first proposed to carry coal to the St Pancras Gasworks as early as 1853, when the company informally tendered to supply 50,000 tons of coal to the works over a three year period (TNA RAIL 236/276/23: 16/08/1853). The offer was declined and coal continued to be conveyed to the gasworks from the northern coalfields by sea and then on to the Gasworks Basin via the Regent's Canal.

Thirteen years later the Great Northern reached an agreement with Johansson and Elliott, owners of the Usworth Colliery near Sunderland, to carry 100,000 tons of coal per annum from York to the St Pancras Gasworks (TNA RAIL 236/144: 246). In November 1866 Seymour Clarke calculated that the viaduct and bridge necessary to carry the railway tracks across the Goods Yard, over the canal and into the gasworks would cost approximately £16,000 to build (TNA RAIL 236/35: 45). The construction of the viaduct was delayed by a disagreement between the railway and gas companies over a levy which the Great Northern proposed to charge on top of the agreed rate of carriage in order to recoup the outlay of the necessary works, which meant that the works were not finally authorised until July 1867 (*ibid.*: 82; TNA RAIL 236/170: 21). The same month the company awarded the contract to construct 'a timber viaduct in the King's Cross Goods Depot and a wrought iron and timber bridge' over the Regent's Canal to Messrs W.H. Bracher & Son of Great Ormond Street for £3,884 (TNA RAIL 236/170: 28). Work on the new viaduct commenced shortly thereafter, and at the end of September the Way and Works Committee authorised the purchase of two weighing machines for the new bridge, the cost of which was met by the gas company (*ibid.*: 49).

The contract between the Great Northern and Johansson and Elliott was renewed four years later and the quantity of coal carried over the viaduct continued to increase until it reached nearly 20,000 tons per annum in subsequent decades (TNA RAIL 236/144: 270).

The construction of the Gasworks Viaduct in 1867

Landscape Feature 24

Whilst no traces of either the tracks leading to the viaduct or elements of the viaduct itself were identified during the various archaeological monitoring exercises that were undertaken within the former Goods Station, certain above-ground elements of the structure were observed during the historic buildings survey. These took the form of supporting timbers that abutted the southern end of the Retaining Wall (Landscape Feature 4), a number of possible tie connections associated with the structure and a pier on the east wall of the northern end of the Coal Drops (Landscape Feature 17).

4.16 The Hydraulic Station, Hydraulic Infrastructure and the Hydraulic Network: the mid-1850s to the 1870s

Landscape Feature 16

It is likely that numerous alterations, extensions and repairs to the hydraulic network were made during the lifetime of the Goods Yard. It appears that many of these modifications were too minor to have been recorded in the records of the various Board Committees that oversaw the company's expenditure, meaning that the documentary resource is an incomplete record in this regard In contrast, a great deal more information concerning the water supply arrangements for the hydraulic network is available, owing in large part to the occasionally fractious relationship between the Great Northern Railway and the Regent's Canal Company.

In November 1852 Joseph Cubitt submitted a request for the installation of an additional two cranes in the 'Departure Goods Shed' (i.e. the Western Transit Shed) to the Executive Committee of the Great Northern (TNA RAIL 236/74). His request was approved and it is likely that the cranes became operational in late 1852 or early 1853. It is likely that these cranes were installed next to the Western Transit Shed's canal branch as shown in Fig. 4.4 (since they break a pattern that is repeated in inverted form in the Eastern Transit Shed; Fig. 3.8). From this location the new hydraulic cranes could have been used very effectively to speed up the unloading of heavy goods from barges for onward transport by rail, thus admitting vessels to the shed's indent more quickly and reducing congestion in the Granary Basin.

As previously discussed, in 1853 Seymour Clarke appealed to the Board for permission to build a new heavy crane and an accompanying track extension on the east side of the Granary Basin to complement the existing crane that was situated to the northwest of that feature (TNA RAIL 236/21: 334; see also Section 4.2; Fig. 4.4).

In addition to the 10-ton crane added to the east side of the Granary Basin in 1853, another 10-ton crane was supplied in the early 1860s. It is likely that this was one of the outdoor cranes depicted on the Humber Plan of 1866 at the northern end of the Goods Shed complex

(Smith 2008: 4; Fig. 3.8). It had been joined by an 8-ton and another 10-ton crane by the time Humber compiled his survey in 1865

In March 1855, Seymour Clarke submitted a report to the Board of the Great Northern concerning the arrangements for the supply of water to the company's facilities at Peterborough and King's Cross Goods Yard and the allocation of responsibility between the two depots for the repair of the pumping apparatus at both (TNA RAIL 236/20: 228). Clarke also drew the Board's attention to 'the slender basis' upon which the arrangements between the Great Northern Railway and the Regent's Canal Company for the supply of water to the King's Cross yard rested. Those arrangements were explained in a letter written by Joseph Cubitt to the canal company in May 1851. When Seymour Clarke was compiling his report nearly four later, he noted that Cubitt's letter represented 'the only information possessed by the Company relative to the arrangements in force' and recommended that the two enter into a 'more formal understanding…than at present exists' (*ibid.*). The Board elected not to 'disturb' this arrangement by regularising it on a more formal basis.

Fig. 4.33 Stanford's Map of 1862 detailing the Hydraulic Station (Landscape Feature 16) and the Horse Provender Store (Landscape Feature 25), scale 1:1,250

In order to meet the growing demand for hydraulic power during the mid-1850s, Walter Brydone was authorised to purchase an additional stationary engine 'with pumping apparatus' in August 1856 (TNA RAIL 236/22: 175). Although it was documented that this was 'to be attached to the existing hydraulic apparatus' the nature of the connection was not described.

4.17 The Horse Provender Store: the mid-1850s to the 1870s

Landscape Feature 25

The Stanford map of 1862 indicated that a new structure had been built against the southern end of the Hydraulic Station in the ten years since Captain Galton had surveyed the Goods Yard for the Board of Trade (Fig. 4.33). Documentary sources indicate that this building was the Horse Provender Store, which was built to process and store feed for the horses that worked in the Yard

In October 1853 the Executive and Traffic Committee of the Great Northern recommended that 'machinery for crushing and stowing Horses' Provender' should be installed in the Granary (TNA RAIL 236/19: 89–90). The proposal was subsequently referred to Seymour Clarke, who recommended that rather than taking over part of the Granary, 'an order be issued for the preparation of a complete and well digested plan for carrying out a system of preparing Provender with proper stowage room for hay and oats with Steam Machinery and steaming apparatus for sweetening the hay, and with access by road, rail and water, with stables in the basement' (*ibid.*). Clarke's proposal for an entirely new building was referred to Joseph Cubitt who approved his colleague's recommendations and prepared draft specifications and estimates for the necessary works. At the beginning of January 1854 Cubitt's plans were approved by the Board and authorisation was granted for the expenditure of £4,300 to build the new Horse Provender Store (*ibid.*).

The structure had been completed by December 1854 when the Board authorised the installation of a new road coal wagon weighbridge nearby (*ibid.*: 112). It is not possible to determine from documentary sources alone if Clarke's plans to build stables in the basement of the building came to fruition. His plan to provide direct communication by rail and water was not realized, the store being accessible by road only.

If plans proposed by Seymour Clarke in November 1855 to convert the Provender Store into stables for Edward Wiggins, the Great Northern's authorised coal cartage contractor had come to fruition, the Horse Provender Store would have been very short lived indeed (TNA RAIL 236/21: 105). In the event, Clarke's suggestion came to nothing, and the building continued to be used to prepare and store horse feed.

It was not until April 1856 that Walter Brydone instigated the construction of an additional turntable and length of railway track to allow wagons loaded with feed to be taken directly to the store (TNA RAIL 236/22: 15). This measure finally provided direct rail access, thereby fulfilling Seymour Clarke's original plan. In contrast, Clarke's proposal to provide access by way of the canal was never realised.

Towards the end of October 1859, Walter Brydone prepared a specification and plans for the extension of the Provender Store and the conversion of the Straw Store to stabling for cartage horses (TNA RAIL 236/85: 314–315). Twelve contractors submitted tenders for the works, the contract for which was awarded to John Jay subject to the condition that the works would be completed within four months 'as it was important that no delay should occur' (*ibid.*: 328). Construction appears to have been approaching completion by the end of April 1860 (TNA RAIL 236/27: 296).

Historical maps suggest that the south side of the canopy that formed part of the Provender Store was extended southwards over the low level roadway to the Coal Depot to meet the curvilinear wall of the higher level Wharf Road viaduct during the five years between 1866 and 1871 (Fig. 4.4; Fig. 4.5). No reference to this undertaking was found in the records of the Great Northern, suggesting that it was probably ephemeral in nature, comprising little more than an external cover that could have been used to shelter carters as they unloaded horse supplies.

Archaeological evidence relating to the construction of the Horse Provender Store in 1854

Landscape Feature 25

The former Horse Provender Store had been demolished and no trace of it survived above ground level. Archaeological evidence of the building was however encountered during a watching brief (KXI07).

The remains of a building orientated northwest–southeast were unearthed in the southwest corner of the Goods Yard in the anticipated location of the Horse Provender Store. The remains were interpreted as the western section of the foundations of that building. The walls that were uncovered appeared to form three rooms and a corridor at basement level and it is reasonable to assume that the rooms were used for the storage of equine supplies. Unfortunately, the internal partitions had been severely truncated to the extent that no noteworthy features survived. In contrast, the northern external wall of the building was better preserved enabling former door and window positions to be determined (Fig. 4.34).

An extant access ramp linked Wharf Road and the bulk of the Goods Yard at the top of the terrace with the low eastern road that ran parallel with the Retaining Wall and the Hydraulic Station at the bottom. It is probable that a similar ramp existed in the mid to late 19th century. If this was the case then the western end of the southwest face of the lowest storey of the Provender Store would have been above ground, whilst the eastern end would have been subterranean as the external surface sloped upwards from west to east. Either way, it is likely that the majority of the Provender Store's lowest storey took the form of a basement since its northern and eastern sections definitely abutted the high ground that was still extant in those locations. It is therefore not surprising that lightwells were observed in the northern façade.

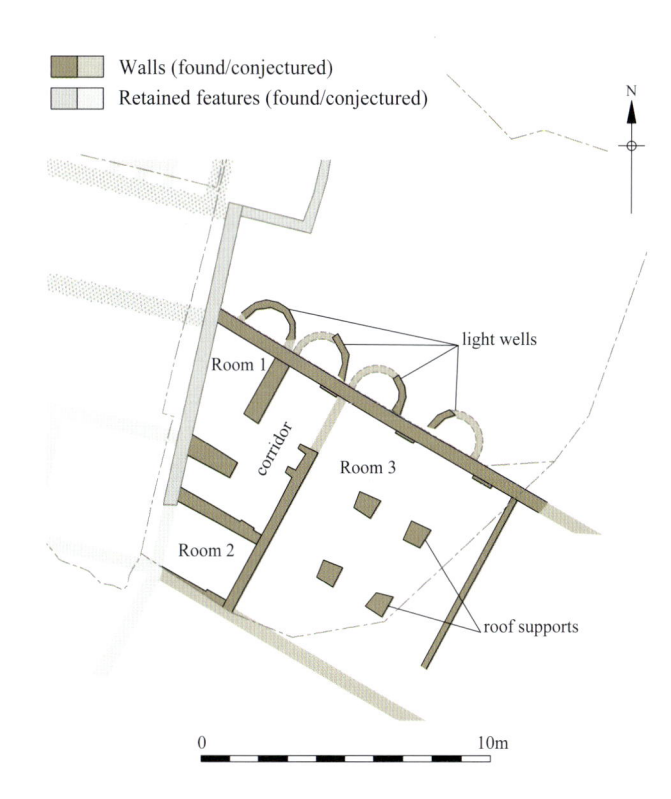

Fig. 4.34 Plan of the western end of the Horse Provender Store (Landscape Feature 25) by 1854 reconstructed from archaeological remains, scale 1:250

4.18 Synopsis

The Great Northern's direct route between the metropolis and the north enabled the company to provide a service that was generally faster and cheaper than those offered by its rivals. However, the new company faced fierce competition during its formative years, when it was attempting to pay off the huge debts that it had accrued during the construction of its network. It could not have endured that difficult period without the revenue it received from the lucrative transport of goods and minerals, particularly coal, to the metropolis.

The efforts of the Great Northern's senior officials placed the company in a far stronger position in the combative railway world of 19th-century Britain by the end of the 1870s than the company had been during the early 1850s. This meant that by the end of that period the company had access to a far wider geographical territory in London itself and the nation at large, which in turn enabled it to attract a larger customer base. Meanwhile, London was expanding rapidly, leading to increased demand for imported goods, as well as increasing the supply of exports that were manufactured by the city's industries. The speed of growth was so great that the period between the mid 1850s and the 1870s saw a rapid increase in the volume of traffic that was handled at King's Cross Goods Yard despite the construction of additional railway goods depots facilities by the company elsewhere in the environs of London. Matters were compounded between 1858 and 1862 by the arrangement with the Midland Railway, although such was the increase in the Great Northern's own traffic that the burden was not significantly eased by the Midland's departure.

The success of the Great Northern attracting London's potato traders to King's Cross from 1852 was a remarkable achievement, although it created its own problems. While the decision to convert the former Temporary Passenger Station into a dedicated potato warehouse was an astute move, the speed with which the conversion was effected meant that many potato traders were forced to work in substandard facilities for many years. Meanwhile increasing congestion in the main Goods Shed, necessitated the provision of additional platform accommodation in the Transit Sheds during the early 1870s.

In an early sign of a problem that would only increase as the 19th century wore on, the Board of the Great Northern somewhat grudgingly approved a number of improvements to the Goods Yard's rather basic road network from the mid-1850s. The external surfaces were originally little more than dirt tracks that quickly became heavily rutted, which prompted a somewhat reluctant campaign to pave them that spanned the entire period from 1853 to 1870.

Despite all of these improvements, it is telling that in 1873, just one year after the Eastern Transit Shed's Platform was enlarged, the Great Northern's General Manager, Henry Oakley, complained about the continuing 'want of further accommodation for the Goods Traffic' at King's Cross. Clogged roads and cluttered platforms seem to have remained a cause for concern, which must have impacted upon the operational efficiency of the complex. What is more, the stockpiling of goods on platforms sometimes posed a significant safety risk as demonstrated by the 'near miss' in one of the Transit Sheds involving the potentially explosive combination of fire and gunpowder that occurred in 1879. While the volume of road and rail traffic increased considerably throughout the period, traffic growth was largely confined to the road and the railway, exposing the long term weaknesses that were inherent in a design that placed such importance on the canal interchange.

The period between the early 1850s and the end of the 1870s saw the Great Northern succeed as a railway goods and coal carrier, something that it could not have been achieved without the effective facilities that it had built at King's Cross. Whilst competitors such as the Midland and the London & North Western companies provided services that directly rivalled the Great Northern's mainline into London, the huge demand that was generated by the needs of the metropolis coupled with the gradual expansion of the Great Northern's national network ensured that the latter company secured a significant market share of London's growing goods traffic for the duration of this period. The company's facilities at King's Cross were not perfect, however, which meant that by the end of the 1870s the railhead was handling more road and especially traffic than it could comfortably handle without the need to invest significant sums to fund improvements. In order to maintain the long term viability of a complex that was pivotal to the success of the company it subsequently became apparent that more significant changes would have to be implemented in order to redress the balance and it is these that will be discussed in the next chapter.

The Shed That Never Sleeps: How King's Cross Fed The City And Fuelled The North

'King's Cross goods...pours out a daily and nightly stream of traffic second to only one or two, if any, of the great London Railway companies. And...it takes in as well as gives out, for up and down, in and out, goods and coal, express trains and 'pick-ups' and return empties, the work goes on 'all round the clock'... All sorts of commodities 'strut their brief hour' upon the stage of an outwards goods shed from a 3-inch 'p.p.' to a gigantic furniture van, or from the 'wings' and 'flies' of a theatrical company 30 or 40ft long to a tiny box of spring flowers requiring the most careful handling, and which Jack Porter, who is a bit of a wag, is apt to speak of as a 'bloomin noosance'...Equally miscellaneous is the 'up' side traffic- bales, packs and trusses of textile fabrics, a small tin of oil here, a hogshead of tobacco there, washing and wringing machines, light castings, bags of nails, pumps, marble mantels, iron bedsteads, huge bales of paper, and a mountainous heap of wicker-work chairs, taking up such loading space, and carried at such generous rates as to promote the Company's dividend much in the same way as a hungry man would fill his stomach with a bottle of soda water'

Reminiscences of Mr. J. Medcalf, former Great Northern Railway Outdoor Goods Manager at King's Cross (Medcalf 1900: 313)

As one of London's principal goods hubs, King's Cross Goods Yard was a bustling, hectic workplace that remained in constant operation, 24 hours a day, seven days a week for more than 100 years. Receiving slow trains and expresses alike from as close as Hertfordshire or as far away as the north of Scotland, the station never slowed or stopped (Erwood 1988: 130-1; Medcalf 1900: 314-8).

Throughout its lifetime, the yard handled a vast array of items, ranging in size, weight and fragility from an enormous gun coil that required a team of 26 horses to move it on by road, to delicate boxes of flowers that were individually carried to waiting carts (Medcalf 1900: 313, 315). Whisky flowed in from the Highlands of Scotland, cotton arrived from the mills of Manchester while paper was brought in by train from the factories of Aberdeenshire (Erwood 1988: 131). A small reminder of the extent of the network established by the Great Northern was discovered during excavations in the Train Assembly Shed, where a North British Railway (NBR)[4] axle box cover, which must have originated in the latter company's locomotive works in Glasgow, was discovered.

One particularly memorable load, the bloated and stinking carcass of a beached whale was shipped from a Lincolnshire beach to King's Cross from where, after much aggravation and unpleasantness for the shunters and porters, it was forwarded by road to the Natural History Museum in Kensington (*ibid.*: 132). But it is those commodities that passed through the yard in enormous quantities that are of particular interest here, since they helped finance the economy of the north, kept London running, bankrolled the Great Northern and its successor the LNER in their turn and dictated the nature of the station's daily routines.

Ever since the early 1850s, London's ravenous appetite for coal meant that huge quantities of that commodity flowed into King's Cross by rail for onward dispatch to the metropolis and beyond. The Great Northern Coal Depot grew from a single set of coal drops and some associated sidings in the

---

[4] The NBR was an Edinburgh based railway company, whose track network covered the east of Scotland and areas in the far north of England. South of Berwick-upon-Tweed, the company had running powers over North Eastern Railway (NER) and Great Northern Railway (GNR) rails along the East Coast mainline, in exchange for NER and GNR access to Scotland. This collaboration enabled the three companies to operate through services between King's Cross and the main Scottish cities of Edinburgh, Glasgow and Aberdeen as well as providing a link with stations even further north. The partnership was solidified as early as 1860, when the three companies established the East Coast Joint Stock (a pool of passenger coaches) and continued until the three partners were absorbed into the LNER in 1923.

1850s, to encompass two sets of drops, an office block and additional tracks by the 1860s. Facilities for the reception and dispatch of coal were further improved later that decade, when the entrepreneurial private coal trader Samuel Plimsoll opened his rival operation in nearby Cambridge Street. A constant stream of horse-drawn drays, tasked with the onward transport of coal to the hungry city would have progressed through the 'dust and din' of the coal yard as rolling stock stationed on the viaducts above discharged their loads (Gordon 1893: 134). All the while the coal stacking ground to the northwest and the mass of sidings that surrounded the various drops would have been a hive of activity as full wagons were taken for unloading whilst the empties were marshalled into their correct positions for their return journeys to the northern coalfields. Although the Great Northern would introduce additional coal reception facilities elsewhere in London as the 19th century progressed, King's Cross remained central to the north London coal trade until the mid 20th century, after which the sidings at Caledonian Road became the focus (Erwood 1988: 132).

Although no single commodity would be able to eclipse coal in terms of volume and profitability, throughout the 19th century huge quantities of stone, bricks and timber were brought into King's

North British Railway axle box cover unearthed in the Train Assembly Shed

This photograph, taken c.1906, shows that in its heyday King's Cross Goods Yard was a hive of activity throughout the night (reproduced from Cassell & Co. 1906: 291 © Camden Local Studies and Archives Centre)

Cross by rail. Until 1899 these items were offloaded in the vicinity of the Coal and Stone Basin and after that in and around the sidings that surrounded the Eastern Goods Shed. The stone arrived from the quarries of Hertfordshire, Lincolnshire and Staffordshire, but above all it was 'imperious' Yorkshire that dominated the trade (Medcalf 1900: 316). Brick traffic represented an increasingly large proportion of the goods brought into King's Cross as the 19th century progressed. Trainloads of White bricks came in from the brickworks of Arlesley (Bedfordshire), whilst cargoes of Fletton bricks were carried from Peterborough in huge quantities (ibid.).

Vast quantities of grain were brought into the yard for storage in the Granary and subsequent distribution to the capital's corn factors and millers (Anon 1851b; Medcalf 1900: 315). Equipped with hydraulic lifts and gravity driven chutes, the Granary was capable of holding upwards of 60,000 sacks of grain. The Granary remained busy until the late 1920s, by which time domestic cereal production had been partially usurped by foreign imports (Atkin 1995: 16–22). This led to the conversion of the structure from a Granary into a bonded sugar warehouse and a depository for goods containers (see Section 9.1).

Textiles, particularly woollen and cotton stuffs from Lancashire and the West Riding, were also imported in bulk, as were metal products from the foundries of the industrial north (Anon 1851b; Medcalf 1900: 315). Whilst the former was brought into the arrivals shed, some components of the latter, namely larger items like girders and boilers that were too big and heavy to be brought inside, were received on external sidings (Medcalf 1900: 315). Wagons carrying such items were presumably directed to the heavy cranes that were dotted around the basins and the station's northern approaches, where crane operators and their crews would have undertaken the difficult and dangerous task of unloading. By 1900 a large trade in hay and straw was also taking place outdoors in an area where locomotives undertook the bulk of the shunting (ibid.: 316).

A short distance to the north of the Goods Station an enormous cattle market at Copenhagen Fields belonging to the Corporation of London was supplied by Great Northern rails from 1855 until the first half of the 20th century (ibid.: 317; Brandon 2010: 85; Kean 1998: 64). Throughout the 19th century, 'countless droves of Durhams, Shorthorns, Teviots' and numerous other breeds arrived by the train load for sale and butchery, along with innumerable sheep, pigs and other forms of livestock (Medcalf 1900: 317). In the first instance these animals were driven to market along York Way, a practice that was presumably reduced by the construction of dedicated unloading pens at Holloway Road Goods Depot after it opened in the early 1860s (ibid.; Nock 1974: 64). Although a shadow of its Victorian self, the cattle market remained operational into the 1930s (Nokes 1938: 124). It is therefore reasonable to assume that livestock continued to be brought in by train until then, although the frequency of arrivals must have fallen drastically, since the sidings that formerly served the market were devoted to the reception of coal by 1938 (Erwood 1988: 132).

After sunset, under the 'naphtha flare' of gas lamps, trains and goods continued to arrive and depart as shunters, checkers, porters and cartage teams carried on with their business (Medcalf 1900: 313). The small hours were skewed towards the reception of perishables that had to reach London's food market before daybreak, making early morning work in areas like the Eastern Transit Shed, the Handyside Canopies and the Potato Market a particularly hectic experience (ibid.: 313). Before the second half of the 20th century, these buildings received train loads of fruit and vegetables and vast quantities of fish traffic from the eastern ports of Grimsby, Hull and Aberdeen throughout the night (ibid.; Erwood 1988: 131; TNA RAIL 1124/127). Indeed, the British love affair with fried fish and chips that began in the latter half of the 19th century was directly driven by urban railway goods depots like King's Cross (Brandon 2010: 85). After the advent of canning, this trade was extended to the reception of tinned foods of all kinds, including large quantities of peas that heralded from the factories of Bardney and Brigg in Lincolnshire (Erwood 1988: 131).

The Potato Market with its 'shabby' stalls was subdivided into 35 to 40 'runs', each of which was the preserve of a single potato merchant (Medcalf 1900: 315). Although subject to yearly fluctuations caused by the weather and the prevalence of pests and disease, during the potato season every 'run' typically received between 100 and 200 trucks daily, each loaded to the brim. Many consignments came from the surrounding counties of Bedfordshire, Hertfordshire and Cambridgeshire, although the majority came from Lincolnshire and Yorkshire, with some originating from as far away as Northumbria and Lothian in Scotland. Freshly picked greens and fruit also poured in throughout

Cattle being loaded into railway wagons for onward shipment to the urban market, no date (SSPL 1996-7038_BTF_92_28 © National Railway Museum and SSPL)

Fish piled high on the quay at Grimsby docks awaiting dispatch by train, October 1961 (SSPL 1995-7233_LIVST_MP_444B © National Railway Museum and SSPL)

Sacks of potatoes stacked in the Potato Market (Landscape Feature 14) at King's Cross in 1951 (SSPL 1995-7233_LIVST_DF_176 © National Railway Museum and SSPL)

the night so as to reach the capital's dinner plates by the following day. When in season, this trade was dominated by rhubarb from Yorkshire, carrots from Bedfordshire and celery from Lincolnshire, accompanied by a vast array of other kinds of vegetables (ibid.; Erwood 1988: 132; Brandon 2010: 87).

In addition to these necessities, certain private companies transported such large volumes of stock via the Great Northern Railway that they saw fit to rent warehouse accommodation in the yard itself. Over the years, notable tenants included the Yorkshire bottle manufacturers, Kilner Brothers, who took up residence in the Midland Goods Shed, Joseph Bagley & Co., another Yorkshire bottle maker, who rented space in the defunct coal drops (after the focus of the coal trade moved to Cambridge Street) and the famous Staffordshire brewers, Bass, who leased a depot elsewhere in the yard for the storage of bottled beer (Medcalf 1900: 313).

Plenty of stuffs flowed outwards in the opposite direction. According to Peter Erwood, who worked for the LNER as a Junior Clerk at King's Cross in the 1930s, 'every day from about 2pm onwards until the small hours, loaded trains left King's Cross carrying the products of London's industry destined for virtually every station on the system, which covered some 6,000 route miles' (Erwood 1988: 131). By Erwood's time, the range of packaging that the yard had to cope with was 'bewildering', as a never ending procession of 'crates, boxes, cartons, sacks, bags, skips, punnets, carboys, drums, barrels... and bales' passed through (*ibid.*).

King's Cross Goods Station was a vital supply artery that was of critical importance to the economies not only the nation's capital city, but also of the farming communities of East Anglia and Lincolnshire and the industrial north. The station simply could not afford to stop, irrespective of the time of year, the day of the week or the hour of the day.

CHAPTER 5

The Introduction of Capstan Shunting in the early 1880s

'I remember there were a lot of derelict buildings ... [that] we used to walk around as apprentices and just explore ... it was like an Aladdin's Cave... one place there was a big tall tower and I don't know what it was for... it was overtaken by doves and it was actually full, like a dovecote and the floor was covered with bird muck ... ankle deep and the smell was terrible....it was a building in the Goods Yard but it's gone now... there were capstans that we used to maintain ... I can't remember where the water supply for them was...it was a pressurised central water supply that went all over the goods yard to operate these capstans; it was quite a historic system'

Former railway mechanic Paul Snushell's description of the relict hydraulic station accumulator tower and some working hydraulic capstans that could be found at King's Cross Goods Yard as late as the 1970s; from an interview by Alan Dein © London Borough of Camden

The Great Northern reached a peak of profitability as early as 1873. Whilst the amount of traffic that it carried would continue to grow after that date, the amount of revenue that it would have to reinvest in its network would also grow, meaning that the company would never better the results of that year (Simmons and Biddle *et al.* 1997: 192). The late 1870s to the early 1880s were therefore characterised by investment across the entire Great Northern network in order to enable the safe passage of an ever growing number of trains.

In the summer of 1881, Henry Oakley (Great Northern General Manager) submitted plans to the Way and Works Committee of the Great Northern for the installation of hydraulic capstan shunting in the Main Goods Shed at King's Cross (TNA RAIL 236/174: 387). Having previously overseen the successful introduction of the technology at the Potato Market (Landscape Feature 14), Oakley proposed an ambitious scheme that would involve the expenditure of more than £13,900 on hydraulic equipment sufficient to make 35 horses and eight men surplus to requirements (*ibid.*; TNA RAIL 236/43: 22, 68; TNA RAIL 236/146: 56; TNA RAIL 236/173: 170). An order worth £8,050 was subsequently placed with Tannett Walker of Leeds for the supply of an accumulator, an additional pumping engine, pipework, capstans, snatch heads and ancillary equipment. Although the exact quantities of the latter items were not specified in the committee's minutes, the order may have covered the cost of approximately 35 capstans and as many as 175 snatch heads (Smith 2008: 8). During the installation of the new technology, Oakley asked for a further £1,510 for 'additional machinery for the hydraulic work at the King's Cross Goods Yard, so as to avoid the necessity for extra horse hire', which was authorised by the Board in October 1882 (TNA RAIL 236/148: 253). It has been estimated that this order may have amounted to a further eight capstans and associated snatch heads (Smith 2008: 8). Three months later a further capstan and five additional snatch heads were purchased from Tannett Walker for £167 (TNA RAIL 236/148: 332; TNA RAIL 236/350: 9). The effects of this extensive technological overhaul on the buildings and railway infrastructure within King's Cross Goods Yard are explored in the pages that follow.

5.1 The Main Goods Shed and Granary Complex: the early 1880s

Landscape Features 7–12

At the end of June 1881 Richard Johnson (who had succeeded Walter Brydone as Chief Engineer on the latter's retirement in 1861) presented a scheme to the Traffic Committee of the Great Northern that he claimed would both 'facilitate the working of traffic and relieve the expenses of working' (TNA RAIL 236/147: 357). Johnson proposed to build a new overhead platform supported by columns at the north end of the Eastern Transit Shed for £5,925, which would store goods 'waiting to be called for'. The proposal was approved by the Board and the platform was built shortly thereafter. It is conceivable that this new structure superseded the platform extension that had been constructed at the opposite end of the shed ten years earlier, a section of which appears to have been removed by the time the 1882 station plan was surveyed (Fig. 5.3). The latter plan suggests that the removal of part of the 1871 platform permitted the turntables that allowed wagons to be transferred from the Arrivals platform to the Granary to be reinstated, although no documentary evidence was found to substantiate this proposition. Although instigated at roughly the same time as the introduction of hydraulic capstan shunting to the shed, the construction of the new platform seems to have formed part of a separate project.

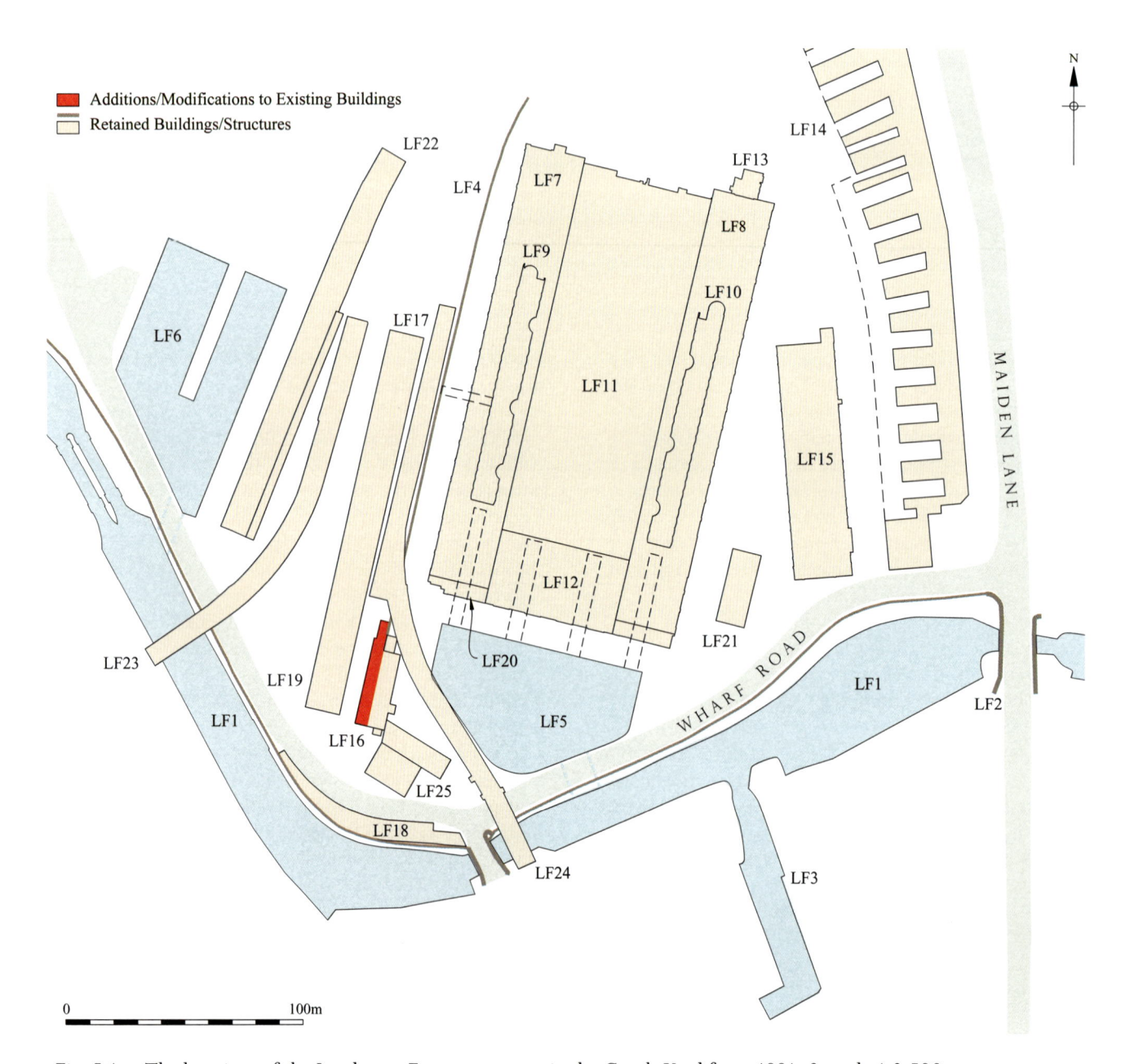

Fig. 5.1   The locations of the Landscape Features present in the Goods Yard from 1881–3, scale 1:2,500

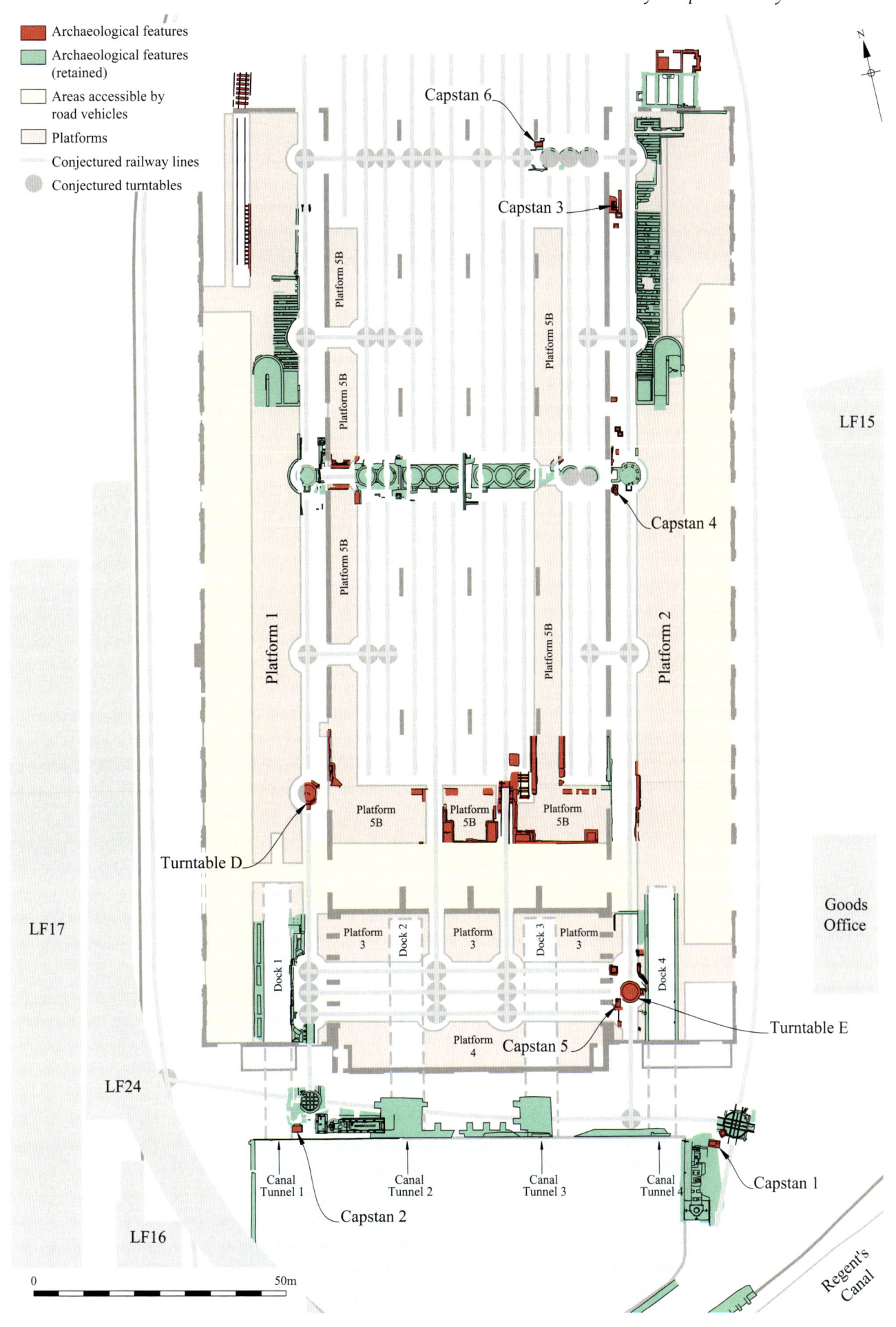

The Main Goods Shed and Granary Complex: the early 1880s 145

Fig. 5.2   The locations of the significant archaeological features dated 1881–3, scale 1:1,000

Historic plans of the station reveal that a considerable quantity of additional platform space was created within the Train Assembly Shed between 1873 and 1882 (Fig. 5.2; Fig. 5.3). Effectively a replacement for Platform 5a (Fig. 4.3; Fig. 5.2), the new feature (Platforms 5B, which comprised several separate short sections) took the form of a large 'U' shaped structure, its 'arms' extending northwards, terminating only a short distance to the south of the northernmost bank of turntables (Fig. 5.3). It was necessary to divide the bank into sections in order to accommodate pre-existing railway sidings. Consequently, the western 'arm' was split into four discrete sections, the eastern 'arm' into two and the east–west 'body' into three (Fig. 5.3). The platform had been inserted at the expense of two of the Train Assembly Shed's north–south railway tracks, one east–west railway track and 26 turntables. Whilst the east–west siding had been completely removed, the north–south examples had been cut short so that they now terminated to the immediate north of the newly constructed 'arms' of the platform. Twenty-five of the 26 affected turntables were removed in their entirety, however the most southwesterly example, Turntable D, was left *in situ* despite the fact that it was, from that point onwards, a redundant feature (Fig. 5.2; Fig. 5.3). New cart access was also inserted between the southern side of the new platform and the north wall of the Granary (Fig. 5.2; Fig. 5.3).

No reference to the construction of the 'U' shaped platform nor the removal of the existing railway infrastructure was found in the minutes of the various Board committees that met during this period, a curious omission given the substantial scale of the alterations. Since the Great Northern took care to record much smaller undertakings, it seems unlikely that these significant alterations were not thoroughly discussed by the company. The most logical conclusion is that the new platforms were built in conjunction with the installation of the first generation of hydraulic capstans, thus forming part of that project and being subject to its budgets, tenders and planning.

The original purpose of the 1882 plan of the goods station (Fig. 5.3) is not recorded; however it was clearly produced by the Great Northern for the use of the company's senior officers. Since the surveyor who prepared it took care to emphasise the Goods Shed complex, the adjacent Midland Goods Shed and the extended Hydraulic Station through the use of thicker, darker outlines than those used to illustrate the other buildings, it is likely that they were the primary focus of the plan.. It is distinctly possible that the document was originally issued to illustrate the scope of the hydraulic extension programme of 1881–2, including the alterations to the Train Assembly Shed's platforms and railway infrastructure. Since the programme of works began in 1881 and the map dates to 1882 it may represents an inventory of recent changes associated with the introduction of capstan shunting rather than a proposal plan, however later sources do suggest that it does not tell the whole story.

The Second Edition Ordnance Survey map of 1894–6 indicates that during the years since 1882 four turntables had been reinstated in their pre-1882 locations, two had been added and another two had been removed from the Train Assembly Shed (including the example that had been rendered redundant in the earlier phase, Turntable D). Two more had been taken out of the southern end of the Eastern Transit Shed allowing the alcove that formed part of the edge of the platform to be modified thus increasing space for goods handling. These changes brought the total number of turntables in the Goods Station to 50 (Fig. 5.4).

Following the installation of the first batch of hydraulic capstans and snatch heads in the Train Assembly Shed in 1881–2, a second consignment was ordered from Tannett Walker in October 1882 and installed early the following year (Smith 2008: 5, 8). It is therefore probable that most if not all of the alterations that appear on the Second Edition Ordnance Survey map of 1894–6 took place in 1883 when this second batch of hydraulic equipment was installed.

Archaeological evidence relating to the introduction of hydraulic capstan shunting to the Main Goods Station and Granary Complex in 1881–2

Landscape Features 7–12

The archaeology encountered strongly supported the content of the station plan of 1882, suggesting that the introduction of hydraulic capstan shunting was indeed accompanied by a dramatic reconfiguration of the platform accommodation in the Goods Station. Evidence demonstrating the existence of the 'U' shaped bank (i.e. Platform 5) as shown on the 1882 Station Plan was found within the Train Assembly Shed (Landscape Feature 11) in the form of 21 discrete masonry fragments.

The platform was clearly a later modification as the archaeology that was uncovered demonstrated that its construction caused two of the Assembly Shed's north–south railway sidings to be truncated, whilst the most southerly east–west siding and all but one of its affiliated turntables were lost. The surviving turntable (situated in the Western Transit Shed) is something of an enigma since the insertion of the new platform cut it off from all but one siding, meaning that it was no longer capable of performing any practical task.

No evidence relating to the cart access that ran through the complex to the immediate north of the Granary (Landscape Feature 12) in 1882 was observed since the feature had been widened and reconstructed in the 1930s. However, remains associated with it were observed in the Granary. The 1882 incarnation of the access point enabled the north face of that building to become accessible to road vehicles for the first time, which in turn provided the opportunity to install more loading points in that elevation. Consequently, the pedestrian doorways that had formerly been used for unloading railway wagons were thenceforth used to manually load horse-drawn carts whilst new chutes for the delivery of grain to waiting vehicles were installed in two window positions thus providing further communication with the first floor. Black circles survived either side of these, within which the numbers '1' and '2' had been written in white paint. Like their counterparts on the southern elevation, it is likely that these numerals formed part of a system that guided carts and goods to the correct positions for loading.

The remains of six hydraulic capstan houses were unearthed archaeologically (Fig. 5.2). These presumably formed part of either the orders that were placed with Tannett Walker by the Great Northern in 1882 or 1883. Capstans 1 and 2 were located outside the Goods Station, beside Turntables A and B, Capstan 6 was found in the Train Assembly Shed and Capstans 3 to 5 were found in the Eastern Transit Shed. It is thought likely that Capstan 5 was part of the 1883 order (see below).

Capstan 6, which was the best preserved, was unearthed beside the northern end of the third 'up' siding from the east in the Train Assembly Shed (Fig. 5.2). It consisted of a hollow metal box capped by three removable metal plates, which permitted access to the interior of the box so that the machinery within could be inspected and maintained. The only *in situ* capstan mechanism that was unearthed during the excavations was found inside this feature, its workings having been protected by the metal lid (Fig. 5.6). The box itself was rectangular, being orientated east–west, its dimensions being 1.65m in length, 0.95m in width and 0.61m in depth, whilst the diameter of an associated turning mechanism was found to be 0.50m at the base, tapering to 0.24m at the top. A deposit of

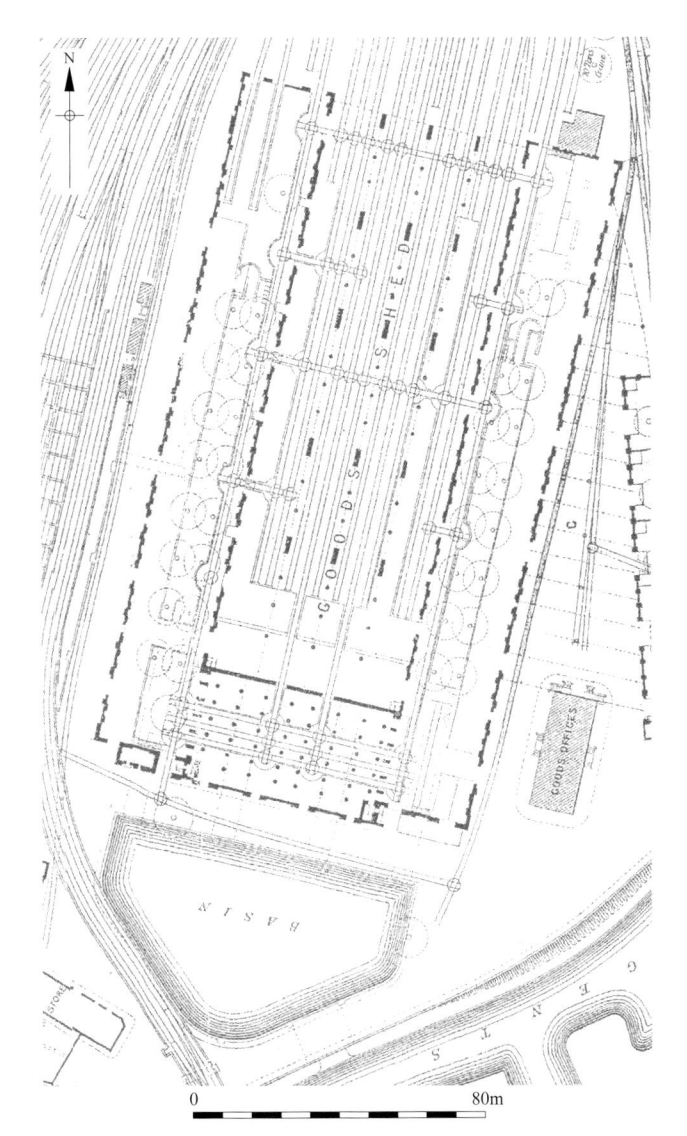

Fig. 5.3 Great Northern Railway Plan of 1882, detailing the Main Goods Station (Landscape Features 7–11), the Granary (Landscape Feature 12) and the Granary Basin (Landscape Feature 5), scale 1:2,000

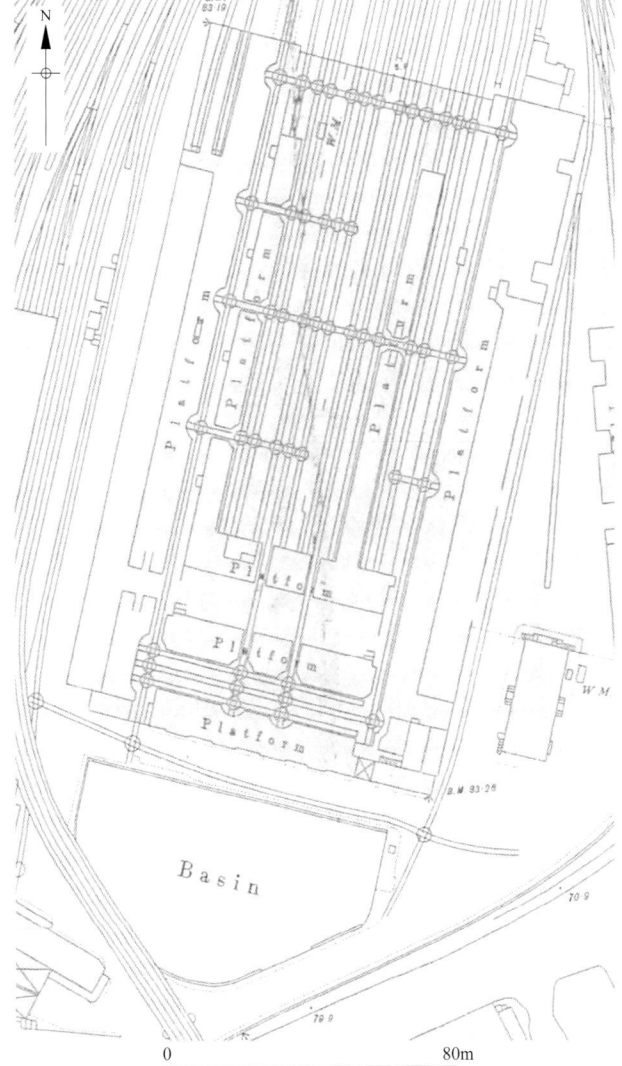

Fig. 5.4 Ordnance Survey Second Edition 1894–6 detailing the Main Goods Station (Landscape Features 7–11), the Granary (Landscape Feature 12) and the Granary Basin (Landscape Feature 5), scale 1:2,000

concrete had been poured into the feature's construction cut in order to cement it in place. Whilst the capstan was largely subterranean, the top of the box being flush with the 19th-century ground surface, the turning mechanism was upstanding to a height of 0.22m. The capstan head would have sat immediately on top of that mechanism.

Analysis of the archaeological remains demonstrated that the Great Northern's capstans, purchased from Tannett Walker between 1881 and 1883, almost certainly utilised the hydraulic motor that was patented by Benjamin Walker in 1882 (Smith 2008: 9).

Walker's patent hydraulic mechanism comprised four cylinders, arranged in two opposing pairs, which faced each other across a gap that was bridged by a 'double ram'. When pressurised water entered one of a pair of cylinders it was exhausted by the other. A forked connecting rod, which acted on the crank of the vertical shaft, took the drive from the double ram, whilst the vertical shaft itself had two cranks for the two pairs of cylinders. The capstan head was mounted on top of this shaft, which it rotated in a clockwise direction. Each of the machine's working valves had three ports: the two end ports were tasked with supplying each cylinder in a pair with high pressure water, whilst the central port conveyed waste water away. The valves were operated by a connecting rod and lever system with a spindle that connected with the slide valve. Cylinder pairs were mounted one above the other in a cast iron box adjacent to the capstan, which was controlled by way of a foot pedal or treadle valve. Walker subsequently modified his hydraulic motor by making the diameter of one cylinder of each pair larger than the other. His new motor also contained rams of different diameters that were connected in pairs as they were in the earlier version of the model. The smaller of the rams was connected to pressure water, whilst the larger ram was alternately connected to pressure and exhaust water, thus simplifying the arrangement of valves (ibid.: 9–10).

Hydraulic capstans were used for both wagon shunting and for rotating turntables, each of which was operated by a team of two, the 'capstanman' and a 'boy'. The job of the former was to operate the treadle that controlled the winding apparatus and hold the hemp rope that would be coiled around the next capstan. The latter would run the rope out from the capstan, attaching it to the wagon that needed to be shunted or turned (ibid.: 10).

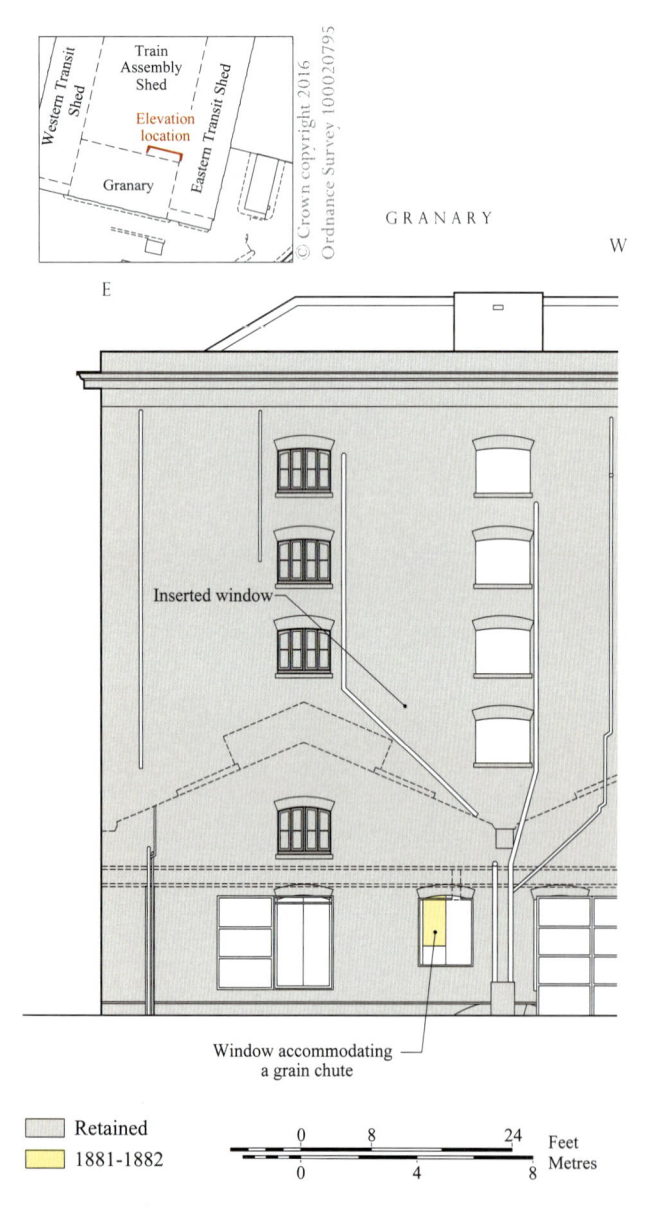

Fig. 5.5 External elevation of the north wall of the Granary (Landscape Feature 12) showing a grain chute that was installed after road vehicle access was granted in 1881–2, scale 1:250

Fig. 5.6 Hydraulic Capstan 6 with *in situ* mechanism, unearthed in the northern end of the Train Assembly Shed (Landscape Feature 11)

It is conceivable that Capstan 3 in the northern end of the Eastern Transit Shed turned the northernmost of the shed's turntables, however the two devices were separated by a distance in excess of 5m so it could only have done this through the use of intermediate snatch heads. It was much better placed to facilitate wagon shunting by hauling wagons from inward-bound trains into the Shed. Further south, Capstan 4 would have been capable of turning an adjacent turntable and shunting wagons further along the length of the shed for unloading. Similarly, the most southerly capstan in the building, Capstan 5, could have been used to haul inward-bound wagons to the end of the structure; it could also have been used to turn at least one adjacent turntable. It is likely that identical capstans were fitted at similar intervals in the Western Transit Shed, although none survived within the confines of the archaeological excavation areas. The one example that was found in the Train Assembly Shed, Capstan 6, was well positioned to turn a number of wagon turntables as well as the adjacent example by means of snatch heads. It may also have been used to haul wagons in or out of the Train Assembly Shed on the adjacent 'Up' line. Externally, Capstan 1 would have turned Turntable A and pulled rolling stock along the adjacent lines, whilst Capstan 2 was responsible for turning Turntable B and hauling wagons along the edge of the Granary Basin (Landscape Feature 5). More hydraulic capstans and associated snatch heads were no doubt installed at regular intervals throughout the Goods Yard but, given the availability of snatch heads, it would not have been necessary to have fitted a capstan beside every turntable.

Archaeological evidence relating to further modifications to the Goods Station in 1883

An archaeological excavation in the Eastern Transit Shed (KXI07) demonstrated that the semi-ovoid alcove that once formed part of the southern end of the Shed's platform was replaced with a smaller semi-circular structure during a later phase, which surrounded Turntable E (Fig. 5.2). Only a 4.50m long fragment of the replacement survived. It presumably originally butted the northern and southern edges of the platform's perimeter wall, a stratigraphic relationship that was lost due to the presence of modern truncations. The new construction was built with red and yellow, predominantly English bonded bricks. It had been erected within a construction cut that truncated two turntable bases, demonstrating that it post-dated these structures. The locations of the turntables in question corresponded with the two examples that were removed from the southern end of the Eastern Transit Shed as shown on the Ordnance Survey map of 1894–6; furthermore, the rebuilt platform edge was identical to the new one that was first shown on that map (Fig. 5.4).

A well preserved metal housing for Capstan 5 was unearthed to the immediate east of the reconstructed alcove and its one surviving turntable base (Fig. 5.2). Since the stratigraphy suggested that its insertion took place when the alcove was remodelled it presumably formed part of the 1883 order from Tannett Walker.

Fig. 5.7 Capstanmen in 1951, photographed in the process of turning a wagon in the King's Cross Potato Market (Landscape Feature 14) through the use of a hydraulic capstan and snatch head (SSPL 1995-7233_LIVST_DF_178 © National Railway Museum and SSPL)

5.2 The Hydraulic Station: the early 1880s

Landscape Feature 16

It must have been immediately apparent to the Senior Offices of the Great Northern that the introduction of hydraulic capstan shunting in the Goods Station would require hydraulic power output to be increased, placing considerable pressure upon the yard's existing power generation facilities. Consequently, enlargement of the Hydraulic Station (Landscape Feature 16) appears to have been integral to the scheme from the outset.

At the beginning of November 1881 a contract for the 'erection of an Engine House for Hydraulic Apparatus' worth £1,931 was awarded to Messrs Wall Bros (TNA RAIL 236/175: 2). Whilst the specification for the works has not survived, the equivalent document for a subsequent extension dating to 1898 has. This document, along with a number of cartographic sources, provides a number of insights into the arrangement and dimensions of the building as it was after the works of 1881–2.

The 1881–2 extension appears to have entailed the construction of a western abutment, which effectively consisted of a new engine room and boiler house that was entirely separate from the originals. The extension must have been approaching completion by the following March, when the Way and Works Committee authorised the expenditure of £600 to purchase a new suction tank and to establish a connection between an existing Armstrong engine and a new 75hp Tannett Walker model that had recently been installed in the new building (TNA RAIL 236/175: 75). It has been suggested that by connecting the new and established hydraulic systems by means of pressure reduction valves, pressure throughout the network could have been raised from 500psi to 650psi (Smith 2008: 11–2). A Goad Fire Insurance plan of 1891 (not illustrated) indicates that the 75hp Tannett Walker engine was accommodated in the new western extension of the Engine House, whilst the old Armstrong 24hp engine was retained in its original location (Goad 1891).

A report prepared by the Great Northern Locomotive Department in 1914 indicates that other engines were present in the Hydraulic Station in the wake of the 1881–2 rebuild. Two small pumping engines, said to be 30 years old when scrapped, were removed in 1914 which in turn suggests that they were installed in the mid 1880s (TNA RAIL 236/216: 271; Smith 2008: 14). Since the age given in the document probably represents an approximation, they could have been installed as early as 1882 thus forming part of the 'refit' of that year.

The Hydraulic Station's accumulator was reported to have worn out by 1881. A replacement was therefore sourced from Sir W.G. Armstrong, Mitchell & Co. for £500 (Smith 2008: 12). With an 18" ram and 20' stroke, the new model was too large for the existing accumulator tower, so it was decided to accommodate both it and the Tannett Walker model that had already been ordered in a new purpose built structure at the northern end of the new extension (TNA RAIL 236/175: 85).

A specification for a programme of later alterations drawn up in 1898, revealed much about the nature of the structure that was created in 1881–2. The document termed the 1851 accumulator tower the 'small Accumulator House', although it no longer performed that function by then. It was described as being '52' girt and 35' high with door and window in same' (TNA RAIL 236/449: 15). It is not clear as to what use this room was put after the accumulators were relocated in 1881–2, but the fact that it was described as being 35' high (16' 6" higher than the adjacent walls), suggests that the tower was left *in situ*. The same document describes the two rooms that housed the engines as 'the Back and Front Engine Houses'. It states that they were separated by an internal wall that was 55' long and 18'6" high to floor level, whilst the wall 'in front of the Front Boiler House and Front Engine House' was described as being 73'6" girt and about 17' high with [glazed] doors and windows in the same' (ibid.: 14–5). The wall between the two Boiler Houses was described as being 56' long and 28' high. The Coal Store was no longer a separate unit, but was mentioned in the context of the 'Boiler House'. The contractor for the 1898 rebuild was instructed to 'take down the present Beams and Coal Track', which was the probable means by which fuel was brought into the Boiler House. The same specification indicates that the coal was discharged into a 'Coal Shoot', although this structure was not described in detail. Access to the Boiler House was apparently gained via 'an old landing and steps'. This statement either refers to an entrance that led down to the Front Boiler Room from an external area to the immediate east or relates to an internal stair and gantry that connected the higher floor of the new 1881–2 Accumulator Tower and Engine House with the lower floors that were present throughout the rest of the building.

The goods station plan of 1882 is the earliest source to depict the Hydraulic Station extension of 1881–2 (not illustrated). In keeping with the documentary evidence (summarised above), it suggested that the extension took the form of a linear structure that abutted the former western external wall of the original Hydraulic Station (i.e. the southern end of the Retaining Wall, Landscape Feature 4). The newly constructed external western wall was straight, running parallel with its earlier counterpart except for the far northern portion, which kinked slightly to the east before continuing to the north. The original Engine House remained extant, the only major alteration being the southerly extension of the Boiler House. Internal doorways between the various rooms are also detailed on the 1882 plan and the distribution of these has been used to aid the ensuing reconstruction.

Archaeological evidence relating to the Hydraulic Station extension of 1881–2

Landscape Feature 16

Remains relating to the Hydraulic Station extension of 1881–2 were revealed during an archaeological excavation (KXO08). The evidence was limited to the northern half of the new extension, as the most southerly room (as shown on the 1882 station plan) had been destroyed when the building was once again modified in 1899. The arrangement of the recorded remnants closely resembled those that appeared on the 1882 station plan.

The Retaining Wall (Landscape Feature 4), which once formed the western side of the original Hydraulic Station, now separated that structure from the newly extended section of the building that could be found to the west (Fig. 5.9). The high ground that abutted the original, eastern section was therefore elevated at a height of approximately 24.00m OD, whilst ground level to the west of the new construction was relatively low at 21.37m OD. The original Hydraulic Station survived unmodified, its working surfaces remaining at the same approximate level as the low ground on the western side of the terrace. The floor of the new Boiler Room was not identified but was probably built at a similar height to its eastern counterpart, whilst the floor surfaces of the newly constructed Engine House and Accumulator Tower were elevated by approximately 1.40m. Such a height differential would explain why no connecting doors seem to have been inserted between the new Engine and Boiler Houses.

The 1882 station plan indicated that the internal pedestrian route from the New Engine House to the New Boiler Room was convoluted. Taken together, the archaeological and cartographic evidence concerning the various floor levels suggests that it would have been necessary to walk through one of the two doorways that linked the old and new Engine Houses, descend a short staircase, walk through a connecting door to the original Boiler House and then pass through another door that connected the old and new Boiler Houses (Fig. 5.9).

The northern half of the outer western wall of the new extension was unearthed 4.64m to the west of the Retaining Wall (Landscape Feature 4). It was 0.52m wide and was constructed of red, frogged bricks, faced with yellow frogged bricks on the external side for decorative effect. At least four stepped foundations continued downwards for over 1m below 19th-century ground level whilst above ground, the wall survived to a height of 24.48m OD, 3.11m above 19th-century ground level on the western (low) side of the terrace. Two windows, which had been bricked up at a later date, were integral to the wall.

After the new outer west wall was built, two internal east–west partitions were erected. The northern example was 4.35m long and 0.58m wide and had been constructed from frogged red bricks. It did not run all the way across the extension, leaving a 0.88m wide gap between it and the external western wall. This gap is thought to represent a doorway between the northernmost rooms. The southernmost internal wall was constructed with identical materials; it presumably separated the two rooms that formed the southern end of the western extension. Beyond that point the 1881–2 Hydraulic Station had been destroyed due to the damaging effects of later building work (Fig. 5.9).

The surviving remains of the smallest and most northerly room of the new extension were 4.32m in width and 8.40m in length. Its floor was paved with stretcher bonded, frogged, red bricks, the tops of which were observed at a height of 22.42m OD. Two octagonal masonry bases were unearthed inside the room. They were 3.14m wide, each side being between 1.33m and 1.30m long. Each consisted of a hexagonal concrete surround (Fig. 5.8), 0.48m wide and 0.24m deep, which retained poured concrete. The concrete had been deposited around a central circular metal ring within which four bolts, arranged in a square pattern, were observed. A circular depression was located at the centre. No doubt these bases supported the weight cases of the accumulators, demonstrating that this room functioned as the new Accumulator Tower (Fig. 5.9). The accumulator bases were positioned off centre relative to the external walls (Fig. 5.9), suggesting that their timber framed guides were positioned against the east wall of the Tower (Tim Smith pers. comm.).

The archaeology that was uncovered demonstrated that the 'new' Engine House could be found to the immediate south of the 'new' Accumulator Tower, directly opposite the original Engine House. It was found to be 5.92m in width by 16.80m in length. The floor was constructed with stretcher red bricks, into which three circular drains were incorporated. The presence of drainage features strongly suggested that, like the floor of the original engine house, the floor of that room also got wet on a regular basis due to condensation and run-off from the steam engines. The top of the surface was found to be 1.39m higher than the original 1850–2 floor to the east, which was not raised at this time. If a means of direct, internal communication between the two Engine Houses existed, as the goods station plan of 1882 suggests[5], this must have been achieved by way of one or more short staircases that were perhaps accompanied by a gantry.

---

5   The 1882 station plan shows two gaps in the dividing wall (i.e. the Retaining Wall, Landscape Feature 4) that separated the Original Engine House to the east from the New Engine House to the west. It is likely that one or more of these represent doorways that enabled internal pedestrian access between the two structures. They do not appear on earlier cartographic sources, suggesting that they were cut into the wall during the 1881–2 extension programme. No trace of them was found during the archaeological excavation, an unsurprising fact since this wall was later torn down to ground level in this location during the 1899 Hydraulic Station rebuild.

A substantial rectangular structure that was oriented north–south was found in the centre of the Engine House, which unmistakably represented the remnants of two or more beds that had been joined together at a later date to form one large construction (Fig. 5.9). The top of the northern end was intact at a level of 23.80m OD whilst the southern end had been truncated horizontally to a depth of 23.05m OD. The separate beds had clearly been built in distinct sections that had been joined together with masonry infill, each element being discernible through the use of different materials and bonding styles.

The northern and central sections of the most northerly example, Engine Bed 4, (Fig. 5.9) were made of York stone blocks, 0.30m thick and of various lengths, which sat on top of seven courses of red, frogged bricks. Ten horizontal rectangular gaps originally appeared to have been present in the eastern and western sides of the structure (five in each), which extended back into the base. These had been blocked at a later date with header bonded, red frogged bricks, one row deep and one row wide. York stone slabs, 0.75m thick, were located above the gaps, bridging them. Seven circular and three rectangular vertical holes were observed directly above in the top of the base; they penetrated the structure until they met the aforementioned horizontal gaps. This was the means by which an engine was secured to the base: bolts would have extended through dedicated fixings in the engine into the engine bed, their entry into the latter being permitted by way of the vertical holes. Nuts were then manually screwed onto these via the horizontal gaps in the main body, which were large enough to admit the hand of a man and the necessary tools. Once this was done, the gaps were blocked with masonry.

A second, similarly sized structure, Engine Bed 5, was found to the south, which was composed of two distinct sections (Fig. 5.9). The northern section was 2.50m long, surviving to a height of 0.87m, and was composed of York stone blocks and frogged, red bricks. Three east–west horizontal voids ran through it, which presumably provided access to the undersides of machine fixings. The southern section was 1.44m long and 1.78m high. It was rectangular with four voids in its southern face and four corresponding voids in its northern face, all of which had been infilled with one row of stretcher-bonded, frogged, red fabric bricks. Whilst it appeared to have been constructed separately to the larger northern section before the two were joined together with abutting masonry walls, it is

Fig. 5.8 Accumulator bases in the 1881–2 Accumulator Tower of the Hydraulic Station (Landscape Feature 16), photograph faces south

The Hydraulic Station: the early 1880s 153

thought that these northern and southern sections are contemporaneous, effectively forming part of the same structure (Tim Smith pers. comm.).

One or the other of Engine Beds 4 and 5 was erected in the early 1880s during the Hydraulic Station extension programme of 1881–2 in order to support the 75hp pumping engine that had been ordered from Tannett Walker in 1881. The other bed may well date to a later period, perhaps having been constructed as late as 1898–9 for a second 75hp Tannett Walker engine that was acquired in 1897. The archaeological and documentary evidence study did not permit a satisfactory conclusion to be arrived at concerning which base was built when so both are considered in this section.

Two small engines, said to be 30 years old when scrapped, were removed from the Hydraulic Station in the 1890s and it is a distinct possibility that they were installed during the Hydraulic Station extension programme of 1881–2. They may have been small ancillary pumping engines or donkey engines, perhaps dedicated to pumping water from the basin. Unfortunately, no clear archaeological evidence regarding their whereabouts within the Hydraulic Station was found.

Archaeological remains relating to the new Boiler House were not discovered. Documentary evidence confirmed that it was situated to the south of the engine house in an area that suffered severe truncation during the Hydraulic Station rebuild of 1899. This later phase of work destroyed the earlier structure in its entirety, so the footprint of the Boiler House has been extrapolated from the cartographic evidence (Fig. 5.9).

According to documentary sources, the Back and Front Engine Houses were topped by two separate pitched roofs that both presumably sprang from the outer walls, coming to rest upon the Retaining Wall (Landscape Feature 4) in the centre of the building. No evidence indicative of the full height of these canopies was found; all that can be said is that they would have been situated at a level that was greater than 24.48m OD at eaves level (i.e. the maximum height of the surviving exterior walls of 1881–2 date) and that they would have been over 2.09m high (based on contemporary floor levels). The new Accumulator House must have taken the form of a tower in order to accommodate the new, larger devices that it was specifically designed to take. It must therefore have been considerably taller than the rest of the Hydraulic Station, requiring a separate roof as a result.

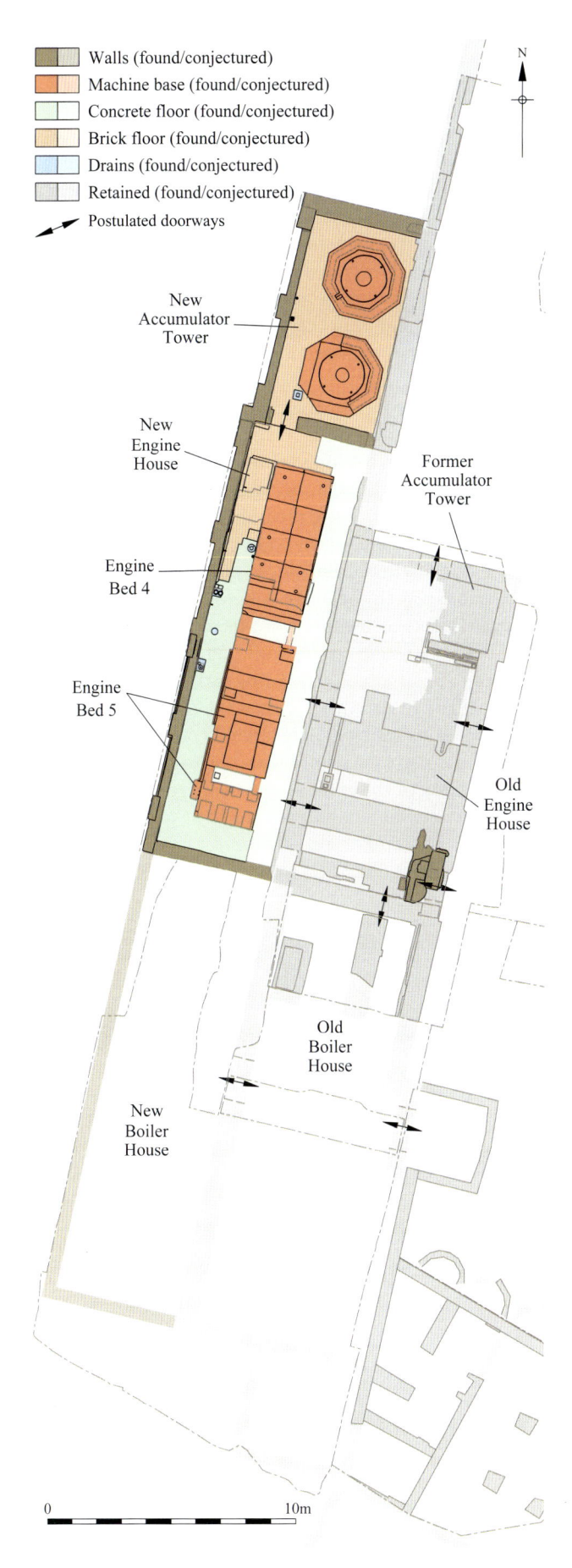

Fig. 5.9   Plan of the Hydraulic Station (Landscape Feature 16) by 1882 reconstructed from archaeological remains, cartographic sources and documentary evidence, scale 1:250

5.3 The Sources of the Building Materials used at King's Cross during the early 1880s

The changes that were made to the fabric of the Goods Yard's buildings during the introduction of hydraulic capstan shunting led to the use of a wide range of different building materials for the first time. Most of these were used during the enlargement of the Hydraulic Station (Landscape Feature 16).

The new walls of the Hydraulic Station (Landscape Feature 16) were constructed in a variety of brick types that included red and maroon stock bricks (fabric types 3032 and 3033), yellow London stock bricks and engineering bricks (3035), whilst the surviving sections of the new brick floor comprised engineering bricks (see Appendix 3 for information concerning brick fabric codes). The new engine bed was constructed of a number of kiln bricks (fabric type 3261) that were manufactured using the Craven process along with numerous examples of (3032) local stock bricks, some of which may have been reused from elsewhere. Substantial York stone slabs capped the masonry, forming the top of the machine base. Both accumulator supports also contained Craven kiln bricks (3261).

In conclusion, the brick and stone types that were used during the introduction of hydraulic capstan shunting in the early 1880s demonstrate that the Great Northern continued to rely heavily upon locally produced materials. These were supplemented by material from Yorkshire, which was probably brought into King's Cross via to the Great Northern's direct rail links with that county.

5.4 Synopsis

Developments that took place at King's Cross Goods Yard during the early 1880s reflect the Great Northern's wider programme of capital investment in new and existing infrastructure. In the case of the goods depot, this was dominated by the introduction of capstan shunting and the remodelling of platform accommodation within the Main Goods Shed. Although this necessitated a substantial financial outlay to purchase the shunting equipment and boost the hydraulic power supply, the changes enabled the depot to meet the demands being placed upon it. Money was therefore saved in the long term since fewer horses and men were required for shunting duties. Mechanisation also sped up the shunting process and the transfer of goods between the road, the railway and the canal.

The installation of additional platform accommodation in the Train Assembly Shed also appears to have formed part of this scheme. This suggests that sections of the building were henceforth devoted to loading and unloading rather than train assembly and disassembly. This was a significant change to the previous traffic management arrangements and a major departure from Cubitt's original design. The provision of the extra platform space would have enabled many more goods wagons to be loaded and unloaded simultaneously, which must have eased rail congestion. Improvements were also made to road vehicle access, most notably through the creation of cart access to the north face of the Granary. This meant that more carts could access the building at any one time thus enabling grain to be shipped onwards to the metropolis more quickly.

These modifications were all designed to enhance the Goods Department's capability to handle the increasing volume of goods that were arriving at King's Cross during the late 1870s and early 1880s. However, as the capital continued to expand and its population's appetite for goods and services grew remorselessly during the decade that followed, it became apparent that these changes offered little more than a temporary respite.

Innovation, Bankruptcy and Reinvention: Samuel Plimsoll and the Great Northern Railway

'The Board of Trade say in their report that 'if the figures be examined, it will be found that about half of the total loss on the coast is to be attributed to over-loaded and unseaworthy vessels of the collier class. For the nine years ending in 1871 this loss is rather more than half'. Now consider, with the information I have given you of the care of Government in other directions, do you not feel justified in asserting emphatically that no vessel needing repairs should be allowed to go to sea until those repairs were properly executed? Do you not feel equally able to say that no vessel should be allowed to be dangerously overloaded? I repeat, Can any valid reason be given against the qualification of any landsman to say, that ships shall not be over-loaded?'

From 'Our Seamen' by Samuel Plimsoll 1872 p.72

Samuel Plimsoll MP (1824–1898) is one of the best known politicians and philanthropists of the Victorian era on account of his political agitation on behalf of British merchant seamen. The scandalous loss of many unseaworthy colliers and their crews in the early 1870s prompted Plimsoll to campaign on behalf of maritime safety legislation. Until Plimsoll's intervention brought the industry's problems to the attention of the public, the vessels that plied the sea lanes between the coal fields of the North East and the port of London were often poorly maintained and dangerously overloaded. His tireless efforts eventually secured a series of reforms that included the invention of the 'Plimsoll Line' that was and still is painted on the sides of ships to prevent overloading by unscrupulous owners.

In contrast to his political campaigning, Plimsoll's business associations with the London coal trade, the Great Northern Railway and its Goods Station at King's Cross are somewhat less well known.

Samuel Plimsoll was born in Bristol in February 1824, the eighth child of Thomas, an exciseman and his wife Priscilla. At the age of 13 Samuel moved with his family to Sheffield, where his father died three years later (Jones 2007: 31). Having started work as a solicitor's clerk at the age of 15, Samuel subsequently found employment at a local brewery, where he remained for ten years. During this period he threw himself into local politics, becoming a stalwart of the local Liberal Club and writing speeches for the Mayor of Sheffield. At the age of 27 he organized the exhibits for the Sheffield Court at the Great Exhibition in Hyde Park. A keen inventor from an early age, Plimsoll ensured that his designs were exhibited in the Sheffield pavilion. Based on his experience of the brewing industry, he applied in 1852 for a patent for the invention of a system of 'more thoroughly and effectually cleansing, extracting, and separating or fining ale, beer, porter, bitter beer, India pale ale, and other malt liquors from the yeast, bottoms, barm, sediment and other extraneous matters and impurities with which it may be in combination' (*London Gazette* no.21382, 19/11/1852: 3130).

Having decided to follow his elder brother Thomas into the coal trade, Plimsoll established himself as an independent 'Coal Merchant, Dealer and Chapman' in Sheffield in the early 1850s. He also acquired an interest in the Great Northern Railway, purchasing a quantity of shares in the fledgling company. Shortly afterwards he hit upon the idea of transporting coal from the South

Yorkshire coalfields to London by rail instead of sea, thereby breaking the monopoly of the North Yorkshire pits (Jones 2007: 34–5). In 1853 he obtained the agreement of colliery owners in Barnsley, Doncaster, Rotherham and Sheffield to ship their coal to London via the Great Northern's lines. Following in his elder brother's footsteps, Samuel moved to London where he opened an office in 1854 in anticipation of the forthcoming trade in South Yorkshire coal. Unfortunately for him, Plimsoll's arrangement with the Yorkshire colliery owners attracted the attention of the Great Northern's General Manager Seymour Clarke, who took steps to protect the interests of the railway company's own in-house coal business, which was then managed by his brother Herbert. In order to use the Great Northern's rails, Plimsoll was obliged to purchase ten coal trucks at a cost of £800 each. Having paid for the wagons, he was unable to use them as Clarke delayed granting the necessary certification. With his business unable to generate an income, Samuel was forced to sell his Great Northern shares in March 1854. This did not however prevent him from turning up at a shareholders' meeting five months later, where he denounced Clarke's machinations and advocated an unrestricted trade in coal (Jones 2007: 36; Thorne 1990: 118). Using the pretext that Plimsoll had been ineligible to attend the meeting on account of having previously sold his shares in the company, the Great Northern proceeded to exact its revenge by allowing his wagons to stand empty and unused. The cumulative delays and outgoings crippled Plimsoll's business, forcing him to file for bankruptcy on 7th February 1855 (*London Gazette* no. 21660, 09/02/1855: 519).

Plimsoll was briefly forced to reside in a lodging house in Hatton Gardens run by the Shaftesbury Society while he worked to discharge his bankruptcy, an experience he recalled in his 1872 book *Our Seamen: An Appeal*. In it Plimsoll revealed that he was chastened by his experience of living in charitable lodgings: 'I went with strong shrinking, with a sense of suffering great humiliation, regarding my being there as a thing to be carefully kept secret from all my old friends. In a word I considered it only less degrading than sponging upon friends, or borrowing what I saw no chance of ever being able to pay' (Plimsoll 1872: 116). Yet the experience also taught him of the selflessness of the men with whom he shared his lodgings, and of the 'splendid patience, fortitude, courage and generosity' shown by those reduced by circumstance to the poorest state in life.

Plimsoll worked hard to remove the taint of bankruptcy. A certificate of bankruptcy was issued on 21st April 1855, following which dividends were paid to his creditors at hearings held at the Council Hall in Sheffield on 2nd June that year and on 26th July 1856 (Jones 2007: 336; *London Gazette* no. 21710, 11/05/1855: 1858; *London Gazette* no. 21898, 04/07/1856: 2368). Having restored his reputation, Plimsoll obtained employment with the mining company Newton, Chambers & Co., the owners of the Thorncliffe Colliery at Chapeltown, near Sheffield. Plimsoll had previously approached John Chambers, one of the partners in the firm, to recruit him to his scheme to carry coal from local pits to London via the Great Northern. Although that scheme failed, Plimsoll impressed both Chambers and his stepdaughter Eliza, whom Plimsoll married in October 1857.

The Cambridge Street Coal Drops in operation c.1905; this image shows the Midland Railway's side of the depot (SSPL 1997-7397_DY_2783 © National Railway Museum and SSPL)

Following their wedding the couple settled in London, where they lived above Samuel's office at no. 32 Hatton Garden, only a few doors away from the charitable lodgings where he had resided a couple of years earlier (Jones 2007: 39). Plimsoll had a second office in the same street at no. 10, from where he traded as a coal merchant and colliery agent in partnership with Robert Mothersill (*London Gazette* no.22401, 06/07/1860: 2563). Plimsoll was also a relentless self-publicist. Having restored his standing with the South Yorkshire colliery owners, Plimsoll succeeded in persuading the London gas companies to use him as their agent (Thorne 1990: 119). In 1858 one of the gas companies agreed to take 52,000 tons of coal supplied by him. In July 1859 he organised an excursion to the Yorkshire coalfields for nearly 200 London coal merchants and their wives, which culminated in special dinner at the Barnsley Corn Exchange, where the guests were treated to a rousing speech from Plimsoll (Jones, 2007: 40). Following the court ruling that brought an abrupt end to the Great Northern's illicit trade in coal on its own account, as well as Herbert Clarke's career as Coal Manager, Plimsoll was one of the first independent coal merchants to move into the Coal Offices in 1860 (TNA RAIL 236/283/4).

There can be little doubt that Plimsoll's prospects were considerably enhanced by his marriage and the patronage of his wealthy father-in-law. Nevertheless, his successful career in business owed much to his own determination, which verged on obstinacy, as well as his enduring talent for invention. Plimsoll believed that the Great Northern's coal drops fragmented his cargo to an unacceptable degree owing to the violence of the descent from wagon to hopper. His dissatisfaction led him to devise an alternative system that allowed coal to be discharged into sacks with minimal breakage and loss, enabling him to maximise the profits from every train load (Thorne 1990: 119). In December 1859 he obtained a patent for a system for 'facilitating the unloading and transferring from railway wagons into carts and barges &c the coals and other matters with which they may be loaded, and for storing the same' (*London Gazette* no. 22343, 06/01/1860: 43). Little over a month later, Plimsoll reached an agreement with the Great Northern that allowed him to build an experimental coal drop on one of the Goods Yard's sidings (TNA RAIL 236/27: 196). The success of this prototype convinced Plimsoll that sufficient demand existed for a new coal handling facility that would rival those provided by the Great Northern. Having established his 'Coal Unloading Company', he set about acquiring the leases of a number of properties at the southern end of Cambridge Street in 1863 so that his plans could be realised (TNA RAIL 236/30: 293). This led to the construction of the Cambridge Street Coal Drops and the associated viaduct and bridge across the Goods Yard and the Regent's Canal in 1865. In order to accommodate the new viaduct it was necessary to demolish the experimental facility in the yard where Plimsoll had perfected his design (TNA RAIL 236/38; TNA RAIL 236/305/20).

It is unclear how much coal breakage was actually prevented by the design of Plimsoll's new hoppers (Thorne 1990: 119). Nevertheless, they were a commercial success, drawing coal merchants away from the Great Northern's Eastern Coal Drops and Coal Offices during the 1870s and 1880s. They remained Plimsoll's property until he eventually sold them to the Great Northern following an agreement struck in February 1891 (TNA RAIL 1189/1424).

CHAPTER 6

Overcrowding at the King's Cross Goods Station during the 1880s and 1890s

'The place is, in fact, like most large goods stations, something of a 'mosaic', additions and alterations having been made from time to time 'here a little and there a little' solely with regard to 'business', viz the convenient manipulation of the traffic, and with little care for any future visits from the 'intelligent foreigner' with a taste for the 'aesthetic'....to the eastward there is the York Road, not to mention the Caledonian Road with its mud, its shabby frontages, and its fish and vegetable stalls, with their naphtha flare lamps at night, but that way madness lies'

Mr. J. Medcalf, former Great Northern Railway Outdoor Goods Manager, describing the appearance of the yard in the late 19th century and the 'madness' of goods handling in the north London area (Medcalf 1900)

While the national rail network continued to grow in extent during the second half of the 19th century, by the 1880s fewer new lines were being opened as competitors abandoned the practice of laying new tracks parallel to those of their rivals (Wolmar 2007: 165, 202). Instead, railway companies increasingly promoted the quality and efficiency of their services in a bid to attract custom. The Great Northern was one of a number of companies that successfully used advertising and publicity stunts to promote their businesses during this era (Nock 1974: 141).

Perhaps the best known of the Great Northern's efforts to court publicity were the famous 'races to the north', which were held in 1887–8 and again in 1895. In conjunction with the company's East Coast allies, the North Eastern and the North British, the Great Northern took on the London & North Western and the Caledonian, pitting King's Cross against Euston as the two consortiums vied to reach Scotland in the shortest time possible (Wolmar 2007: 165–6, 173–4). Improvements were also made to the Great Northern's fleet of steam locomotives under the auspices of the company's celebrated Locomotive Superintendent, Patrick Stirling (Nock 1974: 80–94). The emphasis that the company placed on speed meant that sections of its ageing permanent way had to be replaced during the 1890s in order to prevent derailments, whilst a number of serious accidents during this period obliged it to change operating practices and invest in new and improved signalling and braking arrangements (*ibid.*: 95, 97–111). Despite the stabilization of the railway network, the Great Northern remained keen to seize control of new or existing markets from its rivals. Such an opportunity arose in 1889, when the company acquired the Eastern & Midlands Railway under joint ownership with the Midland, which enabled both companies to gain access to a number of important fishing ports along the Norfolk coast (Simmons and Biddle *et al.* 1997: 192).

Meanwhile the development of the Somers Town Goods Depot by the Midland Railway on the site of Agar Town threatened to challenge the Great Northern's hard-won dominance of the trade in goods between the metropolis and the industrial north (Lewis 2013/2014: 289). The continuing growth in goods traffic to and from the metropolis appears to have insulated the company to some extent from the arrival of its new neighbour, although the development by the Midland of superior potato handling facilities at Somers Town was to cause disquiet among the King's Cross potato traders. When the latter complained about the state of the facilities provided by the Great Northern in the late 1880s, the company responded swiftly and sympathetically.

Rather than representing a crisis period ushered in by the opening of the Somers Town Goods Depot, the 1880s and 1890s instead represented a high point for the Great Northern Goods Department at King's Cross. Goods traffic continued to account for the largest percentage share of the company's revenue during this period as the amount of rail freight carried by the Great Northern reached a peak (Nock 1974: 129–30). Once again, the northern approaches to King's Cross began to suffer from severe congestion as services queued up to traverse

Copenhagen Fields and the Regent's Canal. In order to relieve the problem, the company built a tunnel under the canal in 1892 (*ibid.*: 72–3). By the latter date the Great Northern's London suburban network was approaching completion (*ibid.*: 72).

The increasing volume of goods traffic was largely confined to the road and the railway, as the volume of canal traffic passing through King's Cross declined markedly as the new century approached. At the start of this period, the Regent's Canal continued to carry a significant volume of goods: in 1888 the canal carried 1,009,451 tons of goods, which enabled the owners of the waterway to turn a profit of £45,821 and pay shareholders a dividend of 2.625% (Faulkner 1990: 53). However, the steady decline in canal traffic that had characterised the second half of the 19th century would accelerate throughout the 1890s until by the early 20th century, the Granary Basin at King's Cross Goods Yard would barely be used at all (*ibid.*; TNA RAIL 1189/1426: Brown to Brickwell 08/07/1915).

The number of station and clerical staff that were needed to run the Goods Yard at King's Cross grew in tandem with the volume of traffic received. As the number of station staff increased, the Great Northern was obliged to provide additional office, catering and welfare accommodation for the workforce.

This chapter charts the changes that took place at King's Cross Goods Yard during this period and assesses the impact of these developments upon the fabric of the complex.

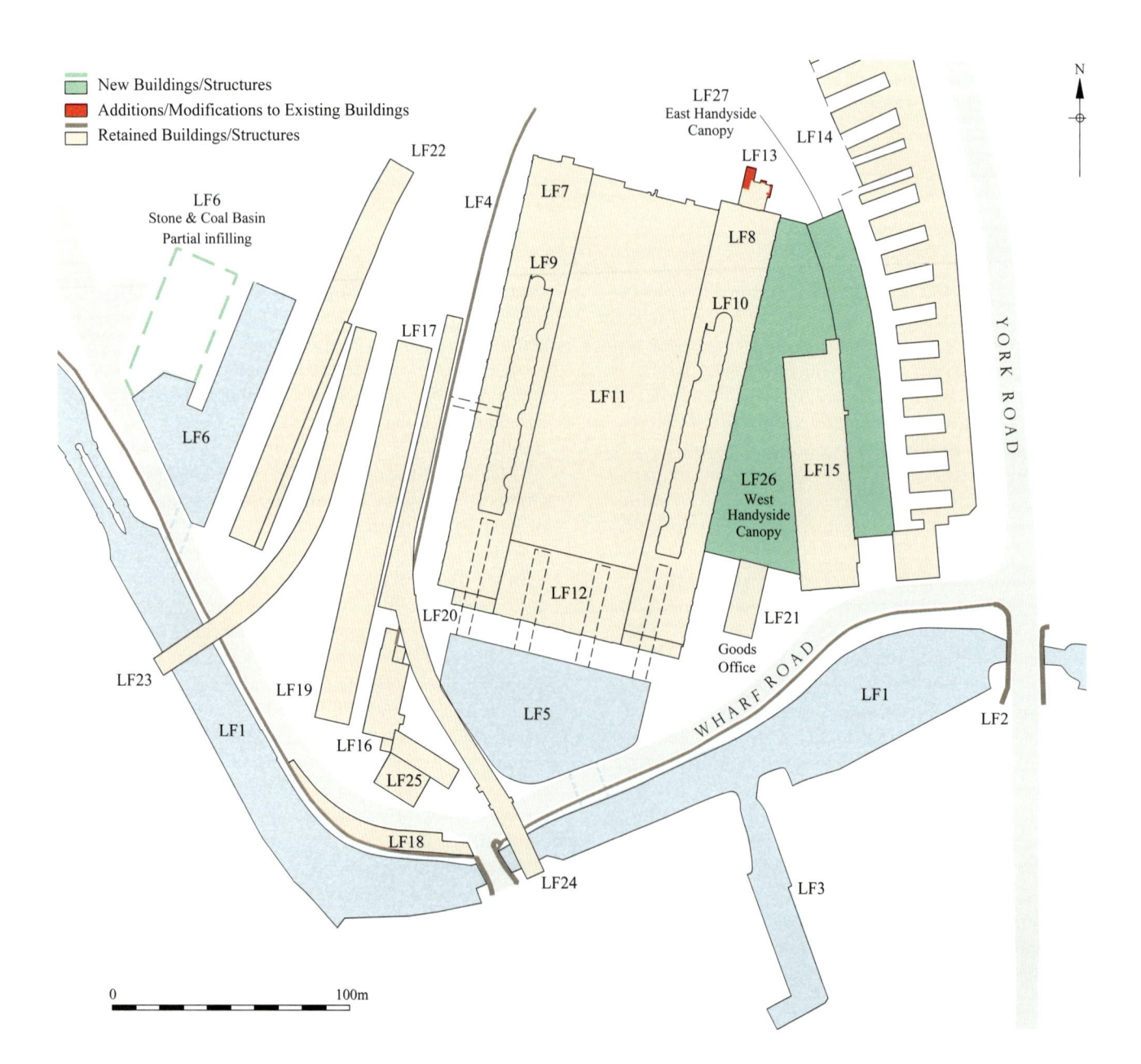

Fig. 6.1   The locations of the Landscape Features that were present in the Goods Yard from the mid 1880s to the mid 1890s, scale 1:2,500

6.1 Road Vehicle Infrastructure: the mid-1880s to the 1890s

Over the course of its first fifty years, the Goods Yard handled a steadily increasing volume of road traffic. By the early 1890s chronic congestion had developed around the vehicular entrances to the yard. Owing to the proximity of the Regent's Canal, the principal means of vehicular access was restricted to the increasingly clogged Wharf Road, which provided a route from York Road (previously known as Maiden Lane), entering the yard via Somers Bridge (Fig. 6.1). It was also possible to gain direct access from an entrance on York Road, however the gradient on either side of this was steep and the bridge narrow thus causing lengthy queues. In 1899 the Great Northern proposed to widen Maiden Lane Bridge (Landscape Feature 2) and to reduce the gradient of York Road, however the proposal foundered and the bottleneck remained at the turn of the new century (TNA RAIL 236/51: 356). Despite the abandonment of the Maiden Lane Bridge scheme, Somers Bridge was rebuilt in 1897 and Wharf Road was widened two years later (TNA RAIL 236/160: 148).

At some point during this period the cart way that ran along the north wall of the Granary (Landscape Feature 12) was transformed into a proper through road, enabling horse-drawn vehicles to traverse the whole Main Goods Shed (Landscape Features 7–11). The construction of this route would have necessitated the removal of a section of Platform 2 in the Eastern Transit Shed. The undertaking would have boosted efficiency at the southern end of the complex by eliminating the bottleneck that must have plagued the turning circle that previously existed. Although direct references to this modification were not found, an unrelated document created in 1897 describes an east–west cart way within the Main Shed, implying that a true road had been created at some point prior to that date (TNA RAIL 236/530: 27).

6.2 The Eastern Transit Shed: the mid-1880s to the 1890s

Landscape Feature 8

Further alterations were made to the interior of the northern end of the Eastern Transit Shed in 1886, after Henry Oakley (GNR General Manager) requested authorisation to spend £46 'for providing an Empties Correspondence office at the north end of the Up Shed and for connecting the present Empties Correspondence and Fish Office into one, to be used as Fish and Pay Office', on the grounds that the existing accommodation was limited (TNA RAIL 236/151: 28). A subsequent request to enlarge the 'Turning-Out Office' (which appears to have been situated in the Eastern Transit Shed) at a cost of £50 was approved in October 1897 (TNA RAIL 236/158: 375). The necessary works must have been carried out in conjunction with the Goods Station enlargement programme of 1897–9 (see Chapter 7) but were subject to a separate plan and budget.

6.3 The Potato Market and the West and East Handyside Canopies: the mid-1880s to the 1890s

Landscape Features 14, 26 & 27

While the Goods Department was busy trying to adapt Lewis Cubitt's mid-century design to meet the demands of the 1870s, the merchants who traded at the King's Cross Potato Market were becoming increasingly restless about the quality of their facilities. Representatives of the traders complained about the conditions that they endured when unloading wagons in the open area between the Potato Market itself and the Midland Goods Shed on several occasions during that decade. Although the Great Northern was sympathetic to the traders' concerns, it was not until the Midland Railway opened a rival market at the Somers Town Goods Depot in the mid-1880s that the company decided to take steps to improve its facilities at Maiden Lane in order to prevent trade being drawn away. The pressure to take action was intensified by a petition signed by 22 of the leading potato salesmen at King's Cross in April 1888. The petitioners requested that the following steps be taken:

'We the undersigned wish to draw your attention to the need of a Vegetable Market in connexion with the Potato Market, at present much of this lucrative traffic is lost, owing to the loss, inconvenience and exposure during the unloading. We beg to suggest that a light roof be placed over the spaces marked yellow. The alteration would add greatly to the attraction of the Market and in inward trade amply repay the Company for the outlay. The saving in ropes alone would be considerable and be an immense boon to us in foul weather' (TNA RAIL 236/362/18, 10/04/1888).

It appears that the traders had already discussed the proposed enlargement of the Potato Market with senior officers of the Great Northern before the petition was delivered to Henry Oakley. The day before Oakley received it, Josiah Medcalf (Outdoor Goods Manager) wrote to Richard Johnson (Chief Engineer) to outline the elements of the proposal that ought to be incorporated in the design of an enlarged Potato Market (*ibid.*: Medcalf to Johnson 09/04/1888). As well as building a canopy over the space between the existing Potato Market and the Midland Goods Shed in order to shelter the potato handlers, Medcalf also asked Johnson to look into the 'cost of...covering in the space between the Midland Shed, west-side and the Main Goods Shed and to be carried northward as the other section' (*ibid.*).

Johnson submitted provisional plans of the proposed roofs to Oakley on 26th April 1888. Although these have not survived, a minute of the Traffic Committee described the proposed works as 'covering the whole of the Market with a Glass Roof and iron columns' (TNA RAIL 236/152: 31). In a letter that accompanied the drawings, he reminded Oakley that the roof of the old Temporary Passenger Terminus (which now housed the Potato Market) still survived and that he therefore intended that the eastern section of the new canopy would be a continuation from the western edge of that old roof (TNA RAIL 236/362/18, Johnson to Oakley 26/04/1888). Johnson divided the works into three blocks or phases:

Area A was 500' long and 50' wide and would cost an estimated £8,066;

Area B was 180' long and 50' wide, the cost of which was estimated to be £3,250

Area C was described as a 'triangular shaped platform...to cover the space between the west side of the Midland Goods Shed and the Main Goods Station', which would cost approximately £9,025 (*ibid.*)

Areas A and B would come to be known as the East Handyside Canopy (Landscape Feature 27), whilst Area C would form the West Handyside Canopy (Landscape Feature 26).

Having received Johnson's plans and estimates, Oakley submitted an outline description of the proposed works to the Board on 2nd May (TNA RAIL 236/362/18, Oakley to the Board 02/05/1888). In an attempt to convince the directors of the long term benefits of the scheme, he made sure to contrast the poor conditions in the Great Northern's existing Potato Market with the 'excellent' rival facilities that could be found at the Midland's Somers Town Goods Depot (*ibid.*). He then went on to present his proposals to the Traffic Committee the following day, which resolved that the Directors should visit the Yard in person to view the proposed scheme for themselves (TNA RAIL 236/152: 31). The next day members of the Board inspected Johnson's plans at the Yard and agreed to approve the application. Johnson was ordered to obtain tenders for 'covering' the Potato Market immediately, in time for the next meeting of the Way and Works Committee (TNA RAIL 236/51: 140). On the last day of the month-the contract for the fabrication and erection of both the east and west canopies was awarded to Andrew Handyside & Co. of Derby for £14,831.16.0 (TNA RAIL 236/148: 253). Construction commenced shortly afterwards (TNA RAIL 236/362/18, A Handyside & Co. to Fitch 08/06/1888).

The specification for the works described how the roof trusses were to be supported by wrought iron lattice girders carried on cast iron columns on both sides, set upon brick corbels (or beams resting on corbels) where they met the east wall of the Midland Goods Shed. To the east they would rest upon the cast iron beams of the Potato Market (TNA RAIL 236/362/18, 05/1888). To the north of the Midlands Shed the trusses were to be supported by wrought iron girders carried by cast iron columns, whilst the roof itself was to be covered with a combination of patent glazing, boarding and slates (*ibid.*).

At the beginning of October Johnson reported that Handyside's men had run into difficulties during the erection of the 'large columns next to the Goods Warehouse', presumably in the area to the immediate north of the Midland Shed Hydraulic Accumulator Tower (TNA RAIL 236/361/4, 02/10/1888). Johnson wrote that excavators had encountered a complex arrangement of drains and hydraulic pressure pipes, the latter almost certainly laid in association with the provision of hydraulic capstans at the Potato Market in the late 1870s (TNA RAIL 236/146: 74; TNA RAIL 236/148: 253).

In May 1889 the Board sanctioned the expenditure of £1,400 for the purchase of hydraulic capstans and snatch heads, which were probably installed in the Handyside Canopies so that the efficiency of the newly improved work space could be boosted (TNA RAIL 236/152: 329; TNA RAIL 236/52: 5). In order to justify expenditure of such a large sum, Henry Oakley pointed out that whilst the cost of annual maintenance was envisaged to be around £156, the capstans would replace seven horses, equivalent to an annual saving of £455 (*ibid.*).

Built heritage evidence relating to the construction of the West Handyside Canopy in 1888

Landscape Feature 26

The West Handyside Canopy was erected between the Eastern Transit Shed (Landscape Feature 8), the Midland Goods Shed (Landscape Feature 15) and the main office in 1888 (Fig. 6.1; Fig. 6.2). It provided the potato traders with a dry area for unloading in an effort to prevent defections to the rival facilities available at the Midland Railway's Somers Town depot.

The western side of the structure extended as far north as the north gable of the Eastern Transit Shed whilst the eastern side extended beyond the end of the Midland Shed where it met the western side of the East Handyside Canopy. Its eastern and western sides were therefore formed by the walls of those buildings, whilst the canopy was fully open at the northern end. The southern extremity was also open, apart from where it met with the north face of the main office (Landscape Feature 21) (Fig. 6.1). Whilst the original canopy covering has not survived, much of the iron roof frame remains extant, including the galvanised steel box guttering.

The roof frame was primarily supported by rows of D-section hollow iron columns cast in three sizes, the variation being necessary because of the differing loads that they were expected to carry. Use of D-section columns enabled the flat parts of those uprights to be set flush against the northern façade of the office and the other walls that the canopy abutted, enabling it to be easily tied into the brick piers of the eastern wall of the Eastern Transit Shed and the western wall of the Midland Goods Shed. Whilst the flat column backs were purely practical, their forward facing sections presented aesthetically pleasing rounded profiles. The base of each was stamped with a small maker's plate that read 'A. HANDYSIDE & CO. LD 1888 DERBY & LONDON'. Stone pads, inserted in the western wall of the Midland Goods Shed, were also used to support sections of the canopy. Where it abutted the Midland Shed's Accumulator Tower, its lattice girder was embedded in the brickwork, which had itself been rebuilt on a slightly different alignment in order to accommodate the canopy. Similarly, the western façade of the Midland Goods Shed (Landscape Feature 15) and the eastern face of the Eastern Transit Shed (Landscape Feature 8) were modified when the West Handyside Canopy was erected against them. In the case of the former, several loading bays and windows had to be infilled, moved or modified. Where the easternmost rafters of the West Handyside met with the west wall of the Midland Goods Shed, the stepped cornice that ran along the top of the latter had to be removed. In order to take the gable of the canopy, the wall also had to be built up above eaves height through the addition of short brick gables in Flemish bond (Fig. 6.3). Similar modifications were made to the Eastern Transit Shed, although in that case timber rather than masonry was used to raise its eastern wall (Fig. 6.4).

Built heritage evidence relating to the construction of the East Handyside Canopy in 1888

Landscape Feature 27

The East Handyside Canopy (Fig. 6.1) formed a narrow, snaking structure that adjoined and ran parallel with the Potato Market (Landscape Feature 14) to the east, abutting the Midland Goods Shed (Landscape Feature 15) and the West Handyside Canopy (Landscape Feature 26) to the west. Like the West Handyside, the earliest roof covering did not survive, although most of the frame and guttering was original.

This canopy comprised closely spaced east–west triangular wrought iron trusses, each spanning *c.*16m, supported on the east side by the cast iron beams of the Potato Market and on the southern and central sections of the west side by stone pads inserted into the brickwork of the Midland Goods Shed (Landscape Feature 15) and the adjacent accumulator house. To the northwest the East Handyside Canopy was integral with its western counterpart, both being carried by north–south lattice girders. On its eastern side, eighteen supporting bays of

Fig. 6.2 The West Handyside Canopy (Landscape Feature 26) abutting the eastern wall of the Eastern Transit Shed (Landscape Feature 8), photograph faces northwest

Fig. 6.3 The West Handyside Canopy (Landscape Feature 26) after the removal of the roof covering abutting the western wall of the Midland Goods Shed (Landscape Feature 15), photograph faces east

Fig. 6.4 Wooden gable added to the east wall of the Eastern Transit Shed (Landscape Feature 8) to enable it to support the upper reaches of the West Handyside Canopy (Landscape Feature 26), photograph faces west

haunched cast iron beams with open spandrels represent the only extant sections of Lewis Cubitt's 1850 Temporary Passenger Terminus (later the Potato Market, Landscape Feature 14).

A scheme to establish a 'telephonic communication' between the Granary and the National Telephone Company's Exchange was proposed in July 1897 by Henry Oakley, who concluded that the annual cost of the connection (£8.10.0) was a worthwhile outlay (TNA RAIL 236/158: 294, 318). Installation was therefore approved by the Board at the end of the following September (*ibid.*: 332). Three disused timber telegraph poles, fixed to the eastern slope of the Handyside Canopy were observed along with a fourth to the east of the Potato Market, an example of which is illustrated in Fig. 6.5. Although subsequently modified, these telegraph poles may originally have formed part of Oakley's early connection.

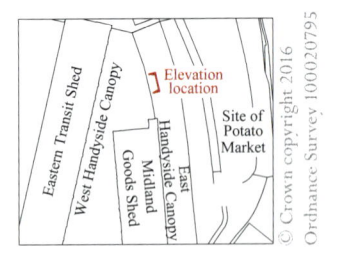

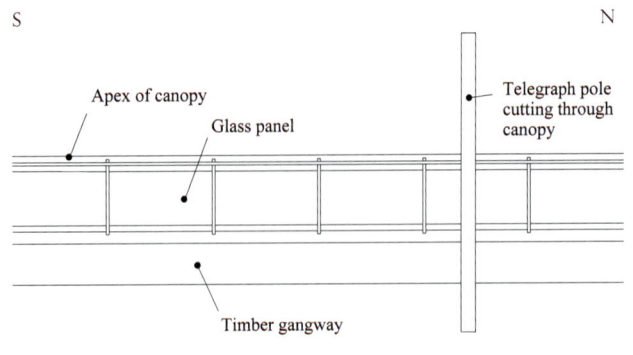

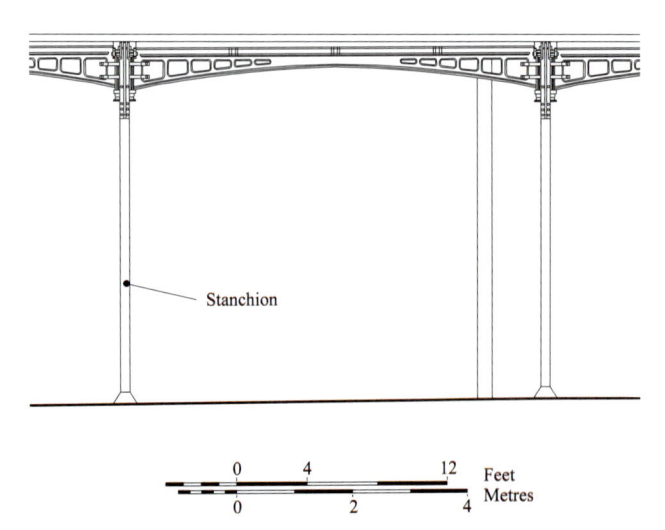

Fig. 6.5 Internal elevation of the east side of the partition that separates the West and East Handyside Canopies (Landscape Features 26 and 27) showing one of the telegraph poles that was probably instated in 1897, scale 1:125

6.4 The Granary: the mid-1880s to the 1890s

Landscape Feature 12

In early 1892 the Great Northern applied to reduce the premium paid for fire insurance cover for the Granary from the existing rate of 12% to 7%, the latter being closer to that paid by rival companies (COL MS 14943/13: 136 1891–2; TNA RAIL 236/54: 159). In order to secure the reduction it was necessary to improve the outdated fire prevention measures in the building and get the changes approved by a surveyor appointed by the Wharf and Warehouse Committee of the London Fire Offices. The inspection was conducted that March by Edwin Crockett, who had previously surveyed the building for the Committee nearly twenty years earlier (Guildhall MS 14943/13: 136 1891–2).

At the conclusion of his visit Crockett presented a report in which he outlined the measures that the Great Northern was required to implement in order to qualify for a reduction of the insurance premium payable on the Granary. These included hanging iron doors in each staircase doorway, fireproofing the upper levels of the staircases and moving the fire main inside the 'north staircase enclosure' from the wall of the Granary (*ibid.*). That November the Board sanctioned the expenditure of £155 for making the recommended alterations (TNA RAIL 236/54: 159), whilst nine months later a further £167 was sanctioned 'for making alteration in the appliances for use in case of fire in the Grain Warehouse at King's Cross…on the understanding that no further requirements are asked for' (*ibid.*).

Built heritage evidence relating to fire prevention improvements in the Granary in 1892–3

Landscape Feature 12

Late 19th to early 20th-century wrought iron plate fire doors had been installed in the access points that led from the Granary stairwells across all floors of the building (Fig. 6.6). Racks for fire buckets, stamped 'GNR', were also observed on the walls of the stairwells at all floor levels, whilst a total of ten fire boxes for the storage of fire-fighting equipment as well as four fire buckets were found. On the east wall of the southwest stairwell at every level, the following had been written upon black painted panels: 'Fire Cock No. 42 to 46'. The brackets, doors and fire cocks were no doubt installed in 1892–3 in an attempt to reduce the station's fire insurance premium in the wake of Crockett's inspection.

Fig. 6.6   Cross-sectional elevation through the Granary (Landscape Feature 12) showing the internal face of its southern wall and the changes of 1892–3, scale 1:500

6.5 The Refreshment Club: the mid-1880s to the 1890s

Landscape Feature 13

During the autumn of 1884 the staff of the Goods Department submitted a request to the Board of Directors for additional canteen accommodation, in order that they might establish separate rooms for the clerks and the men (TNA RAIL 236/49: 10–1). The proposal was supported by Henry Oakley, who estimated that the necessary works would cost just over £370 to complete. By this date the canteen was managed by the Refreshment Club, a voluntary body supported by the subscriptions of its members and underwritten by the company. Approval for the works was granted subject to the condition that the Club paid a nominal rent of £10 to the company each year (*ibid.*).

In the summer of 1888 representatives of the Refreshment Club approached Josiah Medcalf (Outdoor Goods Manager) to request that the company sanction the construction of a new smoking room on the premises (TNA RAIL 236/362/4, Medcalf to Cook 01/06/1888). Medcalf gave his blessing to the proposal, which he argued would 'provide a want calculated to benefit all smoking members, who [had] hitherto been compelled to seek that indulgence in public houses or in the streets' (*ibid.*). The Board sanctioned the expenditure of £200 on the necessary works at the beginning of July (TNA RAIL 236/51: 176).

Drawings of the proposed smoking room prepared by Richard Johnson, indicated that it was to be created on the ground floor of the existing Refreshment Club building by partitioning the scullery and adding a new single bay extension on the north wall (Fig. 6.7). On the ground floor, it would accommodate a larder as well as a new store and office and the coal bunker was to be 'refixed' beneath the new north wall. The first floor of the extension was to be used solely for the Clerks' Smoking Room, whilst the Mens' Smoking Room was to be situated in the easternmost ground floor room of the existing building.

The 1884 and 1888 extensions to the Refreshment Club can be seen on the Second Edition Ordnance Survey map of 1894–6, although it is impossible to determine from the map alone which sections of the structure were erected in 1884 and which were built four years later (Fig. 6.8).

At the end of January 1889 Oakley submitted a further request to the Board for additional accommodation at the Refreshment Club (TNA RAIL 236/152: 249). This proposal was relatively modest in scope, comprising an application for funds to acquire new and improved cooking apparatus for the Club's kitchen for which a quote of £148 had been received. The Board approved the proposal, although it was decided that the Club's rent would be raised from £10 per annum to £15 to contribute towards the cost of the works carried out on the club's behalf over the preceding twelve months. The Refreshment Club continued to flourish during the last decades of the 19th century; by the early 1890s it was reported that it consistently returned an annual profit of no less than £200 (TNA RAIL 236/53: 137).

Archaeological evidence for Refreshment Club extension of 1888

Landscape Feature 13

Archaeological evidence clearly demonstrated that the Refreshment Club was extended twice during the 1880s (Fig. 6.9). The earlier phase of extension consisted of an 'L' shaped yellow brick structure, which abutted the northwest corner of the pre-existing building. Its addition created a small room at ground floor level that by 1888 had become a smoking room (Fig. 6.7; Fig. 6.9). The extension also altered the footprint of the Refreshment Club from an inverted 'L' to a rectangular shape (Fig. 6.9). It is likely that this section of the building was added in 1884 in response to a request from the staff of the Goods Department.

Further north two walls continued beyond the limit of the excavation. The most westerly of the two butted the 1884 extension, suggesting that it was added after that date, whilst its eastern equivalent butted masonry that was added in 1872. The foundations consisted of a mixture of concrete and brick rubble that had been poured into a construction cut. A thin skim of concrete had then been applied, into which a course of header bonded, yellow bricks had been set. It is highly likely that these walls were added in 1888 after Refreshment Club members successfully lobbied for a smoking room. Indeed, the footprint of the archaeological remains closely resembled the footprint of the Refreshment Club as shown on Johnson's proposal plan, supporting this interpretation (Fig. 6.7; Fig. 6.9). The addition of the 1888 extension created a footprint that closely resembled that shown on the Ordnance Survey Map of 1894–6 (Fig. 6.8; Fig. 6.9).

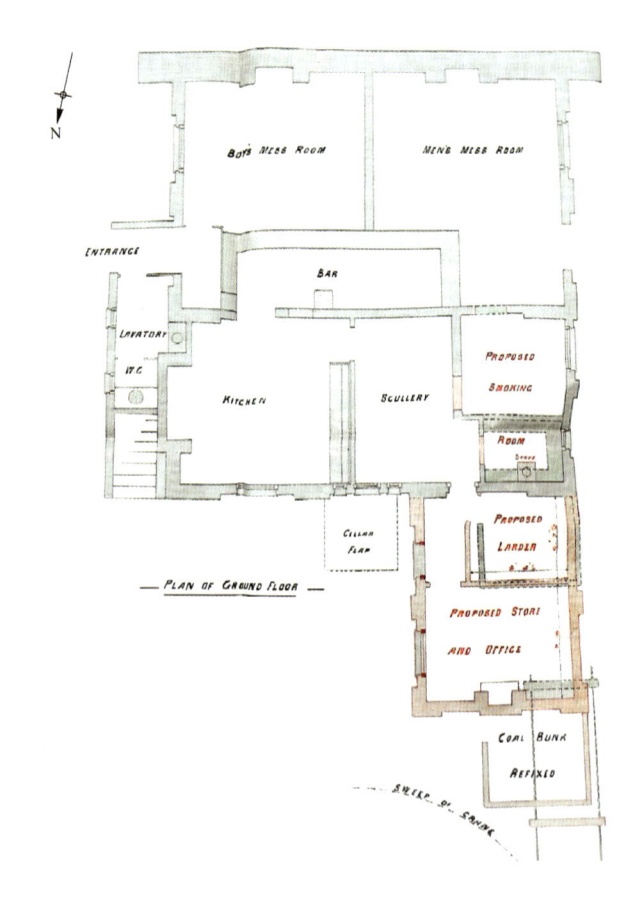

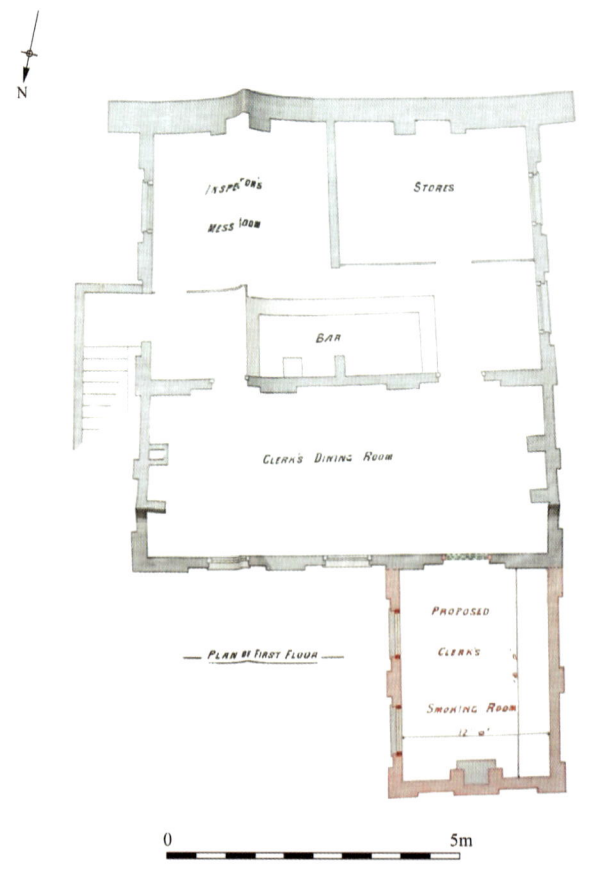

Fig. 6.7 Great Northern Railway Plan of June 1888 showing proposed alterations to the ground and first floors of the Refreshment Club (Landscape Feature 13) by Richard Johnson, scale c.1:125

Fig. 6.8   The Ordnance Survey Map of 1894–6 detailing the modifications that were made to the Refreshment Club (Landscape Feature 13) in order to accommodate the new smoking rooms, scale 1:1,250

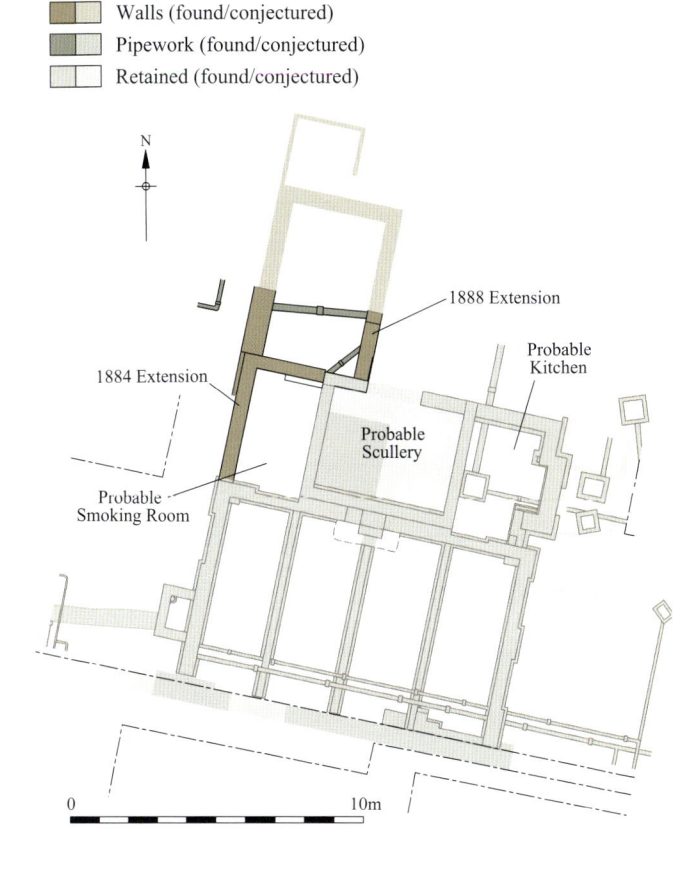

Fig. 6.9   Reconstruction of the Refreshment Club (Landscape Feature 13) by 1888 based on archaeological and cartographic evidence, scale 1:250

6.6 Improvements to Stables in the Goods Yard: the mid-1880s to the 1890s

Landscape Features 9, 10 & 25

The Horse Provender Store (Landscape Feature 25) was an important element of the King's Cross Goods Yard during the third quarter of the 19th century. The number of horses stabled at the Yard increased throughout the period and as the horse stock grew, so did the cost of feeding them. The total cost of provender purchased during the second half of 1873 alone exceeded £2,200 and the price of feed was rising (TNA RAIL 236/145: 15). By March 1880 the company's horse stock had increased to 1,400 animals (TNA RAIL 236/147: 115).

During the 1880s it became increasingly apparent that the capacity of the King's Cross Provender Store was insufficient. This lead the company to build a much larger provender mill at Holloway Road Goods Depot, close to King's Cross (Medcalf 1900: 316). In a report written for the Board in 1883, Josiah Medcalf and Albert Guille (GNR Assistant, later Chief Accountant) found that horses belonging to the Great Northern were more likely to be put out of action by injury or overworking than those owned by rival operators in the capital, a fact that the authors attributed largely to the company's working practices (TNA RAIL 236/149: 148). Concerns about the condition of the stabling at King's Cross resulted in an inspection by Richard Johnson in 1884, who noted that the company's facilities were particularly deficient in drainage and ventilation when compared with those of its rivals. The installation of a 'simple open iron [drainage] channel' in each of the Transit Shed stables therefore featured in his recommendations (TNA RAIL 236/150: 10).

The company failed to fully implement all of Johnson's improvements and it was not until the horses at King's Cross were struck by an outbreak of influenza in 1890 that a sustained effort was made to resolve the deficiencies of stabling at the Goods Yard. In the wake of the outbreak the company commissioned Professor J. Wortley Axe of the Royal Veterinary College to report upon the condition of all of its stables in the capital (TNA RAIL 236/235: 51). In order to prevent future occurrences of the disease, and to improve the general welfare of the horse stock, Professor Axe recommended that the company increase stabling capacity at King's Cross Goods Yard, enlarge the Horse Infirmary and improve access to and ventilation within each of the existing stable blocks (*ibid.*: 31).

On the 14th of May 1891 a Special Meeting of the Horse Committee discussed the improvements that were needed in order to implement Professor Axe's recommendations (*ibid.*: 30). Richard

Johnson submitted plans showing how additional accommodation could be provided for 283 horses and how 'certain improvements could be made to stable ventilation' for a total cost of £25,802. Johnson estimated that the cost of converting the Old Provender Store into stabling for 95 horses would amount to approximately £4,835, while a further £1,000 was required to make general improvements to the existing facilities, including the provision of the additional ventilation as recommended by Professor Axe. The Committee instructed Johnson to prepare further plans and estimates 'of the cost of utilizing the basement of the Forage Store [i.e. the Provender Store] as stables' (*ibid.*: 32–5). Johnson's revised drawings and estimates, which proposed converting the Old Provender Store into a stable to accommodate 120 horses for an outlay of £6,085, were approved by the Committee in mid-June, together with a proposal to improve the ventilation of the company's other stables at King's Cross.

In late July 1891 the Way and Works Committee of the Great Northern awarded the firm of Atherton and Latta the contract for the conversion of the Old Provender Store into stables for £5,775 (TNA RAIL 236/179: 289). It was reported that the contractors were making 'fair progress' towards the end of September (TNA RAIL 236/370/9, Johnson to Way and Works Committee 25/09/1891). One month later, despite inclement weather conditions, Atherton and Latta had already cut out the lower girders and floor and formed the new floor surface (*ibid.*: 30/10/1891). The conversion was finally completed the following February when Mr Poynter (GNR Horse Superintendent) asked Johnson 'to fit up some hydraulic equipment in the Stables recently constructed at the Old Provender Stores... so as to lift provender from the ground to the upper floor' (TNA RAIL 236/235: 68). The Horse Committee approved the proposal on 3rd March, following which it was referred to the Traffic Committee, which sanctioned the expenditure of £130 to provide 'certain Hydraulic Hoisting Apparatus at the Provender Stables... for raising Provender from the ground to the upper floor' the following month (TNA RAIL 236/54: 4).

A plan of the Goods Yard from 1905 (not illustrated), indicated that the Provender Store had been renamed 'Stables' by this time; whilst a plan of 1906 revealed that the main section of the building was divided internally into 22 stalls (Fig. 6.10). The 1905

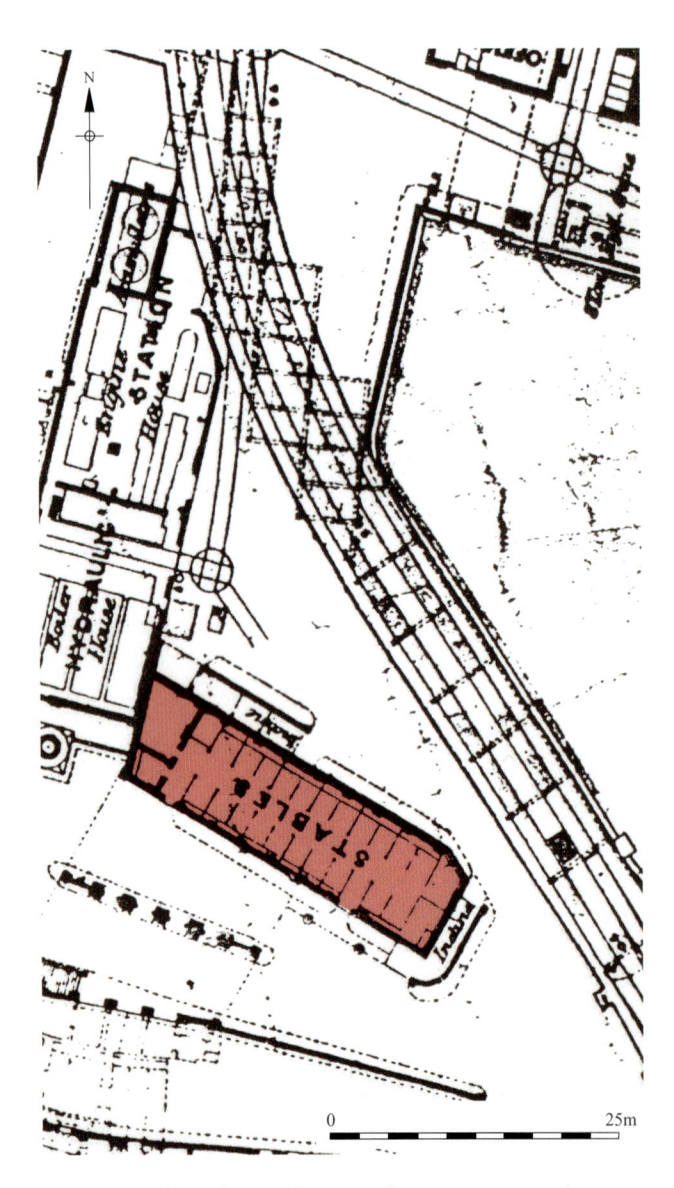

Fig. 6.10 Plan of 1906 showing the Horse Provender Store (Landscape Feature 25) after its conversion to stables, scale 1:625

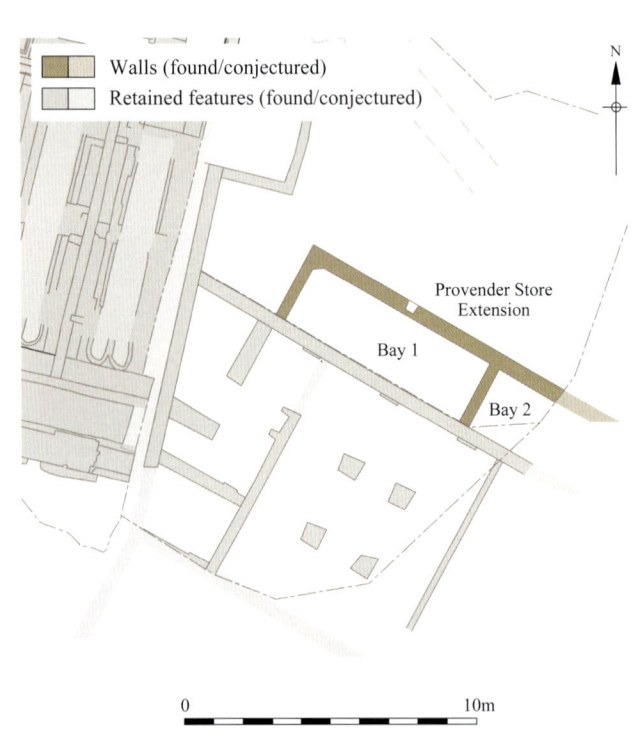

Fig. 6.11 Plan of the ground floor of the Horse Provender Store and Stables (Landscape Feature 25) by *c.*1891 reconstructed from archaeological remains and cartographic sources, scale 1:250

and 1906 plans also depicted a new small rectangular extension abutting the northwest end of the northeast face of the Provender Store, running parallel with the original northeast wall of the building and divided into three equally sized rooms (Fig. 6.10). The addition of the extension and the insertion of these stable blocks probably also occurred in the 1891 when the building was converted from a store to a stable.

Whilst Professor Axe's recommendations concerning the conversion of the Provender Store were implemented quickly, documentary sources indicate that other improvements took longer to materialise. Examples include the construction of an additional exit from the Eastern Transit Shed Stables (which was not given the go ahead until January 1898), whilst permission to make similar modifications to the Western Transit Shed Stables was not granted until October of the following year (TNA RAIL 236/183: 122; TNA RAIL 236/235: 315,375). The contract for the former was awarded to Kirk Knight & Co., who were at that time already engaged in the construction of the Train Assembly Shed's overhead offices. The works cost £370, whilst the additional exit from the Western Stables was to be constructed for of £250 (TNA RAIL 236/235: 315, 375; TNA RAIL 236/183: 122, 158). The latter exit appears to have been additional to the existing subway that led to the Coal Yard, which was integral to the earliest phase of the Goods Station.

Archaeological evidence relating to the conversion of the Horse Provender Store into stables in 1891

Landscape Feature 25

The north-western end of the rectangular extension that was depicted on the 1906 plan was identified during an archaeological watching brief (KXO08). In all probability it was constructed when the Provender Store was converted to stabling in 1891 and was used either for the storage of fodder or tack (Fig. 6.11). Its dimensions were 2.72m northeast–southwest by over 9.92m northwest–southeast, continuing beyond the south-eastern limit of the excavation. Frogged, yellow, English bonded bricks had been used in its construction, five courses of which were visible above ground level. The bulk of the above ground portion had been demolished after the building fell out of use however the foundations of the north-western bay and the westernmost section of an adjoining bay were identified.

Three lightwells, which formed part of the original north-eastern outer wall of the building, were rendered redundant by the extension, which sealed two and truncated a third. The new walls were also not keyed into the main body of the former Provender Store because they were added later.

In order to convert the building into a stable, it was necessary to construct a new first floor. The partitions that separated the stables from one another as shown on the 1906 map may therefore have been located at first floor level, which would explain their absence from the archaeological record. This interpretation is supported by documentary sources, which suggest that some of the ground floor continued to be used for the storage of feed after the building was converted into stabling.

Built heritage evidence of alterations to the Western and Eastern Stables in 1884–98

Landscape Features 9 & 10

The granite setts that formed the floor of the Western and Eastern Stables (Landscape Features 9 and 10) were partially relaid in order to accommodate two north–south drainage channels built from reused railway lines that ran the length of each stable block (Fig. 6.12). They were presumably inserted in response to Richard Johnson's recommendation of 1884 to insert an 'open iron channel' in the subterranean stabling areas.

The emergency exit that was added to the Eastern Transit Shed Stables in 1898 was identified during an historic building recording survey (KXF07). It consisted of a subterranean 'L' shaped ramp that connected the below ground stables with the West Handyside Canopy area (Fig. 6.12). No such addition was found in the Western Stables, suggesting that the plan to construct a new ramp in that block was abandoned.

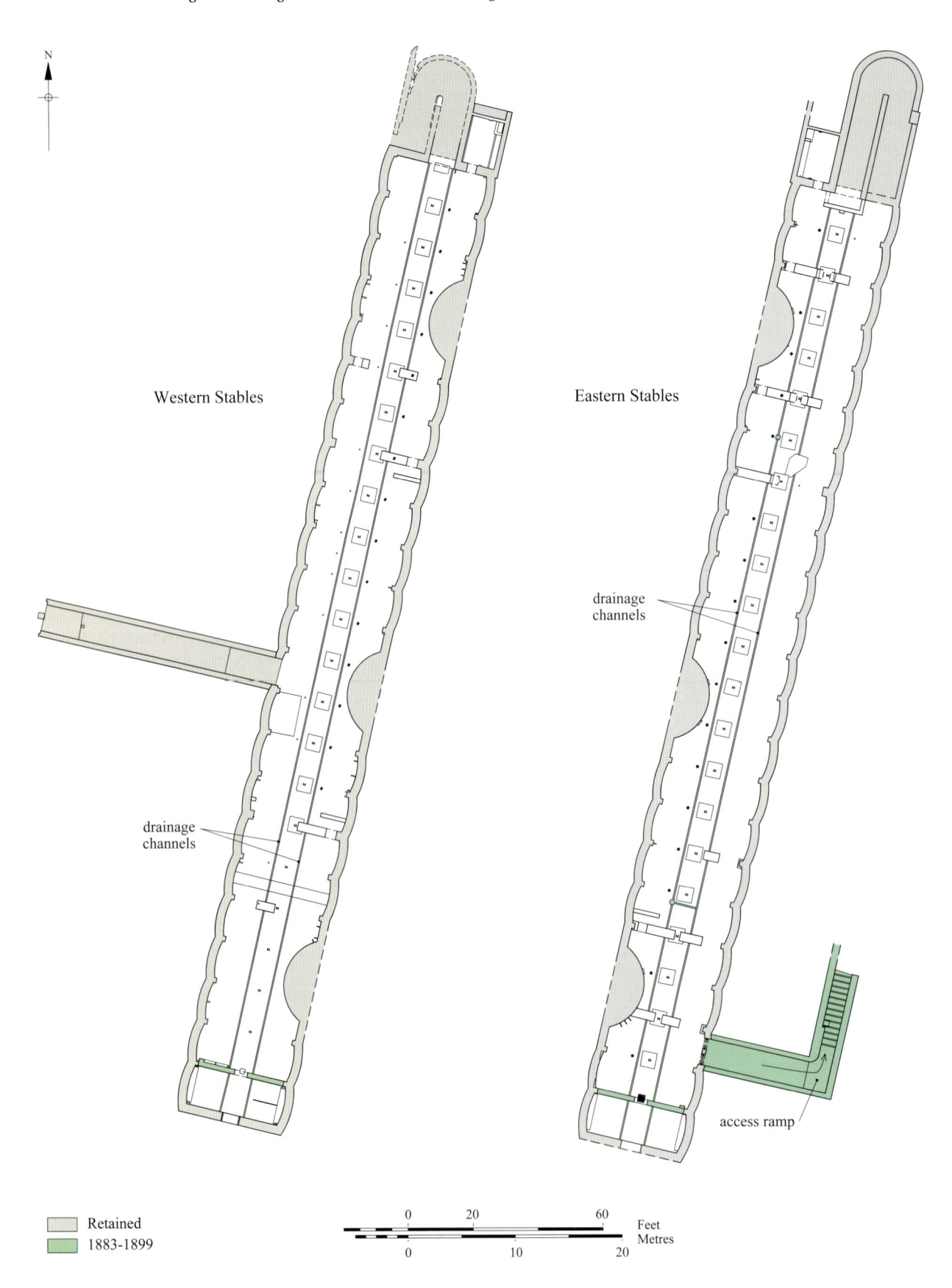

170 *Overcrowding at the King's Cross Goods Station during the 1880s and 1890s*

Fig. 6.12 Plan of the Western and Eastern Stables (Landscape Features 9 and 10) by 1898 reconstructed from built heritage and archaeological evidence, scale 1:500

6.7 The Relocation of the Coal Trade to the South of the Canal between the mid-1880s and the late 1890s

Landscape Feature 18

During the 1870s and 1880s the focus of the coal trade at King's Cross had moved away from the old Coal Drops (Landscape Features 17 and 22) towards Plimsoll's technologically superior facilities in Cambridge Street (Landscape Feature 23). As a result, the original Coal Offices (Landscape Feature 18), which had been designed to operate in conjunction with the adjacent Great Northern Railway Coal Drops, declined in importance.

In 1882 the Great Northern considered plans to erect an office block upon the vacant plot of land in front of the Passenger Terminus, prompting several of the coal merchants based in the original Coal Offices (including both Herbert Clarke & Co. and Rickett Smith & Co) to express an interest in leasing office space in the proposed development (TNA RAIL 236/147: 309, 357; TNA RAIL 236/172: 261, 360, 374). Unfortunately for them, the railway company decided against building the proposed block, which meant that the merchants were obliged to stay put. Following the abandonment of the scheme, Clarke and his fellow merchants became increasingly frustrated by their isolation from the centre of the King's Cross coal trade. The following year Clarke demanded that the rent he paid be substantially reduced on the grounds that his offices were 'not now available for the purpose of the Trade' in contrast to 'other Merchants [who had] …offices outside the station, where orders are given' (TNA RAIL 236/148: 139). By the late 1890s, the focus of the coal trade at King's Cross had relocated to Cambridge Street (Landscape Feature 23), although greatly reduced quantities of coal were still being received in the remaining drops at the north end of the Eastern Coal Drops (Landscape Feature 17).

In order to retain its share of the lucrative coal trade, the Great Northern acquired the northern half of the Cambridge Street site from Samuel Plimsoll in the early 1890s (TNA RAIL 236/53: 50)[6]. In the autumn of 1896 it was decided that the existing drops should be extended and that a new range of offices for the use of the coal merchants should be built at the junction of Cambridge Street and Wharf Road (TNA RAIL 236/158: 72, 98). Although separated from their namesake by the Regent's Canal, the latter premises were subsequently frequently referred to as 'the Coal Offices' in company records (e.g. TNA RAIL 1189/1424).

The decision to construct a new administrative block on Cambridge Street encouraged those merchants who had remained in the original Coal Offices to move out. By 1890 only a handful remained, although these included two of the largest firms, Charrington Sells & Co. and Herbert Clarke & Co. By the mid-1900s however, only John Irving & Sons, who were based on the first floor of Block 3, rented office space in the building (Kelly & Co. 1904: 446).

The coal traders were gradually supplanted by hay and straw merchants, the first of whom established premises in the former Coal Offices in 1892 (Kelly & Co. 1892: 371). The arrival of these salesmen was probably with a consequence of the conversion of the nearby Provender Store into stabling (TNA RAIL 236/53: 136; TNA RAIL 236/54: 3; TNA RAIL 236/235: 28, 35, 68). In 1898 the hay and straw traders were joined by the clerical staff of the Great Northern Horse Department, which had recently been enlarged following the appointment in 1894 of Andrew Moscrop to the post of Horse Superintendent (TNA RAIL 236/235: 321; TNA RAIL 236/600). By early 1900 the Horse Superintendent's department occupied the principal block of the former Coal Offices (Block 2), whilst the three adjacent offices (Blocks 3 to 5) stood vacant (TNA RAIL 236/235: 390; TNA RAIL 236/600).

By the turn of the 20th century, Moscrop was responsible for the company's stock of 1,300 horses in the capital, 1,000 of which worked in the Goods Department at King's Cross (Wade 1900: 208). As the Department grew, moves were made to consolidate several of the workshops, offices and stores scattered throughout the Yard, so in April 1898 it was agreed

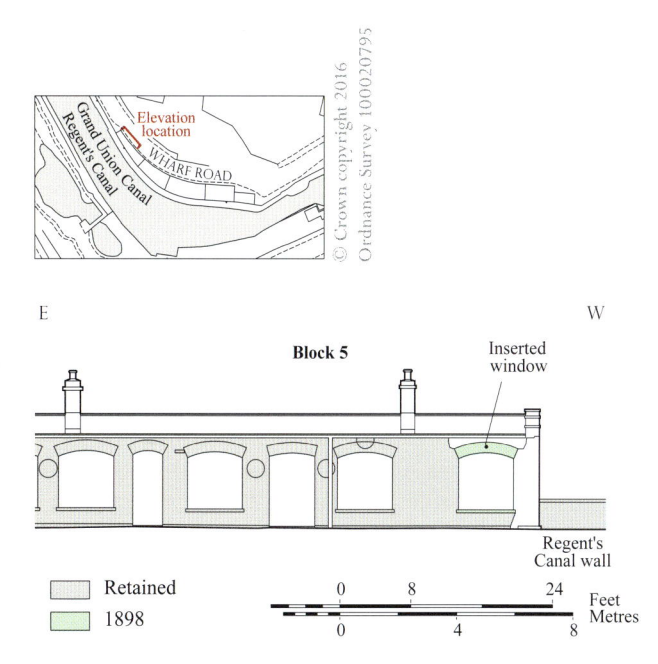

Fig. 6.13 External elevation of the north wall of Block 5 of the Coal Offices (Landscape Feature 18) showing the changes of 1898, scale 1:250

---

6 The north side of the drops was fed by GNR rails, the opposing side having been fed by the Midland Railway.

that alternative accommodation be found for the harness making workshops, which were then located in premises used by the Locomotive Department (TNA RAIL 236/235: 328). In February 1900 R.H. Twelvetrees (Great Northern Goods Manager) proposed that the three unoccupied Coal Offices adjacent to Moscrop's office should be adapted in order to accommodate the workshops (*ibid.*: 390). In June of that year, Twelvetrees submitted plans of the necessary works, which he estimated could be carried out at a cost of £70 (TNA RAIL 236/236: 3). Once established in the former Coal Offices, the harness workshops took on a number of apprentices and purchased new machinery, the latter including a number of advanced electrical stitching machines (Grinling 1905: 28). One of the units was still known as the Tack Room as recently as the mid 1980s (DEGW *c*.1987; Gilbert *c*.1985: 38).

Built heritage evidence for the Coal Offices and the creation of the Tack Room in 1898

Landscape Feature 18

Internal connections between the various units in Blocks 3 to 5, including the former toilet block at the western end, had been established by the time that the built heritage survey took place. It is reasonable to assume that the connection was created in 1898, when they were converted from separate coal offices rented by different companies into facilities occupied by the Great Northern Railway Horse Superintendent's Department. Once these internal connections were created, it was possible to do away with one of the doorways that formerly communicated with Wharf Road. It was converted into a window, perhaps in order that the workshop could be better illuminated. The former toilet block that had been positioned at the western end of the structure may also have been converted into part of the Tack Room at this time, hence the insertion of an internal doorway in the wall that formerly shuttered it off from the rest of Block 5 and the creation of a large window for the purposes of illumination (Fig. 6.13). This hypothesis is supported by brick fabric analysis, which found that the structural elements in question were formed from mid to late 19th-century bricks (fabric types 3032 and 3034), suggesting that the changes probably predated the 20th century (see Appendix 3 for further information on brick coding).

6.8 The Midland Goods Shed: the mid-1880s to the 1890s

Landscape Feature 15

Kilner Brothers continued to use the former Midland Goods Shed as a bottle warehouse for the remainder of the 19th century. During that time, the company undertook a series of extensive improvements to the building, the most significant of which was the installation of electric lighting in 1895/6 (TNA RAIL 236/181: 387). Under the terms of an agreement reached by the company and the Great Northern, the railway company agreed to supply the cabling and arrange the connection to the public supply, while Kilner Brothers were responsible for providing their own lamps (*ibid.*). Other alterations undertaken by the company during this period included a number of unspecified 'alterations to the sanitary arrangements... in accordance with the requirements of the Sanitary Authorities' approved in April 1899, and the purchase of a hoist for £226 in October 1901 (*ibid.*: 368; TNA RAIL 236/160: 95; TNA RAIL 236/185: 73).

The construction of the Handyside Canopies in 1888 necessitated a number of alterations to parts of the Midland Shed. Contractors for the construction of the canopies were instructed to cut 'through the corner of [the] Accumulator House for the tee, struts and rafter of roof principal, average 2' 3" long and make good after same' (TNA RAIL 236/362: 18 Estimate 12). This as alteration was observed during the building recording exercise, and is described in Section 6.3.

At the end of October 1889 Richard Johnson proposed that a new two track tunnel be constructed alongside the existing Gasworks Tunnel to increase access for express passenger, suburban and fast goods trains to King's Cross Passenger and Suburban stations (TNA RAIL 236/364/3: 30/10/1889; TNA RAIL 236/178: 279–80). Johnson estimated that the tunnel would be approximately 528 yards in length and would take around eighteen months to complete (TNA RAIL 236/178: 361). The route proposed by Johnson would take the new tunnel under the Regent's Canal, the Midland Goods Shed and the Potato Market (TNA RAIL 236/364/3: 30/10/1889). The contract for the works was awarded to Messrs H. Lovatt & Co. of Wolverhampton, who commenced work in the summer of 1890 (*ibid.*: 03/06/1890; TNA RAIL 236/179: 11). By the end of September the contractors were underpinning the walls of the Midland Shed (TNA RAIL 236/364/3: Johnson to Way and works 26/09/1890). Five months later Johnson reported that while good progress had been made with the underpinning of the east wall of the Midland Goods Shed, the progress of tunnelling northward beneath the roadway on the southern side of the shed was comparatively slow (*ibid.*: 28/02/1891). Although the contractors' tunnel had reached the Midland Goods Shed by the end of March,

Johnson subsequently reported that progress under the shed and the Potato Market remained slow, owing to the need to exercise 'great care' while working beneath the buildings (*ibid.*: 24/03/1891, 27/06/1891). It was not until the New Year that Johnson was to report that tunnelling in the vicinity was finally approaching completion (*ibid.*: 06/01/1892).

6.9 Synopsis

By the end of 1892 the Great Northern had overcome the engineering challenge of digging a third tunnel under the Regent's Canal and completed its programme of improving rail access to the Goods and Passenger Stations at King's Cross (Nock 1978: 72). While this represented a considerable and lasting achievement, it had not been matched by the somewhat half-hearted efforts to relieve the chronic congestion that affected road vehicle traffic entering and departing the yard, nor by any major rail infrastructure changes inside the Goods Yard itself. Records suggest that the introduction of capstan shunting in 1882 and the changes that were made in its immediate wake were probably sufficient to ensure that the complex would continue to function tolerably well in the years that followed, whilst the arrival of the Midland Railway's sizeable depot at Somers Town in 1887 no doubt absorbed some of the trade that would otherwise have gone to King's Cross. However the company fully appreciated the financial consequences of losing business to its rivals, which explains why it responded so promptly to the concerns of the potato traders in the late 1880s. Had the Midland not attempted to entice the potato traders away from King's Cross however, it seems unlikely that the Great Northern would have been as accommodating to their needs as it eventually was. This contrasted with the lack of attention shown to the coal merchants, who remained isolated in the Coal and Fish Offices on the north bank of the canal for several years after the focus of their trade had migrated to the Cambridge Street Coal Drops on the opposite bank.

Similarly the company's response to the inadequacies of cramped and sometimes insanitary stabling in the Goods Yard, which had been highlighted by its own senior officers as early as 1883, was initially hesitant and only really gained traction following the influenza outbreak amongst the horse stock in 1890. The construction of the provender mill at Holloway Road during the preceding decade fortunately enabled the company to convert the old Provender Store into stables, while the recently vacated Coal and Fish Offices provided a new base for the enlarged Horse Department. Despite these welcome changes, corporate records create the impression that the Goods Yard was becoming increasingly overcrowded as the 1890s progressed. The Great Northern also made no attempt to alter the yard's canal infrastructure, which took up a great deal of valuable space but was growing increasingly under-used, though any changes in this area were contingent upon the cooperation and goodwill of the owners of the Regent's Canal.

As the 1890s drew to a close, it became apparent that the King's Cross Goods Yard was struggling to cope with the increasing volume of road and railway traffic. The Yard's hydraulic infrastructure was also ageing and in need of modernisation. These challenges coincided with the arrival of a new rival which was about to enter the London railway goods market. In order to face these and other challenges, it became increasingly necessary for the Great Northern to make sweeping improvements at the King's Cross Goods Station.

Vital Cogs In A Vast Machine: The Labour Force Remembered Through History And Archaeology

'...the goods department's interest began once wagons had arrived at the reception sidings and the train engine had been detached. First of all the number-takers listed the wagon numbers, checked the labels to see that they were actually for King's Cross (occasionally this proved not to be the case!), and removed the yellow envelopes containing the invoices (i.e. the railway's own transit documents) from the clip on the wagon solebar. In rainy weather, these were often sopping wet, and their subsequent treatment in the office partook something of the nature of unravelling the Dead Sea Scrolls! The number takers having finished, the shunters took over, and as tracks in the shed became cleared, loaded wagons were shunted up in rakes of about 15 for unloading...Control was then in the hands of the shunters, and it seemed a marvel to me that there never seemed to be any mishaps. The only time I ever saw one was when a 20-ton brake van became derailed by some over-enthusiastic fly shunting. It blocked three tracks leading into the Inwards Shed for about five minutes until it was re-railed by methods which in my later Army days were usually referred to as brute force and bloody ignorance!....Muscle power ruled, and the skill, panache even, with which heavy and awkwardly shaped packages were got out of wagons and on to the delivery vehicles was marvellous to behold...Once the outbound wagons were loaded and labelled, they were drawn out of the shed and marshalled for the appropriate train by the shunters, after which the Operating Department took over, and the Goods Department had no further interest in the matter until they arrived at their destination'

From 'Memories of King's Cross Goods' by Peter Erwood, a Junior Clerk at King's Cross in the late 1930s (Erwood 1988: 133–4)

The buildings and the infrastructure that made up the yard were vital to its success, but it was the people who worked there who enabled it to function. By the mid-1930s the workforce employed in the Goods Yard numbered upwards of some 2,000 individuals, with yet more employed by the various private merchants and businesses which traded from premises around the depot (Erwood 1988: 131). This section describes a small number of the many occupations that brought the shed floor and the sidings to life, highlighting the importance of manpower to the effective working of the Goods Yard.

A visit to the Goods Station at any time of the day or night would have bombarded the senses with innumerable sights, smells and sounds. The clanking of wagons, the shouts of the checkers and porters, the whinnying of horses and the clack of their hooves would have punctuated the drone of the hydraulic machinery. All the while, chugging locomotives would have been working the outdoor sidings, whilst a thunderous rumble would have been audible every time a wagon discharged its load into the coal drops. Until its closure in the early 20th century, the stench of the adjacent gasworks would have wafted across the canal whenever the wind blew in a northerly direction, whilst anyone visiting the Handyside Canopies would have been assailed by the smells of the fish and vegetable sidings, complete with the occasional wagonload of spoiled goods. In wet weather, unpaved areas of the station would have quickly turned to mush, while in dry spells the cartage teams would have kicked dust up into the air as they trundled past. Plumes of steam from the chimney of the hydraulic station and the locomotives that worked the rails would have been a constant feature, whilst more often than not the entire area would have been shrouded in smog from the hearths and factories of the city. To the unaccustomed visitor the experience would have the potential of being quite overwhelming, but to the staff of the Goods Station it would simply have formed part of their usual routine.

When a goods train arrived at the station's reception sidings, the locomotive would be detached and its rolling stock would be divided by the employees of the arrivals area, whilst men stationed in the departure zone would be responsible for forming newly loaded wagons back into trains for onward dispatch to the north. The many oversized spanners that were discovered in and around the Train Assembly Shed were presumably used to assemble and dismantle goods trains. A fragment of vacuum hose, a vacuum hose end and an associated clamp from a wagon[7] were also discovered in the shed, and it is likely that these items were broken and lost when trains were being assembled or dismantled.

After incoming wagons and vans had been separated, checkers and number takers would be on hand to list their identification serials and to read the transit documents that were attached to each. These were intended to confirm that the shipment was destined for King's Cross and detail its onward path through the station and into the world that lay beyond (ibid.: 133).

Teams of horses, men and boys were responsible for guiding empty carts and wagons laden with all kinds of goods through the main shed and the various covered areas within the yard's precinct. Until the advent of hydraulic capstan shunting in the early 1880s, this was achieved inside the shed by horse and man power alone. Even after the introduction of hydraulic power in the main Goods Shed, capstanmen remained a common sight throughout the station, although the

---

7   These items formed part of the braking system of a train. Flexible pipes ran down the entire train so that brakes could be applied to every piece of rolling stock.

An employee at King's Cross manually commences the shunting process through the use of a 'pinch bar'; when inserted between the wheel and the rail the lever on the tip of this tool could be used to move wagons over short distances, 1953 (1995-7233_LIVST_DF_201_A (1995-7233_LIVST_DF_201_A © National Railway Museum)

A selection of tools discovered during archaeological excavations in the Train Assembly Shed

Hook discovered in the Main Goods Shed

Shunters' poles recovered from excavations in the Main Goods Shed

175

number of horses in use did fall. Until the late 1930s, the shunters would have use of the many turntables that were dotted throughout the Transit and Train Assembly Sheds in order to turn wagons across the complex at right angles. An assortment of items accidently dropped into the wells of the turntables was found during the archaeological investigation, including glass, pottery sherds and broken clay pipes, a fork, a spoon and somebody's door key, which were found in the turning mechanism of Turntable A.

Shunting with or without the help of hydraulics was a potentially hazardous undertaking, as a large railway cart running at speed on a goods shed siding was more than capable of killing a man and inflicting life-threatening wounds upon a cart horse (Anon 1851b). Indeed, official figures from 1912 suggest that an incredible one in ten shunters were either killed or seriously injured whilst at work (May 2003: 30). To reduce the danger, the men used shunters' poles, which comprised long wooden sticks fitted with a metal hook that could be used for coupling and uncoupling rolling stock without walking between wagons. Some idea of the importance and prevalence of this simple tool can be gleaned from the fact that six complete examples were unearthed during the course of the archaeological work that was undertaken in the Main Goods Shed, along with the metal tips of a further five. Also discovered was a broken section of the metal links that ran between the wagons, onto which the poles were hooked.

In certain outdoor areas, where the risk of a conflagration was comparatively low, tank engines and their footplate crews could be seen working the sidings (Medcalf 1900: 315). By the 1930s, this job fell to about six LNER Class J52 0-6-0 saddle tank locomotives, a modified variant of a design by the famous Great Northern locomotive engineer Patrick Stirling. The LNER had inherited a fleet of these robust locomotives when it took possession of the Goods Yard at Grouping in 1923 (Erwood 1988: 133).

Loading and unloading was the job of the bank staff. Typically they were formed into small gangs that included a checker (i.e. the supervisor), a 'caller off' who read or attached the address labels and administrative papers to the various consignments and about four goods porters. Although the latter were expected to do the lifting, in practice this was usually done by the whole team (*ibid.*: 134). Assistance was provided by a network of mechanised cranes, hoists and jiggers controlled by machine operators, as well as a system of manually operated pulleys and winches. This machinery no doubt accounts for the many small broken wheels, pulley blocks and associated hooks that were unearthed throughout the yard during the archaeological excavations.

From the earliest days, power for this equipment would have been provided by the Hydraulic Station. It is reasonable to assume that the number of men working there would have grown after the station was enlarged in 1882 in order to cope with the increased demand that the introduction of mechanised capstan shunting brought. In all likelihood, this happened again in 1899 after the Western Goods Yard opened to traffic and the hydraulic station was further improved.

Despite the assistance provided by hydraulic machinery, a large amount of energy was still expended by the porters in order to successfully transfer all kinds of loads between wagons, road vehicles and, in the early decades, the canal. Even by the late 1930s, after a very substantial modernisation programme had taken place, extensive manual handling remained common practice (*ibid.*: 131). Often aided 'by nothing more sophisticated than hand barrows and cargo hooks with occasional help from a crowbar or two', the bank staff were expected to manhandle all kinds of commodities in and out of open wagons and box vans alike (*ibid.*). Indeed, many barrow and trolley wheels appear to have perished under the strain of the daily grind, since seven examples suffering from varying degrees of warping and buckling were unearthed within the confines of the Eastern Goods Shed.

By LNER days, the easternmost bank in the Inwards Shed (identified here as Platform 2) had been marked out so as to divide it into smaller sections that corresponded to specific districts of London, whilst a facility in the Outwards Shed was divided in much the same way but by northern destinations and stations on the line out of King's Cross (*ibid.*: 134). This proved to be a reasonably effective technique which aligned railway wagons with corresponding road vehicles in order to limit congestion and reduce lateral manual movement along the banks (*ibid.*).

The colder months presented the Good's Station's labourers with a wide range of challenges. In snowy or frosty conditions, wagons were sometimes unable to depart whilst pick-ups could be delayed, leading to 'thousands of trucks', loaded with commodities, standing in and about the station (Medcalf 1900: 315). This caused problems for the shunters, who would then be asked to extract one specific wagon for loading or unloading that was, more often than not, marooned within a mass of rolling stock (*ibid.*). Such occurrences were commonplace throughout the year, but were exacerbated by poor weather (*ibid.*: 315–6).

In the weeks leading up to the festive season, the intensity of incoming goods traffic would dramatically increase in order to meet the demands of the Christmas table. Activity in the Metropolitan Cattle Market would pick up pace, whilst huge numbers of 'specials' carrying rhubarb from Yorkshire would pour into the Potato Market and the Handyside Canopies, typically between the hours of 7pm and 10am, all of which added to the workload of the shunters and the bank staff (*ibid.*: 318).

Like all sections of permanent way, the station's railway sidings had to be maintained in order to enable shunting to take place successfully. This task was entrusted to plate layers, who may well have been responsible for dumping a number of lengths of redundant rails under the Transit Shed banks, along with old fixings from chairs and broken sleepers, which were discovered during the archaeological investigations. Two rakes and a fork were also discovered in the Eastern Transit Shed, which were probably used to smooth the ballast.

While horses and their handlers were absolutely central to operations in the Goods Station, other animals could be found at work in the yard. Feral cats were a constant presence and they were actively encouraged into areas like the Granary to control the rodents which infested the building. During the tenure of the LNER, the need to control pests resulted in the creation of one of the most unusual jobs in the yard, that of 'The Cat Man'. Employed by the railway company directly, he put out milk and food for the cats and collected the corpses of rats that they had killed. Although his income was modest, he was able to supplement it by selling kittens 'for sixpence a go, mostly to the female staff of the accounts office' (Erwood 1988: 137). The cats and 'The Cat Man' himself appear to have fascinated other company employees who worked at the Yard. One of the Yard's strays, the formidable and battle scarred 'George', became so well known that the LNER chose to publish his likeness on a postcard (Christopher 2012: 95).

As might be expected, the Goods Yard's labour force was overwhelmingly drawn from the communities that lived around the King's Cross area. The vast majority of those who worked at the yard came from the local working class population, although many members of the clerical and

George, the battle-scarred Goods Station cat

managerial staff would have been from slightly more prosperous backgrounds. In addition to the predominantly white workforce, the yard employed a handful of employees who came from minority groups, who are worthy of special mention here as it would otherwise be very easy to forget that they were ever there at all.

Following his retirement from the Great Northern's service in 1897, Josiah Medcalf (1832–1932) wrote a piece for the Railway Magazine in which he reminisced about his long service with the Great Northern at King's Cross (TNA RAIL 1156/12/2; Medcalf 1900). Writing about the Potato Market, Medcalf described a railwayman who was known to his colleagues as 'Black Jem' on account of his ethnicity. Jem was evidently a valued member of the workforce, whose aptitude, work ethic and intelligence were praised by Medcalf, who described him as 'one of the best capstanmen in the country'.

The recruitment of women to perform traditionally male jobs during the First World War proved that they could successfully undertake a wide variety of roles in the railway industry. By the end of the war, more than 68,000 women were employed by the railway companies, although the end of hostilities and the return of men to their peacetime jobs meant that the number of women working on the railways had fallen to 20,000 by 1920 (Earnshaw 1990: 36). Documentary evidence indicates that a number of women worked as porters in the Western Goods Shed during the conflict (TNA RAIL 236/190: 39). Despite the dramatic fall in the numbers of female employees after the war, women continued to occupy a variety of roles, mainly clerical, throughout the interwar period. It was still surprising however, to learn that in the 1930s a small contingent of women were working as porters at King's Cross. Although employed by one of the Goods Station's tenants rather than the LNER, the gang of women unloaded up to 30 wagons of bottles per day on behalf of their employer, the glass manufacturer Joseph Bagley & Co. of Knottingley, West Yorkshire. The women evidently suffered a great deal of prejudice, being viewed with suspicion and even trepidation by many of their male colleagues. Even some of the relatively well-educated clerks found their presence troubling, one of whom described the girls as 'fearsome and muscular' (Erwood 1988: 133). According to the latter source, the prevailing view in the yard was that the women were little better than prostitutes, a line of work in which they were rumoured to indulge when not at work unloading their wagons (ibid.: 132–3). This opinion doubtless owed less to the moral standards of the women themselves than to the prejudices of those who objected to them working in a traditionally all-male occupation.

CHAPTER 7

The King's Cross Goods Station at the turn of the 20th Century: Investment and Enlargement

'A year ago, when prognostications were rife as to the loss of traffic which the Great Northern Company would suffer when the new Great Central line was opened, it might have been thought prudent on the part of the former not to embark on a costly scheme for enlarging its goods accommodation in London; but the management took an opposite view, and their policy has been abundantly justified by the steady manner in which their traffic has continued to increase week by week, notwithstanding the commencement of the new competition. Thus it has come about that the new terminal accommodation for goods traffic at King's Cross, which was brought into use on Saturday, 8th July, is available at a time when it is pressingly wanted, besides being welcome as a weapon for strengthening the Great Northern Company's position in London under circumstances of keener competition than has hitherto been experienced'

A journalist describes his visit to the newly enlarged Goods Station at King's Cross for the publication 'Transport: The Weekly Review' (Anon 1899)

The ten years either side of 1900 were perhaps the closest that Britain's railways ever came to something approaching a 'Golden Age'. The nation's track network was almost complete, the quality of the rolling stock was undergoing a dramatic transformation for the better, locomotives were becoming increasingly powerful and safety standards were improving (Wolmar 2007: 164, 169, 185). Most companies therefore chose to plough resources into improvements in order to attract customers rather than attempt to enlarge their existing networks (*ibid.*: 194). With the internal combustion engine in its infancy at the end of the 1890s, it would be several years before long distance road transport could compete with the steam train.

The Diamond Jubilee of the Great Northern Railway in 1910 therefore coincided with a period that has been described as its zenith (Nock 1974: 129). The company's express passenger services were fast and frequent whilst its suburban services continued to extend into the London commuter belt (*ibid.*: 71; 129). Yet goods traffic of an ever increasing and varied nature continued to turn the greatest profit and account for the greatest number of services, and the Great Northern was able to attract a great deal of this custom due to a number of pioneering innovations not available to the company's rivals (*ibid.*: 129, 141; Simmons and Biddle *et al.* 1997: 192). The Great Northern was the only carrier that used fast, fully braked goods trains capable of running at very high speeds (Simmons and Biddle *et al.* 1997: 192). One such express covered the distance between King's Cross and Manchester in just five hours and 50 minutes, an extraordinary feat for an Edwardian goods service (Nock 1974: 130). The company also custom built a range of wagons specifically tailored to carry different freight types, such as covered vans that were used to transport bananas from the docks at Liverpool and the high capacity bogie wagons that conveyed the huge volumes of brick traffic carried by the company's trains (*ibid.*: 129). In order to further boost efficiency, the Great Northern also began to greatly improve the quality of their goods handling facilities, modernising and enlarging the marshalling yards at Doncaster, Colwick, Peterborough and Ferme Park (*ibid.*: 30).

Despite all this, passenger rather than freight services were still given priority upon Britain's railway network, with goods trains occasionally becoming lost in the system, sometimes for hours, as they waited for a 'go' signal (Wolmar 2007: 193). In 1907, the Midland Railway introduced a system that involved installing telephones in signal boxes so that the passage of all trains throughout their network could be charted, thus making it harder for mistakes of that nature to arise (*ibid.*). Whilst the Midland was the first British company to pioneer this innovative system it was the Great Northern that really embraced it, introducing telephones in the Leeds district in 1912, the Western District in 1913 and across the remainder of its network during 1914 (Nock 1974: 131).

Despite these welcome development, the railways faced many challenges at the turn of the new century.

180 *The King's Cross Goods Station at the turn of the 20th Century: Investment and Enlargement*

The demands of an ageing track network, competition between carriers, an increasingly discerning public and the continued implementation of Government regulations regarding ticket pricing and freight haulage costs meant that the railway companies were no longer able to generate huge surpluses (Wolmar 2007: 205). The dawn of the Edwardian period would also see railway employees demand better terms and conditions as the workforce grew increasingly unionised (Simmons and Biddle *et al.* 1997: 342). Although the latter development was fiercely resisted by the railway companies, a number of strikes during the 1900s would eventually lead to the introduction of various contractual improvements, all of which would have the effect of lowering profit margins (*ibid.*; Wolmar 2007: 200–5). Taken together, these factors impacted upon the dividends that were paid to the Great Northern's shareholders. Though the value of the company's dividends remained competitive, they were worth less than they had been in earlier years (Nock 1974: 129).

The British economy began to slow as it entered the 20th century, though that trend was defied by London which was one of the few areas in the country that continued to enjoy economic expansion (Scott 2007: 36–9). Unlike most of the rest of the nation, its local railway network therefore continued to grow due to the demands of the city's industries and its new commuter population (Wolmar 2007: 185). The expansion of its suburbs had by now taken the form of ribbon development along its railway network as improved services, a wider range of public transport and the widespread provision of workmen's fares enabled

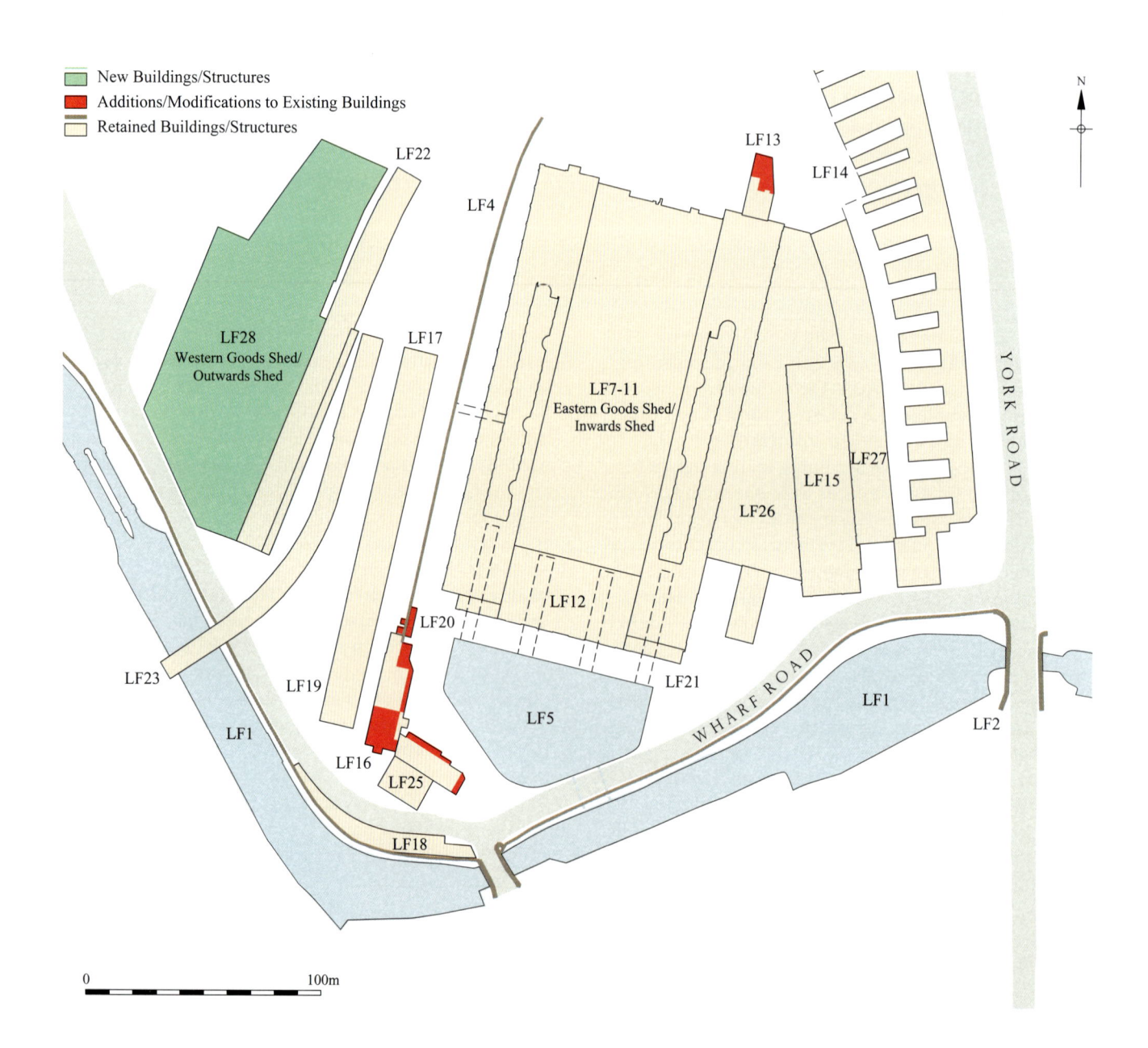

Fig. 7.1   The locations of the landscape features present in the Goods Yard from 1897–1913, scale 1:2,500

employees to vacate the overcrowded centre and move further out (Brandon 2010: 66). The continued growth of the capital's economy and its population, driven increasingly by the railways themselves, meant that the demand for goods and services remained very high.

Ever since the early 1850s the Great Northern Railway had consistently attracted a significant share of that traffic, however the status quo was about to be challenged by a potent new competitor. Under the auspices of its ambitious chairman, Sir Edward Watkin, the Manchester Sheffield & Lincolnshire Railway was rebranded as the Great Central Railway, which applied for a London extension in 1897. Having won the right to enter the capital, the Great Central set about building the last of London's major railway termini, the Marylebone Passenger Station and Goods Yard, which were completed in 1899 (Simmons and Biddle *et al.* 1997: 189). Watkin envisaged that his new concern would ultimately connect with his other company, the South Eastern, via the Metropolitan Railway, enabling its trains to continue to Normandy via a tunnel under the English Channel, thus providing a direct link between Manchester and Paris. While the formation of the Great Central 'caused a stir in every railway boardroom in England', it was the Great Northern that faced the greatest threat from the new arrival (Nock 1974: 114). The new company's mainline from Manchester to London would not access any areas of the country that lacked railway lines, and would instead exclusively rely on the poaching of traffic from rival companies (*ibid.*; Simmons and Biddle *et al.* 1997: 189). The Great Northern was therefore faced with the potential loss of all goods shipments west of Sheffield, the company's great fear being that the Great Central would retain all of the London-bound goods traffic that it had previously taken from the Manchester Sheffield & Lincolnshire Railway (Nock 1974: 114–5). Given the already congested state of the King's Cross Goods Yard, the arrival of a new and technologically advanced depot at Marylebone must have caused deep concern. The situation must have been particularly alarming given that the Great Central intended to make most of its money from goods rather than passenger services; indeed, a massive 67% of their receipts did eventually come from this sector (Simmons and Biddle *et al.* 1997: 189).

This chapter explores how the Great Northern attempted to resolve the problem of congestion at King's Cross Goods Yard and the company's response to the threat posed by the Great Central. The impact of other improvements carried out at the yard during the Edwardian period, such as improvements to the yard's welfare facilities, are also considered.

7.1 The Inwards Goods Shed: the late 1890s to the early 1900s

Landscape Features 7–11

The late 1890s ushered in a phase of extensive and much needed change at King's Cross. In May 1897 the Board of the Great Northern approved a scheme that finally promised to end the delays and congestion that had afflicted the yard throughout its existence, by separating outwards and inwards goods traffic and doubling handling capacity (Anon 1899: 6; TNA RAIL 236/158: 208, 369). The scheme was intended to disentangle 'down' (i.e. north-bound) from 'up' (i.e. south-bound) lines by establishing independent Inwards and Outwards Goods Depots. The centrepiece of the latter was to be constructed over two levels on the site of the largely redundant Coal and Stone Basin. Once the new Shed had been completed and the Western Coal Drops had been converted into a subsidiary Outwards Goods Shed, the Granary, Transit Sheds and the Train Assembly Sheds would handle inward-bound freight only. That complex of buildings was henceforth termed the 'Inwards' or the 'Eastern Goods Shed' within the 'Eastern Goods Yard' whilst the new complex became known as the 'Outwards' or 'Western Goods Shed' (Landscape Feature 28) within the 'Western Goods Yard' (Fig. 7.1). The latter depot lay beyond the limit of the area encompassed by this study and is not described here.

When the 'Western' or 'Outwards' Goods Shed opened in March 1899, the purpose of the former Outwards platform in the Western Transit Shed changed almost overnight from dealing with outgoing to incoming goods. Similarly, the sidings and platform space that had been added to the area immediately north of that shed in the 1870s 'in order to facilitate the transfer of Traffic for the north coming from the City' were rendered redundant since they had not been designed to receive incoming commodities. Historical maps and plans indicate that they were removed at some point between 1894 and 1905 (Fig. 7.3), most probably around March 1899 when their significance dramatically decreased. A plan of the Goods Yard surveyed in1905 also suggests that a central bank extension to Platform 5 was added in the Train Assembly Area, thus increasing the amount of space that was available for unloading (Fig. 7.3).

The conversion of the Transit Sheds and the Train Assembly Shed to the Inwards Goods Shed impacted upon those buildings in a number of ways. In December 1897 it was decided that additional office space should be provided for both the Inwards and Outwards Sheds for an estimated total cost of £11,640 (TNA RAIL 236/183: 91). Whilst the greater part of this sum was to be spent on new accommodation at the Western Goods Shed, the demand for additional office space at the Inwards Goods Shed prompted the Board to approve the construction of a new range of offices in the Train Assembly Shed. The contract for these works was awarded to Kirk Knight & Co. at the beginning of February 1898 for the sum of £2,023 (TNA RAIL 236/530: 'Indenture Kirk Knight & Co. and GNR', 04/02/1898). The contractors were instructed to take down part of the original roof over the Train Assembly Shed and to build high level offices upon compound steel girders with new cast iron columns beneath. The section of roof taken down was then to be reinstated atop the new offices. Access to the offices would be gained via the Granary, where a new corridor was to be constructed on the second floor between the staircase in the southwest corner and a doorway cut out of the north wall of the building (*ibid.*: 'Specification, Bills of Quantities & Tender 12/1897' 14–9). Bricks used in the construction of the new offices were to be of the same make and quality as those of existing structures, with the exception of piers under the girders, which were to be made of bundled Staffordshire blue bricks (*ibid.*: 27). At the point where the compound girder met the north wall of the Granary the contractors were instructed to 'shore-up the brickwork and roof over the large arch to the [east–west] Cartway [and to] cut out the arch and make good the piers for the taking of the compound girder'.

Each office was to be fitted with a large mahogany topped desk, together with a cast iron fireplace manufactured by the Teal Fireplace Company of Leeds (*ibid.*: 41). Construction of the new offices was underway by the end of March (TNA RAIL 236/183: 158). Although the exact date of completion is unknown, the offices had been in use 'for several months' by the spring of 1899 (TNA RAIL 236/530: Ross to Latta 12/04/1899), by which time a number of departments had relocated there. While the lower level was occupied by the Goods Delivery Office, the upper level was allocated to the Shipping Office and the Inwards Correspondence Office (Erwood 1988: 135).

The entire enlargement programme was fast approaching completion by the spring of 1899, by which time the two yards had been working independently of one another for several months. Owing to the decision to run inward-bound traffic into both Transit Sheds the delays caused by the previous system were largely eliminated, although the new arrangements highlighted the inadequacy of existing road vehicular access (Medcalf 1900: 314). In order to improve the road connections between the platforms, Henry Oakley proposed in May 1899 that a roadway be constructed across the southern end of the Train Assembly Shed, linking the two Transit Sheds together more effectively (TNA RAIL 236/160: 144). Oakley's recommendation was accepted and it was agreed that the road would be built for an outlay of £215. The contract for the construction of the road was awarded to Messrs Wall & Co. the following month (*ibid.*: 146).

Fig. 7.2 The western end of the newly constructed 'Western Goods Shed' (Landscape Feature 28) fronting the north bank of the Regents Canal (TNA ZLIB 3/19 © The National Archives) Reproduced from The Illustrated handbook of the Regent's Canal and Dock Company published in 1911

Built heritage evidence relating to the construction of the Overhead Offices in the Train Assembly Shed in 1898–9

Landscape Feature 11

The majority of the original circular cast iron columns that upheld the roof of the Train Assembly Shed were removed when the building was re-roofed in the 1950s. The built heritage survey (KXF07) found that eight of these columns remained in situ, where they had previously supported the floor of the Transit Shed's Overhead Offices from 1899 into the 1950s.

During the construction of the Overhead Offices in 1898, these eight circular columns were joined by three I-beam examples that were inserted at the expense of two sections of original masonry. Eleven structural supports, consisting of cast iron brackets and riveted steel plate girders resting on padstones, were embedded just below eaves height in a rebuilt section of the east wall of the Western Transit Shed (Fig. 7.4). The girders were originally crossed at right angles by three rolled steel joists that had been inserted in the south wall of the Granary (Landscape Feature 12). Together, these structural elements supported the floor of the Overhead Offices.

At the southern end of the complex, three rolled steel joists were stamped with the maker's mark 'BURBACH 45 NP', indicating that these elements had been forged by the Burbach steelworks in Germany.

The offices had originally been surmounted by a pitched roof, indicated by the scar that was left by the structure on the west side of the north elevation of the Granary (Landscape Feature 12; Fig. 7.6). The construction of the roof had necessitated infilling one window and partially blocking two others. A doorway providing direct communication with the second floor of the Granary was then inserted (Fig. 7.6).

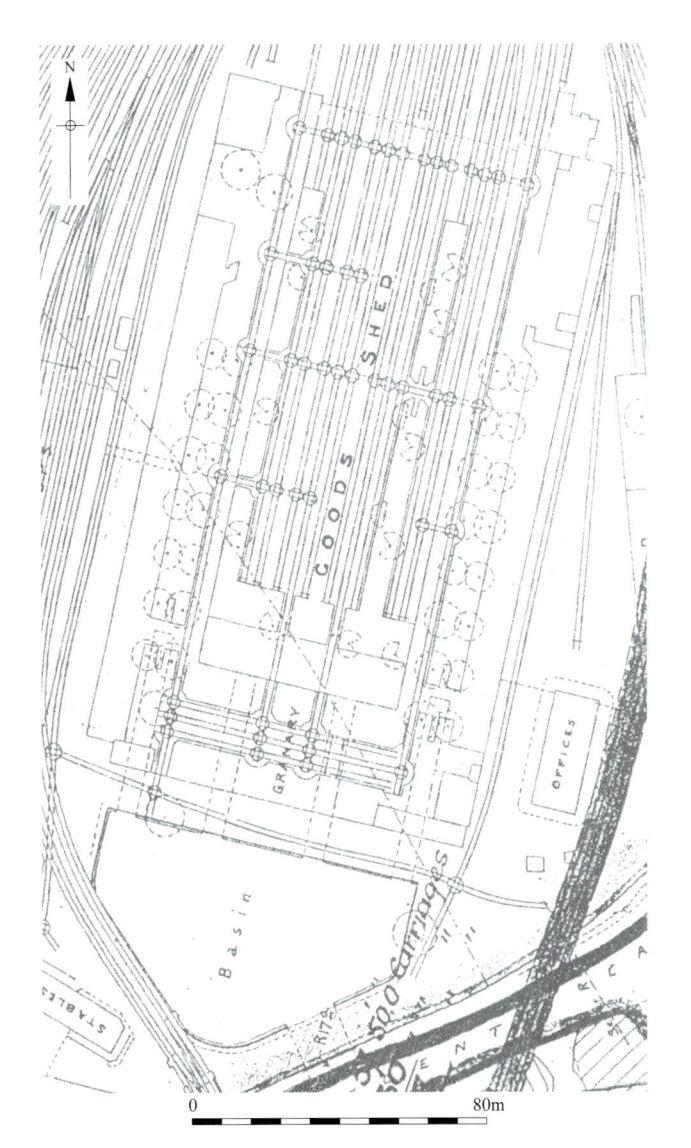

Fig. 7.3 The Great Northern Railway Plan of 1905 detailing the Eastern Goods Shed (Landscape Features 7 to 11), the Granary (Landscape Feature 12) and the Granary Basin (Landscape Feature 5), scale 1:2,000

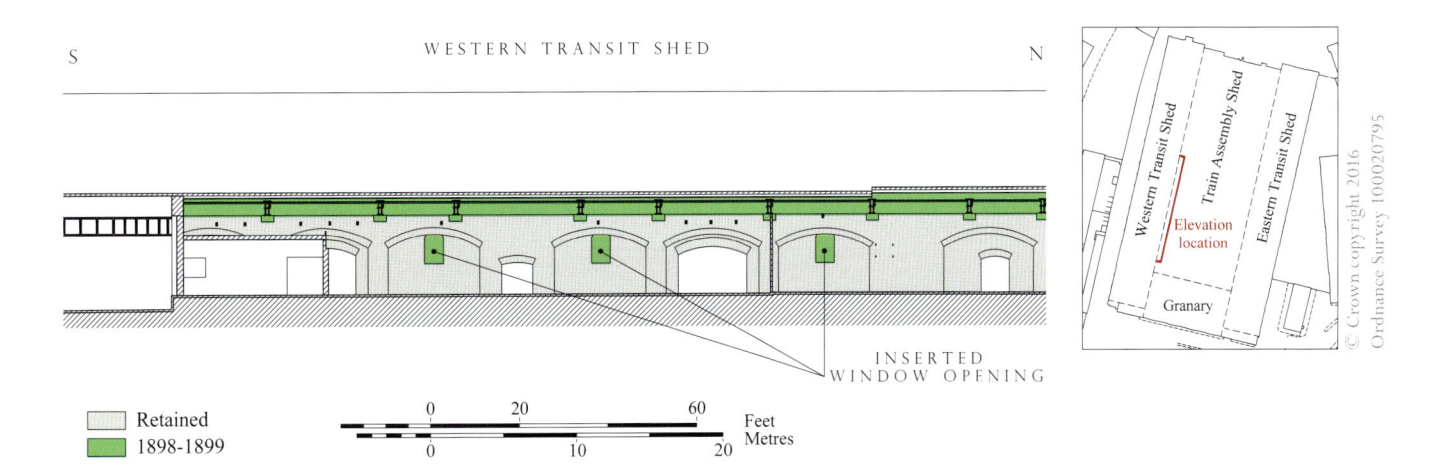

Fig. 7.4 Cross-sectional elevation through the Train Assembly Shed (Landscape Feature 11) showing the external face of the east wall of the Western Transit Shed (Landscape Feature 7) and the changes of 1898–9, scale 1:500

Before the Train Assembly Shed's overhead offices were constructed, the Western Transit Shed was predominantly lit from above by its partially glazed roof. Subsequently the shed was lit by several new window openings, inserted towards the southern end of the Western Transit Shed's eastern wall (Fig. 7.4).

The upper, eastern section of the shed's northern elevation (around and above the opening that formerly accommodated the shed's railway siding) has been rebuilt at some point in the past. The nature of the brickwork that was used suggested that this occurred during the insertion of the Assembly Shed Offices in 1898–9 (Fig. 7.5). Why such an extensive rebuild was undertaken in that location at this time remains uncertain, as the north end of the shed was far removed from the new structure. A likely explanation is that the arch that surmounted the reconstructed opening had started to subside. This certainly occurred to the arches of the equivalent openings in the Eastern Transit Shed, which had to be reinforced with rolled steel joists (RSJ) in the latter half of the 20th century.

A plan of the station surveyed in 1905 suggested that a central bank extension had been added in the Train Assembly Area during its conversion in 1899 (Fig. 7.3). No archaeological evidence relating to it was found, so it is reasonable to assume that the feature was either never present or had not survived. One possibility is that this central area was used for stockpiling certain goods but was never formerly turned into a raised bank. Alternatively the 1899 structure may have been replaced in its entirety by a later platform in the 1930s.

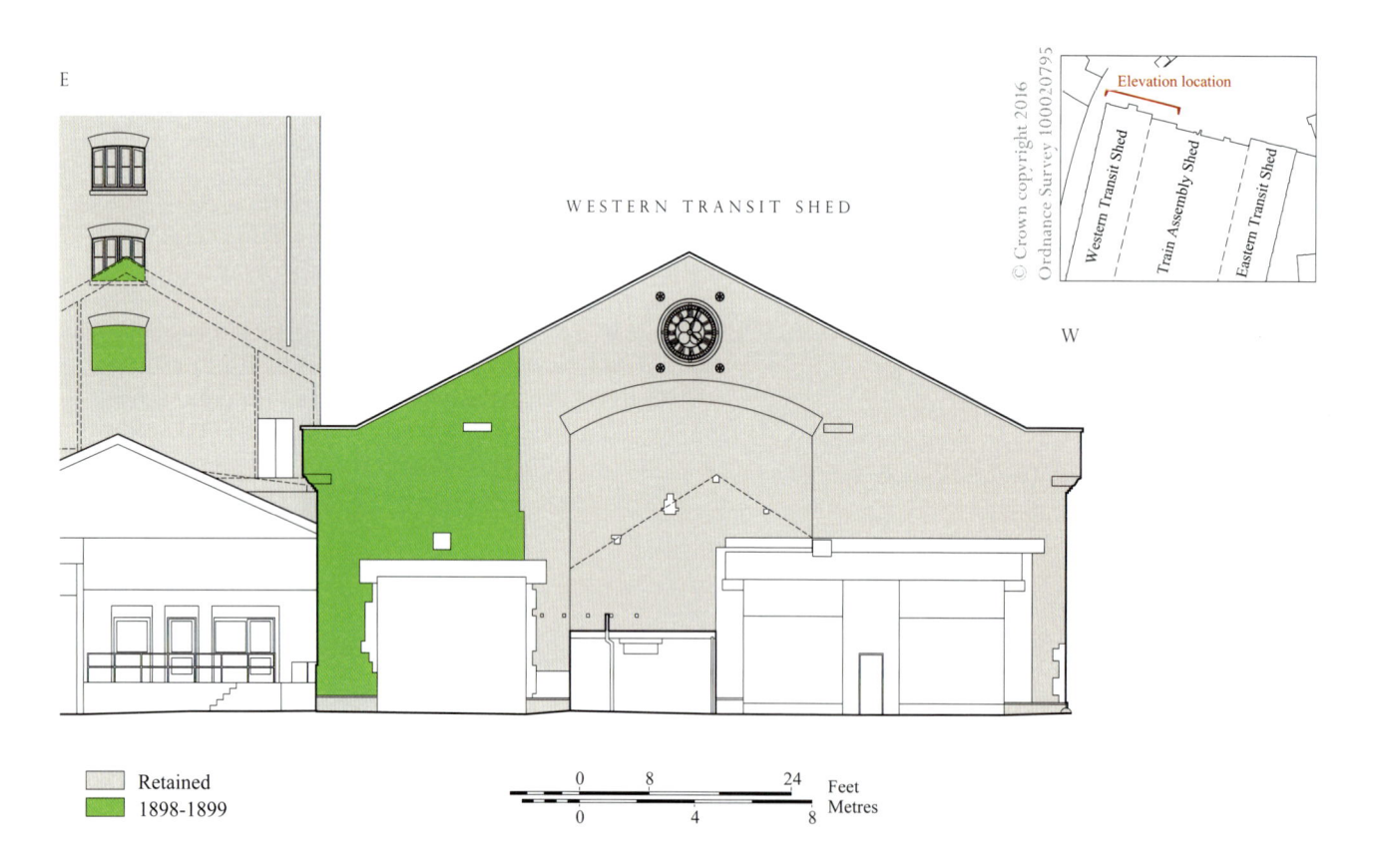

Fig. 7.5 External elevation of the north wall of the Western Transit Shed (Landscape Feature 7) showing the changes of 1898–9, scale 1:250

7.2 The Granary: the late 1890s to the early 1900s

Landscape Feature 12

While the 1897–9 conversion programme had a significant effect upon the way that goods were handled in the Eastern Goods Shed, the Granary appears to have remained largely unaffected by the changes. Any alterations to the building that were made during this period were a result of the construction of the Overhead Offices in the Train Assembly Shed's (Landscape Feature 11).

This work entailed removal of part of the roof over the Train Assembly Shed and construction of level offices upon steel girders with new cast iron columns beneath. Access to the new offices was via the Granary, where a new corridor was to be constructed on the second floor, illuminated by window openings cut into the west wall (TNA RAIL 236/530: Specification Bills of Quantities & Tender 12/1897 14–9, 29). The east wall of the corridor was to be built between the existing cast iron columns upon RSJs, while both the floor and ceiling were to be made fireproof in accordance with the recommendations made by Crockett six years earlier (*ibid.*: 14; COL MS 14943/13: 136 1891–2). The fireproof floor was to be constructed upon RSJs, and covered with coke breeze and Portland cement concrete, the latter 'to be finished smooth to take 'Duffy's' or other approved deal block flooring 1½" thick' (TNA RAIL 236/530: Specification Bills of Quantities & Tender 12/1897 30). Bricks were to be of the same make and quality as those of existing structures, with the exception of piers under girders, which were to be made of bundled Staffordshire blue bricks (*ibid.*: 27). For the north elevation of the Granary, the contractors were instructed to 'shore-up the brickwork and roof over the large arch to the [east–west] Cartway [and to] cut out the arch and make good the piers for the taking of the compound girder', using Staffordshire bullnosed blue bricks to construct the piers.

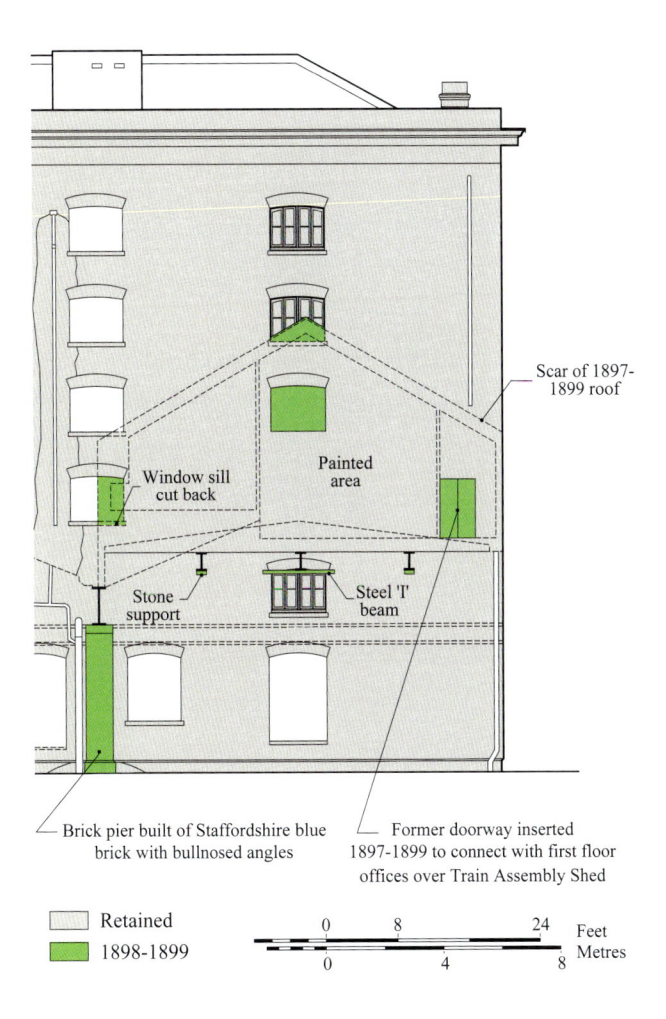

Fig. 7.6 External elevation of the western side of the north wall of the Granary (Landscape Feature 12) showing the changes of 1898–9, scale 1:250

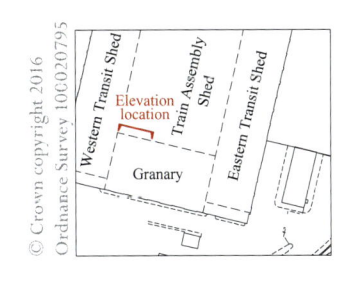

Built heritage evidence of modifications to the Granary during the construction of the Train Assembly Shed Overhead Offices in 1898–9

Landscape Features 11 & 12

The evidence that was observed during the field work corresponded with the available historical information. Three second floor window openings surmounted by segmental arches were inserted in the west elevation of the Granary, just above the eaves of the Western Transit

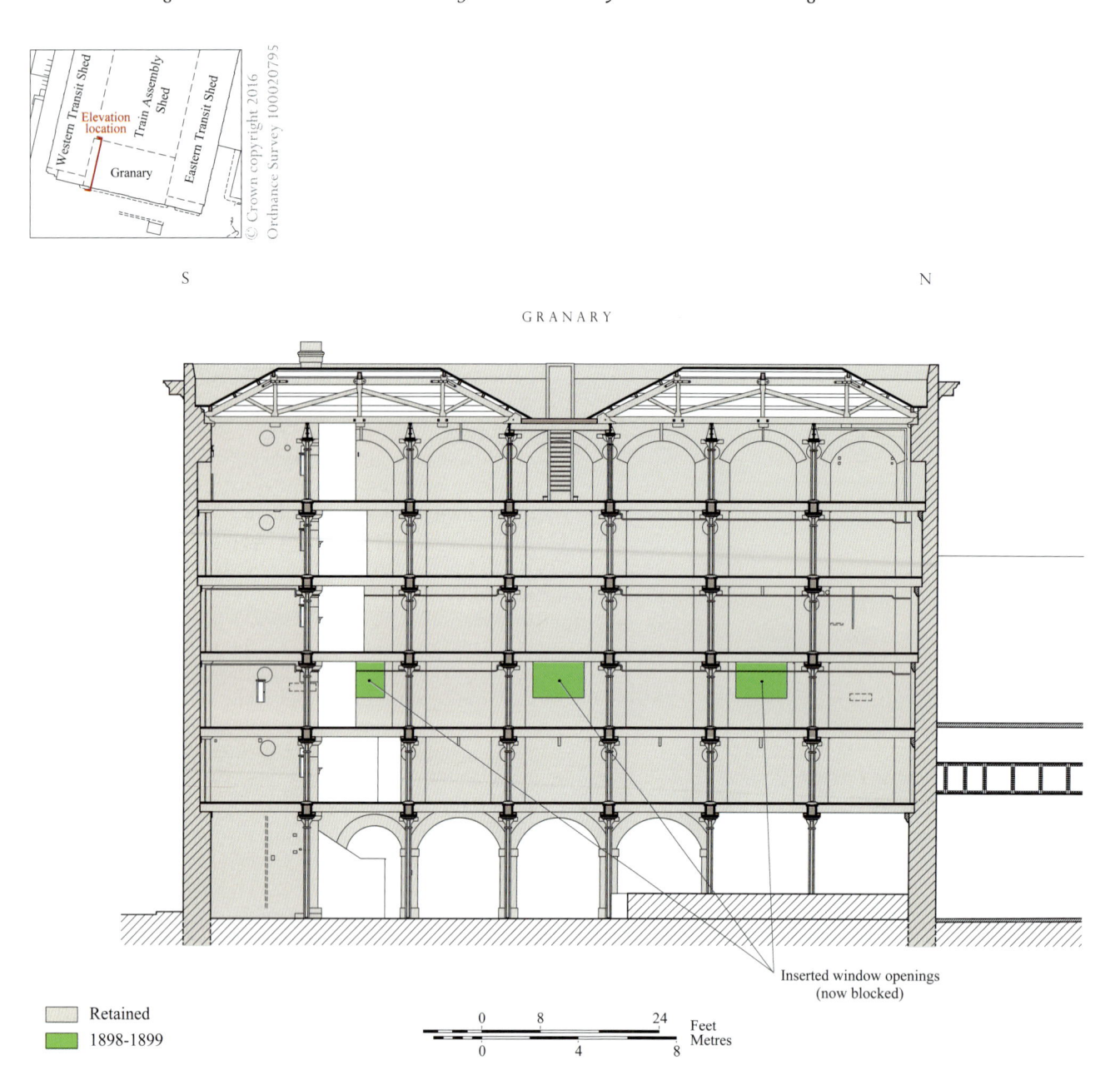

186   *The King's Cross Goods Station at the turn of the 20th Century: Investment and Enlargement*

Fig. 7.7   Cross-sectional elevation (looking west) showing the changes of 1898–9; note the windows (now blocked) that were inserted in the west wall of the Granary (Landscape Feature 12) to illuminate the corridor that connected the building with the Train Assembly Shed's Overhead Offices (Landscape Feature 11), scale 1:250

Shed (Fig. 7.7). They were no doubt added between 1897 and 1899 to illuminate the new passageway that led from the Granary to the Train Assembly Shed's Overhead Offices (Landscape Feature 11).

Where the Overhead Offices met the north elevation of the Granary, a third floor window was bricked up, whilst it was necessary to partially infill the window that sat directly above it as well as the second floor window in the adjacent bay (Fig. 7.6). In order to retain a first floor window an east–west steel beam was inserted across the upper part of the opening, which supported a north–south RSJ that formed part of the new offices.

7.3 The Western and Eastern Goods Offices: the late 1890s to the early 1900s

Landscape Features 20 & 21

Following the conversion of Cubitt's Goods Shed to receive incoming traffic in the late 1890s, the Outwards Goods clerks decamped to the high level offices at the new Western Goods Shed. This must have freed up a great deal of space in the Western Goods Offices although it is not entirely clear as to what use this was put at the time. Additional accommodation had also been provided for the clerks of such departments as the Inwards Correspondence Office following the construction of the new high level Goods Delivery Office in the Train Assembly Shed. This must have made the Western and Eastern Goods Offices a good deal less cramped, greatly improving the working conditions for the clerks therein.

7.4 The Hydraulic Station: the late 1890s to the early 1900s

Landscape Feature 16

Less than fifteen years after the Hydraulic Station had last been enlarged, Locomotive Engineer Patrick Stirling informed the Locomotive Committee in the autumn of 1895 that the hydraulic pumps were 'practically worn out and require renewal' (TNA RAIL 236/211: 55; Smith 2008: 12). Furthermore, it was noted that three of the five boilers at the Hydraulic Station were 25 years old and were in need of replacement (ibid.).

Despite the acquisition of a pair of 'modern pumps and cylinders' from Tannett Walker of Leeds, it was clear that the output of the Hydraulic Station was not sufficient to meet the needs of the growing depot. It certainly would not be able to power the proposed Western Goods Yard in its existing state. This prompted Stirling's successor Henry Ivatt to seek permission to acquire an additional steam pumping engine in June 1897 (TNA RAIL 236/211: 213). Ivatt's request was approved and a tender from Tannett Walker to supply an engine fitted with 14' diameter cylinders by 21' stroke, with an output of 160–80 gallons per minute was referred to the Board in late July. This engine offered the added attraction of being able to be installed 'without alteration to the building' (ibid.: 220). It has been suggested that this engine was therefore likely to have been a direct replacement for the old 25hp Armstrong model, which was to be scrapped (Smith 2008: 12).

While the requirement for additional engine capacity had been addressed for the time being, the old boilers were clearly approaching the end of their working lives. A temporary solution was found by mounting two locomotive type boilers outside, but the following March a tender submitted by Daniel Adamson of Hyde to supply three new 7' diameter boilers at £322 each was approved by the Locomotive Committee of the Great Northern (ibid.). In May 1898 this decision was revoked in favour of replacing all five existing boilers with four 8' diameter Lancashire examples. The decision to purchase the Lancashire boilers necessitated the enlargement of the Hydraulic Station, which was too small to accommodate them. It was also necessary to build a taller chimney in order for the boilers to function correctly.

In September 1898 a new pumping engine 'capable of delivering 300 gallons of pressure water per minute', was ordered from Tannett Walker in order to increase power output in advance of the completion of the new Outwards Goods Depot (i.e. the Western Goods Yard) (TNA RAIL 236/212: 58). Following its purchase, the owners of the Regent's Canal permitted the Great Northern to take an additional quantity of water from the canal basin for condensing purposes. A payment of £50 per quarter for the water taken was agreed, all of which was to be returned to the basin after use; the railway company were also to be responsible for maintaining all of the necessary pipework (TNA RAIL 1189/1426: 'Agreement for the Supply of Water at King's Cross' 02/10/1916).

Since neither the new boilers nor the newly acquired steam engine could be accommodated in the existing Hydraulic Station it was clear that the existing structure would have to be remodelled. A specification for the requisite works was issued by the Engineer's Office in June 1898. The roofs, the central (dividing) wall and the external front (east) wall of the Boiler and Engine Houses were to be taken down and new roofs and walls for both were to be erected. The existing c.80' tall chimney was to be demolished and replaced with a 104' tall shaft in a new location (TNA RAIL 236/449 Specification: 14). Other parts of the building that were to be taken down included the walls of the 'small Accumulator House' and the workshop, although the later Accumulator House was to be left alone. Although the roofs of the Engine and Boiler Houses were to be removed, the contractor was told to leave 'the large timber beams under the roof' of the Engine House in place and to take particular care when taking down the rest of the structure so that no damage would be caused to the machinery within (ibid.: 27–8). It was necessary to keep the engines operational throughout the rebuild, suggesting that they were powered by the two locomotive type boilers that had been installed outside the Hydraulic Station whilst the Boiler House was out of commission (ibid.).

Certain walls, entrances and openings were either to be retained or to be relocated in the new building. For example, the contractor was instructed to take down the 'old landing and steps down to Boiler House Floor and refix [the] same in [an] altered position as will be directed' (ibid.: 18). The rebuilt Hydraulic Station was to have a new single gabled roof, framed with 12' x 4' tie beams in a single length. The roof principals would comprise 8' x 6' Queen Posts, 9' x 6' principal rafters and 9' x 4' purlins, clad with North Welsh Countess slates (ibid.: 28).

The earliest plan of the Goods Yard showing the Hydraulic Station after it had been remodelled in 1898–9 confirmed that the facility had indeed been enlarged by rebuilding the northern end of the building (Fig. 7.8). The plan also suggested that the section of the Retaining Wall (Landscape Feature 4) that formerly divided the 'back' and 'front' Engine and Boiler Houses had been demolished in order to create a single, larger engine house and one sizeable boiler room.

A plan of unidentified origin, apparently dating to 1906, offered some schematic internal details concerning the post-1899 arrangement of machinery inside the Hydraulic Station (Fig. 7.9). Annotations on the plan make it clear that the unaltered northern room continued to function as the Accumulator House whilst the central room housed the engines and the southernmost room contained the boilers. Two circular structures are shown within the unmodified Accumulator House, which represent the two accumulators that were acquired back in 1881–2, whilst four parallel rectangular structures appear side by side within the Boiler House, each of which must represent one of the four new boilers that were ordered in 1898 (Fig. 7.9).

The arrangement of machinery within the Engine House was harder to interpret with confidence. Seven structures of varying shapes and sizes were shown within its confines on the 1906 plan. As demonstrated by a later plan, compiled for insurance purposes, post-1899 the two 75hp Tannett Walker engines, acquired in 1891 and 1897, were instated on the large beds in the western half of the building, whilst the 100hp Tannett Walker condensing engine, purchased in 1898, was installed upon one of the beds in the eastern half of the building, along with a 40hp machine (Fig. 7.11). A reference in a Locomotive Department report regarding the removal, scrapping and replacement of two small pumping engines in 1914 proves that at least two small engines were present in 1899 (TNA RAIL 236/216: 271; Smith 2008: 14). Removed in 1914, they were said to be around 30 years old when they were scrapped, which suggests that they were installed in the mid-1880s or thereabouts (perhaps during the 1881–2 enlargement

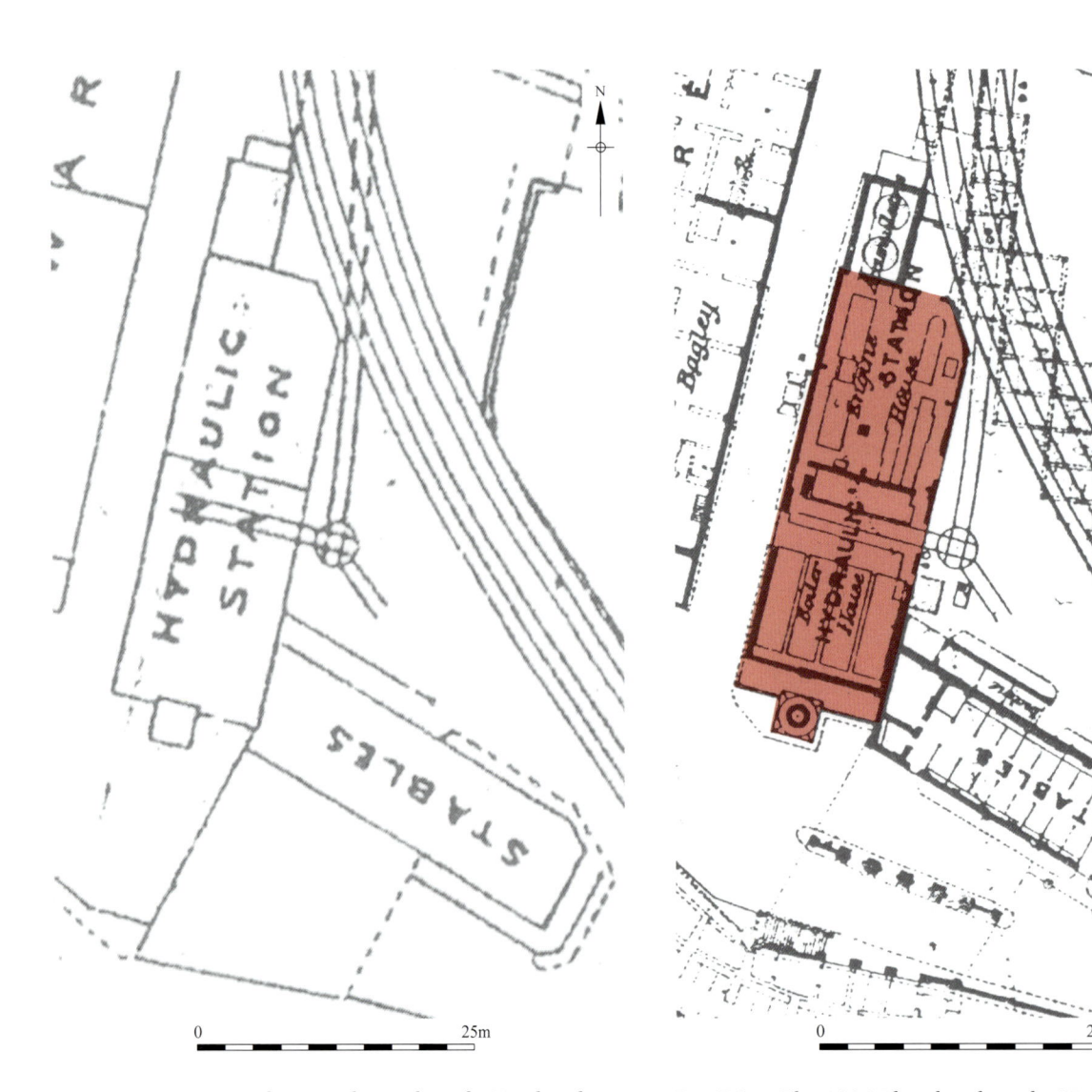

Fig. 7.8   Great Northern Railway Plan of 1905 detailing the Hydraulic Station (Landscape Feature 16), scale 1:625

Fig. 7.9   The 1906 Plan detailing the Hydraulic Station (Landscape Feature 16), scale 1:625

The Hydraulic Station: the late 1890s to the early 1900s 189

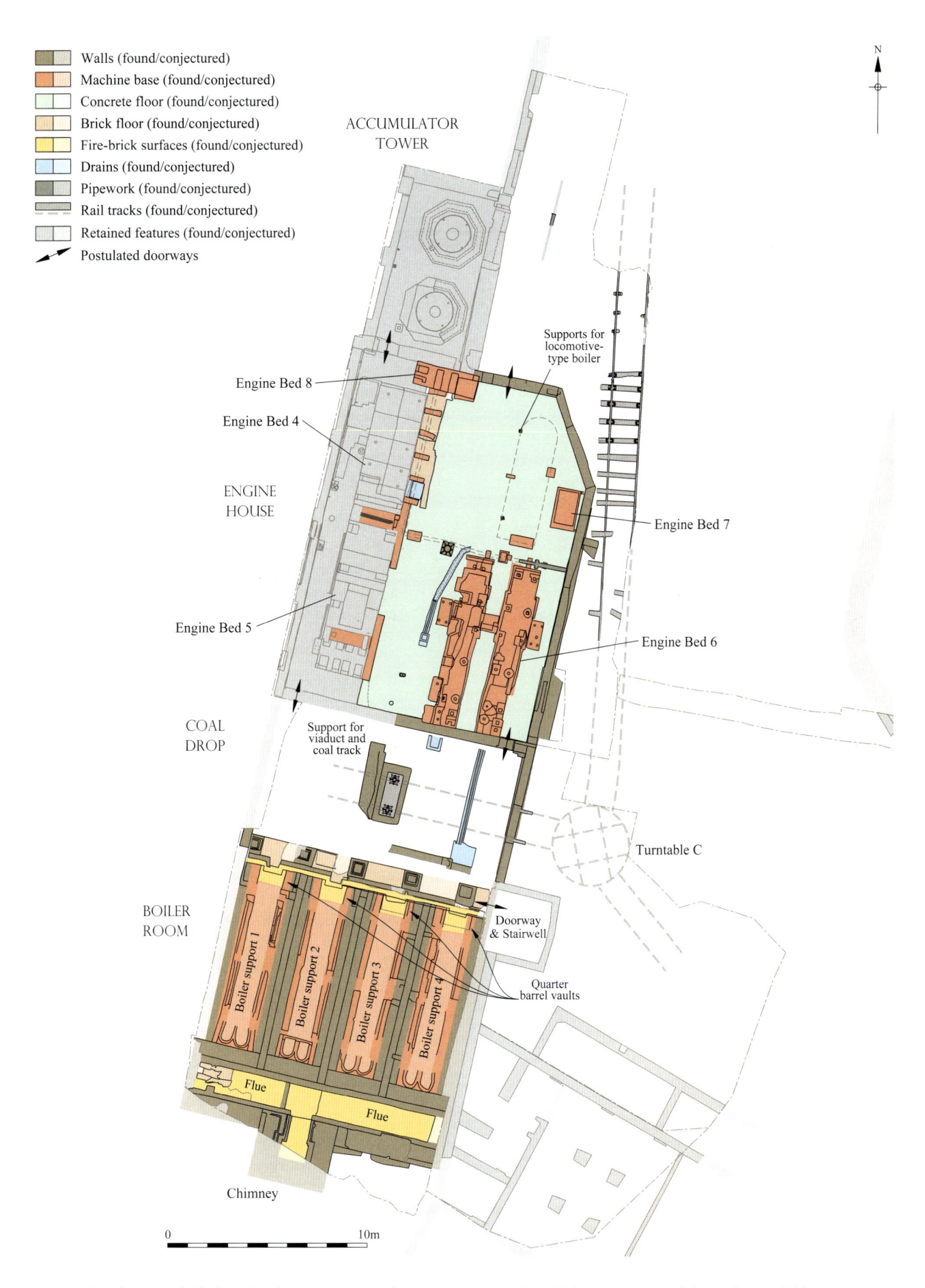

Fig. 7.10 The remodelled Hydraulic Station (Landscape Feature 16) by 1899, reconstructed from the available archaeological and cartographic evidence, scale 1:250

works). Since they were not removed until 1914, they must have survived the Hydraulic Station remodelling programme of 1898–9. The presence of these smaller engines within the Engine House post-1899 may therefore account for a further two of the structures that are shown on the 1906 plan.

In summary, at least five engines are known to have been installed in the Hydraulic Station's Engine House by 1899, although seven structures of varying sizes are shown within that room on the 1906 plan. With regard to the distribution of the engines, it was impossible to state with absolute confidence which model was situated where on the basis of the documentary evidence, with the exception of the two 75hp examples. Fortunately, archaeological evidence revealed a fuller picture (see below).

Historical maps and plans indicate that the Hydraulic Station was extended once more by 1905 (Fig. 7.8) and again between 1905 and 1906 (Fig. 7.9). The earlier phase of expansion involved the addition of a very small rectangular abutment to the northern side of the Accumulator Tower (Fig. 7.8). By 1906, this was joined by two slightly larger structures, one of which abutted the eastern side of the Retaining Wall, whilst the second abutted the Engine House and the Accumulator Tower (Fig. 7.9). Documentary evidence alone gives little clue as to the purpose of these enigmatic additions.

Archaeological evidence relating to the Hydraulic Station enlargement of 1898–9

Landscape Feature 16

The physical remains of the Hydraulic Station confirmed that it was indeed modified in 1898–9 and archaeological evidence provided additional details concerning the extent of the rebuild and the internal arrangements of the machines. The archaeology that was encountered also showed that part of the north wall, all of the south wall and much of the eastern side of the pre-existing Hydraulic Station were entirely rebuilt with red and yellow frogged bricks set in indurated lime mortar at this time. The rebuild was therefore more extensive than cartography alone suggested (Fig. 7.10).

In accordance with the specification for the rebuilding works, both the western side of the building and the 1881–2 Accumulator House were left unaltered whilst the old chimney was replaced with a new one that was found to be integral to the newly erected southern wall of the Boiler House. The Old Accumulator Tower was demolished and the dividing wall between the former Back and Front Engine and Boiler Houses (i.e. a section of the Retaining Wall, Landscape Feature 4) was torn down as specified, although the archaeology that was encountered demonstrated that the lower reaches of it survived below the new floor of the remodelled building.

A slender remnant of this wall also survived above floor level, forming part of a new internal dividing wall that separated the remodelled Engine and Boiler Houses. As such, by 1899 that partition was a complex multi-phase structure that incorporated elements of masonry dating to 1850–2, 1882 and 1898–9. It is reasonable to assume that some work would have been required in order to make the conglomerate presentable, which may explain why the contractor was asked to 'face up the old walls at both ends of the same' after taking down the former dividing wall between the two Engine Houses (TNA RAIL 236/449: Conditions of Contract 18).

The Engine House of 1898–9

The north-eastern section of the remodelled Engine House ran at an oblique angle relative to the rest of the building in order to accommodate a railway siding, thus forming a five sided structure, the internal dimensions of which were 16.80m north–south by 12.36m east–west (Fig. 7.10).

A layer of rubble and clay was deposited inside the eastern half of the completed structure sealing the 1850–2 brick floor of the first incarnation of the Hydraulic Station as well as the remains of the partially demolished wall that formerly divided the Back and Front Engine Houses. The ground-raising material created a new floor at a level comparable with the 1881–2 surfaces to the west, that were retained post-1899. This obscured the brick infill of a doorway that formerly linked the old 'Back and Front Engine Houses' with the Old Boiler Room to their south, proving that the doorway fell out of use at this time. A concrete bedding layer, 0.20m thick, sat above the ground raising material and the remaining upstanding elements of the dividing wall. It was capped by stretcher bonded bricks which formed the new floor surface in the eastern half of the Engine House. At a height of 22.39m OD, the 1881–2 surface in the western half of the building was now a mere 0.27m higher than its new equivalent to the east. A 0.60m long concrete ramp was therefore installed to the south of Engine Bed 2, which eliminated the trip hazards that would otherwise have been caused by this slight height differential.

A rectangular metal plate, 0.64m north–south by 0.56m east–west, was found embedded in the approximate centre of the floor of the newly remodelled Engine House. A circular void was observed in the centre of the plate, which was 150mm deep. It once supported a central, circular columnar roof support (Fig. 7.10) that must have upheld the new gabled canopy that spanned the entire east–west extent of the Hydraulic Station. Further weight is added to this interpretation by the inclusion of a central, circular feature in approximately the same location as the metal plate on the 1906 plan (Fig. 7.9). It was coloured in an identical way to the surrounding masonry walls of the Hydraulic Station demonstrating that it was a structural element. The plan also showed two similar circular features, centrally placed against the eastern and western walls of the engine house along with another in the centre of the southern wall. Together these formed a cross shape, the centre of which coincided with the location of the metal plate. This suggests that a total of four columns once supported the 1898–9 Engine House roof.

The distribution of the engine beds that were unearthed archaeologically (Fig. 7.10) closely resembled the arrangement shown on the 1906 plan of the Hydraulic Station (Fig. 7.9), although some anomalies were encountered.

Only one earlier engine bed (either Bed 4 or 5) was retained during this phase (Fig. 7.10), when it was joined by a virtually identical counterpart (again, either Bed 4 or Bed 5).

In the eastern half of the building, two parallel, sub-rectangular masonry beds, bridged at a right angle by a third rectangular brick feature that was integral to the other two elements, were erected. The north–south structures are clearly shown on the 1906 plan as two separate constructions (Fig. 7.9), however the archaeology that was encountered suggested that they were actually part of one structural entity, henceforth identified here as Engine Bed 6 (Fig. 7.10).

The most easterly of the two north–south masonry elements that formed Engine Bed 6 was 1.76m in width. It was roughly rectangular with irregular sides, 0.76m of it being visible above the 1898–9 floor level. The external faces were built with English bonded red bricks held together with indurated light grey mortar. The brickwork was a minimum of four courses thick and hid an internal core that was composed of ad hoc materials that included bricks, stone blocks and reused circular millstones that had been bonded together with concrete. Five circular and four rectangular holes were observed in the top (Fig. 7.10). They did not truncate the structure but had been intentionally created when it was built and clearly did not represent an afterthought. The rectangular holes were between 110mm and 140mm wide, whilst their circular counterparts (formed by the centres of the millstones) were 60mm to 72mm in diameter. It is likely that a total of fourteen voids originally existed, although the probable locations of five of these had been destroyed by later building activity (their locations are extrapolated in Fig. 7.10). A series of bricked up openings were identified in the eastern and western faces of the bed. These were between 1.40m and 0.32m long and between 0.28m and 0.42m high. Each had been bricked up at a slightly later point and were aligned with one or more of the holes that were observed in the top of the structure. The western masonry element was an approximate mirror image of its eastern equivalent and it is likely that it originally had an identical arrangement of openings and bricked up voids (Fig. 7.10). As detailed in Section 5.2, these holes were used to secure the engine to the bed.

The 1898–9 concrete surface butted the sides of Engine Bed 6 because the elevated floor would have been incapable of supporting such a heavy structure. It was necessary to throw its weight down to ground level by positioning it upon the remains of the 1850–2 machine beds (i.e. Engine Beds 1, 2 and 3; see Fig. 3.63), which remained extant immediately below Engine Bed 6.

It has been suggested that Engine Bed 6 was the correct size and shape to support a compound steam pumping engine. An engineer's report suggests that the 100hp Tannett Walker engine that was ordered for the 1898–9 Engine House was a tandem-compound, yet the nature of this engine base suggests that it was actually a cross-compound (Tim Smith pers. comm.) Another anomaly can be found on the Goad Insurance Plan of 1921, which suggests that the engine was 120hp rather than 100hp (Goad 1921; Fig. 7.11).

It seems that the 75hp Tannett Walker engine that had been installed in the western half of the building during the 1880–2 extension works was left in its original position, either on Bed 4 or Bed 5 (only one of which was in existence in the early 1880s as discussed in Section 5.2). Which of the two beds was the older remains unknown, however it can be said is that both were extant from 1898–9 onwards after which they both accommodated 75hp Tannett Walker engines, variously purchased in the early 1880s and in 1897. The logic for this is as follows.

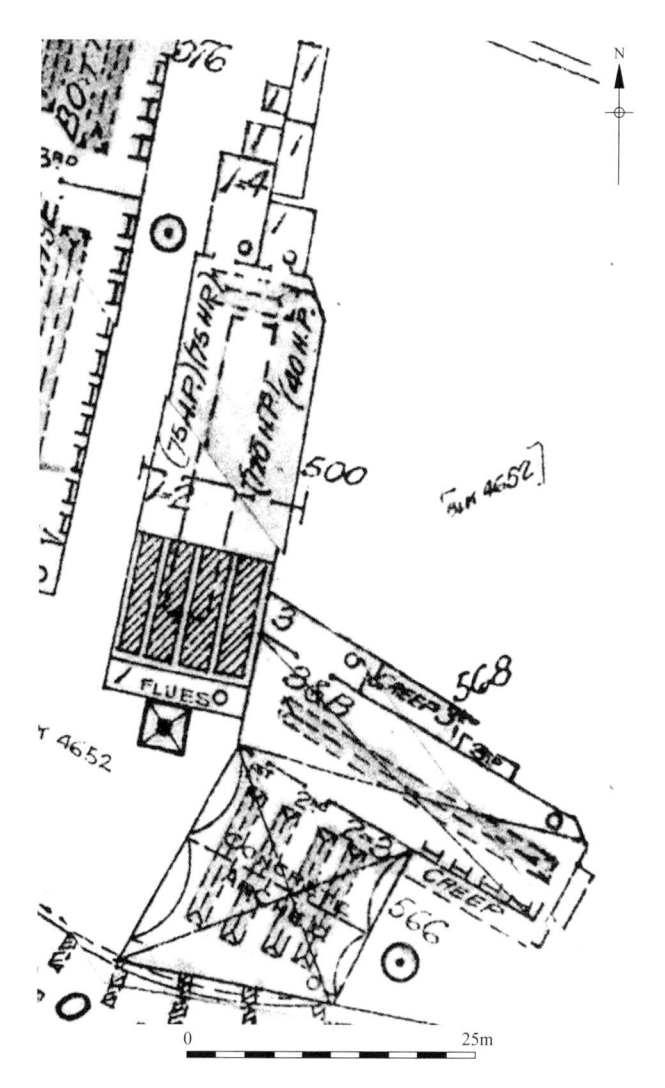

Fig. 7.11 The Goad Fire Insurance Plan of 1921 detailing the Hydraulic Station (Landscape Feature 16), scale 1:625

It is known that a 75hp Tannett Walker engine was installed on Engine Bed 1 in the Old Engine House one year before the 1898–9 programme of works began, since a Locomotive Department report noted that it could be 'got into the site now occupied by the old engine [which was worn out] without alteration to the building' (TNA RAIL 236/216: 271; Smith 2008: 14). The 'old engine' in question was the 24hp Armstrong example that had been installed back in 1850–2. The 1898–9 rebuilding works resulted in the loss of Engine Bed 1 after the construction of Engine Bed 6 for the new 100 or 120hp machine, meaning that the 75hp example, which was virtually new, would have to be moved. Consequently, by the time that the Goad Insurance Plan of 1921 had been compiled, it had been reinstalled in the western half of the Hydraulic Station, alongside the 1882 example, either on Engine Bed 4 or 5 (Fig. 7.11; Fig. 7.10). In all likelihood, this rearrangement occurred during the remodelling works of 1898–9.

The two small engines that were purchased in the first half of the 1880s and sold for scrap in 1914 were retained during this phase. Like the 1897 Tannett Walker machine, they were probably moved from their original positions during the 1898–9 programme of rebuilding. Their likely positions within the remodelled building could not be conclusively determined from documentary material alone, however the archaeological evidence has provided some clues.

A small rectangular structure, Bed 7, was unearthed close to the eastern wall of the Engine House. It was oriented north–south and was composed of red bricks that were capped by one large York stone slab (Fig. 7.10). The structure was 1.78m long and 0.48m wide, 2.06m of it being visible above the level of the 1898–9 floor surface. The Goad Insurance Plan of 1921 indicates that a 40hp engine was situated upon this base by that time (Goad 1921; Fig. 7.11). In all probability this was the 40hp example that was brought to King's Cross from Farringdon Street Goods Depot as a replacement for the two 1880s engines that were scrapped in 1914. However, the base itself definitely pre-dates this event as it appears on the 1906 plan of the Hydraulic Station (Fig. 7.9). It was probably constructed during the 1898–9 programme of works and certainly cannot pre-date that event since its eastern side was situated beyond the footprint of the earlier version of the Hydraulic Station. This suggests that Bed 7 was not originally constructed for the Farringdon engine; instead, it probably accommodated one of the early to mid 1880s engines after the rebuild of 1898–9. The other early to mid 1880s engine may have been positioned in the Engine House on Bed 8, another small base that abutted the dividing wall between that room and the Accumulator House (Fig. 7.10). The extant remains demonstrated that Bed 8 was erected during the 1898–9 rebuild since the eastern side of it sealed the partially demolished remains of the dividing wall that formerly separated the 'Back' and 'Front' Engine Houses.

The north-easterly 'bed', which is depicted as a bullet-like shape on that map of 1906 (Fig. 7.9), was not found. In its stead, three metal stanchions with 'H' shaped profiles were embedded in the concrete floor, along with two smaller masonry bases to their east. Together they may represent the so called missing 'bed' that is shown on the 1906 plan. It has been suggested that they may have helped to support a locomotive type boiler that has since been removed as machines of this kind were commonly mounted above ground rather than being fixed to a masonry base (Tim Smith pers. comm.). It is therefore possible that one or more of the two locomotive type machines that were mounted outside the confines of the Boiler House during the 1898–9 rebuild were situated in this location. They presumably played a pivotal role in keeping the hydraulic network operational whilst the Boiler House was being reconstructed. The fact that they appear on a map dating to 1906 suggests that they were retained either to provide additional power when necessary or to act as a 'back up' system should any of the main boilers require maintenance or repair.

The small rectangular structure shown on the 1906 plan against the west-central section of the southern wall of the engine house was not found archaeologically. Given that no trace of it was unearthed it is reasonable to assume that this enigmatic structure was relatively insubstantial and did not support an engine.

The Boiler House and Coal Drop (Coal Shoot) of 1898–9

Like its earlier counterpart, the floor of the new Boiler House was built at a similar level to the low ground on the western side of the terrace. Earlier versions of the structure permitted access and egress from the high ground to the east via a sunken doorway that was situated at the bottom of a stairwell that accommodated a staircase. This entrance was retained post-1899 in a narrowed and modified form, the southern side of the entrance being rebuilt using bull nosed, brown glazed bricks, whilst yellow and red bricks continued to form the northern side (Fig. 7.12). These masonry elements had been inserted within the larger 1850–2 doorway so that the outline of the earlier structure could still be discerned. The 1898 specification stated that an 'old landing and steps down to the Boiler House floor' should be lifted and refixed in an 'altered position as will be directed' (TNA RAIL 236/449: Conditions of Contract 18). This may relate to the narrowing of this entrance or alternatively could refer to an internal gantry and stairway that may have provided an internal link between the elevated floor of the Engine House and the lower level Boiler House.

The reconstructed and extended Boiler House of 1898–9 was rectangular in shape, its internal dimensions being at least 12.96m east–west by 17.92m north–south. The lower reaches of its northern, eastern and southern walls were unearthed archaeologically, whilst its western wall was most probably positioned just beyond the western limit of the excavation area (Fig. 7.10). Unlike the newly reconstructed Engine House, the floor of the new Boiler House was not raised. At 21.46m OD, it was respectively 0.66m and 0.93m lower than the floors that were present in the eastern and western sides of the 1898–9 Engine House. In contrast it was virtually flush with the external 19th-century ground surface at the base of the terrace to the west but was 2.54m lower than the top of the terrace to the immediate east.

The Boiler House was divided into two sections by an internal partition, which took the form of a row of identical rectangular column bases that were aligned east–west (Fig. 7.10). These were formed by rectangular concrete bases capped by square brick plinths that were composed of at least seven courses of grey bricks. The partition separated the remodelled Coal Drop to the north from the Boiler Room proper to the south (Fig. 7.10).

Two metal fixings, 1.43m apart, were observed in both faces of the external eastern wall of the Coal Drop. Although sealed by later masonry, their lower reaches had clearly been built into the wall when it was created in 1898–9 and were therefore contemporary with that phase of rebuilding. As demonstrated archaeologically and by the 1906 plan, they were aligned with a section of external railway track situated to the immediate east of the Hydraulic Station, which formed part of a spur that ran into the Boiler House (Fig. 7.9; Fig. 7.10). The features were therefore interpreted as a railway viaduct in cross-section, supported internally by a rectangular concrete base that upheld a pair of struts (Fig. 7.10). At a height of 24.12m OD, the viaduct was roughly level with the external 19th-century ground surface to the east and approximately 2.66m higher than the internal floor of the Coal Drop.

A reference in the specification of 1898–9 suggested that the area where coal was discharged was known as the 'coal shoot' whilst the viaduct itself was variously known as the 'Coal Track' and 'Truck Road' (*ibid.*). It was an essential component of the new facility as it enabled the bottom opening 7½-ton capacity wagons that the Great Northern used to discharge their loads directly into a hopper. It is therefore reasonable to assume that a similar mode of operation occurred here, albeit on a smaller scale. The Hydraulic Station 'chute' therefore probably consisted of a simple 'coal drop' that was located directly below the viaduct, the coal storage area being positioned to the immediate north.

The floor of the Coal Drop and coal storage area was characterised by a compact dirt surface resembling beaten earth, which was discovered at a level of 21.46m OD. Two drains had been cut into it (Fig. 7.10), the presence of which suggests that the room got wet on occasions, perhaps due to condensation from the boilers that could be found to the south or because this area was partially open to the elements so as to admit the railway viaduct. It is therefore highly probable that the so called earthen 'floor' was originally capped by a hard surface otherwise it would have become muddy and rather impractical when damp.

The boilers themselves were situated to the south of the Coal Drop, housed in four parallel, elongated masonry supports (Fig. 7.10). Each was 8.82m in length and 1.94m wide, the four separate positions being delineated by three internal dividing walls. The features were evenly spaced being between 2.75m and 2.78m apart. They presumably supported the four Lancashire Boilers that were ordered during the 1898–9 works.

Boiler Supports 1 to 4 (Fig. 7.10) were each composed of a rectangular masonry case that was formed by upstanding, red, frogged brick walls bonded with lime mortar. Curved 'W' shaped alcoves, between 0.50m and 0.54m in width and 0.74m and 0.84m in length, formed the most southerly sections of the structures. The main bodies were composed of header bonded bricks although shorter stretcher bonded sections were occasionally observed on the outer faces. The internal sides were lined with one course of stretcher bonded firebricks that included curved examples in the semi-circular alcoves. These fireproof linings appeared to have been set in a very soft layer of clay. Perhaps this clay was used to accommodate the movements of the boilers and associated bases resulting from their slight expansion and reduction in size as they heated up and cooled down. In some places the bricks appeared to have been dry-walled, but it seems more likely that the high temperatures that they were exposed to destroyed the clay and the mortar in those areas. The firebricks were not bonded into the main body of the external walls and were therefore freestanding, which would have enabled them to be replaced easily if damaged by heat. The northern ends of the probable boiler supports were each formed by two quarter barrel vaulted brick structures with concrete cores, their curved faces being on their internal sides. The bricks were crumbly to the touch as a result of heat or water damage. To the north, quarter barrel vaults were observed, which butted up against an east–west wall that incorporated four 'U' shaped kinks that were aligned with the boiler positions (Fig. 7.10). Its southern side was faced with firebricks, suggesting that it was frequently exposed to heat emanating from the boilers. The 'U' shaped sections may have upheld or formed fire boxes into which fuel was shovelled, each of which was given additional structural support by pairs of quarter barrel vaults.

Soon after the external walls of the boiler supports were complete, the gaps between each of them were infilled with light reddish grey, angular, pebble-sized clay fragments, presumably to improve their structural integrity. Being composed of poured concrete, the basal internal bedding for the floors of Boiler Supports 1 to 4 must also have been created at this time as these deposits were retained by the surrounding walls of each boiler position. In each instance, the concrete bedding layers supported brick floors that were composed of stretcher bonded firebricks. In the alcoves, they sloped gently downwards from a level of 21.18m OD in the south to 21.09m OD in the north. The remaining sections were virtually flat, being observed at heights that ranged between 20.99m OD and 21.09m OD.

To the immediate south of the four boiler positions a probable flue was observed on an east–west alignment, running across the entire length of the Boiler House. It consisted of two parallel walls composed entirely of firebricks that were between one and two courses wide and 0.24m high. The flue itself was substantial being 1.28m wide. An internal floor was also present, which consisted of at least two courses of fireproof bricks, the top of which was found to be at a level of 21.58m OD. A 1.09m wide gap was present in the centre of the south wall of the flue, which led into the remodelled chimney. A fire brick floor was unearthed inside the latter structure, the top of which was found to be at a height of 21.41m OD thus effectively forming an unbroken continuation of the floor of the flue. It is likely that waste gases were expelled from the boiler casings into the flue before being drawn up the chimney. Exactly how this was achieved remains unknown, however a network of above ground pipes seems the most likely explanation. These were probably removed after the building fell out of use, perhaps at the same time as the boilers.

Archaeological evidence relating to later alterations to the Hydraulic Station in the early 20th century

Landscape Feature 16

Excavations exposed the northern and eastern walls of an extension that abutted the eastern wall of the Accumulator House, as shown on the 1906 map (Fig. 7.9; Fig. 7.14). These elements formed a rectangular structure, the internal dimensions of which were 3.92m by 6.18m. The new room appears to have been divided into two small sections that took the form of an 'L' shaped corridor and a small rectangular room that housed a small concrete machine base, which was surrounded by a stretcher bonded brick floor.

The lack of documentary evidence pertaining to this structure precludes a definitive interpretation of its function, however a clue may be found in its southern wall (i.e. the former outer wall of the Hydraulic Station's Engine House), which subsequently contained a doorway that offered communication between the Engine House and the new abutment. Since the floor of the earlier structure to the south was found at a height of 22.12m OD, 2.03m lower than the high ground to the east and the floor of the new extension, it must have been necessary to pass between the two via a flight of stairs. This suggests that the two rooms shared more than a mere physical connection since it is unlikely that efforts to link them internally in this way would have been undertaken unless they had interconnected functions that required easy and rapid pedestrian movement between them. This in turn suggests that the new extension housed supplementary apparatus that was associated with the Engine House, effectively forming yet another extension to the Hydraulic Station.

To the north, the footprint of the second structural element that was added between 1905 and 1906 was identified (Fig. 7.14). Strictly speaking, it was not a true extension to the Hydraulic Station and is more accurately described as an outbuilding. Abutting the Retaining Wall (Landscape Feature 4) to the west, the north, east and south walls of the new structure were erected in one continuous build. Together, they created a rectangular building, the internal dimensions of which were 5.49m by 2.65m. Due to serious horizontal truncation, the locations of any obvious doorway positions were not observed in the exterior, however topographical constraints meant that it must have been entered from the high ground at the top of the terrace upon which it sat.

A mass of metal pipework was found inside, which consisted of eight pipes, two of which were linked to a circular tank-like structure in the northwest corner. A brick base, 1.07m long, 0.69m wide and 0.75m tall, supported the tank to the north, east and south, whilst a newly created sub-rectangular alcove that had been cut into the Retaining Wall supported it to the west. Whilst the function of the pipework remains uncertain, it is suspected that it was most probably associated with the hydraulic system (Tim Smith pers. comm.).

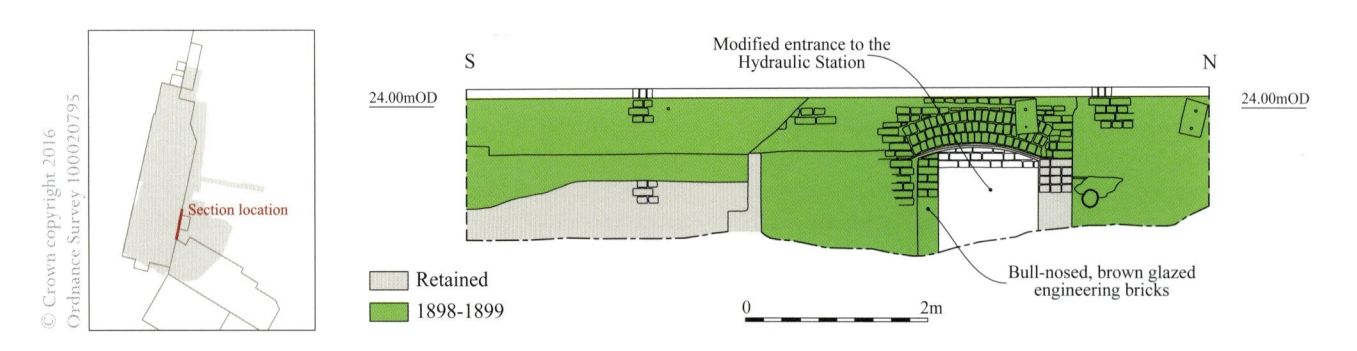

Fig. 7.12 East facing section showing the partially rebuilt external wall of the Hydraulic Station (Landscape Feature 16) in the location of the door and stairwell showing the changes of 1898–9, scale 1:80 (Area of Excavation is grey in inset)

Fig. 7.13 The Hydraulic Station's Boiler House (Landscape Feature 16) of 1898–9 under excavation, photograph faces east

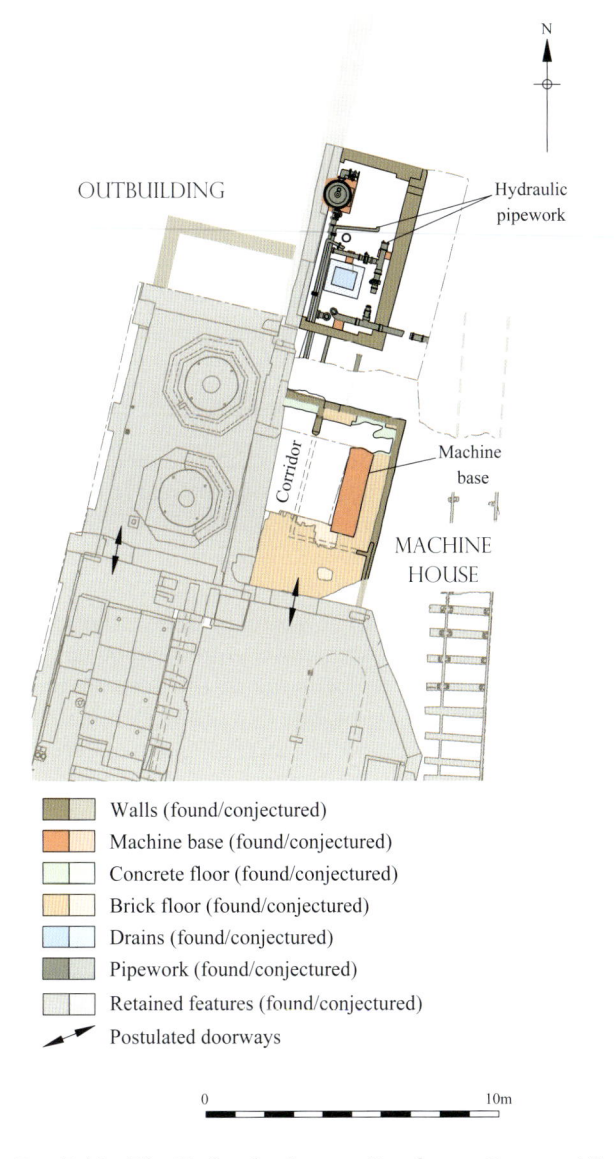

Fig. 7.14 The Hydraulic Station (Landscape Feature 16) and associated structures in plan by 1906 reconstructed from archaeological and cartographic evidence, scale 1:250

7.5 The Gasworks Viaduct: the late 1890s to the early 1900s

Landscape Feature 24

In February 1904 the Gas Light and Coke Company (GLCC), which had formed from the merger of the Chartered and Imperial Gas Light and Coke Companies in 1876 announced that it had 'decided experimentally' to close the St Pancras Gasworks (TNA RAIL 783/110: Grinling to Bury 03/02/1904, Brooks to Grinling 13/02/1904). Given that railway coal traffic to the works had ceased the previous month, it is unlikely that the news came as a complete surprise to the Great Northern, which was keen to take down the redundant Gasworks Viaduct (TNA RAIL 783/112: Brooks to Milne-Watson 06/01/1904). The gas company, on the other hand, was uncertain about the future of the works, which it continued to man with a skeleton workforce (LMA B/GLCC/126: 04/02/1904).

The senior officers of the GLCC had yet to make up their minds regarding the fate of the works by January 1908, when Alexander Ross (GNR Engineer) pressed the Board of the Great Northern to take steps to acquire the premises. Ross maintained that if the site was secured, the railway company could dispose of a number of unwanted 'assets' at King' Cross, including the Gasworks Viaduct and Bridge, the Congreve Street Bridge and the Battle Bridge Road Bridge, each of which he described as being 'in an advanced state of dilapidation'[8] (TNA RAIL 783/110: Ross to Brickwell 27/01/1908). Ross envisaged that once cleared of the remaining plant and machinery, the gasworks site could be used to accommodate either a further set of coal drops or an additional carriage shed (*ibid.*: 25/07/1910).

In July 1910 the gas company finally announced that it was prepared to sell 5½ acres of the gasworks site to the Great Northern for £55,000 (*ibid.*: Milne-Watson to Bury 20/07/1910). Having gained possession of the site the following spring, the Great Northern set about building a wall around the site, part of which was retained by the gas company as a gasholder station. In early 1912 the railway company invited contractors to clear the site of building materials and reduce the brickwork to ground level (*ibid.*: Brickwell to Bury 18/03/1912).

It is not entirely clear when the Great Northern demolished the Gasworks Viaduct, although most of the structure in the Goods Yard must have been taken down before the Third Edition Ordnance Survey Map was surveyed in 1914. Owing to a dispute between the Great Northern and the Regent's Canal and Dock Company

---

8 The GNR was liable for maintaining the Congreve Street Bridge, which spanned the mainline tracks to King's Cross Passenger Station between the gasworks and York Road (TNA RAIL 783/110: Ross to Bury, 16/03/1906).

(RCDC) over the ownership of property on the south bank of the canal, permission to demolish the piers and girders of the bridge over the canal was not granted until June 1917[9] (TNA RAIL 1189/1423: 01/06/1917). Following the resolution of the disagreement, the Great Northern was advised to secure the services of Messrs G. Cohen & Co. of Commercial Road to demolish the bridge, which was eventually taken down the following year (*ibid.*; TNA RAIL 1189/1425/3).

7.6 The Sources of the Building Materials used at King's Cross during the late 1890s and the early 1900s

Specialist analysis revealed that a number of building materials that had not previously been seen at the Goods Yard were used during the enlargement of the latter in 1897–9 (for further information on brick fabric types and codes pertaining to this period see Appendix 3).

Stock bricks in the 3038 and 3032 fabrics continued to be used in various machine bases within the Train Assembly Shed (Landscape Feature 11), whilst Glasgow (type 3033) stock bricks were identified in some hydraulic inspection chambers and the new Train Assembly Shed platforms (Landscape Feature 11). Fabric 3035 type bricks, possibly originating from the Midlands, were also identified in the platform edges, whilst Engineering bricks were occasionally made use of. Fletton bricks, many of which were stamped 'LBC' (London Brick Company), were found in the redesigned platform arrangements in the Eastern Goods Shed and Granary Complex (Landscape Features 7–12), indicating that they probably originated from the London Brick Company's works at Peterborough.

A large assemblage of bricks was retrieved from the excavation of the Hydraulic Station extension (Landscape Feature 16) of 1897–9. The new walls of the building had been built of kiln bricks (3261), some of which bore the Craven stamp, London Brick Company (LBC) bricks (3038), purple machine frogged bricks (3032) and a number of non-local bricks of a fabric type that looked as though it had been produced from glacial clay. The new machine bases and engine beds included kiln bricks (3261), some of which were stamped 'Craven', LBC bricks (3038) and purple stock bricks (3032). Reused millstones (formed from Millstone Grit) were also identified in the core of one of the engine beds, which may have been imported from Derbyshire or Yorkshire or recycled from a more local mill. Engineering bricks had been used to construct a series of plinths that separated the Boiler House from the Coal Drops. A combination of those bricks and LBC bricks

(3038) had also been used in the construction of the supports that upheld the fire boxes of the Lancashire Boilers. The main bodies of the masonry boiler supports were predominantly made from purple bricks (3032) although some red, machine pressed examples (3033), yellow stocks (3035), LBC examples (3038) and some yellow stamped bricks made by Cliff & Sons of Wortley, Leeds were also present. Each boiler support was lined with kiln bricks (3261), some of which were curved, in order to protect the rest of the structure from the intense heat to which it would be exposed. Unsurprisingly, kiln bricks had also been used to build the floors of the flues and the base of the chimney, whilst engineering bricks and stock bricks (3035) formed the walls. Some kiln bricks bearing the stamp 'CRAVEN' were also recovered from the chimney. The base of the foundations of the boiler positions were made of concrete, although in some cases that layer was capped by North Welsh slate.

Whilst local fabrics continued to be used, many other building materials used at King's Cross appear to have come from further afield. Transportation from these far flung areas would have been easy for the Great Northern since most of the non-local masonry that was found hailed from areas that the company or its partners served. It would therefore have been possible to ship them to King's Cross at cost price thereby dramatically reducing transport expenses. Bricks and stone therefore continued to be brought in from Yorkshire, which had long been well connected to the Great Northern's major trunk lines, while these imports were joined in this phase by a great deal of material from Peterborough. Before the arrival of the railway, that city was not a manufacturing centre, having long being overshadowed by its larger neighbour, Stamford (Simmons and Biddle *et al.* 1997: 376). Since the arrival of the Great Northern, Peterborough had grown rapidly, as had the local brick manufacturing industry along the length of the main line from Fletton to Yaxley (*ibid.*; Goslin 2002: 17). Brick traffic therefore expanded accordingly to the extent that it soon began to form a substantial proportion of 'up' goods traffic. The Great Northern had also been making inroads into parts of Derbyshire and the Midlands since the late 1860s, so it is not surprising that materials from quarries and factories in those regions were also encountered at King's Cross (Nock 1974: 58–9).

Given that the Great Northern's routes were generally concentrated in the eastern half of the country, the use of bricks from the Glasgow area may be unexpected. However their presence can be explained by the complex and often fluid alliances formed by the railway companies. The east coast collaborators, the Great Northern, the North British and the North Eastern, had long since realised the power of their alliance and had been running joint stock along the length of their network as far back as 1860. Their reach into Scotland was then further increased five years later after a series of political manoeuvres resulted in an amalgamation of the North British and the Edinburgh & Glasgow Railways (Simmons and Biddle *et al.* 1997: 139). This

---

9 The canal company had been renamed the Regent's Canal and Dock Company in 1900.

meant that the Great Northern was in league with a company that had access to Glasgow, meaning that bricks could be brought to King's Cross from that city with relative ease.

The presence of slate is noteworthy, as the quarries of North Wales were far removed from the Great Northern's network. Many had however, long been connected to the railway by narrow gauge lines, with special tracks being laid right up to the quarry edges so that the material could be transported from remote valley mines to mainline stations or Welsh ports (*ibid.*: 453). Throughout the 19th century, most slate was transported to London by sea, though rail exports were beginning to exceed those that were carried by ship by the early 20th century (*ibid.*). This meant that in 1898–9 when the Goods Yard at King's Cross was being extended, slate was being brought to London by rail and by sea in approximately equal measures. However, Wales was situated beyond the Great Northern's geographical territory, London-bound slate instead being predominantly carried by the London & North Western and the Great Western into Camden and Paddington respectively (*ibid.*). Consequently it seems most likely that the Welsh slate that was found at King's Cross came from a London based middleman, suggesting that the final leg of its journey to the yard was made by road.

7.7 Synopsis

The stability and financial security of the Great Northern Railway Company during the late Victorian and Edwardian periods finally provided it with a chance to tackle head on the persistent problem of rail traffic congestion at the King's Cross Goods Yard. The company decided to construct a sister depot which opened in 1899, the centrepiece of which was the Western Goods Shed. The erection of the new depot and the conversion of the original into an inwards shed was a costly and time-consuming project that also necessitated the enlargement of the Hydraulic Station. The implementation of the scheme resulted in the most substantial operational changes that had been made to the complex since its inception. However the investment paid off; the capacity of the yard was practically doubled, reducing rail congestion throughout the complex at a stroke whilst streamlining and rationalising the movements of incoming and outgoing wagons.

The newly built Western Goods Shed replaced the largely redundant Coal and Stone Basin, which had represented a significant waste of space given the fact that the volumes of canal traffic entering the complex had declined considerably. On the other hand, the

Fig. 7.15 The largely redundant Granary Basin standing empty (TNA ZLIB 3/19 © The National Archives) Reproduced from *The Illustrated Handbook* of the Regent's Canal and Dock Company published in 1911

Granary Basin in the Eastern Goods Yard survived the improvement programme, despite the fact that it was little used. This meant that the Eastern Goods Yard's road vehicular infrastructure remained largely unmodernised as carts continued to negotiate the perimeter of the underused basin. Although a small amount of extra land would be made available through the subsequent demolition of the disused Gasworks viaduct, the basin itself took up a great deal of space that might have been put to a better use. However given the often fractious relationship between the Great Northern and the senior officers of the canal company at this time, it is difficult to envisage how the railway company could have achieved this without the cooperation of their neighbour. Whilst plans to rectify this situation were in preparation during this period, the outbreak of the First World War in 1914 would mean that they would be delayed even further. The clogged highways that characterised the approaches to the Eastern Goods Yard throughout the 1800s therefore remained a feature of the complex as it entered the early 20th century.

The construction of the Western Goods Yard also did much to improve the lot of the depot's employees. Most obviously, the substandard offices in which the station's clerical staff worked were ameliorated by the creation of abundant new office accommodation. Working conditions were therefore improved for many staff as a result of the enlargement scheme. Staff welfare would be further improved in 1908 through the extension of the Refreshment Club (discussed in the following vignette) and it is probably no coincidence that this occurred at a time when railway employees were becoming increasingly unionised, organised and vocal concerning pay and conditions.

It is conceivable that the improvements of 1897–9 were made, in part at least, in response to the threat to the Great Northern posed by the opening of the Great Central's goods depot at Marylebone. Thanks to the extensive overhaul that accompanied the construction of the Western Goods Shed and the vast amount of additional handling space that it created, the Great Northern had acquired the means and the capacity necessary to hold its own against its new and technologically advanced rival.

In the event, the advent of the Great Central was not as problematic for the Great Northern as the company might have anticipated. This was in large part the consequence of a series of errors on the part of the Great Central. Sir Edward Watkin's ambitious programme of expansion greatly overstretched the Great Central's finances, creating a deficit from which the company would never fully recover (Wolmar 2007: 181–5). In fact the Great Central's finances were so strained that from 1889 onwards no ordinary dividends would be paid to their shareholders for the remainder of the company's existence (Simmons and Biddle et al. 1997: 189). The Great Northern's Board also decided to abandon its confrontational stance towards the Great Central's London extension, in favour of a more pragmatic approach which saw the company dropping its objections to the Great Central's Bill in return for continued running rights into Manchester (Nock 1974: 115). From then on the two companies managed to maintain a cordial relationship to the extent that they would often work together in mutually beneficial ways during the years that followed.

In conclusion, whilst the construction of the Western Goods Shed entailed a huge financial outlay, earlier attempts at improving the efficiency of the Goods Sheds and eliminating the problem of goods traffic congestion at King's Cross had demonstrated that it was probably impossible to achieve the desired results without considerable expenditure. Moreover the Great Northern decided to proceed with the enlargement of the station at a time when it could comfortably afford to absorb the cost of doing so. Since the carriage of goods to and from the capital was pivotal to its profit margins, the company's decision to invest in its London goods head was a prudent and well-timed measure. The implementation of the enlargement and improvement programme of 1897–9 therefore meant that the Great Northern's London goods railhead was able to meet the demands of the new century, whilst the Great Central was deprived of the technological advantage that it might otherwise have possessed.

Finds from The Refreshment Club and The Extension of 1908

Berni Sudds

A snapshot of a bygone institution has been provided by the recovery of a small but interesting collection of china originating from the King's Cross Goods Station Refreshment Club (Landscape Feature 13). The broken remains of 24 mugs, of two different sizes, and a least one plate from the same utilitarian refined white earthenware service were retrieved from demolition debris sealing the remains of the Club buildings. The service is simply decorated with maroon slip bands and a black transfer printed logo reading 'KING'S + GOODS STATION REFRESHMENT CLUB' in a double lined circular belt with 'G. N. R.' in the centre.

The Refreshment Club was a voluntary body, underwritten by the Great Northern and funded by members' subscriptions, which came into being sometime between 1872 and 1884 in the then Refreshment Room, which stood at the eastern end of the Eastern Transit Shed (Landscape Feature 8). Welfare facilities of one form or another appear to have existed in this location from the 1850s onwards, but with the aim of 'deterring absenteeism and detrimental levels of alcohol consumption' with staff going home to eat meals, or frequenting public houses in the vicinity, an on site canteen was opened in 1872 (see Section 4.7 for further details). The Club enjoyed considerable success and the facilities were subsequently extended on more than one occasion during the last two decades of the 19th century, and more significantly during the early years of the 20th century.

The service can be included within the category of 'institutional wares', specifically commissioned from pottery manufacturers by a variety of organisations. Similar institutional wares are known to have been made to order for the armed forces, hospitals, hotels and dining establishments (Jarrett and Thompson 2012). The service would have been ordered from a specific pottery, probably in a

Mugs from the King's Cross Goods Station Refreshment Club (two sizes 3⅛"–¼: 3⅞")

large batch, but no makers' marks were identified. It is tempting to suggest that it is related to the large scale 1908 extension and improvement of the Club (further details of which are given below) but this cannot be proven. Given that it is marked 'G. N. R.', it certainly pre-dates 1923, when the railway merged with other lines to form the LNER

In March 1908 the Traffic Committee authorised the expenditure of £2,100 for the extension of the profitable Refreshment Club and a contract for the construction of a new kitchen and messroom was awarded to Messrs W. Pattinson & Sons Ltd for £1,893 at the beginning of the following month (TNA RAIL 236/187: 261, 272). This amount represented a substantial investment in the fabric of the building, which later maps and plans confirm was considerably enlarged during the early decades of the 20th century. Such expenditure is made less surprising given the profitability of the enterprise and its importance to the functioning of the yard, providing on site facilities for drinking, eating and smoking. The building remained in the hands of the Refreshment Club in 1942, although it is not clear whether the Club continued in its existing form after nationalisation in 1948. The building was depicted on the 1953 Ordnance Survey Map, although it had been demolished by 1970. A small fraction of the southeast corner was discovered archaeologically, however the rest of the 1908 structure was situated beyond the limit of the excavation.

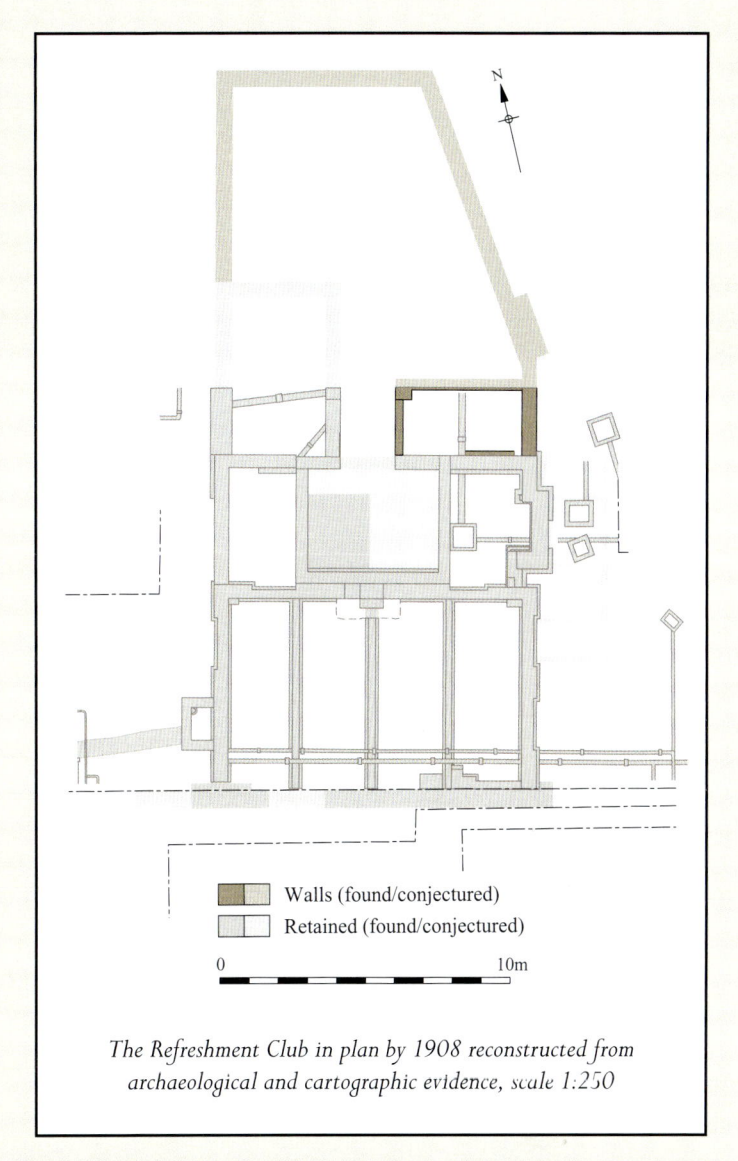

The Refreshment Club in plan by 1908 reconstructed from archaeological and cartographic evidence, scale 1:250

CHAPTER 8

The King's Cross Goods Station during the First World War: 1914–1918

'One air raid, under the arch there [i.e. at St Pancras chambers], a load of soldiers in uniform were going back to where they was going... [then the Germans] dropped a bomb [and] killed them... I remember when they brought the first Zeppelin down, I remember seeing it in the sky, all the flames, looking towards Charrington Street, [I will] always remember it.... Mum worked on the railways... She worked at the Somers Town depot, loading them up. She used to go to work in trousers, first time I saw ladies in trousers. Just across here'

Excerpt from an interview with local resident Jack Capper, who was born in Somers Town in 1911. By Alan Dein © London Borough of Camden

By the time that the First World War broke out in August 1914, the governments of the rival Great Powers had long anticipated that the railways would play a pivotal role in the coming conflict. The primary function of the railways, especially at the outbreak of war, was to facilitate the rapid mobilisation of troops, although they were to fulfill many other vital roles as the war progressed. In fact railways had figured in military preparations for war since the American Civil War of 1861–5. That conflict had demonstrated how they could be used to move troops to the battlefront, sustain an army's insatiable demand for food and ammunition and remove the wounded from the battlefield in relative comfort.

The earliest efforts to harness the railways for the purposes of national defence in Britain took place in the 1860s, with the formation of the Engineer and Railway Volunteer Staff Corps. The Corps was entrusted with drawing up the mobilisation timetables that the railway companies would follow in order to transport troops to their ports of embarkation when war was declared (Earnshaw, 1990: 5–6). This body was succeeded in 1896 by the Army Railway Council, which in turn became the War Railway Council in 1903. The Council set out to resolve how the Government could control the railways during wartime, whilst ensuring that the sometimes competing needs of the railway companies, the armed forces and the civilian population could be met. Out of this body emerged the Railway Executive Committee, which was formed in November 1912 in order to bring together all of the railways under a single authority in the event of war. Comprised of senior representatives of the ten major railway companies, the Executive ensured that the railway system operated as a single entity in the national interest, enabling the through passage of trains anywhere in mainland Britain and directing resources to where they were needed most. Chaired by the London & South Western Railway's (LSWR) General Manager, Herbert Aschcombe Walker, the Railway Executive Committee took control of Britain's railway companies on 5th August 1914, one hour after war had been declared at 11pm on the 4th. The Executive's powers were however, strictly limited. The railway companies retained their peacetime management structures, administration and traffic patterns. Those changes that did take place were a direct consequence of the prioritisation by the Executive of military and government traffic over scheduled civilian services. The Executive also took upon itself the right to set wages, fares and levels of compensation, powers which were to have a profound and ultimately deleterious impact on the railway companies following the Allied victory in 1918.

The first responsibility of the Executive was to ensure the smooth running of the mobilisation plans that had been drawn up before the war. General mobilisation came into effect on 5th August 1914 and took place over a period of fourteen days. Between the 9th and 17th August, 334 trains carried the British Expeditionary Force (BEF) of nearly 70,000 troops, 21,523 horses, 166 field guns, 2,446 vehicles and 2,550 tons of store to Southampton (*ibid.*: 12). The initial mobilisation was just the beginning however, and the next four years would see the railways moving millions of troops across

203

Fig. 8.1 Wagon used for carriage of gunpowder and explosives. The First World War exposed a nationwide shortage of wagons suitable for the carriage of gunpowder and explosives. While some companies converted existing rolling stock, the Great Northern built a batch of 8-tons vans at Doncaster Wagon Works in 1915. This example bears the instruction 'To be returned to Kings +' on its side

the national network. The Great Northern served a number of major military depots, including Barrowby near Grantham, Clipstone, Ripon and Catterick Garrison, together with the Royal Naval Air Service (RNAS) Training Establishment at Cranwell aerodrome. By 1916 troops passing through King's Cross could make use of rest and relaxation facilities with sleeping accommodation and quiet rooms at King's Cross Passenger Station, laid on by the Great Northern and manned by the YMCA and the Salvation Army (TNA RAIL 236/189).

Although the mobilisation of August 1914 briefly interrupted regular services, civilian passenger traffic returned to normal for the remainder of that year (Earnshaw 1990: 16). Before war broke out, passenger trains usually received preferential treatment over freight traffic, which was often delayed so that passenger services could run promptly (Wolmar 2007: 193). All of this was to change over the course of the year that followed, as the Government insisted that freight be given top priority. During the course of 1915 the volume of troop trains, goods and coal trains and military traffic increased dramatically. By 1918 the volume of freight traffic heading southward on Great Northern rails from industrial centres in West Yorkshire, Teesside and Tyneside to London and the southern ports had more than doubled (Nock 1974: 150). In order to overcome delays on the lines into the marshalling yard at Ferme Park, the Great Northern began running a continuous service of longer, heavier goods trains into the yard, equally spaced in time so as to minimise congestion, along a line that was 'loaded to the hilt'. In contrast, fewer passenger services were scheduled whilst those that remained were run in closely spaced groups to minimise disruption to freight haulage.

Following the failure of the British offensive at Neuve Chapelle in March 1915 owing to a shortage of high-explosive artillery shells, the Government established the Ministry of Munitions to increase the output of guns and ammunition. Within days of its formation in June 1915, the Ministry ordered the Railway Executive to provide sufficient capacity at railway workshops for the manufacture of 2,250 shells per week (Earnshaw, 1990: 41). Workshops such as the Great Northern Locomotive Works at Doncaster were converted into facilities for reforming and refilling spent shells and cartridge cases, which were transported to and from the factories by rail in vast quantities. Explosives, and the chemicals used to make them, were carried by specialist rolling stock such as the 8-ton gunpowder vans built by the Great Northern's Doncaster Wagon Works in August 1915. A few months after it was created, the Ministry of Munitions established a Transport Branch in order to coordinate the huge increase in munitions traffic carried by the railways. While munitions trains were restricted to certain routes and usually directed away from urban areas for obvious reasons, the concentration of munitions factories in the London area forced such traffic through built up areas day and night. In addition to the massive amount of freight passing through King's Cross on its way to the Channel Ports, the station handled trainloads of explosives destined for the British forces fighting in France (Wragg 2012: 12).

Whilst the Royal Navy ensured that Britain faced a negligible threat of invasion by sea, ports along the east coast of England were vulnerable to hit and raids by the German Navy, which bombarded Scarborough and Hartlepool in December 1914. These raids and other German naval activity in the North Sea had the unanticipated effect of diverting East Coast shipping traffic onto the railways, placing additional strain on the network. Although the threat from aerial bombing must have appeared vanishingly remote at the start of the war, by early 1915 German aircraft and airships had already launched attacks from the air against the British mainland. On 19th January 1915 a Zeppelin inflicted considerable damage on the railways at King's Lynn docks (Earnshaw 1990: 24). Zeppelin raids increased in intensity throughout the spring and summer of 1915, and on 7th September 23 bombs fell around Euston, Holborn and Liverpool Street Stations. The following January Zeppelins bombed the headquarters of the Midland Railway at Derby, causing severe damage and disruption. The tide turned against the Zeppelins in September 1916, prompting the German high command to switch to the use of Gotha bombers. Gothas first attacked the capital during a daylight raid on 13th June 1917, dropping a number of bombs on Liverpool Street Station. On 7th July, St Pancras Station sustained a direct hit during a raid, while attacks in September

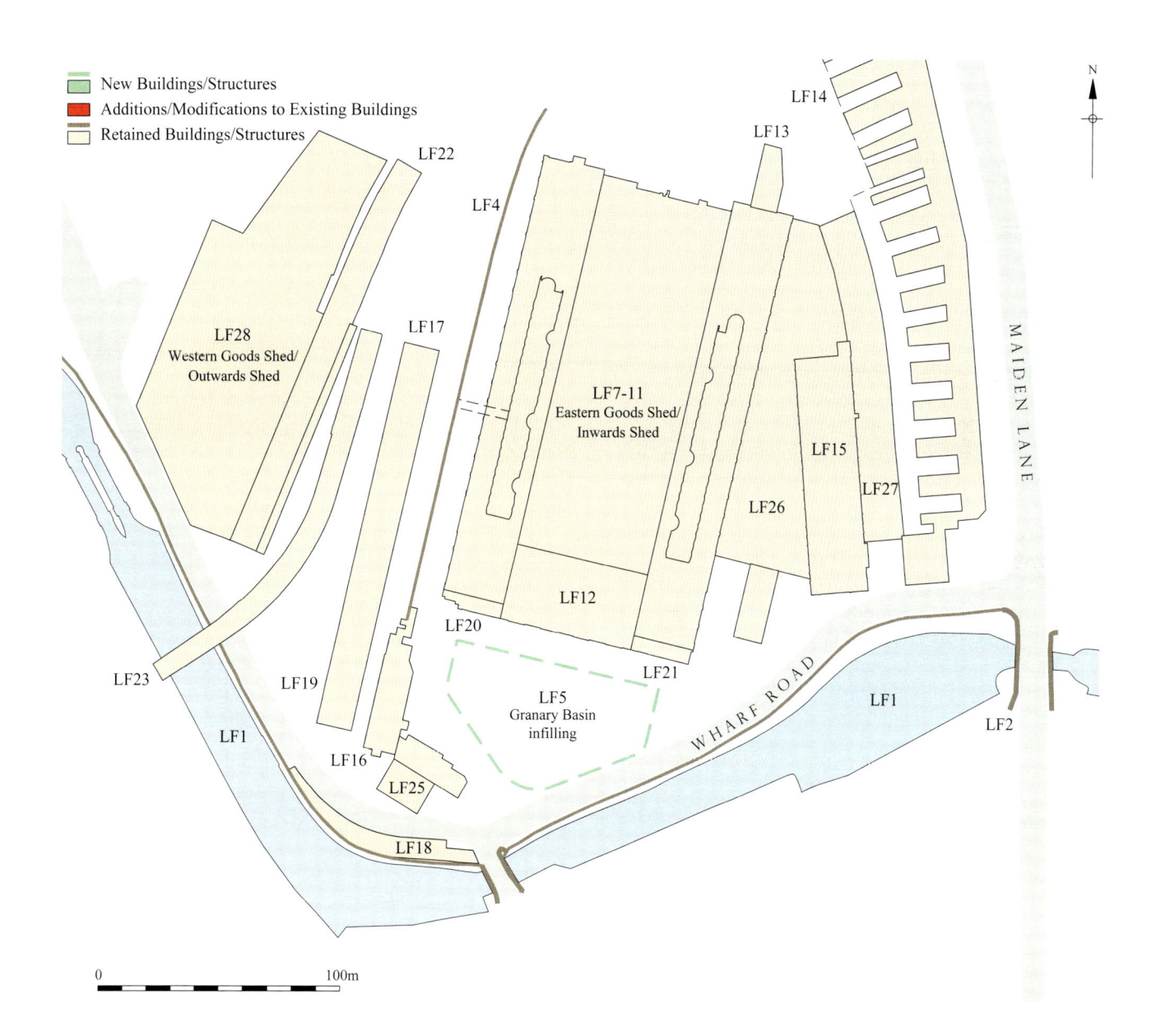

Fig. 8.2   The locations of the landscape features present in the Goods Yard from 1914–8, scale 1:2,500

damaged buildings at Hornsey, Paddington and the Stratford Locomotive Depot. On 17th March 1918 St Pancras was hit again, when five bombs fell on the station and the Midland Grand Hotel, killing 20 people, most of whom were sheltering in a carriage drive leading from the hotel to the main booking hall. By the latter date the Great Northern had adopted the precaution of using the Gasworks Tunnel at King's Cross as a shelter for mainline passenger trains whenever enemy aircraft approached (Wragg 2012: 12).

Approximately 5.7 million British men served in the armed forces between 1914 and 1918 and their recruitment had a significant impact upon the labour market in Britain. Although railway workers were exempt from conscription due to the essential nature of the roles that they performed, huge numbers joined up voluntarily. Railwaymen who belonged to the reserve forces were mobilised automatically, whilst other sections of the workforce, such as cleaners, clerics, administrators, porters and station staff, were not exempt from compulsory military service. Altogether, 184,475 railwaymen of military age enlisted during the First World War, representing 45% of the workforce (Wragg 2012: 14). The departure of so many workers created a large number of vacancies, which were increasingly filled by women and less-skilled men. Before war broke out, the railways employed 13,046 women, the majority in domestic grades such as cleaners, laundresses, canteen workers and hotel staff (Earnshaw, 1990: 34). In 1915 the Railway Executive decided that women could handle most of the clerical work, ticket collection, and carriage cleaning duties, in addition to certain porterage, engineering and maintenance roles. By the end of the war women accounted for 16% of the entire railway workforce. Women were prevented from working in the most physically demanding jobs, such as engine driving and firing, as well as from specialist roles such as manning signal boxes. Their employment was also resisted by the railway unions, who insisted that women were employed on a temporary basis in order to protect their members' jobs in the long term. Despite these restrictions, women were employed in a variety of physical jobs such as engine cleaning. An all female team at the Great Northern's New England depot at Peterborough kept the company's fleet 'remarkably smart and clean' throughout the conflict (Nock 1974: 152). Records indicate that women worked in a variety of roles at King's Cross previously reserved for male staff.

The following section describes the alterations that were made to the fabric of the Goods Yard during the war years, and considers to what extent they were a consequence of the conflict.

8.1 The Construction of Goods Way and the Infilling of the Granary Basin: 1914–1918

Landscape Features 29 & 5

By the turn of the 20th century it was evident to the senior officers of the Great Northern that the majority of goods entering and exiting the Goods Yard by means other than the railway did so by road, whereas the importance of waterborne transport had clearly declined over the preceding decades. The decline of this traffic was particularly apparent in the Granary Basin, which was by then used for little more than the extraction and discharge of water from the hydraulic network.

In 1911 the Great Northern purchased the disused St Pancras Gasworks. This acquisition finally provided an opportunity to transform vehicular access from York Road, with a view to making the acute traffic congestion of previous decades a thing of the past. An entirely new route providing access to the Yard from both Wharf and York Roads was therefore proposed (TNA RAIL 236/182: 260–1, 348; TNA RAIL 783/110). The company sought the necessary Parliamentary powers in 1912, which were granted by the Great Northern Act of 1913 (TNA RAIL 1189/1423; TNA RAIL 1189/1428: Brickwell to Directors 31/10/1917). The Act permitted the Great Northern to construct a new road that would run along the south bank of the canal before crossing that waterway via a new bridge, thus replacing the awkward approach to Somers Bridge and the old bridge itself. On the north bank of the canal the new road would cross the southern part of the Goods Yard, joining York Road at the old Maiden Lane entrance. The scheme could not commence immediately, however, as it was necessary to infill the Granary Basin first

The Great Northern went to some length to ascertain whether it could close the Granary Basin without reference to the canal company, relations with which were at a low ebb. Having satisfied themselves that the power to close the connection between the Granary Basin and the canal resided solely with the railway company, the senior officers of the Great Northern took the decision to close the redundant Basin in the spring of 1915 (TNA RAIL 236/469). Records indicate that the process was already underway by early July (TNA RAIL 1189/1426: Brown to Brickwell 08/07/1915). It was decided that tipping would take place in phases in order that the intake and outlet pipes to and from the Hydraulic Station could continue to function without interruption until they were relocated (TNA RAIL 1189/1426: Brickwell to Gresley 19/01/1916). Although Charles Brown, the Great Northern's Chief Engineer, envisaged that the replacement pipes would join the canal at a point immediately to the east of Somers Bridge, no agreement had been reached regarding their relocation by late September, by which time tipping had already advanced to the intake on the west side of the Basin (TNA RAIL 1189/1426: Brickwell to Gresley 21/09/1915, 24/09/1915).

Tipping was still taking place in January 1916 when A.J. Brickwell, the Great Northern's Chief Surveyor, noted that the dumped material was getting close to the overflow pipe situated a short distance to the south of the hydraulic intake (*ibid.*: 19/01/1916, 21/01/1916). The following month, Brown submitted a new plan for the diversion of the pipes from the Basin to the canal (TNA RAIL 236/189: 298). Whilst this plan was promptly approved by the Board, it was still necessary to make arrangements to find an alternative supply of water for the Hydraulic Station. The most obvious solution was to use the canal, which resulted in a new agreement between the Great Northern and the canal company that October (TNA RAIL 1189/1426: 'Agreement for the Supply of Water at King's Cross' 02/10/1916). The terms of the agreement stipulated that the pipes that connected the infilled Granary Basin with the Hydraulic Station be removed and replaced with two 9" intake and outlet pipes, which would meet the canal adjacent to the westernmost of the Coal and Fish Offices (*ibid.*: 'King's Cross Goods Yard: water supply from the Regent's Canal, mains etc.' 16/10/1916; Fig. 8.4). The agreement also allowed for the closure of the well (accessible from a manhole on the west side of the Basin beside the former outlet pipe) and the 8" suction pipe used to top up the water supply to the Hydraulic Station. It was decided to retain the 5" pipe that supplied water from the Hydraulic Station to the tanks in the roof space of the Granary, ensuring the continued survival of the latter structure's original hydraulic hoists for the time being.

Although no reference to the closure of the canal docks beneath the Granary has been identified during the course of this research, accounts of the infilling process indicate that tipping progressed in a southward direction from the buildings towards the canal. The final stage of the works (blocking the mouth of the former tunnel between the Regents Canal and the Basin) did not take place until 1921, after the necessary permissions had been received from the canal company (NRRG 138812 LNE: 08/12/1920; TNA RAIL 1189/1425/3). This evidence suggests that the canal docks were probably infilled at an early stage of the process, perhaps as early as 1915.

The fact that this scheme was carried out during the financially lean war years was a consequence of timing of the 1913 Act and the failure of the Great Northern and the Regent's Canal and Dock Company to reach a settlement at an earlier date. Having discovered from contracts drawn up 65 years earlier that it was entitled to close the Granary Basin without reference to the canal company, the Great Northern set about filling it in during the summer of 1915. Having previously spent a substantial sum acquiring the site of the St Pancras Gasworks, it is not surprising that the Great Northern pressed ahead with the scheme, which promised relief from the road traffic congestion that had long afflicted the roads entering the Goods Yard. What is more, the scheme did not involve significant expenditure since the Basin

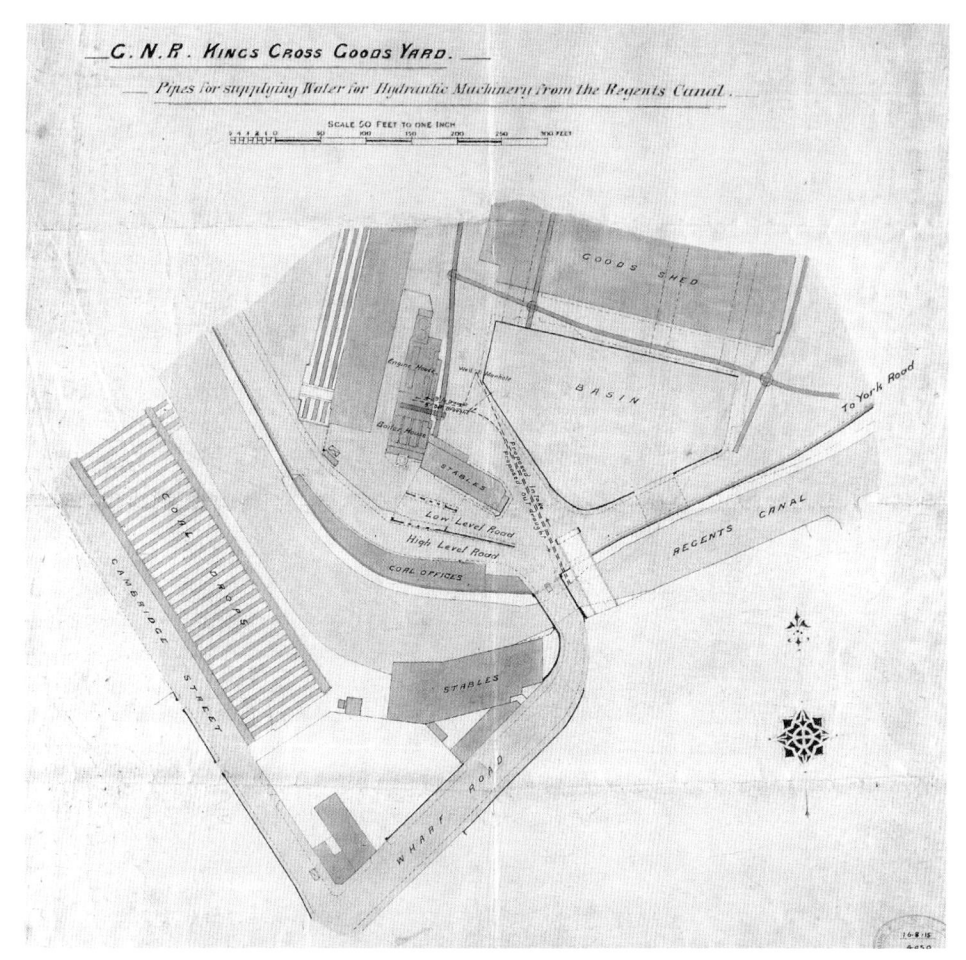

Fig. 8.3 Great Northern Railway Plan of August 1915 detailing the proposed alterations to the water supply from the Regent's Canal, scale 1:2,500 (TNA RAIL 1189/1425/3)

could be filled with waste material. Construction of the roads did not commence until the conflict was drawing to a close, by which time the Great Northern was already looking towards the future.

8.2 The Hydraulic Station, 1914–1918

Landscape Feature 16

In 1914 a steam pumping engine that had previously been used at the Farringdon Street Goods Depot was installed in the Hydraulic Station. The machine was brought in to replace two smaller engines, said to have been 30 years old at the time of their replacement (TNA RAIL 236/216: 271; Smith 2008: 14). It has been suggested that the 'new' engine was a 40hp model that was installed on Bed 7, as shown on the Goad Insurance Plan of 1921 (*ibid.*; Fig. 7.10; Fig. 7.11; Tim Smith pers. comm.).

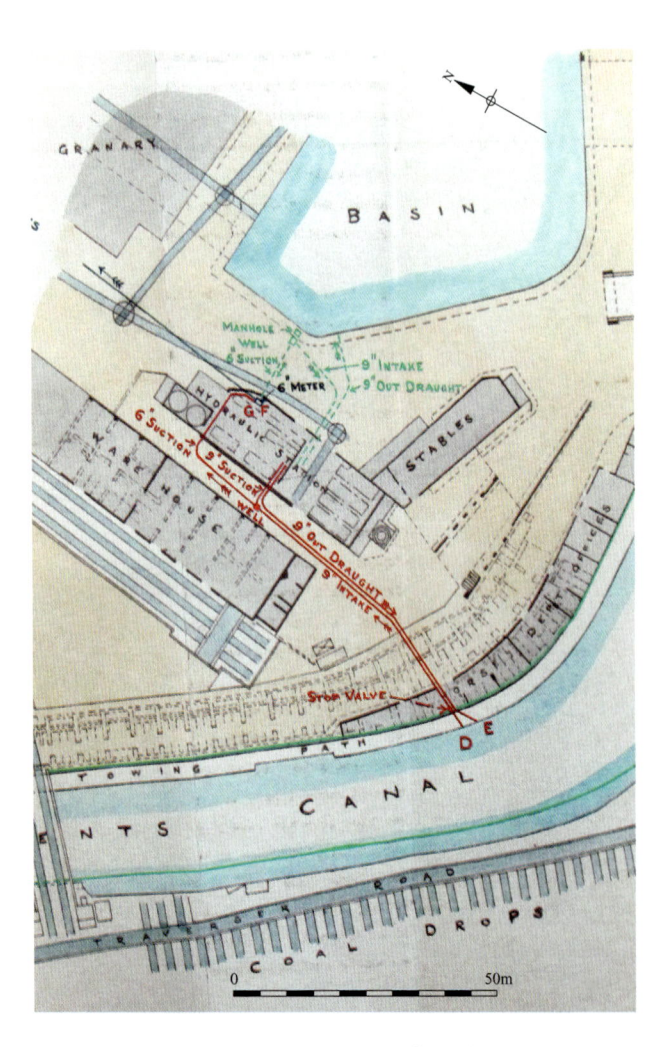

Fig. 8.4 Great Northern Railway Plan of October 1916 detailing alterations to the water supply from the Regent's Canal, scale 1:1,250 (TNA RAIL 1189/1426: 'Agreement for the Supply of Water at King's Cross' 02/10/1916)

The introduction of the 'new' machine ensured that the Hydraulic Station could continue operating without significant interruption at a time when great demands were being placed upon it by additional wartime traffic.

The second significant development which impacted upon the Hydraulic Station during the war years involved the diversion of the hydraulic intake and outdraught mains from the redundant Canal Basin, which was being infilled, to the canal itself. Documents associated with those works reveal how the decision affected the arrangement of machinery in the Hydraulic Station.

The relocated intake main discharged into a new well that was sunk adjacent to the southwest corner of the Hydraulic Station's Engine House, from which a 6' suction pipe led through a duplicate 6' meter to a pump that propelled water via a 5' supply pipe to the tanks in the roof of the Granary[10] (Fig. 8.3; Fig. 8.4). A 9' suction pipe carried water from the well to the condenser of the 100hp or 120hp engine, whilst the waste water from the latter was returned to the canal via a 9' outdraught pipe (TNA RAIL 1189/1426: 'Agreement for the Supply of Water at King's Cross' 02/10/1916).

8.3 The Midland Goods Shed, 1914–1918

Landscape Feature 15

The construction of the Western (Outwards) Goods Shed (Landscape Feature 28) between 1897 and 1899 greatly increased goods handling capacity at King's Cross, however by the spring of 1915 it had become apparent that it was necessary to make available additional accommodation for both incoming and outgoing goods traffic. Whether this was a consequence of the demands placed upon the station by additional wartime traffic is not specified in the records of the Great Northern, although the substantial sum budgeted for the works (£24,000) and the company's apparent haste to see them completed suggests that it may have been an important consideration (TNA RAIL 236/189: 190, 206). The scheme would comprise the extension of the Western (Outwards) Goods Shed to provide 'additional berthing accommodation for wagons', the provision of a 'shunting spur' in the sidings in the Western Goods Yard and the conversion of the Midland Goods Shed (Landscape Feature 18) from a bottle warehouse to a facility 'for the reception of general Inwards Traffic' (TNA RAIL 236/189: 206). The latter could be achieved for the relatively modest sum of £1,996 plus the funds necessary to compensate Kilner Brothers for the relocation of their

---

10   The new well replaced an earlier well situated on the west side of the Canal Basin close to the old 9' intake pipe which was closed when the basin was infilled.

Bottle Warehouse to premises in the Eastern Coal Drops (*ibid.*: 213; Goad 1921, not illustrated). Shortly after the various claims for compensation from tenants affected by the scheme were settled, drawings of the alterations required to convert the southern end of the shed back to Goods use were issued. Though the surviving example of the contract drawing is in poor condition, it is possible to discern the new cart entrances in the southern end of the east and west elevations, as well as the configuration of the girders that surmounted the new openings and the design of the new dormers in the roof (NRRG DMFP: 00026298).

The company's willingness to relocate one of its longest established tenants, and the apparent acquiescence of Kilner Brothers in their removal from the Midland Goods Shed appears to suggest that there was an urgent need to accommodate additional inbound traffic by the spring of 1915. It is therefore reasonable to assume

Fig. 8.5 External elevation of the west wall of the former Midland Goods Shed (Landscape Feature 15) showing the changes of 1915, scale 1:250

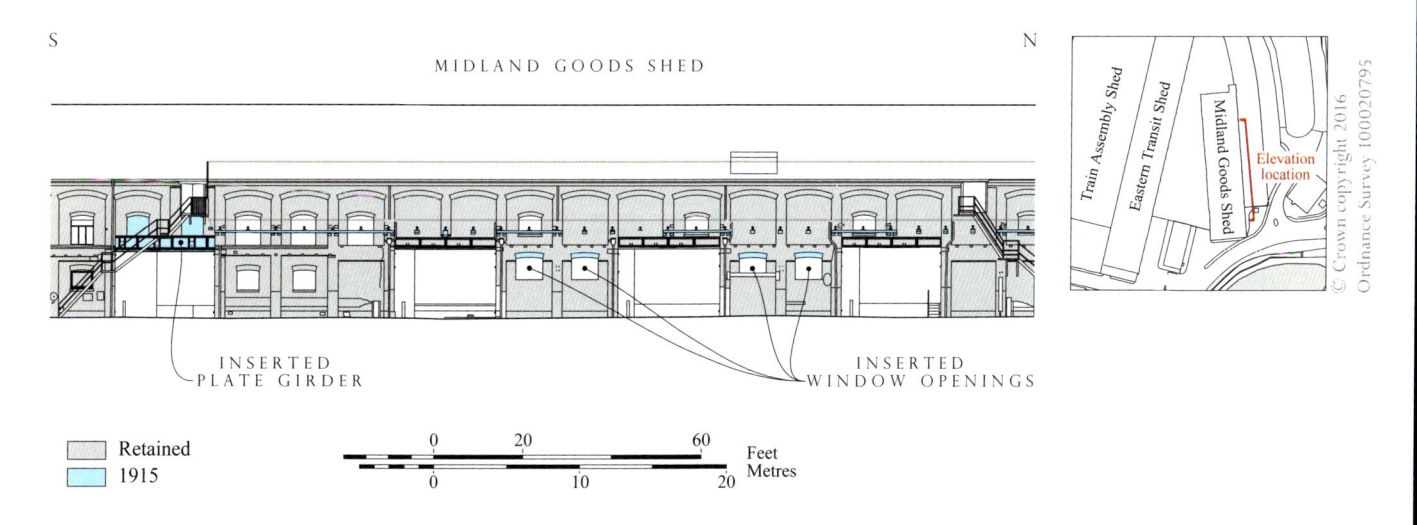

Fig. 8.6 External elevation of the east wall of the former Midland Goods Shed (Landscape Feature 15) showing the changes of 1915, scale 1:500

that the Midland Goods Shed was appropriated in order to better accommodate the increased volume of traffic entering the yard during the First World War, although further research into the nature of that traffic remains to be undertaken.

Built heritage evidence of the conversion of the former Midland Goods Shed to inwards use c.1915

Landscape Feature 15

The former Midland Goods Shed (Landscape Feature 15) effectively became part of the Inwards Goods Depot in 1915. In order to prepare the building for inwards use it was necessary to make several alterations to the fabric of the structure. In the western elevation two loading bays for the use of road vehicular traffic were reinstated (Fig. 8.5). Their insertion meant that two former window positions had to be blocked, before the large metal girders that formed the lintels were erected above heavily braced sandstone blocks. An identical opening was also created in the eastern elevation (Fig. 8.6). Two windows were then added in the opposing side of the building, along with another four in the eastern wall (Fig. 8.6). They were stylistically different to their earlier counterparts, being surrounded by dark blue, bullnosed engineering bricks that typified the early 20th century elsewhere on site.

8.4 Alterations to the Inwards Goods Shed and the Granary, 1914–1918

Landscape Features 7–12

At the beginning of August 1917 the Way and Works Committee of the Great Northern authorised the provision of an emergency exit from the Inwards Correspondence Office for an estimated cost of £102 (TNA RAIL 236/190). This exit appears to have consisted of a vertiginous wooden staircase which led to the westernmost platform of the Train Assembly Shed (Erwood 1988: 130; NRRG LNER: 95-K-42).

The records of the Great Northern do not state the reason behind this decision, though it is interesting to note that the work was commissioned just one month after neighbouring St Pancras Station was bombed by German aircraft. Whether the Committee's decision was influenced by recent events is unknown, however it is conceivable that the emergency staircase may have been commissioned as a precautionary measure in the event of future air raids.

8.5 Staff welfare facilities, 1914–1918

Following its enlargement in the early 1890s, the Refreshment Club (Landscape Feature 13) consistently turned an annual profit. By the mid-1900s however it was apparent that the club's facilities were in need of an overhaul. In April 1908 a contract worth nearly £2,000 for the construction of a new kitchen and messroom was awarded to Messrs W. Pattinson & Sons Ltd of Sleaford (TNA RAIL 236/187: 261, 272). While the Refreshment Cub remained reasonably up-to-date throughout the First World War, the same could not be said about other welfare facilities at the yard, in particular the toilets (e.g. TNA RAIL 236/190: 39).

The recruitment by the railway companies of large numbers of female employees during the First World War has already been mentioned. Female clerical staff were already working in both the Outwards and Inwards Goods Offices at the goods yard in June 1915, when the Traffic Committee recommended that women's lavatories should be provided in both locations for a total outlay of £184 (TNA RAIL 236/189: 224). In October 1916 the same committee called for the expenditure of a further £149 for temporary lavatory accommodation for female staff at the yard, although the precise location of these facilities is not recorded (*ibid.*: 356). Whereas the majority of women employed by railway companies before the war had worked in 'domestic' roles, as the war progressed a much wider range of positions was made available to female staff. By the end of 1916, around 900 women were employed by British railway companies as goods porters, a number of whom worked at the King's Cross (Earnshaw, 1990: 36). In January 1918 the Traffic Committee recommended that new messroom and lavatory accommodation be provided for the Outwards Section of the Women's Porterage Staff for the sum of £450 (TNA RAIL 236/190: 39). Given that outwards goods traffic was handled by the Western Goods Shed by this date, it is probable that this reference relates to the construction of new facilities in the Western Goods Yard. References to female employees of the Great Northern during the decades that preceded the First World War in the company's records are few and far between; the majority of those at the King's Cross end of the line appear to have worked as cleaners and at the Great Northern Hotel, although there were some female clerks working in the Western Goods Shed as early as 1899. By the end of the war however, women were working in a wider variety of clerical and porterage roles in both the Inwards and Outwards Goods Yards.

8.6 Synopsis

The imposition of state control on the railways at the outbreak of the First World War represented something of a double edged sword for the railway companies over the years that followed. The Railway Executive was generally quick to finance measures considered essential to the war effort; within weeks of the start of hostilities it authorised four major new connections on the national rail network to improve the flow of traffic between lines belonging to different companies, including a 'defensive link' between the Great Northern and Midland lines at Peterborough. The cost of the extension of the Western Goods Shed and the conversion of the former Midland Goods Shed at King's Cross was initially met from the Great Northern's own funds, though the company would have been compensated by the Government if those works were considered to be of benefit to the war effort, as they almost certainly were. However the state's concern for the wellbeing of the railway companies and their infrastructure came a poor second to its overriding interest in ensuring that they were effectively utilised in the national interest for the duration of the war. This disparity between the short-term priorities of the state and the long-term financial interests of the companies meant that weaknesses in the system that emerged during the war, such as the bottleneck at Ferme Park, were resolved by altering running arrangements, rather than by investing in infrastructural improvements.

The volume of freight carried southward along Great Northern rails to London and the south more than doubled from its pre-war level during the four year conflict. The company responded by adapting existing facilities at King's Cross, including the conversion of the former Midland Goods Shed to goods use in 1915, part of a wider programme of works that also saw the enlargement of the Western Goods Shed and improvements to the shunting arrangements in the Goods Yard. It is not known whether the Midland Goods Shed was adapted in order to receive munitions trains, although King's Cross is known to have handled cargoes of explosives during the war.

Although aerial bombardment was in its infancy at the beginning of the war, the destruction wrought by German air raids taught the railway companies that their assets were key targets in this new form of warfare. At King's Cross, the Great Northern responded by sheltering trains in the Gasworks Tunnel during air raids, and it is likely that the provision of an emergency exit in the Inwards Goods Shed represented a rudimentary air raid precautions (ARP) measure taken in the aftermath of a raid on neighbouring St Pancras Station.

The enlistment of nearly half of the railway workforce in the armed forces during the First World War created a massive number of vacancies, which were increasingly filled by female employees for the first time the history of the railways. At King's Cross women worked in the clerical and porterage departments of both the Inwards and Outwards Goods Departments, necessitating the construction of at least three sets of women's lavatories between 1915 and 1918. Whilst the female porters were replaced by men after demobilisation in 1918/19, women continued to work in clerical roles for the remainder of the history of the Goods Yard.

Ultimately, the First World War had a disastrous impact upon the finances of the railway companies. In 1913 they had made profits of £45 million; by 1921 they were running at a loss of around £9 million, in large part thanks to the cost of manpower (Wragg 2012: 17). Whilst wages had accounted for £47 million in 1913, they had risen to £160 million by 1920. When state control came to an end in 1921, the railway companies were in a perilous financial position, which the government of the day proposed to resolve by merging them into four large groups. In 1923 the Great Northern became part of the London & North Eastern Railway (LNER), which inherited the assets and liabilities of seven major railway companies and 27 subsidiaries.

A Short History of the Horse at King's Cross Goods Yard

'There are 300 horses working out of the Great Northern King's Cross depot alone; the Midland and North Western have almost as many, and all the railway coal stations are tended by a numerous herd of distributors, so that we shall be well within the mark in allotting 1,500 fairly good horses to the leading London coal merchants. Some of these horses are shires, some of them Clydesdales, some of them would more than satisfy the Leicestershire people as not being eligible for entry in any stud-book. You can see theta [sic] of all sorts, good, bad, and indifferent, at work in dozens up that curious thoroughfare- though it looks like a cul-de-sac- which runs out of Pancras Road under the arches by Battle Bridge, round by the gasworks and between the Midland and Great Northern Railways. There you will find coals to the left of you, coals to the right of you, volleying and thundering. In every arch is a platform; on every platform are two weighing machines; over each weighing machine is a shoot [sic] which delivers into the sacks on the scales, and from which the coal stream is cut off with a lever much as you turn off your water at a tap. Overhead are the waggons [sic]; down the shoots the coal roars, and booms, and hisses in a cloud of dust, as sack after sack fills up and is run out on the hand truck into the vans, in the shafts of which stand the horses gently bobbing their nosebags and utterly indifferent to the dust and din'.

Extract from 'The Horse World of London' by W. J. Gordon 1893: 134

Cart and shunting horses were instrumental to the success of King's Cross Goods Yard throughout most of its history. Until the mid-20th century, horses were relied upon for the transportation of incoming and outgoing commodities to and from London and, in the days before the introduction of hydraulic capstan shunting, they were largely responsible for manoeuvring wagons in the station's internal areas where steam locomotives were not permitted. The horse was therefore of considerable

A railway horse and his master engaged in wagon shunting, 1953 (SSPL 1995-7233_ LIVST_DF_200_A © National Railway Museum and SSPL)

importance throughout the tenure of the Great Northern Railway and during the early years of its successor, the London & North Eastern Railway. This section therefore explores the lot of the horses that lived and worked at King's Cross.

As previously discussed, stabling for horses was integral to Lewis Cubitt's original design for the Goods Station. Subterranean stables were incorporated in the foundations of the Eastern and Western Transit Shed platforms, along with stabling for about 120 animals below the Wharf Road Viaduct (TNA RAIL 236/273; Thorne *et al.* 1992: 101). Beyond the Goods Shed stables, there were a number of other stables dotted around the yard with several situated in the north-west corner in the vicinity of the Locomotive Department. As many as 500 horses were kept at the yard during the 1850s, housed in five or six separate stable blocks (Thorne *et al.* 1990: 105; Anon 1859).

During the earliest years of the Goods Station's existence, the Great Northern awarded contracts to independent cartage contractors to bring goods and coal into and out of the yard for onward distribution by road or rail. The contract for the cartage of goods was won by a Mr Sherman, whilst the company awarded the contract for the onward transport of coals to Edward Wiggins (TNA RAIL 236/71: 164, 198; TNA RAIL 236/273; TNA RAIL 236/275/23). The railway company let offices and stables in the complex to both companies; Sherman's beasts were briefly housed in the Western Stables, for which he paid an annual rental of £200 in January 1851, although it was subsequently decided to move them to the newly completed Eastern Stables (TNA RAIL 236/71: 164. 198; TNA RAIL 236/17: 228). Sherman's tenure was short lived, and it is likely that the railway company used the intervening period to build up its own horse stock. When Sherman's contract came to an end the following December, the Great Northern decided to moved its own animals into the recently vacated Eastern Stables (*ibid.*). When it was proposed in February 1851 that Wiggins' animals should reside 'at the end of the Coal Offices', Joseph Cubitt urged caution, pointing out that the site was inconveniently placed and that the proposed accommodations were too small for the size of Wiggins' stock (TNA RAIL 236/273: 07/02/1851). Cubitt's warning was heeded and the majority of Wiggins' horses were instead housed in the Western Transit Shed Stables after Sherman vacated them in October 1851 (TNA RAIL 236/17). Nevertheless a rent return dated September 1852 indicated that a small number of his animals were also kept in the vicinity of the Coal Offices, presumably in the location that Cubitt had rejected as being unsuitable (TNA RAIL 236/275/23).

Following the Great Northern's decision to bring coal and goods cartage services in-house after the expiry of Sherman's and Wiggins' contracts, the number of horses stabled at King's Cross Goods Yard continued to increase. Nine years after the yard opened, the number of animals working there had swelled to 800, whilst by the turn of the 20th century the numbers dealing with deliveries alone amounted to 1,500 (Thorne *et al.* 1990: 105). Consequently, by the year 1900 up to 1,000 heavy horses could be seen working simultaneously in and around King's Cross Goods Yard throughout the morning, noon and night, accompanied by over 300 cartage contractors (Holden 1985: 95–6). The role of these cartage teams was described in a short memoir by Peter Erwood, a Junior Clerk at King's Cross in the late 1930s:

'Once loaded, the van driver and his vanguard (usually a fourteen or fifteen year old boy, and frequently the son of a railway employee) departed on their round, checking out at the Cartage Office by the main gate. There, they might be instructed to make collecting calls as well, and in any case they kept an eye open for cards displayed in shop windows or at factory gates marked 'LNER Carman to Call'. [To] the busier and more industrialised districts, certain road vehicles were sent out empty on a regular pick-up round. Returning vehicles stopped at the Cash Office by the main gate to hand over monies received for goods sent 'Carriage Forward'...and then, if necessary, went round to the Outwards Shed for unloading' (Erwood 1988: 134).

All loaded wagons would have to pass over a weighbridge before leaving the yard. The machine would measure the weight of the loaded dray in order to track the passage of goods in and out of the station. The reading could then be used in combination with the distance that would be travelled and the number of deliveries that needed to be made in order to calculate the timespan within which a cartage team would be allocated a bonus upon their return for quick and speedy work. Crucially for the railway company, it could also be used to make sure that no embezzlement took place. Before departing, carters would also be issued with a standard kit that

Carters and their drays return and set out on their rounds via the Goods Yard's Maiden Lane entrance c. 1900 (© Neil Parkhouse Collection)

was designed to help them make their rounds and attend to their animal's needs. It included basic welfare items for the horse such as a nosebag and a bucket for feeding and watering as well as a tool kit that was designed to help manoeuvre the dray and facilitate the delivery and collection of heavy loads. Typically this included a sack, a pulley roller, a 'scotch'[11], skids[12], a rope, a chain and packing as well as cart lamps for use after dark (Holden 1985: 101). All these items remained the property of the railway and the carters were responsible for their safe return (*ibid.*). Commonly, the more expensive items, such as the cart lamps, were exchanged for tokens so that their whereabouts could be more accurately monitored (*ibid.*).

The huge number of horses and drays that were required at larger Goods Yards meant that the railway companies tended to employ farriers, wheelwrights and tack-makers directly (*ibid.*: 100). Indeed, a visiting journalist reported as early as 1859 that a large farriers shop and a sizeable tack workshop were already present at King's Cross Goods Yard by then (Anon 1859). These services were subsequently brought together under the aegis of the Great Northern Horse Superintendent's Department, which appears originally to have been based in the north-west corner of the yard, close to the Locomotive Department. The Horse Superintendent and his staff took over part of the recently vacated Coal Offices (Landscape Feature 18) in the 1890s, before moving the yard's tack room into the same block in June 1900 (TNA RAIL 236/235: 328; TNA RAIL 236/236: 3).

A large number of artefacts associated with the care and maintenance of working horses were unearthed during excavations in the Goods Station. For example, within the modern backfill of the Transit Shed Stables a blacksmith's punch of some antiquity was discovered, along with an awl, possibly for repairing tack, and a number of simple tools that may have been used to maintain road vehicles, such as spanners, a wrench, a chisel, an oil can and a pair of pliers as well as a device for nipping the noses of horses to keep them under control. Some broken components from traces (used to fasten a harnessed horse to a dray) were also unearthed in the Eastern Stables.

Working conditions for horses at King's Cross were considered to be exemplary by mid 19th-century standards; indeed, a reporter who visited the yard nine years after it opened wrote a glowing account of the lives of the horses that lived there. Describing the underground transit shed stables, he declared that 'the cleanliness and order in which they are kept is remarkable'

---

11   a device that helped the cart to break when descending steep inclines.
12   a ladder-like item, usually carried underneath the dray; heavy parcels were often removed from the cart by sliding them down this vital piece of equipment.

(Anon 1859), however the conditions he described left much to be desired by 21st-century standards. The crowded stables were 'a complete vista of racks, bins and horses heads'; no natural light was admitted, so they were illuminated by 'innumerable jets of gas'. Because the animals worked alternate twelve hour stints, this meant that many did not see daylight for considerable periods of time. The stablemen suffered alongside their animals, working twelve hour shifts that alternated on a monthly basis between nocturnal and diurnal. Despite the 'cleanliness and order', the reporter admitted that the atmosphere in the underground stables 'with the greatest care and attention, cannot be of the purest', suggesting that even he found the conditions in the poorly ventilated environment oppressive (ibid.).

Various schemes to improve the lot of the horse at King's Cross were instigated after the yard opened for business. In the early days the animals were given drinking water that was drawn directly from the Regent's Canal. The dirty and polluted condition of this water led Edward Wiggins to complain to the railway company about the impurities of his animals' water supply. In response, Seymour Clarke proposed that Wiggins' animals be watered from the New River Company's mains instead, an arrangement that was agreeable to both Wiggins and the Board of the Great Northern (TNA RAIL 236/27: 120). Having approved Clarke's proposal, the Board asked Clarke to investigate the possibility of providing an improved supply of water for the company's own horses.

The chief threats to the health of the animals that worked at the Goods Yard were accidents and disease. The Great Northern built a horse infirmary near the northern end of the Yard early in its existence, spending considerable sums upgrading its facilities as the 19th century progressed. By the turn of the 20th century the infirmary was treating injured animals from across the entire Great Northern network (ibid.).

Of the many workplace injuries suffered by horses in the Goods Yard, being struck by railway wagons appear to have been particularly common and potentially life-threatening. A reporter who visited the yard in 1859 provided a graphic description of the results of these injuries:

'We were shown several poor creatures who had lately met with severe accidents. One, a fine large carthorse, a day or two previous, had had cut out of its hind-quarter, a lump of solid flesh as large as a dinner-plate through the unfortunate creature having come into contact with the corner of a railway truck when in motion. The poor animal appeared to be in great agony, as the thread with which the surgeon had sown it into its place had come loose, and the flesh hung from the carcase' (Anon 1859).

After regaling the reader with an account of a collapsed horse tormented by flies that it 'had not the strength to shake off' and another rendered lame after having its hoof torn from its foot, the reporter reiterated that 'it was pleasing to feel that [the horses] were so well cared for'.

It appears that the Great Northern only made a coordinated effort to address the problem of unsafe working practices in the mid-1880s, in response to a to review of the Horse Department's requirements for new stock conducted by Josiah Medcalf and Albert Guille in October 1883. At the time that the report was commissioned the company had more than 1,650 horses stationed across the capital, having risen from 1,250 ten years earlier (TNA RAIL 236/145: 117; TNA RAIL 236/149: 120). The report concluded that rival operators experienced fewer horses put out of action by injury or overworking than the Great Northern, a fact that the authors primarily attributed to the company's working practices in London (TNA RAIL 236/149: 148). It appears that this report led to the establishment of a panel of Directors known as the Horse Committee. The authors of the report also expressed concern about the condition of the stabling at King's Cross in particular, which prompted the Horse Committee to ask Richard Johnson to review stabling at the Goods Yard in late 1884 (TNA RAIL 236/150: 10). Johnson reported that the accommodation was deficient in both drainage and ventilation and recommended that the former could be bettered through the installation of a 'simple open iron channel' in each Transit Shed stable, whilst air quality would need to be improved considerably in order to reach the standards of the stabling that was provided by the Great Northern's rivals (ibid.).

Evidence of improvements to the drainage in the Transit Shed stables made in accordance with Johnson's directions was found during the archaeological investigation. It is possible that a block drain lifter, discovered on the floor of the Western Stables, was used to open the grate that was instated as a result of his advice. However, the Great Northern failed to fully implement his recommendations regarding ventilation. In fact, it appears that the company they did not make any concerted attempt to resolve the deficiency until after 1890, following an outbreak of influenza among the horses stabled at King's Cross. In the immediate wake of the epidemic, Professor J. Wortley Axe of the Royal Veterinary College was commissioned to report upon the condition of all the company's stables in London (TNA RAIL 236/235: 51). In order to prevent future occurrences of the disease and to improve the general welfare of the horses, the professor recommended that the company increase stabling capacity at King's Cross, enlarge the Horse Infirmary and improve access to and the ventilation within each of the existing stable blocks at the Yard (*ibid.*). Following the enlargement of the Horse Infirmary at considerable cost, the company eventually turned its attentions to improving conditions at the cramped stables at the Goods Shed. In January 1898 and October 1899 respectively, the Board therefore approved expenditures of £380 and £250 for the provision of additional exits to the Eastern and Western Transit Shed Stables (*ibid.*; TNA RAIL 236/183: 122; TNA RAIL 236/235: 315). Archaeological evidence suggested however, that this decision may have been subsequently revised. An additional ramp was added to the Eastern Stables only; the Western Stables remained unmodified, perhaps because they already possessed a second exit.

Many railway employees were evidently very fond of the animals with which they worked. Accounts of the good temperaments, strengths and abilities of individual railway horses abound (Holden 1985: 82, 115, 117–9, 126–7, 135). Most handlers trained their animals through the use of elaborate commands, provided them with titbits and treats and took great care regarding grooming and appearance (*ibid.*: 82, 115, 127, 138). Indeed, it was not unusual for an urban railway horse to be entered into a stable competition or to take part in a carnival or a parade, in which

Fragment of an ornamental plaque in the shape of a horse (head missing) recovered from the Eastern Stables

case animals and handlers alike would don their finery and sport the livery of their railway company with pride (*ibid.*: 82). This sense of pride is captured in an artefact that was recovered from the Eastern Stables at King's Cross. Completely superfluous from a functional perspective, it is an ornamental plaque in the shape of a horse that may have once adorned the harness of one of the animals.

The latter half of the 1890s and the early years of the new century coincided with the appearance of the first petrol driven vehicles on London's streets. Indeed, the role of the horse in mass public transport had already been eroded as London's well established suburban and underground railway networks were joined by electrified tramways in the early 1900s (Wilson 2007: 57–61). While the earliest motorised vehicles were not yet capable of challenging the supremacy of the cart horse as a haulier, as the 20th century progressed it must have become increasingly apparent that the future lay with mechanised traction. During the first half of the 20th century the transition would take place very gradually, so when war broke out in 1914 horses remained of huge importance to the transportation requirements of soldiers and civilians alike.

During the First World War huge numbers of horses in civilian use were requisitioned for military service. At first railway horses were also subject to impressment, although the Railway Executive Committee (which directed Britain's railways between 1914 and 1921) soon objected to the War Office, which instructed the Army to put an end to the practice for all railway horses, other than those already in the Armed Horse Reserve (Holden 1985: 20). Independent cartage contractors, cabbies and other operators who frequented the nation's railway termini were offered no such protection and many of their animals continued to be taken into military service as a result.

The decline of the cart horse at King's Cross began in earnest after the London & North Eastern Railway inherited the Great Northern Railway Company's assets at Grouping in 1923. Soon after the new company came into existence, Ralph Wedgwood, the LNER's first Chief General Manager, conducted an investigation into the company's horse stock in the capital. He discovered that the company had inherited a huge number of cartage horses from the various railway concerns that it controlled in the capital, amounting to almost 2,500 animals. In contrast, the LNER's London-based contingent of motorised trucks totalled just 22 (TNA RAIL 390/58: Minute 165 04/10/1923). The company was therefore aware that it needed to modernise its cartage arrangements but insufficient funds were made available to make a rapid change. Horses therefore remained ubiquitous at King's Cross throughout the cash-strapped 1920s, with at least one independent hay and straw merchant continuing to trade out of the Coal and Fish Offices (Landscape Feature 18) until as late as 1930 (Kelly & Co. 1930).

Although the LNER's financial woes offered the company's horse stock a temporary reprieve, the wheels of mechanisation had been set in motion. By the autumn of 1928, the Southern Area of the LNER had built up a sizeable fleet of motor vehicles, many of which were stationed at King's Cross on the site of the newly infilled Granary Basin (TNA RAIL 390/704). The 1930s and the 1940s would therefore see the cart horse phased out at the yard as its dominance finally came to an end.

CHAPTER 9

The King's Cross Goods Station during the Interwar Period: Grouping, Depression and Transposition

'GOOD NEWS: ...the General Strike is over... The emergency train services will be amended and increased as rapidly as the circumstances permit... FREIGHT TRAFFIC: Great progress is being made in the movement of goods traffic. Meat trains were run yesterday from Aberdeen to King's Cross and from Liverpool to North Eastern England... LOYALTY: Messages of loyalty have been received from various points on the LNER...GRIT: An unemployed seaman stoker walked from Manchester to King's Cross to volunteer as a fireman. His boots were badly cut, but he has been supplied with a new pair. MESSAGE TO STRIKERS: At the conclusion of the strike the number of staff whom the Company can employ will be materially reduced. The effect of the Strike upon the trade of the country must be to diminish substantially the tonnage of traffic to be handled, and it will necessarily take a considerable time for trade to recover. The company wish it to be understood that at the conclusion of the strike they will give preference for employment to those of their staff who have remained at work or who offer themselves for re-employment without delay...Please show this to as many people as possible in order to give the widest publicity'

The LNER's response to the General Strike of 1926 as published in the company newsletter, the LNER News, on May 13th of that year (TNA RAIL 393/321)

The celebrations that marked the Allied victory in the First World War could not mask the serious socio-economic problems that confronted the Government as it sought to establish the 'land fit for heroes' promised by Prime Minister David Lloyd-George. Roughly 10 million Britons, half of them in the armed forces and the remainder in civilian trades, had until then been employed in war-related work. The Government therefore faced the daunting task of assimilating these people into a peacetime workforce (Greaves 2007: 132).

In an attempt to avert the social unrest that had convulsed Continental Europe after the Bolshevik Revolution and the defeat of Germany, the British Government decided that in order to stimulate the economy, fiscal policy would not be tightened. This led to a very brief financial boom that was accompanied by a rapid growth in employment, although this was unsustainable at a time when many industries were battling war-induced bottlenecks in supply. This resulted in too much money and not enough goods, a combination that generated rapid inflation (Greaves 2007: 133). The only recourse for the Government was to pursue a policy of fiscal tightening that had caused considerable economic contraction by the end of 1920, which was accompanied by soaring unemployment. Coupled with dire international economic circumstances and Britain's crippling war debts these domestic problems had a profoundly negative impact upon the British economy, hitting coal, manufacturing and heavy industries in the north of the country with particular force (*ibid.*; Price 2007: 149).

Whilst the Government had initially reverted to a *laissez-faire* economic policy at the end of the war and allowed private companies to regain autonomy over their businesses, the railways were treated as a special case, remaining under State control until 1921. Before the First World War broke out, Britain's privately owned railway companies were in the habit of undercutting each other to the extent that many were running at a loss. Temporary state control during the conflict put an end to these practices, so it was feared that a return to 'business-as-usual' during the turbulent period following the war could have a devastating impact upon the transport network. Government control was therefore extended for a further two years by the Transport Act of 1919, which also saw responsibility for the railways taken out of the hands of the Board of Trade and given to the newly established Ministry of Transport. The new Ministry was led by Sir Eric Campbell Geddes, a dynamic former railwaymen who had held important posts during wartime in the Ministry of Munitions and latterly as Director General of Military Railways (Bonavia 1983: 2–4).

Sir Eric had ambitious plans for the future of the British transport system, which included the wholesale reorganisation of the railways. Although nationalisation was briefly considered, the Ministry opted instead to reduce the large number of small companies by grouping them into four large concerns, which it was hoped would encourage greater coordination and less competition. The grouping process was authorised by the Railway Act of 1921 (also known as the 'Grouping Act'), which created 'Southern', 'Western' 'North Western, Midland and Scottish' and 'North Eastern, Eastern, and Eastern Scottish' regional groups. The Great Northern was squeezed into the latter grouping, which was named the London & North Eastern Railway (LNER), the second largest of the 'Big Four' railway companies formed in January 1923 (the other three being the London, Midland and Scotland or LMS, the Great Western and the Southern). The post-grouping companies absorbed a multitude of smaller concerns, which were in varying states of financial wellbeing. Whereas the Great Northern's shares had paid a healthy dividend of 4.37% in 1912, and the North Eastern had paid an impressive 6%, other members of the LNER group had been doing rather less well before the war, when Great Eastern shares had paid only 2.5% and Great Central shares had paid nothing at all (Wragg 2012: 23). There were other flaws in the arrangement, not least the fact that despite the stated goal of reducing competition between operators, there was a great deal of overlap in the system. The two largest companies, the LMS and the LNER competed on two key routes; firstly

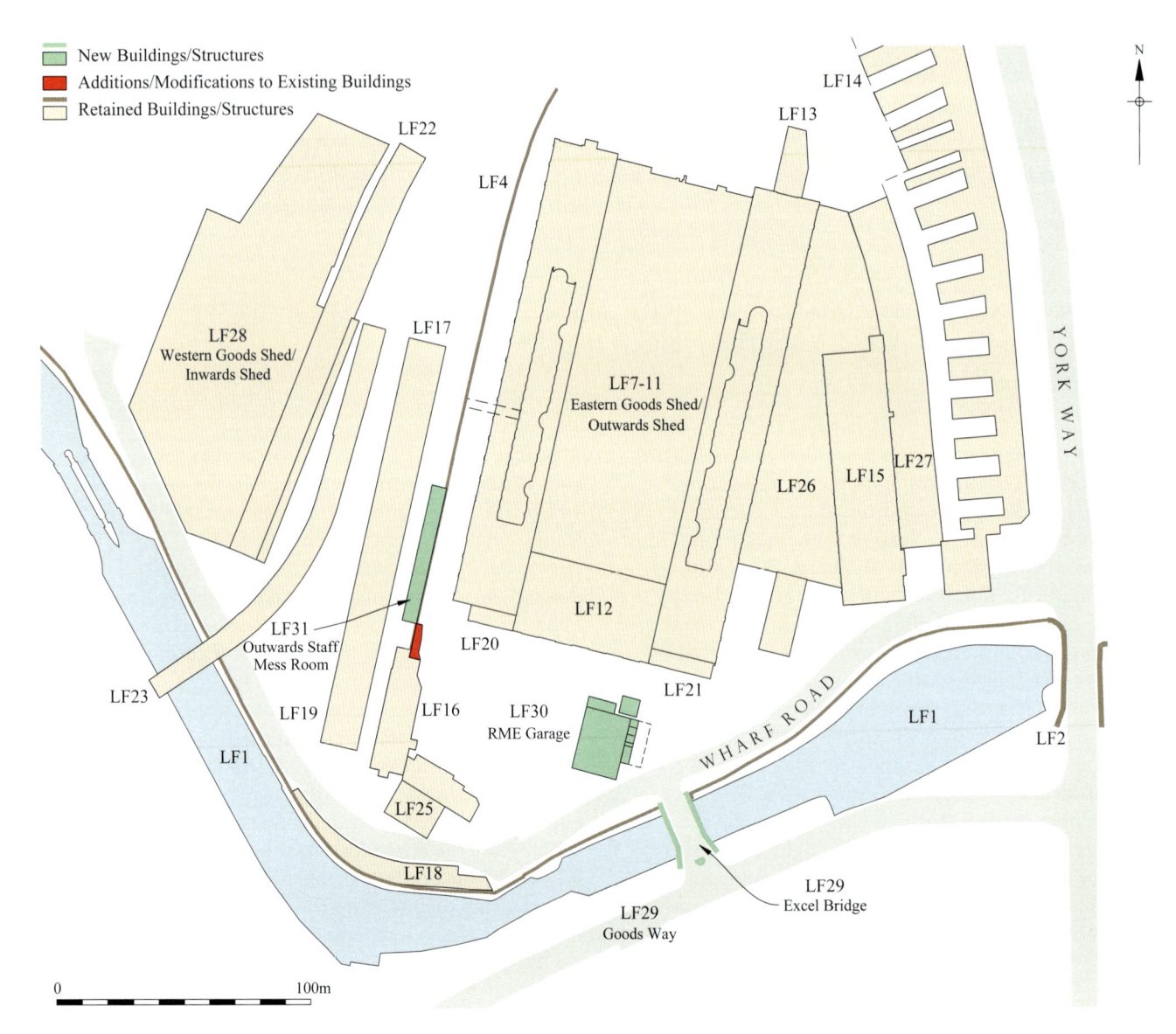

Fig. 9.1 The locations of the landscape features present in the Goods Yard from 1919–38, scale 1:2,500

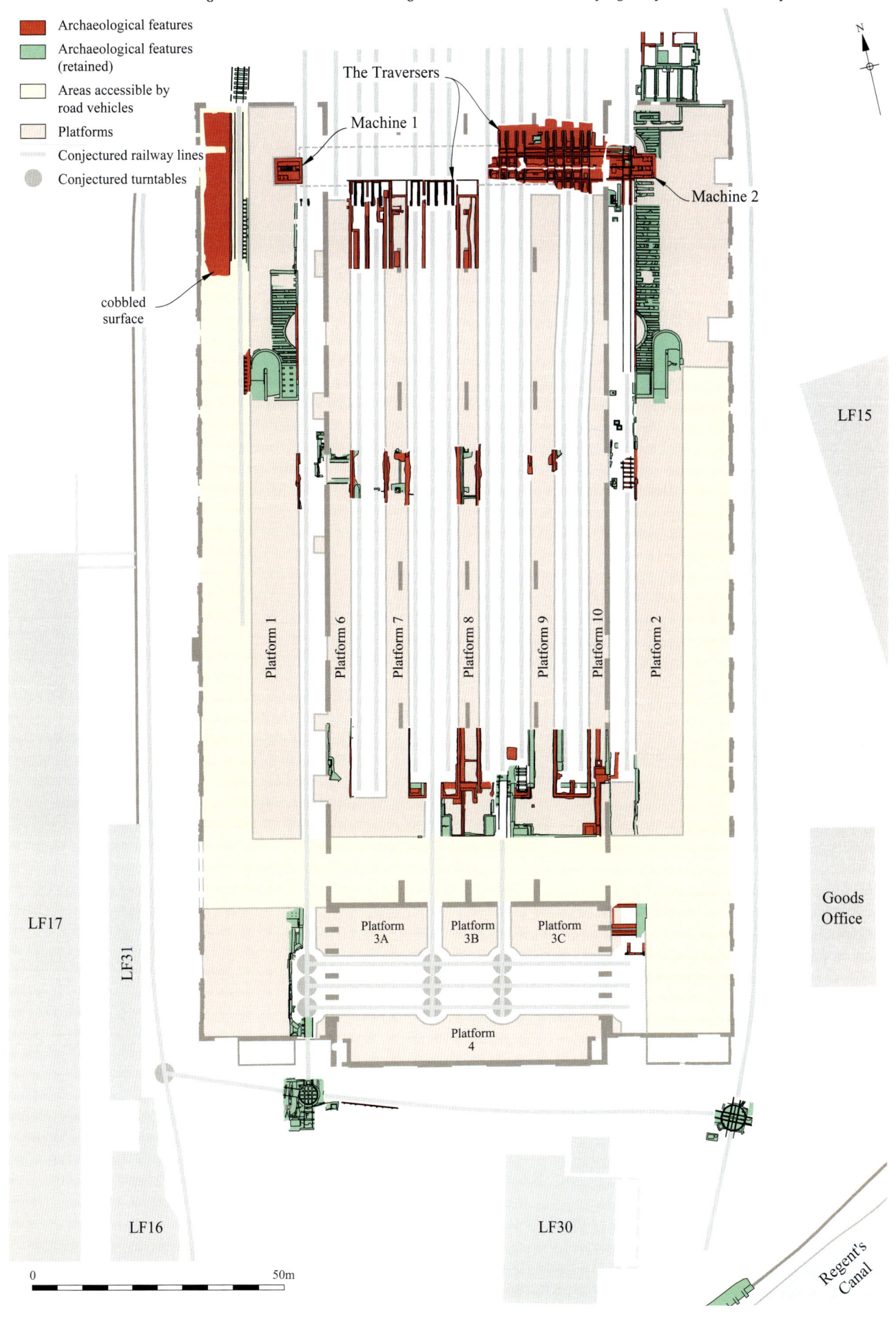

Fig. 9.2　The locations of the significant archaeological features dated from 1919–38, scale 1:1,000

between London and Manchester and secondly on their long distance lines between London and Scotland.

The competition that developed between the LNER and the LMS on their London to Scotland routes came to define many people's perceptions of interwar railway travel. Shortly before grouping took place, the Great Northern built its first 4-6-2 Pacific locomotives at Doncaster. Designed by the company's Chief Mechanical Engineer Nigel Gresley, the locomotives were designated Class A1 (later upgraded to Class A3) by the LNER after grouping. The new company appreciated that the performance of these locomotives was a powerful marketing tool, which it used to great effect during the period often described as the 'Golden Age' of steam. This began in earnest on 1st May 1928 with the historic, record breaking, non-stop run of the LNER's Flying Scotsman from King's Cross to Edinburgh Waverley. The feats of the A3s, and their successors the streamlined A4 Class captured the public imagination as they vied with the LMS Coronation Class Pacifics to become the fastest steam locomotives in the world, a title that was won by the A4 Mallard on 3rd July 1938, a record which still holds today (Williams 2013: 117–23). Yet the glitz of the company's main line Anglo-Scottish express services contrasted with the lack of investment elsewhere on its network, notably its branch and suburban lines. Behind the glamour of the Flying Scotsman and the A4s lay the inescapable fact that the LNER was, by a considerable margin, the most impoverished of the 'Big Four' (Whitehouse and St John Thomas 2002: 10).

The company was disadvantaged from the moment that it was formed due to the awkward combination of lines that it inherited. Although second only to the LMS in terms of overall size, and employing the equivalent of approximately three quarters of its rival's staff on a network that was just 7% smaller, the LNER's annual traffic receipts were one third less (Bonavia 1982: 1). The company had also inherited numerous small and little-used local branch lines in depressed areas in the north and east of the country, which further added to its financial woes (Williams 2013: 41). What is more, since so many of its lines serviced Britain's industrial heartlands, two thirds of the company's profits came from freight haulage, substantially more than any other railway company (Whitehouse and St John Thomas 2002: 29; Bonavia 1982: 1). The poor health of the LNER's finances was therefore further compromised by the general decline in the volume of goods transported across its entire network during the depressions of the 1920s and 1930s, when heavy industrial production fell into a rapid decline (Whitehouse and St John Thomas 2002: 10, 37).

None of the railway companies were immune to the social unrest and labour disputes that arose during the 1920s, and the entire railway network was brought to a halt during the General Strike of May 1926 (Greaves 2007: 140). Although the strike lasted just nine days, the railway companies lost substantial sums of money as a result (Wragg 2012: 22). Of greater long-term concern to the Directors and shareholders of the LNER was the state of the British coal industry. Having been a significant exporter before the First World War, the industry had been hit successively by the collapse of exports during the war, the worldwide recession that followed it and then by the miners' strike that triggered the General Strike. While the rest of the striking workforce returned to work after that dispute ended, the miners remained on strike until late November. With the global supply of coal already exceeding demand before the strike, British coal exports collapsed as overseas customers switched to coal from mines in Germany and elsewhere. The collapse of international coal exports had an immediate effect upon the revenues of the LNER, its coal receipts dropping from £14.7 million in 1923 to just £7.9 million in 1926 (Whitehouse and St John Thomas 2002: 10, 38). Taken together, these factors meant that from the year that it was founded the LNER's profits would decrease annually, being many millions of pounds less near the end of the interwar period than at the start.

Although the carrier was undoubtedly hit hard by the recessions and depression of the 1920s and 1930s, some respite was offered by the revenues earned from carrying foodstuffs and agricultural produce. Agricultural and fish specials increased during the period as the company cornered a majority share of that traffic. Advances in packaging and production of food produce, including canning, the increasing transfer of milk in tankers rather than churns and, most strikingly, the sudden and rapid growth of domestic sugar production, had the welcome effect of boosting revenues (*ibid.*: 34). This helped the LNER to maintain an operating profit, which meant that the company was rarely in danger of actually going bust. However, the company's net return out of which dividends were paid, which was inadequate from the start, fell from £13 million to just £7 million between 1923 and 1932, causing considerable unrest among shareholders (*ibid.*: 13–4). In comparison with rival companies little money was available for modernisation, despite the increasingly dated facilities that the LNER had inherited.

The LNER inherited a number of large goods stations in the capital from its constituent companies. In addition to the King's Cross Goods Station, these included the London goods depots of the Great Eastern Railway at Bishopsgate and Spitalfields and the Great Central Railway's goods station at Marylebone. During the pre-grouping era, each of these depots had adapted over time to handle specific types of cargo carried by their operators. In contrast to King's Cross and Marylebone, both of which handled traffic which originated mainly within the British Isles, Bishopsgate Goods Yard also received a great deal of continental traffic from Harwich. This included large numbers of containers designed to carry ferry passengers' luggage, a service pioneered by the Great Eastern. By the middle of the 1920s it had become apparent that Bishopsgate was growing increasingly congested, a situation that was exacerbated in the summer of 1925 by the destruction by fire of the nearby Spitalfields Goods Station (TNA RAIL 390/547, 30/07/1925). It was therefore decided to divert the heavy continental traffic from Bishopsgate to King's Cross, where it would be stored prior to onward distribution by road or rail (TNA RAIL 390/59: Minute 545 30/07/1925). The yard was ill-prepared for the ensuing influx of containers and cased goods that it had not been designed to handle. In order to accommodate this new and increasingly common form of cargo it was therefore necessary for the LNER to find the capital necessary to modify existing arrangements.

Little money had been invested in updating the Goods Yard's buildings during the 1900s and 1910s following the massive expenditure of the late 1890s. Similarly, the yard remained reliant upon hydraulic technology to shunt wagons and move goods through the complex, at a time when newer facilities elsewhere were increasingly being adapted to use electrical power. The LNER had also inherited an army of horses that were distributed across London at a time when motor vehicles were becoming a quicker and more efficient short-haul option, in part thanks to improvements made to the quality of the capital's arterial roads by Government work schemes. Modernisation was therefore needed on many fronts in order to enable the LNER's goods handling operation in the capital to remain competitive. In contrast to the declining fortunes of other areas of the country served by the LNER network, London's economy continued to thrive during the interwar period. Fuelled by 'new' light industries, a growing consumer culture and an expanding contingent of office workers, London's economy continued to expand throughout the 1920s and 1930s making it one of the few areas where a demand for goods and raw materials was sustained (Inwood 1998: 708, 724–33).

Owing to the continued impoverishment of the LNER, the company's freight operations and its less prestigious passenger services were modernised 'painfully slowly' during the 1920s and 1930s (Whitehouse and St John Thomas 2002: 11). This chapter considers how the LNER attempted to overcome the challenges that faced the Goods Station at King's Cross during a period when the money needed for investment was nearly always in short supply.

9.1 The Modernisation of Goods Handling at the Granary during the Interwar Period

Landscape Feature 12

The goods handling techniques and equipment that were used at King's Cross Goods Yard in the late 1920s had been devised to handle loads carried by traditional goods wagons and vans. By the beginning of the following decade the increasing use of containers had led to difficulties being experienced with existing equipment, not least the fixed cranes. A meeting of the Traffic Committee in January 1930 heard that difficulties were encountered when using the two fixed cranes at the southern end of the Granary[13] to unload containers, owing to their lack of power (TNA RAIL 390/61: Minute 1498 09/01/1930). The Committee was told that the position of the cranes necessitated that a large amount of shunting take place, during which delays occurred 'owing to the approach [rail] road to the cranes being occupied by wagons at which the traders' teams [were] employed loading or unloading' (*ibid.*). To increase crane lift capacity a 6-ton petrol-electric mobile crane was purchased from Ransomes and Rapier of Ipswich, which would permit loading or unloading of trucks to be undertaken almost anywhere in the Yard whilst eliminating the need to shunt wagons to and from fixed cranes. The new machine would be controlled by a single operator, thereby permitting the company to dispense with the services of one of the two existing crane attendants, as well as two goods porters, one van setter and a horse, enabling a net saving of £251 per annum.

The volume of container traffic continued to increase throughout the 1930s. By the middle of the decade two 6-ton mobile cranes were stationed at King's Cross Goods Depot, followed by a third in 1936 (TNA RAIL 390/64: Minute 2929 23/07/1936). The growth of containerisation accelerated the demise of the heavy fixed hydraulic cranes through their replacement with more agile and cost effective modern mobile equivalents. Although the former Granary Basin cranes survived *in situ* until after the Second World War, another 10-ton hydraulic fixed crane was replaced by a Ransomes and Rapier 6-ton petrol-electric standard mobile crane in February 1943 at a cost of £1,245 (TNA RAIL 390/1974: Minute 3175 18/02/1943).

The diversion of the Harwich continental traffic from Bishopsgate to King's Cross in 1925 led to a backlog of loads being stored in wagons at the Inwards

---

13   These cranes had lifting capacities of 8 and 10 tons respectively and had been retained despite the earlier infilling of the Granary Basin in order to unload goods from wagons shunted to their locations (TNA RAIL 390/61: Minute 1498, 09/01/1930).

Shed, which led to further delays and congestion (TNA RAIL 390/59 Minute 545 30/07/1925). A solution to the problem was found in the Granary, the top three floors of which then stood empty. Although the original hydraulic lifting apparatus remained *in situ*, it was not suitable for lifting the cased goods that were beginning to arrive from the Continent. Plans were therefore prepared to replace the hydraulic hoists with an electric lift that would take goods to and from the top three floors and for electrically operated jiggers that would allow goods to be moved from the Granary to road vehicles for onward distribution (TNA RAIL 390/547: Memorandum to the Traffic and Locomotive Committees 30/07/1925).

Tenders were invited for the installation of two 10cwt and one 15cwt electric landing hoists, one 30cwt electric goods lift and a 50 KVA AC air-cooled transformer in November. The contract to install the hoists and the lift was awarded to S.H. Heywood & Co. of Reddish, Greater Manchester in January 1926 for £1,152. The Specification for the hoists stipulated that the supporting steelwork was to be fitted between the existing cast iron girders and attached to the 'front wall', the latter by means of rolled steel channels supplied and fitted by the railway company (*ibid*.: 'Specification for Electric Landing Hoists' 10/1925). In order to carry the 30cwt electric goods lift a new lift shaft was to be inserted beside the north wall of the building, in the seventh structural bay from the west. The cage of the new lift was to be not less than 8' wide by 10' deep by 7' high, steel framed, fitted with collapsible steel gates and carried by no fewer than four flexible steel wire 'ropes' (*ibid*.: 'King's Cross Goods Yard Granary: Arrangement of 30cwt Electric Lift' 10/1925). The contractor indicated that it would not be possible to construct a cage of the dimensions specified, and offered to provide one that was 7'8" wide by 9'10" deep with a 7'8" clear opening, of the same type as that the company had supplied to the Great Western Railway Goods Depot at South Lambeth. This offer was accepted, and both shaft and lift were installed in 1926. Although the new electric lifts supplanted the hydraulic hoists in the upper levels of the Granary, no effort appears to have been made to remove the overhead water tanks, which continued to occupy the roof space as late as 1942 (Goad 1942).

The principal commodity stored in the upper levels of the Granary during the second half of the decade was sugar, which was kept on the second, third, fourth and fifth floors of the building by the English Sugar Beet Corporation from 1926 (TNA RAIL 390/61: Minute 1462 24/10/1929). The corporation paid the LNER more than £1,000 per annum to accommodate these goods, which had previously been carried by water and stored in Port of London Authority (PLA) warehouses elsewhere in the capital. In the autumn

Fig. 9.3 A Ransomes and Rapier mobile crane in operation at King's Cross Goods Yard in 1949 (SSPL 1995-7233_LIVST_RF_341 © National Railway Museum and SSPL)

of 1929 the English Sugar Beet Corporation asked the railway company to permanently adapt the four floors as a bonded warehouse, in return for which the former was prepared to pay up to £1,600 *per annum* for the remaining two years of its tenancy. Keen to retain a guaranteed source of income in the midst of a recession, the LNER accepted the sugar company's proposals and agreed to make a number of alterations to the fabric of the building in order to comply with Customs regulations. These modifications included the provision of new office and lavatory accommodation, the disconnection of the grain chutes and the closure of the openings on the second floor, together with a number of minor (unspecified) alterations on the remaining floors. The Traffic Committee authorised the expenditure of £327 to complete these works on the 24th October 1929.

In addition to handling large volumes of previously unfamiliar commodities such as sugar and bananas[14], these alterations enabled King's Cross to receive an increasing quantity of goods containers. As early as 1928 the company forwarded 8,330 container loads carrying 12,462 tons of goods by rail, the containers being a mixture of types designed to be carried onward by road and sea (*ibid*.: Minute 1279 21/02/1929). By the spring of 1929 the LNER was planning to introduce a range of containers designed for the carriage of commodities as diverse as soft fruits (for delivery to jam factories), meat and bricks (*ibid*.: Minute 1318 25/04/1929). Within a year King's Cross alone was handling as many as 430 containers per month (*ibid*.: Minute 1498 09/01/1930).

Archaeological evidence relating to the conversion of the Granary into a shipping container reception facility and sugar warehouse in 1926–9

In order to install the electric lift in the Granary in 1926 it was necessary to block one vertical row of windows in the north elevation, which were situated in the location of the new lift shaft (Fig. 9.5). Whilst elements of the shaft must date to the 1920s, it was widened when the lift was replaced in the 1950s. Consequently, most of the extant remains that were observed were not original elements.

---

14 King's Cross started receiving bananas from Jamaica via the East India Docks in April 1929 (TNA RAIL 390/60: Minute 1250, 03/01/1929).

Fig. 9.4 A sugar beet lineside storage yard. The interwar period saw domestic sugar beet production skyrocket, a factor that caused the Granary to be converted into a bonded sugar warehouse. The commodity was transported by rail on an industrial scale as shown in this image (SSPL 1996-7038_BTF_29_22 © National Railway Museum and SSPL)

The building recording survey (KXF07) noted several changes that were no doubt made when four floors of the Granary were turned into a sugar warehouse in 1929. All the grain chutes that were observed had been blocked and it seems likely that this took place at that time so that the building would comply with bonded warehouse customs regulations. The same explanation may account for the blocking of the various other openings that formerly criss-crossed the structure. Presumably this was also around the time when the words 'Flour and Sugar 1st Floor' were painted upon the southern door jamb of the ground floor entrance to the eastern stairwell, which suggests that the building continued to receive flour during this period albeit in smaller quantities.

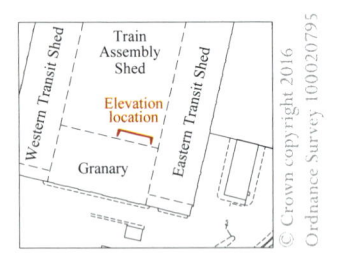

Fig. 9.5 External elevation of the eastern side of the north wall of the Granary (Landscape Feature 12) showing the windows that were modified when the lift was inserted in 1926, scale 1:250

9.2 The Midland Goods Shed during the Interwar Period

Landscape Feature 15

At the beginning of the 1920s the Goods Yard's Telegraph Office was still based on the ground floor of the offices at the southern end of the Midland Goods Shed. Since the introduction of telephones at King's Cross in the late 1890s the volume of telegraphic traffic had declined, and it was therefore decided to replace the telegraphic connection between the Goods Yard and the Passenger Station with an alternative mode of communication (TNA RAIL 236/158: 294, 318). Five months before grouping, the Great Northern's Way and Works Committee had approved a proposal to install a pneumatic tube via which messages could be transmitted in either direction between the Midland Goods Shed telegraph office and its counterpart on the main departure platform at King's Cross (TNA RAIL 236/191: 56). In March 1923 the newly founded LNER approved its predecessor's decision to close the office and install the pneumatic connection, thereby saving approximately £600 in wages and equipment costs per annum (TNA RAIL 390/58: Minute 34 15/03/1923; TNA RAIL 390/296: Memorandum from CGM to Traffic & Works Committee 10/03/1923). The tube was installed by Messrs T. Cooke & Sons of York in June 1923 (TNA RAIL 236/191: 56; TNA RAIL 390/296: Memorandum of the Works Committee 04/06/1923).

Despite having been an element of the Inwards Goods complex since 1915, the Midland Goods Shed was left out of a scheme to transpose the functions of the existing Outwards and Inwards Sheds that was proposed in July 1935 (TNA RAIL 390/63: Minute 2709 25/07/1935; see below).

9.3 The Hydraulic Station during the Interwar Period

Landscape Feature 16

Railway companies started using electrical power to drive machinery at goods and locomotive yards shortly before the turn of the 20th century. Whilst the adoption of electrical power at older yards was often a gradual process, newer yards such as the Great Western Railway's locomotive depot at Old Oak Common (completed in 1906), relied entirely upon electricity for lighting and the operation of machinery (Thompson and Gould 2010). Having decided to take advantage of the new power source, many railway operators at first chose to generate electricity themselves; Old Oak Common's supply was drawn from the Great Western's power station at Park Royal, whilst the Great Northern built a power station for its own use at Holloway in the 1890s. This situation persisted into the second decade of the 20th century, after which time electricity increasingly began to be drawn from local municipal suppliers.

The electrification of plant and machinery at King's Cross Goods Yard gained momentum after the end of the First World War, and continued to accelerate following Grouping in 1923. In December 1925 Nigel Gresley, the Chief Mechanical Engineer of the LNER warned that the boilers used to supply steam for the Engineer's workshops in the Locomotive Depot were worn out and recommended that instead of repairing or replacing them, electricity should be substituted for steam (TNA RAIL 390/68: Minute 843 07/01/1926). The Goods Department was also keen to switch to electricity, and in November of that year tenders were invited for the supply of electrically operated lifts and jiggers that would replace the hydraulic lifting apparatus in the Granary (TNA RAIL 390/547: Memorandum to the Traffic and Locomotive Committees 30/07/1925). The same month the Locomotive Committee approved expenditure of £7,359 for electrically driven pumps for the hydraulic system (TNA RAIL 390/30: Minute 425 26/11/1925) so three electrically driven automatic three-throw ram pumps were duly ordered from Hathorn Davey & Co. of Leeds at a cost of £3,732.10s.0d (TNA RAIL 390/31: Minute 586 06/01/1927). Since the LNER had decided to shut down the Holloway generating station in the mid-1920s, the necessary electrical power required by the machines was drawn from the St Pancras Borough supply.

The introduction of electrically-driven pumps rendered the boilers in the Hydraulic Station redundant. These boilers had been removed by September 1929, when permission was given to move the District Electrical Repair Shop and Stores into the empty Boiler House (*ibid.*: Minute 1078 26/09/1929).

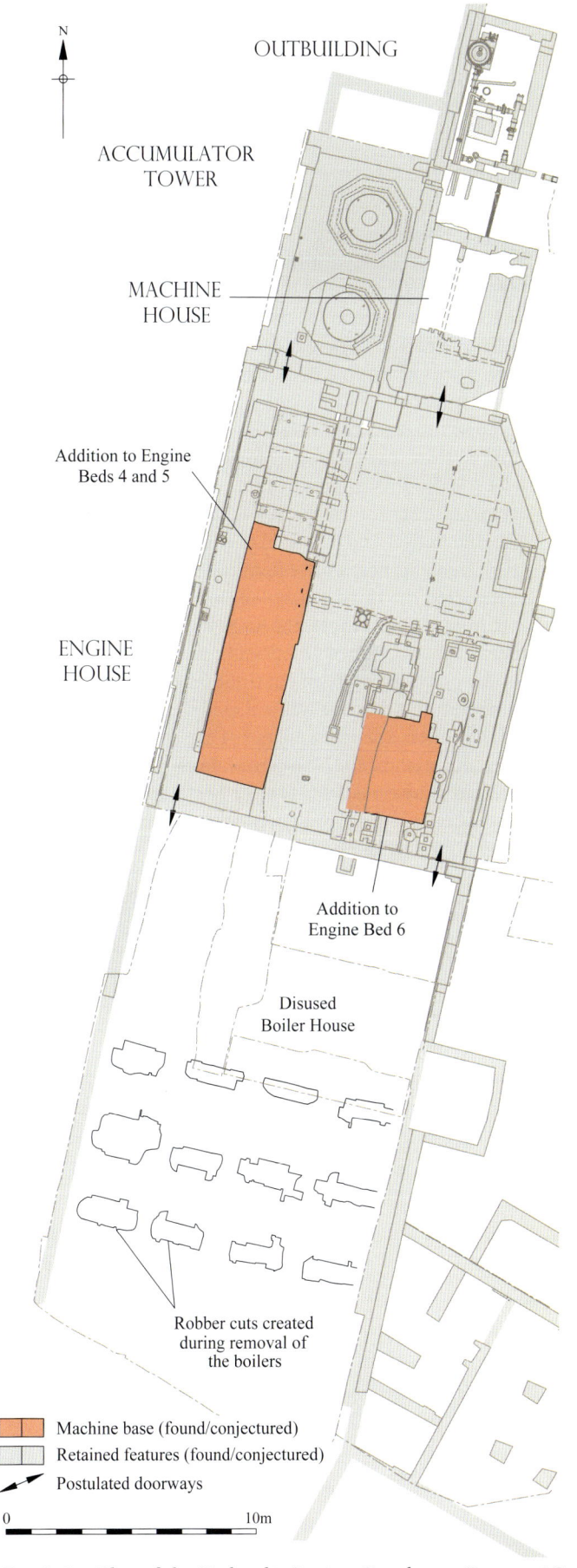

Fig. 9.6 Plan of the Hydraulic Station (Landscape Feature 16) after electrification in 1927 reconstructed from cartographic and archaeological evidence, scale 1:250

Archaeological evidence relating to the electrification of the Hydraulic Station in 1927

Landscape Feature 16

Whilst the archaeological evidence suggested that the footprint of the Hydraulic Station was not modified or extended during the 1920s, poured concrete additions, 0.50m thick, were made to Engine Beds 4, 5 and 6 (Fig. 9.6). The first of these sat directly on top of the southern two thirds of Engine Beds 4 and 5, thus turning the two separate bases into one structure. The second capped the central and south central sections of Engine Bed 6. These caps probably formed bases for the three electrically-driven pumps that were installed in the Hydraulic Station in the mid-1920s. The cap that sealed Engine Beds 4 and 5 was large enough to accommodate two such engines, whilst Engine Bed 6 probably supported the third example (Tim Smith pers. comm.).

In the Boiler House, the masonry boiler supports had been raggedly truncated in twelve places so that each boiler position had been cut three times (Fig. 9.6). Most probably, this damage was inflicted during the removal of the boilers after they fell out of use at the point of electrification in 1927. The truncations may have been created to insert jacks below the heavy boilers so that they could be removed more easily prior to being scrapped or reconditioned for reuse elsewhere.

9.4 The Coal, Fish and Cartage Department Offices and the Road Motor Engineering Garage during the Interwar Period

Landscape Features 18 & 30

The former Coal Offices (Landscape Feature 18) continued to house the Great Northern Railway's Horse Department until at least the end of the First World War (TNA RAIL 1189/1426: 16/10/1916; TNA RAIL 236/600). In addition to the department's staff, the offices also continued to be used by independent hay and straw merchants, with at least one firm of salesmen trading from the building as late as 1930 (Kelly & Co. 1930).

The functions of the former Horse Department Offices were not significantly affected by Grouping in 1923, following which the newly formed LNER retained them for the use of its own Cartage Department until the 1930s and possibly beyond.

Shortly after Grouping, the LNER set out to identify areas where duplicated or excessive expenditure could be reduced. Cutting the size of the cartage departments of its constituent companies presented a significant opportunity to make savings. An investigation conducted in the autumn of 1923 by Ralph Wedgwood (later Sir Ralph), the Chief General Manager, found that the company had inherited nearly 2,500 cart horses in London, compared with a motor vehicle fleet of only 22 trucks (TNA RAIL 390/58: Minute 165 04/10/1923). Over the years that followed the company gradually reduced the horse stock in the capital, replacing them with *ad hoc* purchases of motor vehicles of various types. Consequently, the horse-drawn vans that had carried exports from King's Cross Goods Yard during the early years of the LNER were gradually superseded so that by the mid-1920s, a significant proportion of the LNER's fleet of motor vehicles was instead stationed there (TNA RAIL 390/704).

By the autumn of 1928 it had become apparent that facilities for the repair and maintenance of motor vehicles in the Southern Area of the LNER network were 'quite inadequate' to meet the needs of the company's still expanding vehicle fleet (*ibid.*: 23/10/1928). That November proposals were approved to build a covered repair garage and workshop at King's Cross Goods Yard; however the following year it became necessary to defer the scheme 'in view of the urgent need for economy' (*ibid.*: Works Committee 03/02/1929). By the end of 1933, however, the number of motor vehicles in the fleet at King's Cross had risen to 200, prompting Nigel Gresley to revisit earlier plans to establish a garage within the depot (*ibid.*: Memo to Locomotive Committee 07/12/1933). Although the site of the old St Pancras Gasworks had previously been earmarked for the new facility, it was decided instead to use the infilled Granary Basin, which had recently been paved for use as a car park (TNA RAIL 390/71: Minutes 2692 27/07/1933, 2704 28/09/1933). Tenders for the construction of the repair shop were invited in January 1934 (*ibid*.: Minute 2782 04/01/1934). That April the contract for 'the provision and erection of [a] New Repair Shop and Garage and protected corrugated sheeting to sides and roof' was awarded to Fairfield Shipbuilding and Engineering Co. Ltd for £2,252.17s.7d (*ibid*.: Minute 2877 26/04/1934; TNA RAIL 390/704: 27/04/1934). The Road Motor Engineering (RME) Garage and the adjacent motor vehicle parking area (Landscape Feature 30) were first depicted on a plan of the Goods Station produced by the railway company in 1933, which appears to have been surveyed before the building was fully completed (Fig. 9.7).

The LNER Cartage Department continued to maintain a significant presence at King's Cross Goods Station and still kept a large number of horses in stables across the yard. The Department therefore continued to occupy the former Coal Offices (Landscape Feature 18) throughout the 1930s, using Block 5 for harness repairs until at least 1942, a continuation of the tack manufacturing and repair shop established there in the late 19th century (NRRG: DMFP00026218; Goad 1942; not illustrated).

In addition to housing the administrative staff of the Cartage Department, part of the former Coal Offices (Landscape Feature 18) appears to have reverted to its original use during this period. The London coal trade enjoyed a brief revival between the wars, and a number of coal merchants, including Tyne Main Coal Co. Ltd, J.L. Davies & Co. Ltd and Range Brothers traded out of the building as late as 1945 (Kelly & Co. 1940; 1945). Evidence from post-war trade directories suggests that independent coal merchants continued to trade from the Offices throughout the 1940s until around the time of nationalisation, when the building was ceded in its entirety to the Goods Department of British Railways.

Today, Landscape Feature 18 is frequently referred to as the 'Coal and Fish Offices' (or indeed the 'Fish and Coal Offices') although it is not entirely clear when it became associated with the administration of the fish traffic at King's Cross. Because this traffic was handled 'in-house' by the Goods Departments of the Great Northern and the LNER, historical directories do not provide any clues (Robbins 1935: 25). Whilst the quantity of fish arriving at King's Cross had increased considerably from the late 1880s onwards, the trade at King's Cross was smaller than that handled by the Great Eastern Railway's London goods terminus at Bishopsgate (TNA RAIL 1124/127: 4–6). It was only after Grouping in 1923 that imports started to increase more rapidly. Prior to this, a great deal of the fish that arrived at King's Cross came from East Anglia, however during the LNER years this traffic would be redirected to Bishopsgate, meaning that the trade at King's Cross was instead dominated by northern ports such as Aberdeen, Grimsby and Hull (Robbins 1935: 25; Erwood 1988: 132). Nearly 14,000 'packages' of fish were despatched daily from King's Cross, the majority of which were sent to Billingsgate Market, although large quantities were also transferred to the Southern Railway (SR) for onward transit to stations on that network (Robbins 1935: 25).

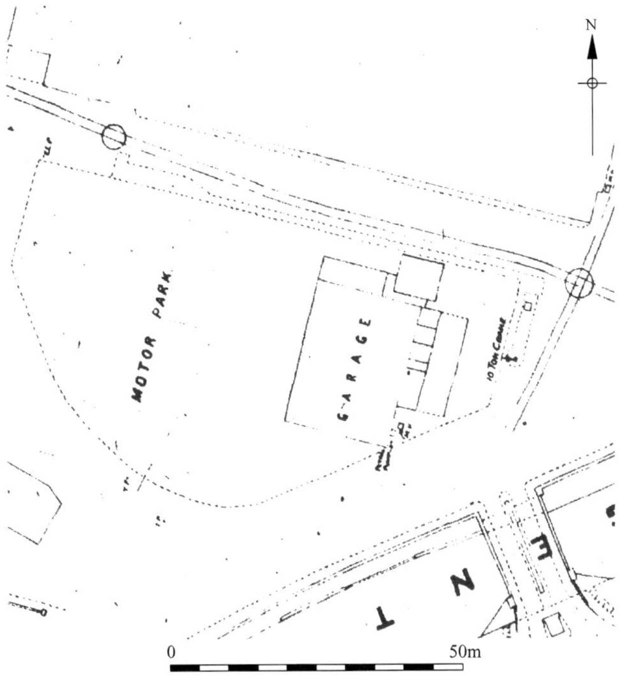

Fig. 9.7 LNER Plan of King's Cross Goods Station of 1933 detailing the RME Garage (Landscape Feature 30), scale 1:1,250 (NRRG DMFP 00026218)

9.5 The East and West Handyside Canopies and the Fish Trade during the Interwar Period

Landscape Features 26 & 27

Fish traffic was initially received in the Eastern Transit Shed (Landscape Feature 8), however by the late 1880s the growing volumes of that highly perishable commodity appear to have posed a number of logistical problems for that facility. The erection of the West Handyside Canopy (Landscape Feature 26), which covered the inward goods siding and roadway between the Main Goods Shed and the Midland Shed from 1888 onwards provided an ideal alternative location that enabled fish to be rapidly transferred into waiting road vehicles for immediate onward distribution. By the interwar period the original 1870s siding and a later short line of track had come to be known as the 'Long and Short Fish Road' (Bussell and Tucker 2004b: 2).

The late 1930s would see the East Handyside Canopy (Landscape Feature 27) and the adjacent Potato Market (Landscape Feature 14) continue to handle vast quantities of vegetable traffic. As many as 100,000 tons of potatoes flowed into the Potato Market every year, in addition to substantial amounts of seasonal fruits and vegetables, including up to 160 wagons of peas per day, plus multiple wagonloads of celery and rhubarb from the forcing sheds of Yorkshire (Roberts 1938: 635).

9.6 The Eastern Goods Shed and the Granary during the Interwar Period

Landscape Features 7–12

Towards the end of the 1920s the LNER announced that it intended to rebuild the locomotive depot (known as 'Top Shed') at King's Cross (Townend 1975: 38). The scheme was immensely costly and was one of only a handful of major works carried out by the company at a time when capital expenditure was tightly restrained. In contrast, little money was available for the renewal of the buildings of the Goods Yard, which were increasingly showing their age. In March 1935 the Goods Department complained to the Board about the condition of the roofs of the Inward and Outwards Goods Sheds, the Potato Market and the Midland Goods Shed (TNA RAIL 390/72: Minute 3110 28/03/1935). It was claimed that water had penetrated the canopies of these structures causing damage to the goods below, prompting the department to recommend that the company embark upon 'a general systematic overhaul of the roofs' over the next four years. The department also proposed that the roofs of the West and East Handyside Canopies be replaced, in 1935 and 1936 respectively (*ibid.*). Although the plan that once accompanied the report has not survived, a draft version, showing the phased roof refurbishment has been preserved at the Network Rail Records Group in York (NRRG: GNR [LNER] King's Cross Goods Station *c*.1935). This drawing indicated that refurbishment of both the Eastern and Western Transit Sheds comprised the final phase of the programme as originally envisaged, and was scheduled to be carried out by the company's own staff in 1939 for a total estimated cost of £16,350 (TNA RAIL 390/72: Minute 3110 28/03/1935). In the event the roofs in question were not replaced, almost certainly as a consequence of the Second World War. The necessary repairs could not be undertaken until after the end of that conflict for obvious reasons, by which time they were in an advanced state of dilapidation.

In the wake of the recessions of the 1920s and the Great Depression of the early 1930s, a number of initiatives designed to promote economic growth and relieve the burden of mass unemployment were implemented by central government. In 1929 Winston Churchill, then Chancellor of the Exchequer in Stanley Baldwin's Conservative Government, abolished Railway Passenger Duty, on condition that the resulting sums were used to modernise the railways (Wragg 2012: 25). One of the most significant measures used by Government to stimulate the economy after the Great Depression was the Guarantees and Loans Act 1934, which was designed to encourage private capital investment in major infrastructural works. Large companies were invited to design projects of proven public utility, the finances of the successful bids being guaranteed by the Government on the proviso that the 'winning' firms hired otherwise unemployed labour. The Railway Finance Corporation was set up in order to provide loans at a highly competitive interest rate of 2.5%. In 1935 the Ministry of Transport asked each of the 'Big Four' railway companies to supply a list of improvement schemes that they considered to be economically viable and of ultimate benefit to their shareholders for inclusion in the proposed Government Works Schemes. While both the Great Western and the Southern Railways applied for funding to improve mainline passenger services, the LMS and LNER sought to finance improvement to their freight operations. In a memorandum to the Board written in July 1935, Sir Ralph Wedgwood (LNER Chief General Manager) argued that King's Cross Goods Station warranted inclusion on the company's list of potential improvement schemes. In his report to the Board, Wedgwood highlighted the significant technological and operational inefficiencies of the Inwards and Outwards Sheds at King's Cross Goods (TNA RAIL 390/63: Minute 2709 25/07/1935). He described the existing facilities as 'inadequate and uneconomical', observing that while the existing Inwards Shed could accommodate nearly 220 wagons, only 35 of these were accessible to road vehicles at any one time. Given that approximately 210 wagons were unloaded daily, access restrictions meant that the process of unloading had to be repeated on six separate occasions throughout each working day. Time and money were wasted as a result, with shed staff standing idle whilst shunting operations took place. Further time was lost by the continued use of obsolete 19th century hydraulic capstans and turntables for shunting purposes.

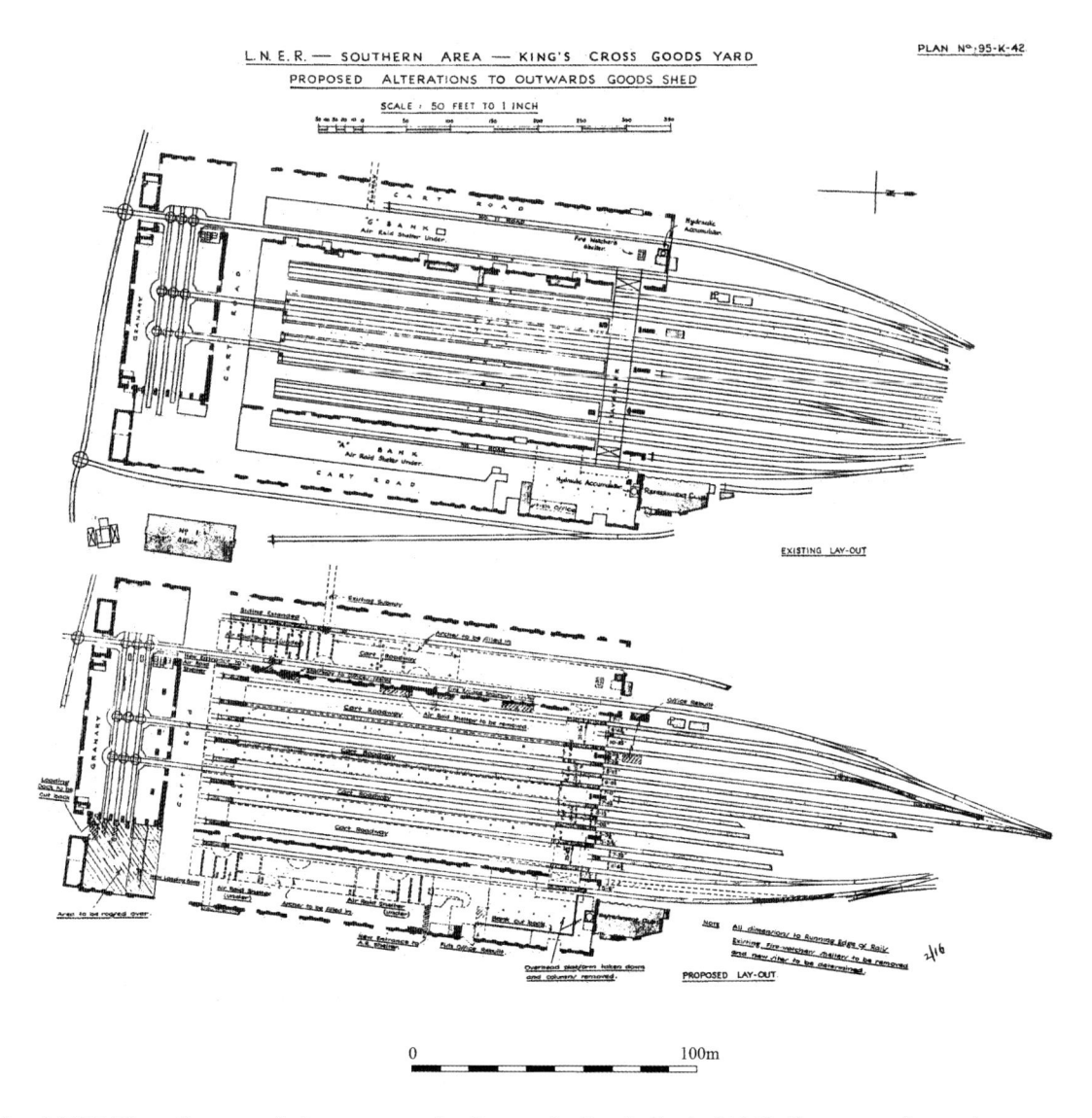

Fig. 9.8 LNER Plan of proposed alterations to the Outwards Goods Shed of 1942 illustrating the track arrangement in the Eastern Goods Shed (Landscape Features 7 to 11) and the Granary (Landscape Feature 12) in the late 1930s, scale 1:2,500 (NRRG LNER 95-K-42)

Wedgwood's solution was a scheme for 'transposing the functions of the Inwards and Outwards Sheds and remodelling and equipping the depot to meet modern standards'. If implemented, it promised to completely transform and modernise the LNER's goods handling facilities at King's Cross for an estimated cost of £37,295 (TNA RAIL 390/72: Minute 3223g 25/07/1935). It was calculated that if the scheme would reduce pilot shunting time by more than 12,000 hours across the two sheds *per annum*, save approximately 2,000 hours in delays to trains entering King's Cross Goods and effect a net saving of £13,379 (TNA RAIL 390/63: Minute 2709 25/07/1935). Having been persuaded of the merits of the scheme, the Board approved its inclusion in the package of measures for which the company sought financial assistance from the Government. Wedgwood's ambitious 'Transposition Scheme' was far-reaching in scope and ultimately resulted in the creation of a fully modernised and efficient goods handling depot.

In order to adapt the former Inwards Shed for Outwards use, it was necessary to re-lay eight of the ten sidings in the Train Assembly Shed (Landscape Feature 11) and remove the majority of the turntables and hydraulic capstans in the Transit and Train Assembly Sheds (Landscape Features 7, 8 and 11). Wagons were to be transferred from siding to siding by two new electrically operated traversers, which would ensure that they would be available for marshalling as soon as loading was completed (*ibid.*; Roberts 1938: 634). To make space for the new traversers, openings were to be cut into the north ends of the east wall of the Western Transit Shed (Landscape Feature 7) and the west wall of the Eastern Transit Shed (Landscape Feature 8) thus enabling the direct transfer of wagons between the three buildings. It was decided that the turntables associated with the Granary would be retained, as the installation of another traverser in that location would have been difficult without making substantial changes to the structure of the building (Fig. 9.8).

Two 22-ton electrically operated traversers were ordered from Cowans Sheldon & Co. Ltd of Carlisle, plans of which were issued in September 1936 (NRRG DMFP 00026358; reproduced here as Fig. 9.9). A photograph taken for an article published in Railway Gazette magazine in 1938 showed one of the two in action. A goods van was shown passing beneath a newly inserted girder spanning the brick piers at the north end of the Train Assembly Shed. The photographer captured the shed foreman operating the machine's wall-mounted push button control unit, which was situated a short distance from a hydraulic capstan head that had somehow survived Transposition intact (Fig. 9.10).

In addition to the insertion of the traverser it was decided that the toilets and other installations that had

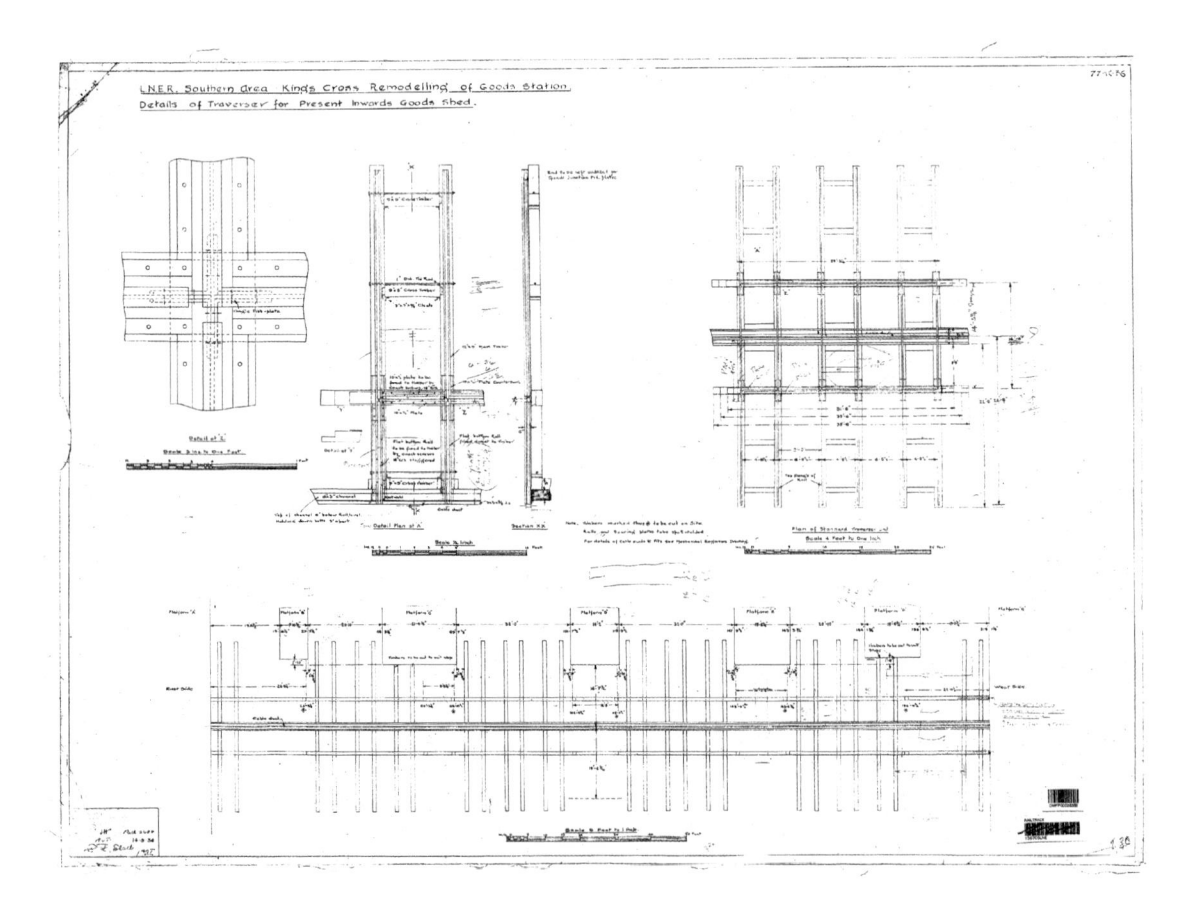

Fig. 9.9 LNER Plan of the Cowans Sheldon electric traverser for the Outwards Goods Shed (Landscape Features 7–11) of September 1936, various scales (NRRG DMFP 00026358 LNER)

been built over sections of the Cart Road at the south end of the Train Assembly Shed should be demolished in order to create wider openings in both the eastern and western elevations of the Transit Sheds. The aim of this exercise was to improve vehicular access to the loading banks (TNA RAIL 390/63: Minute 2709 25/07/1935; NRRG LNER King's Cross Goods Shed Alterations to lay-out 1935).

Outline plans of the Transposition Scheme were published in the LNER's in-house magazine in 1938. A more detailed plan of the works was produced in 1942, which was compiled in association with various alterations undertaken during the Second World War. The latter plan is reproduced here as Figure 9.8.

The use of electricity as a power source was becoming ever more prevalent at King's Cross during the interwar period, and Transposition necessitated the installation of several pieces of electrically operated machinery. Hydraulic power was not to be banished altogether from the new Outwards complex however, as it was decided to retain 27 hydraulic cranes on the platforms of the Transit Sheds (TNA RAIL 390/63: Minute 2709 25/07/1935).

The final act of the Transposition Scheme took place over the first weekend of March 1938, when the staff of the Goods Delivery Office and the Eastern Goods Offices (Landscape Feature 21) moved to the former Outwards Goods Offices at the Western Goods Shed (Erwood 1988: 135; Roberts 1938: 635). Amongst the departments that were forced to give up their accommodation in the Western Goods Shed were the Outwards Counter Office, which moved to the ground floor of the Eastern Goods Offices and the Outwards Labelling Office, which was relocated to the Western Goods Offices (Erwood 1988: 135; NRRG: 66-ME-520 02/1966).

Fig. 9.10 One of the Cowans Sheldon Electric Traversers in operation in the Train Assembly Shed (Landscape Feature 11) in 1938 (© Railway Gazette International)

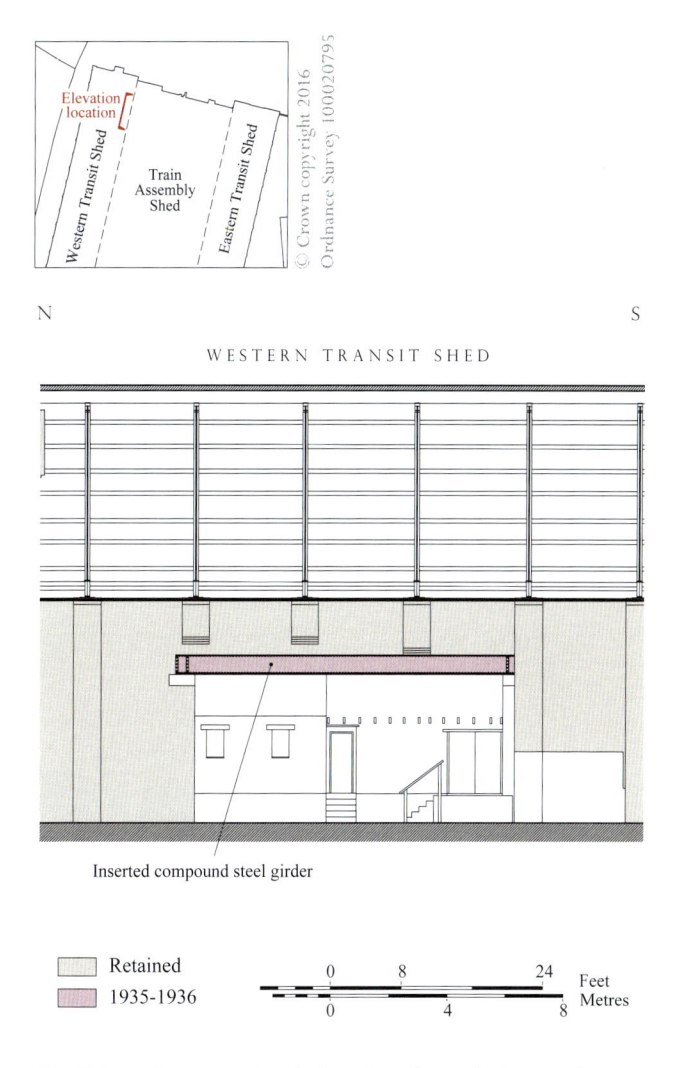

Fig. 9.11 Cross-sectional elevation through the northern end of the Western Transit Shed (Landscape Feature 7) showing the internal face of the east wall and the changes of 1935–6, scale 1:250

Built heritage and archaeological evidence relating to the traversers installed in the Eastern Goods Shed during Transposition in 1935–8

Landscape Features 7–11

In order to install the new wagon traversers, it was necessary to make a number of structural changes to the fabric of the former Inwards Shed. The most obvious of these involved the widening of the set of railway wagon openings that formerly accommodated the most northerly railway siding (Fig. 9.11). Compound steel girders ('RSJs') formed the new lintels that were respectively inserted in the eastern and western walls of the Western and Eastern Transit Sheds. They bore the rolling marks 'Appleby-Frodingham 24' x 7½" and 'British Steel'.

The arrangement of roof supports in the Train Assembly Shed (Landscape Feature 11) also had to be modified to make way for the traversers. Three of the brick piers that had been erected in 1850–2 were replaced by three vertical twin British Standard Beams, battened together, and three vertical single examples. They upheld the southern ends of a series of north–south riveted compound girders that were strong enough to span the traversers in their entirety, their northern ends being embedded in the piers that formed part of the Train Assembly Shed's northern elevation.

Excavations in the Train Assembly Shed (Landscape Feature 11) and the Eastern Transit Shed (KXI07) revealed the remains of a large brick and concrete structure (Fig. 9.12; Fig. 9.2). The main body of the feature consisted of a series of more than 22 rectangular masonry arms arranged in 11 pairs that were oriented north–south along with a further five that were oriented east–west. Together, they interlocked to form a grid of rectangular voids surrounded by upstanding masonry that was composed of red bricks arranged in English bond. This was set in a bedding layer of concrete that had been poured into a construction cut which truncated the northernmost bank of former turntable bases thus rendering them redundant. A 0.23m thick concrete cap sealed the entire structure, the top of which was found at a level of 24.01m OD. Pairs of rail-like 'beams' had been set into the concrete cap along the length of each east–west 'arm'. Where these had been removed, imprints in the concrete demonstrated their former presence. The structure was interpreted as part of the foundations that once supported the traversers (Fig. 9.12).

A complex piece of machinery, resembling a small engine with a wheel-like structure to its immediate east, was found to the west of the main body of the traversers' foundations in the Western Transit Shed (Fig. 9.13). Henceforth termed Machine 1, it was oriented east–west and sat upon a rectangular concrete base (Fig. 9.2). The structure was surrounded by a larger rectangular perimeter wall that presumably once formed part of a small house that encased the

Fig. 9.12 Remnants of the traverser support of 1935–6 exposed by excavation, photograph faces south

Fig. 9.13 The 1935–6 electric motor that powered the traversers in the Western Transit Shed (Landscape Feature 7), photograph faces southeast

machine. An internal concrete floor, found at a level of 23.11m OD, was unearthed within the footprint of the machine house. Six wooden chocks, arranged in two parallel east–west rows, were set into the top of this in the southwest corner. A virtually identical structure, Machine 2, was unearthed in a similar position within the Eastern Transit Shed (Fig. 9.2), although the arrangement of the components within the house, including both the machine itself and the wooden chocks, were inverted relative to those in the Western Transit Shed.

The main bodies of these machines were interpreted as electric motors that powered the traversers, whilst the wheel-like structures that were offset towards the traverser support may have played a part in controlling the winch mechanisms that pulled the transfer tables, which were responsible for carrying the rolling stock to and fro. The roles of the sets of wooden chocks remain enigmatic, one possibility being that they supported mechanisms that were associated with the winding gears.

Archaeological evidence relating to the platform modifications in the Eastern Goods Shed and the Granary made during Transposition in 1935–8

Landscape Features 7–12

The remains of five linear structures on a north–south alignment were unearthed during excavations in the Train Assembly Shed. Each was composed of wall fragments that had been constructed with reddish purple bricks set upon coarse, poured concrete foundations. The upstanding brick sections of the walls were between 0.38m and 0.46m wide, their southern sections butting up against and encasing the northern side of the east–west sections of Platform 5. Together they formed the perimeters of five hollow, upstanding structures that were oriented north–south. The masonry that was observed strongly suggested that they had been erected in one continuous build as the brick perimeters of at least four of them effectively formed one long, continuous, snaking wall. The five rectangular structures each represent a new loading area within the new Outwards Shed, termed Platforms 6 to 10 in this document from west to east. Worthy of note is the fact that Platforms 6 and 10 encased remnants of the north–south 'arms' of Platform 5, which had been inserted earlier. A series of small 'U' shaped structures were observed at the southern end of the platforms, each coinciding with the end of a railway siding. They probably once enclosed buffer-stops (Fig. 9.2).

Since all the remaining turntables in the Train Assembly Shed (Landscape Feature 11) had been removed when the traversers were installed, all the surviving masonry niches that formerly accommodated turntables in the east–west 'arms' of the surviving sections of Platform 5 were blocked when these structural elements were encased in Platforms 6 and 10. The remaining recesses in Platforms 1 and 2 in the Western and Eastern Transit Sheds (Landscape Features 7 and 8) were also blocked with identical masonry (Fig. 9.2).

Although the installation of the traverser resulted in the loss of all the turntables in the Main Shed (although the nine that were associated with the Granary

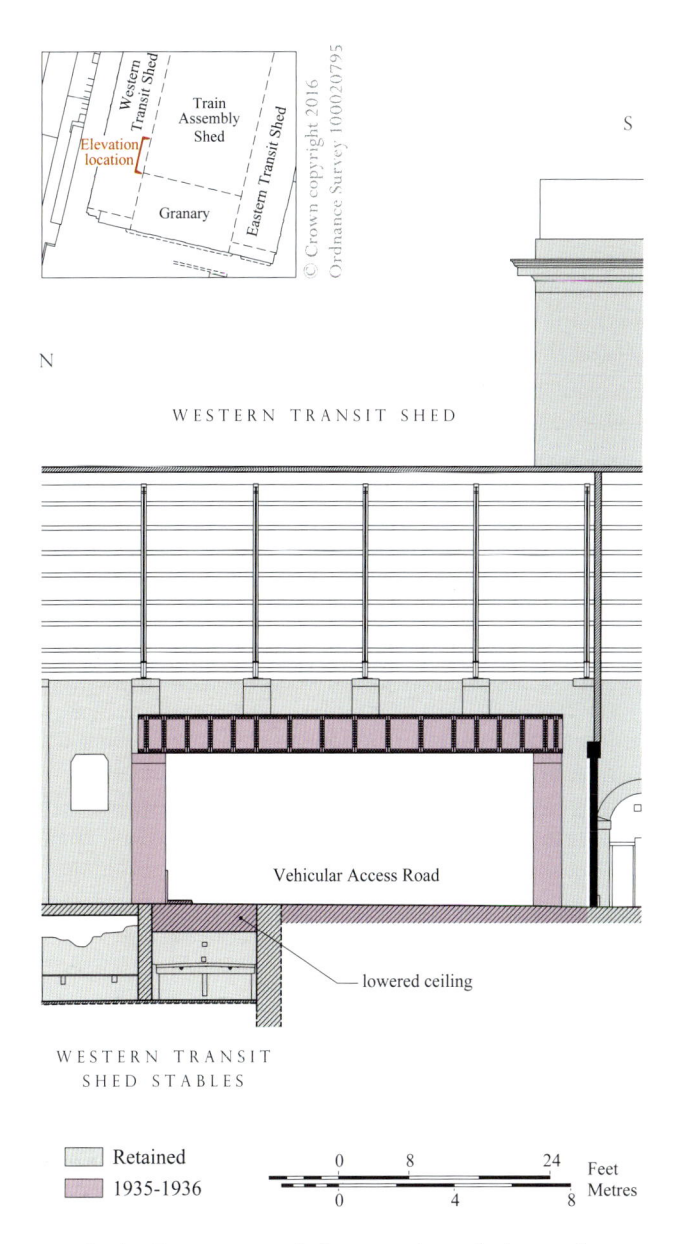

Fig. 9.14 Cross-sectional elevation through the southern end of the Western Transit Shed (Landscape Feature 7) and the Western Transit Shed Stables (Landscape Feature 9) showing their east walls and the changes of 1935–6, scale 1:250

survived), the locations of most of the railway sidings did not change, with the exception of the alterations described in the following paragraph.

The recorded remnants of the traversers suggested that it would have been necessary to shift the northern sections of the two easternmost sidings in the Train Assembly Shed slightly eastwards in order to link them with the rails that ran onto the north–south sections of those new mechanisms. This inference was supported by the available cartography and also explained the sub-rectangular footprints of Platforms 9 and 10, which had clearly been erected with this track alteration in mind (Fig. 9.2; Fig. 9.8).

The remodelling of the access road in the Eastern Goods Shed during Transposition in 1935–8

Landscape Features 7–11

The east–west cart road that had been created during the late 19th century was enlarged during Transposition so that it was broad enough to enable road vehicles to pass each other in both directions. This necessitated the widening of the access points that had previously been cut through the southern ends of the walls of the Transit Sheds (Landscape Features 7 and 8) and the insertion of new structural supports. The new openings were rectangular in shape with lintels that were formed by compound steel girders encased in concrete supported by sandstone pads (Fig. 9.14). The jambs were edged with black bullnosed engineering bricks and were protected by guard stones at the base on either side. These new openings had to be inserted in the eastern and western walls of both Transit Sheds so that the enlarged road could run through the entire complex.

The enlargement of the road meant that the most southerly rooms in the former Transit Shed Stables (Landscape Features 8 and 10) had to be modified. Their ceilings needed to be lowered, the replacements being formed by concrete shuttered with timber (Fig. 9.14).

9.7 Improvements to Staff Welfare Facilities during the Interwar Period in the Western Goods Offices and the Inwards Staff Messroom (later the Outwards Staff Messroom)

Landscapes Features 20 & 31

Documentary evidence reveals that the first floor of the Western Goods Offices had been turned into a messroom by the 1960s (NRRG: 66-ME-520 02/1966), but this most probably took place considerably earlier. After the First World War the Board of the Great Northern authorised expenditure on a number of new mess facilities in the Goods Yard, and it is plausible that the conversion happened around that time (TNA RAIL 236/190: 39, 203).

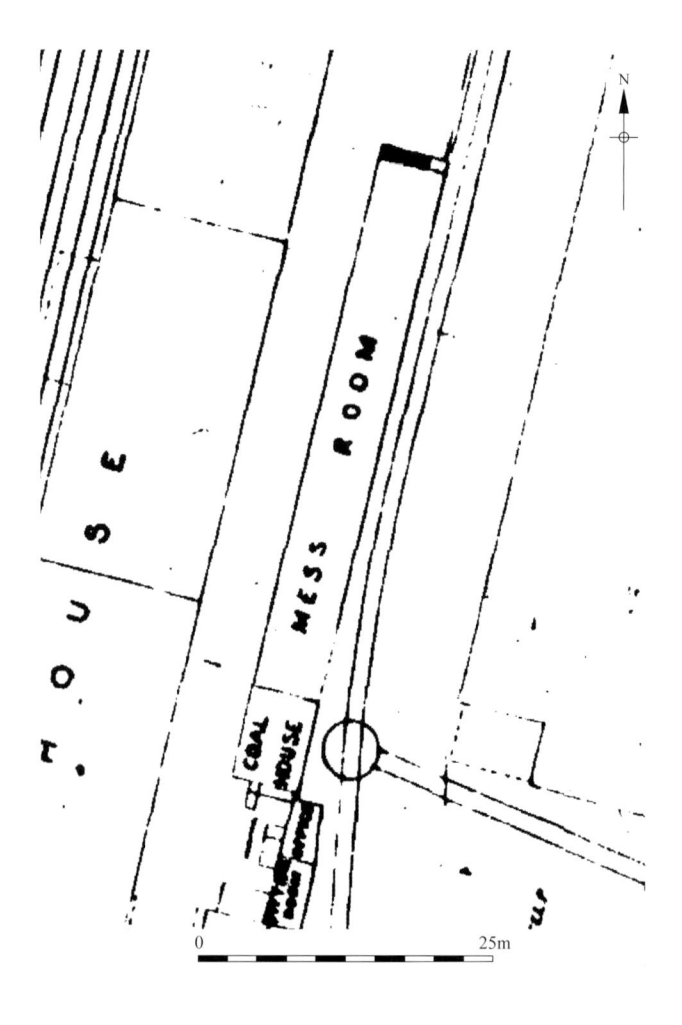

Fig. 9.15 LNER Plan of King's Cross Goods Station of 1933 detailing the Inwards Staff Messroom (Landscape Feature 31), scale 1:625 (NRRG DMFP 00026218)

At the end of May 1924 the District General Manager of the Southern Area of the LNER advised the company's Traffic Committee that the staff of the Inwards Goods Department were pressing for improvements to their existing mess facilities (TNA RAIL 390/58: Minute 345 29/05/1924). Staff members were particularly dissatisfied with the Inward Messroom, which was used by as many as 380 men, yet contained seating for only 30. The Committee approved a proposal to build a new messroom for the Inward Goods Shed 'fitted with ranges, tables etc.', capable of accommodating 170 men at one sitting thus providing 'ample facilities for washing purposes and the drying of clothes'. The cost of the endeavour was estimated to be in the region of £2,460. At some point between the beginning of June and the end of October a contract to build the new messroom was awarded to Messrs Ekins & Co. for £2,709 (TNA RAIL 390/67: Minute 487 02/10/1924). These references relate to the construction of the long narrow building, located a short distance to the north of the Hydraulic Station that has been known since the 1990s as the Laser Building (Landscape Feature 31). The earliest cartographic depiction of the new facility can be found on a plan of the Goods Yard surveyed in 1933 (Fig. 9.15).

The Transposition programme required that the entire bank and clerical staff of each shed swap workplaces with each other. This in turn meant that it was necessary to reorganise existing mess facilities to meet the needs of their new users (Erwood 1988: 135; Roberts 1938: 635). The same month that Transposition was announced it was decided to provide separate messroom and washroom space for the cartage staff, the majority of whom had previously used the Traffic Department Staff Room. In order to provide sufficient accommodation for these employees it was decided that 'the low level of the existing Inward Shed staff messroom be adapted accordingly' for an estimated cost of £1,588 (TNA RAIL 390/63: Minute 2709 25/07/1935). While the Inwards Staff Messroom (Landscape Feature 31) was being adapted for the Outwards Staff, mess facilities for the relocated Inwards Staff were also being

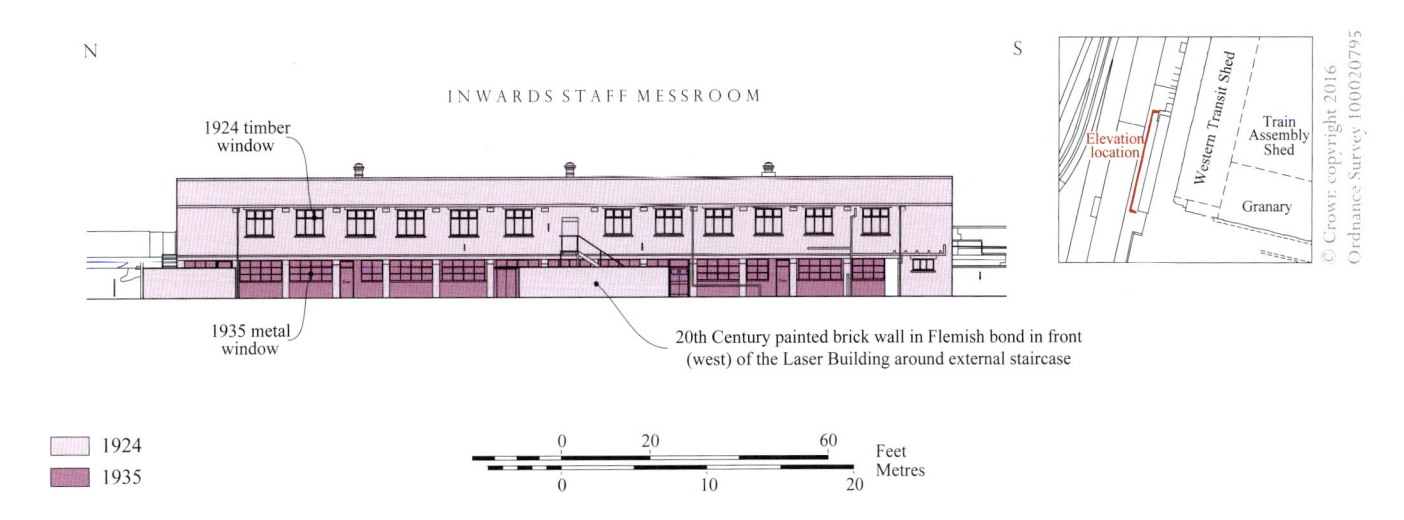

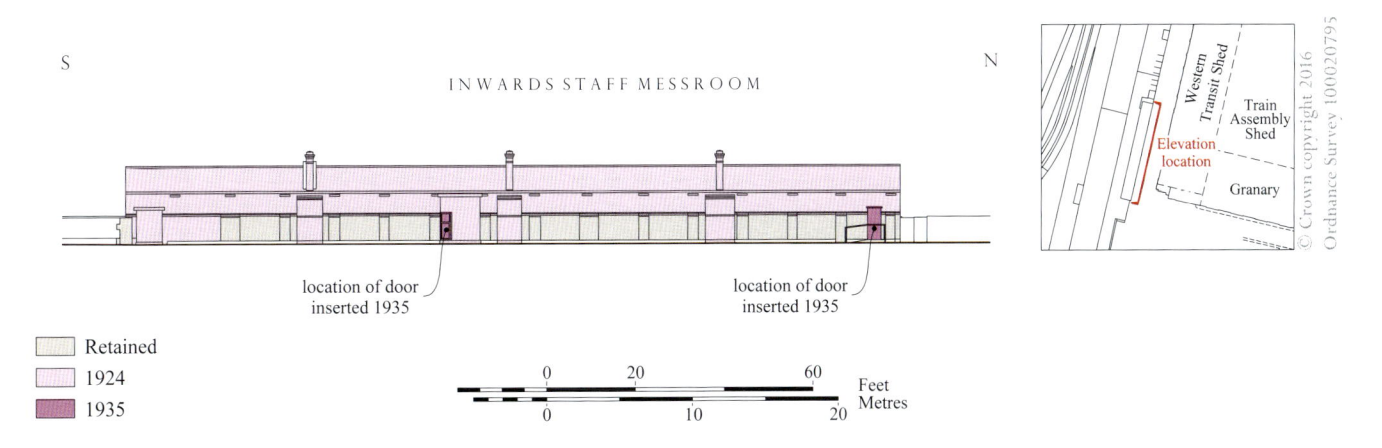

Fig. 9.16 External elevation of the western wall of the Inwards Staff Messroom, later the Outwards Staff Messroom (Landscape Feature 31) showing the changes 1924–36, scale 1:500

Fig. 9.17 External elevation of the eastern wall of the Inwards Staff Messroom, later the Outwards Staff Messroom (Landscape Feature 31) showing the changes of 1924–36, scale 1:500

improved. In late July 1937, approximately eight months before Transposition took place, the Traffic Committee approved a proposal to enlarge the existing Outwards Messroom to provide seating for 188 men of the Inwards Goods Department and lockers for 321 (TNA RAIL 390/64: Minute 3169 22/07/1938).

The resumption of peacetime working conditions at the end of the First World War meant that the majority of the women taken on by the railway companies as temporary employees during the conflict were replaced by men. With the exception of certain female clerical grades, this appears to have been the case at King's Cross Goods Station. No evidence of new or additional welfare facilities for female staff at the station have been found in the records of the LNER, suggesting that they remained a small minority in the yard's workforce during the interwar period.

The construction of the Inwards Staff Messroom in 1924

Landscape Feature 31

The Inwards Staff Messroom comprised a two storey brick structure in Flemish bond that was twelve bays long (Fig. 9.1; Fig. 9.15; Fig. 9.16; Fig. 9.17). It was erected towards the western edge of the Eastern Goods Yard on the low ground at the base of the terrace. The structure abutted the west side of the Retaining Wall (Landscape Feature 4) so that a large proportion of the east side of the building was formed by that structure. This meant that most of the east wall pre-dated the structure, having been created in 1850, whilst the uppermost section of that building along with the other three sides of the structure were erected when the Messroom came into being in 1924 (Fig. 9.16; Fig. 9.17). Due to its topographical position only one storey could be seen from the top of the terrace to the east (Fig. 9.17) whereas two were visible from the bottom of the

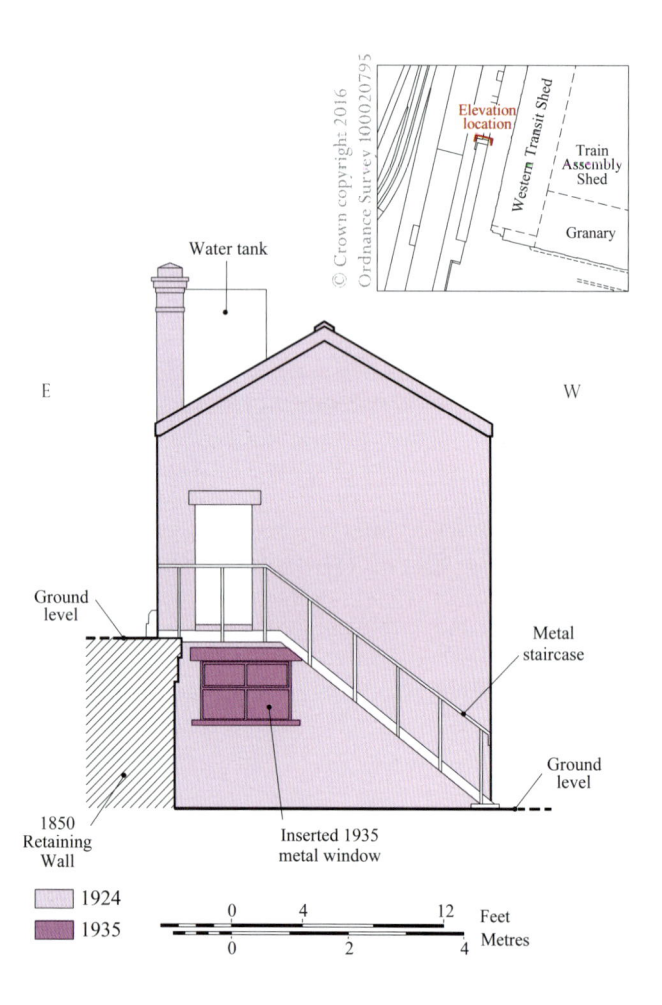

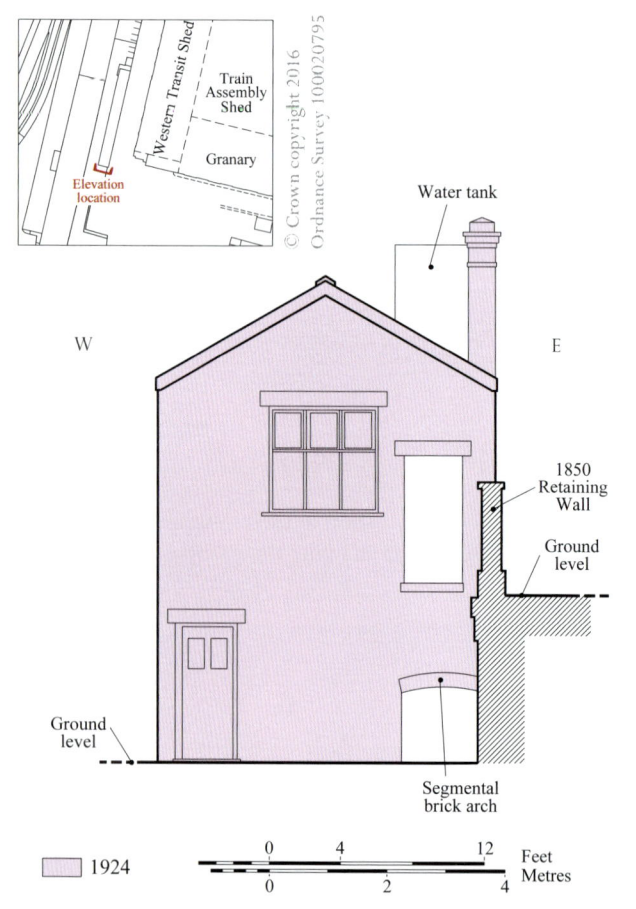

Fig. 9.18 External elevation of the north wall of the Inwards Staff Messroom, later the Outwards Staff Messroom (Landscape Feature 31) showing the changes of 1924–36, scale 1:125

Fig. 9.19 External elevation of the south wall of the Inwards Staff Messroom, later the Outwards Staff Messroom (Landscape Feature 31) showing the changes of 1924–36, scale 1:125

terrace on the western side (Fig. 9.16). The building was surmounted by a pitched, slated roof with gabled walls at both ends, its upper floor being heated by fireplaces that fed into three chimney stacks.

In the north elevation a threshold was present at first floor level, access from the ground being gained via a metal staircase that ran up the outside of the building (Fig. 9.18). The staircase also provided a means of communication between the base of the terrace and the high ground to the east of the Retaining Wall (Landscape Feature 4). A similar doorway and staircase arrangement was observed in the southern elevation (Fig. 9.19) and in the approximate centre of the western elevation (Fig. 9.16). In the southern elevation, an original doorway was also present at the base of the staircase at ground floor level, next to a small arched opening that originally housed a coal chute that fed the building's coal store (Fig. 9.19). The eastern elevation was originally predominantly blank with the exception of one original doorway in the most southerly bay (Fig. 9.17).

Illumination was provided at first floor level by twelve original windows in the western elevation (Fig. 9.16) along with another in the southern wall (Fig. 9.19). The majority of these 1920s windows comprised three tall panes surmounted by three smaller examples encased in timber frames, although the only original ground floor window in the western façade was somewhat smaller, containing just three panes (Fig. 9.16).

The appearance of the building suggested that when it was first constructed in 1924, staff welfare facilities were confined to the upper storey. At ground floor level, the fabric of the structure indicated that it originally consisted of 14 'stalls' that were open to the elements on their western sides (Fig. 9.16), although it may originally have been possible to close them off in some way through the use of shutters. The stalls may have been used as storage areas or, more probably, for the 'stabling' of motor vehicles or horse-drawn carts.

The conversion of the Inwards Staff Messroom to the Outwards Staff Messroom in 1935

Landscape Feature 31

When this building was enlarged in 1935 for the use of the Outwards Staff it was necessary to convert the storage or stabling area on the ground floor into a mess facility in order to increase capacity. This required that the former 'stalls' in the western element be fully bricked up and that additional windows and doors be inserted at ground floor level (Fig. 9.16). A further two doorways, which have been severely modified in modern times, were also cut into the eastern elevation at first floor level, either at this time or during a subsequent phase of work (Fig. 9.17).

The ground floor was not originally heated, so a major component of the conversion scheme involved the extension of the chimney flues downwards so that new hearths could be installed. The toilets appear to have been situated at the southern end of the building on both floors, the rest of the structure being taken up by drying rooms and mess areas.

9.8 The Sources of the Building Materials used at King's Cross during the Interwar Period

Fletton bricks from Peterborough continued to be widely used at the Goods Station during the interwar period and many were used in the Eastern Goods Shed during its conversion to outwards use. Kiln bricks (fabric type 3261) and frogged bricks bearing the stamp 'CRAVEN' were also noted, as were unfrogged stock bricks (3034) and machine pressed examples (3035) (see Appendix 3 for further details regarding brick fabric types and codes).

9.9 Synopsis

At the end of the First World War Britain's railway companies were in a considerably worse financial situation than they had been at the beginning. Four years of restricted funding, rising wage bills and maintenance arrears had taken their toll on the companies' finances and on the country's railway infrastructure. When the state relinquished control of the railways in 1921, it was apparent that the *status quo ante* could not continue. The solution devised by the Ministry of Transport, which involved amalgamating smaller underperforming companies with larger and more successful competitors into four large regional groups, was intended to encourage greater coordination and less competition. Though it represented a sensible response to the challenge, grouping was not without its problems. From the outset, the LNER was a prisoner of its inheritance. Although second only to the LMS in terms of overall size, the LNER's annual traffic receipts were one third less than those of its rival. The company's reliance on freight haulage, particularly coal, made it uniquely vulnerable to the economic turbulence of the 1920s and 1930s. The LNER was always the poorest of the 'Big Four' railway companies and its profits decreased year-on-year throughout the major part of the period. In 1938 the company couldn't even afford to pay a dividend on its shares.

Yet the outlook was not entirely bleak for the LNER, while several of the developments that helped the company to maintain an operating profit throughout the period also drove investment at the King's Cross

Goods Station. The increasing use of goods containers led to a number of innovations, including the acquisition of a number of modern mobile cranes and investment in the previously under-used Granary. As the carriage of sugar beet became an increasingly important source of revenue for the LNER, a large part of the same building was converted into a bonded sugar warehouse. Investments in motor vehicles were made at the expense of the cart horse and a garage facility was constructed on the infilled Granary Basin, thus dragging the road transport arm of the company into the motorised age. Other developments included the somewhat overdue electrification of the hydraulic station and an overhaul of the yard's communication network.

The workforce of the Goods Yard also benefitted from improvements in welfare facilities during the period, including the provision and subsequent enlargement of a brand new building (Landscape Feature 31) specifically for that purpose. It is possible that these developments took place in response to the increasing power of Trades Unions, which gave the workforce a louder voice and greater representation during the interwar period.

The most significant investment in the infrastructure of the Goods Yard of the interwar period took place in the second half of the 1930s, when the State-sponsored Transposition programme transformed the functions of the Eastern and Western Goods Sheds. The National Government's adoption of Keynesian economic policies in response to the Great Depression gave the railway companies a huge incentive to modernise their infrastructures at greatly reduced risk to themselves and their shareholders. The success of Sir Ralph Wedgwood's ambitious scheme meant by the late 1930s the LNER was the proud owner of a fully modernised goods yard. The vast undertaking resulted in the most revolutionary series of operational changes since the construction of the Western Goods Yard in 1899.

In conclusion, despite the economic turmoil of the 1920s and 1930s and the LNER's inherent structural and financial weaknesses, the company successfully achieved its goal of modernisation at King's Cross during the interwar period. This enabled the busy yet inefficient depot to continue to play a pivotal role in the supply of both domestic and overseas commodities to the nation at large and the lucrative London market in particular. Thanks to the implementation of the Transposition programme the efficiency of the King's Cross Goods Yard had improved considerably by the end of the 1930s, by which time the British economy was clearly on the mend. In outer London and the south-east of England a housing boom had been underway since the early 1930s, as economic recovery and the increased availability of mortgage finance fuelled a vast wave of suburban expansion. The recovery was intensified by the Government's policy of rearmament in the years preceding the Second World War, providing a much needed boost to the economies of many of those areas worst hit by the Depression (Price 2007: 152). The growth of Britain's GDP from that point onwards was exceptional relative to the larger democratic economies of the day, bringing the long period of recession and depression to a long-awaited end.

'No Human Activity Can Be Carried On Without Paper': The Working Lives Of The Goods Station's Clerks

'No human activity can be carried on without paper and railway work was no exception. There were reams of it, mountains of it; and when I first walked through the door of the King's Cross Goods Delivery Office on the morning of the 16th August 1937, as a newly joined Probationary Clerk, the first thing that I saw was a huge table running the entire length of the office- some 40ft- around which about twenty men were sitting, each completely surrounded by pieces of paper in all shapes, sizes and colours. From time to time, large, perspiring and usually unshaven and generally dishevelled men burst through the door, bearing clipboards with yet more wads of paper, which they detached and dropped on the table, seized fresh piles placed ready for them, and rushed out again, the whole operation being carried out to a monotonous obbligato of obscene language in which the usual word figured largely, if not almost exclusively'.

From 'Memories of King's Cross Goods' by Peter Erwood, a Junior Clerk at King's Cross in the late 1930s (Erwood 1988: 133–4)

During the 19th and 20th centuries the workplace at King's Cross was strictly hierarchical. At the top were the managers, most of whom had spent their entire working lives on the railways. At the bottom were the shed workers, porters and other manual labourers, while the middle ground was occupied by the vast contingent of administrators, who were themselves demarcated by a complex system of grades that was almost impenetrable to outsiders (TNA RAIL 236/21: 126; TNA RAIL 236/171: 254). A handful of reminiscences written by former members of the station's administrative staff reveal much about the conditions and culture of the station's offices, which appear often to have been little more genteel than the shed floor. The foul mouthed and overworked occupants of these offices suffered their own share of discomforts as they faced a huge bureaucratic mountain on a daily basis. Peter Erwood, who was a member of the station's clerical staff in the late 1930s, recalls the noisy and sometimes stifling environment of the offices:

'Above the general hubbub, telephones rang constantly. It seemed like Bedlam, but it was not. Everyone knew what he was doing, but much of the work had to be done against the clock, and the office was too small, uncomfortable and generally inconvenient to make for easy working...the huge table left little room for free movement, and one's progress down the office was a sort of obstacle race, or rather crawl. It was stuffy in summer, cold and draughty in winter, and the coal burning antique Victorian fireplace...did not warm more than the three or four people sitting nearest to it... The catalogue of inconveniences was completed by the lack of anything much in the way of washing facilities- one hand basin, cold water only, against the wall, adjacent to an antique and explosive gas ring, complete with a two gallon kettle. There was no in-house loo, the nearest being a sinister subterranean affair fifty yards or so across the yard towards the main gate, and obviously not designed to encourage people to withdraw there for a quiet smoke!....The routine paperwork was enormous... each consignment required three documents; the consignment note, made out by the sender, which was the basis of the legal contract between him and the railway company; the invoice, which was the railway's own transit and accounting document, prepared at the sending station; and the delivery sheet, which when signed by the consignee, concluded the transaction. The consignment notes made out by the larger firms were usually legible and often typewritten, but others were sometimes very much the reverse. At country stations, especially, farmers' hands were more used to speeding the plough than handling pen or pencil, and they needed a lot of convincing that 'Mr. John Smith, London' was not a sufficient address!' (Erwood 1988: 134).

Erwood was just one of the huge complement of administrators that made the Goods Station function. Different sub-sections were tasked with managing the accounts, monitoring the flow of commodities through the station in terms of volume and type and ensuring that shipments were accompanied by the correct paperwork so that they would reach their intended destination in the right amount of time. In LNER days it was the clerks of the Goods Delivery Office who were responsible for implementing the convoluted pricing system that had evolved over time. This required members of staff to establish the weight of the consignment, ascertain the category of the goods in question (through the use of a complicated classification system) and the mileage that they would travel. In order to determine the weight of consignments, large quantities of iron weights were kept in the offices, numerous examples of which were unearthed during archaeological excavations in the Eastern Goods Shed. Having ascertained the weight and classification of the consignment, it was necessary to determine whether or not the consignment would be collected and delivered door to door by the railway company itself, whether the sender and the receiver would take responsibility for transit to and from the departure and arrival stations or whether the railway would undertake one or the other of those legs of the journey on behalf of their client (ibid.: 136). Of course, things did not always go to plan, and clerks were required to deal with the consequences of lost or delayed shipments. A common, aggravating and avoidable problem encountered by the clerks of the Goods Department was described in detail by Peter Erwood:

'The invoice was supposed to accompany the goods, in an envelope clipped to the wagon solebar. However, when the invoice arrived at the destination station, all the aforesaid gubbins had to be copied yet again, by hand, in pencil, on the delivery sheet. In an effort to save time, the larger stations often sent invoices ahead by passenger train, care of the guard. This was all right in theory, but at King's Cross the invoices sometimes lay around in the passenger station waiting for a messenger to carry them up the road to the goods station, with the result that the consignment sometimes arrived before its invoice, thus provoking the 'Found on hand, no account' procedure, which involved much eventual clerical labour to match the consignment up with the invoices marked 'Not received' (ibid.).

Despite the mountain of paperwork that Erwood and his colleagues faced, during the cash strapped days of the LNER the yard's offices were under resourced. Erwood maintained that pencils were substituted for typewriters and mental arithmetic for adding machines, though the company did use sizeable numbers of both (ibid.: 135). The artefactual assemblage that was retrieved from the Goods Yard reflected the clerks' reliance upon traditional office supplies since it included nothing more technical than ink bottles (one of which was marked GNR Stationery Stores) and graphite sticks.

Like the shed floor, the offices were also worked around the clock. Indeed, by 1937 more Inwards clerks were employed in the arrivals area by night than they were by day, since the hours between 7pm and 7am were the busiest in that section of the station. In contrast, departing services were skewed towards the late morning and the early evening, meaning that the situation was reversed for the Outwards clerks. At least the offices were illuminated by electric lighting, which had been installed as early as 1900 (Medcalf 1900: 318).

Throughout the life of the yard, staff responsible for the administration of incoming traffic were based in and around whatever buildings were functioning as the inwards shed at that point in time, whilst a similar set-up could be found in the vicinity of the outwards shed. After its construction in 1899, the new offices that formed part of the Western Goods Shed were considered to have been greatly superior to their older equivalents to the east, which were by then dingy, musty and overcrowded with 'smoke-grimed ceilings' owing to the use of gas lamps at night (ibid.). Soon after the roles of the Eastern and Western Goods Yards were transposed in the late 1930s, the Inwards and Outwards administrators were obliged to swap offices. 'What the Outwards chaps thought of the piggery that they ended up in, I never knew, but I doubt that they were very chuffed about it!' commented one of the Inwards clerks in the wake of the move (Erwood 1988: 135).

Nearly forty years before Peter Erwood joined the LNER, Josiah Medcalf published his vivid and sometimes humorous recollections of life in the Goods Yard. Medcalf recalled an occasion when a clerk dared to admire his surroundings through the window of his second floor office. 'Upon my word, it's quite like Venice', the unnamed man announced whilst gazing at the Regent's Canal, replete with [in Medcalf's words] 'its top dressing of London slime' (Medcalf 1900: 319). The disparity between his colleague's somewhat wistful observation and the grim reality of the canal evidently struck a chord with Medcalf, who saw fit to preserve it for posterity.

Although the working environment was invariably dominated by men, female office workers could also be found in railway goods stations. Before the First World War compelled the railway companies to employ women in grades formerly occupied by men, the number of women in the workplace was modest, although a small contingent worked at King's Cross Goods Yard as administrators, secretaries and typists as early as the late 1890s. It is also possible, given trends observed elsewhere, that others were employed as telegraph operators during the pre-war period (May 2013: 28). In the years after the completion of the Western Goods Yard in 1899, a room in that depot was set aside for the exclusive use of 'the lady clerks' (Medcalf 1900: 319). It appears that the management took a rather patriarchal view of the female members of the workforce, believing that it would have been both detrimental to productivity and morally improper to integrate the sexes. Male workers were therefore prohibited from entering the female office unless the visit was 'strictly necessary for the purposes of business' (*ibid.*). The employment of women in a wide range of roles during the First World War appears to have done little to dilute the prevailing male-dominated culture of the goods offices. Peter Erwood recalled that as late 1938, when a single female clerk was employed alongside the men of the Goods Delivery Office, she soon found herself isolated from her colleagues through the use of a fire screen (Erwood 1988: 135). It was claimed that she eventually left on account of the foul language that was liberally thrown around by the men, though her treatment appears, to modern sensibilities at least, to have been akin to a form of solitary confinement.

Fragment of a Great Northern Railway ink pot marked 'TO BE RETURNED TO G.N.R. Stationery Stores'

Lady clerks at work in the 'Abstracting Office' at King's Cross Goods Yard during LNER days (reproduced from Roberts 1938: 635 © LNER Magazine)

CHAPTER 10

The King's Cross Goods Station during the Second World War: 1939–1945

'[Air Raid Precaution] people would be designated on a rota to go up onto the roof on a vantage point just to look out to see if they could see any bombers coming ... at the beginning of the War...it was the droning of aircraft coming towards you that let you know ... As soon as the War broke out the Control Office at King's Cross was evacuated to Knebworth in Hertfordshire ... they were linked up through the Air Ministry and if there was raids imminent or something ... they used to get the information and they would ring out to all the signal boxes, all the marshalling yards with a message of London Central Colours ...if it was daytime and there was an 'air raid possible' you would get a London Central Yellow which would tell you to be on your guard .... but once you got the London Central Reds the people that weren't essential to keeping the railways running ... I mean maybe office staff and that ... they would immediately go to the air raid shelters but if you were an operator man or you was in a signal box or a driver or you was on a train or a shunter, you just had to keep working'

Excerpt from an interview with George Case, a former railway employee at King's Cross during World War II. By Alan Dein (© London Borough of Camden)

As the threat of conflict in Europe grew during the mid-1930s, the British Government began to make preparations for the protection of the national infrastructure and the civilian population against attack from the air. In response to growing public anxiety about the country's readiness for war, Parliament passed the Air Raid Precautions (ARP) Act 1937, which compelled local authorities to take measures 'for the protection of persons and property from injury or damage in the event of hostile attack from the air' (TNA HO 45/17610). Although the Act established state funding for ARP schemes, its provisions did not apply to public utility and transport undertakings, which were expected to fund such measures at their own expense.

In 1937 a 'Railway Technical Committee' formed of representatives from the 'Big Four' railway companies, the Ministry of Transport, the Home Office and the London Passenger Transport Board (LPTB) met to discuss the arrangements necessary for the protection of railway property and personnel, the provision of emergency repairs and lighting restrictions during wartime (Wragg 2012: 43). In a report issued in June 1938, the committee estimated that the cost of providing the ARP measures necessary to protect the railways and ensure the continuing movement of traffic would be in excess of £5 million, a heavy burden upon the already depleted finances of the railway companies. The Government subsequently offered to meet the major proportion of this sum in the form of a grant of £4 million, the largest single portion of which was allocated to the LPTB. The total sum allocated to the individual railway companies was approximately £3 million, of which the LNER received £764,950 (*ibid*.: 44). The remainder was pooled by the 'Big Four' in order to enable them to stockpile materials for repairs and maintenance, as well as spare parts for locomotives and rolling stock. The grant also funded the relocation of the headquarters of the companies from London to the comparative safety of the countryside. The LNER relocated its headquarters to The Hoo, a mansion near Hitchin, Hertfordshire.

The Munich Crisis of September 1938 further highlighted deficiencies in Britain's preparations for war. In response, the Government brought forward the Civil Defence Act 1939 the following spring. The Act compelled all employers to provide ARP shelters in the workplace and obliged the railway companies to make special provision to protect both 'vulnerable points' (VPs) and 'Specified Areas' on the railway network, the latter including goods stations and marshalling yards (TNA RAIL 390/1165: 28/09/1939). Contracts for the construction of air raid shelters in the King's Cross District were awarded to two companies, W. & C. French

Fig. 10.1 A naval gun is conveyed by special train from Catterick Bridge to Woolwich on 24th July 1946 (SSPL 1995-7233_LIVST_FE_157 © National Railway Museum and SSPL)

and Pitchers Ltd, in the spring and early autumn of 1939 (TNA RAIL 390/1148: Minute 4078 30/03/1939; TNA RAIL 390/1165: 28/09/1939).

At the height of the Munich Crisis in September 1938, the Ministry of Transport advised the 'Big Four' and the LPTB that the Government intended to take control of their operations in the event of war. The same month, the Railway Executive Committee (REC) was reconstituted, albeit in a different form to its First World War predecessor (Bonavia 1983: 15; Wragg 2012: 46). Whereas the earlier Executive had been under the control of the Board of Trade, the 'new' REC came under the control of the Ministry of Transport, which assumed control of the railways on 1st September 1939, two days before Britain declared war on Germany. From that date, the railway companies were obliged to hand over all revenues to the Government, in return for which they were allocated shares from a pool of around £40 million, proportionate to their average net peacetime revenues (Wragg 2012: 53).

As in the previous war, steps were taken to improve connections between lines that were considered to be of strategic importance to the war effort (Bonavia 1983: 15, 28). In the London area, this involved adding extra lines so that links between the various 'Big Four' companies could be maintained in the event of a strike upon an existing connection such as the Metropolitan or the East London line. Those affecting the LNER included the single line spur that was built at Haringey in order to link the Great Northern main line with the Tottenham and Hampstead Joint Line and a connection that was inserted at Gospel Oak so that the LNER could forward traffic to the Southern Railway via the West London or the North & South West Junction. Another section of track was laid down at Bounds Green so that traffic from the north could be diverted to the Great Eastern or the Great Central main lines via the Hertford Loop should the Great Northern mainline become incapacitated south of Langley Junction at Stevenage.

During the First World War the railways had suffered as a consequence of the loss of the huge numbers of personnel who had enlisted in the armed forces. In 1939 the railway companies could expect to lose those members of staff who belonged to the reserve forces and those who had signed up for the Civil Defence and ARP services. In accordance with the National Service (Armed Forces) Act 1939, every man between the ages of 18 and 41 (excluding apprentices) was liable to be called up. Certain grades of railway personnel were categorised as reserved occupations, exempt from call-up unless the state required them to operate trains for the armed forces.

By 1945, a total of 98,000 men and about 4,000 women had been released from the railways to serve in the armed forces (Williams 2013: 328). As in the First World War, many of the vacancies were filled by women. Female recruits were far more numerous during the Second World War than they had been during the earlier conflict, and women and girls were allowed to occupy a much wider range of positions than previously. By the end of the war the LNER had employed some 10,000 women, not only as station staff and as office workers but also in physically demanding roles that had hitherto been the preserve of men, such as signallers, guards, welders, plate layers, road vehicle delivery drivers, engine cleaners and crane operators (Williams 2013: 332, 335, 344). Despite the enlargement of the female workforce, manpower shortages continued. These compelled the Government to issue the Essential Work Order in October 1941, which prevented railway staff from changing jobs without official permission. This did not go far enough, so from May 1942 the Government agreed not to call up staff from operating grades within the LNER without the company's consent (Bonavia 1983: 27). Remarkably, when traffic reached its peak just before D-Day in 1944, fighting men were actually returned from the armed forces to the railways in order to enable them to meet the additional demands placed on the system by preparations for the invasion of Europe (*ibid.*).

In order to protect the railways from attack from the air at night, a blackout was imposed across the national network at the beginning of the war. No lights could be shown externally, locomotive fires were concealed by canvas screens, carriage windows were screened and colour signals were hooded. The imposition of the blackout created numerous safety problems for the railway companies, and it was blamed for many as 230 minor accidents within the first two months of the war (Wragg 2012: 47). Goods trains moved at night in order to keep the lines clear for faster passenger traffic during daylight. The LNER's peacetime reliance upon revenues from freight traffic meant that the company was affected

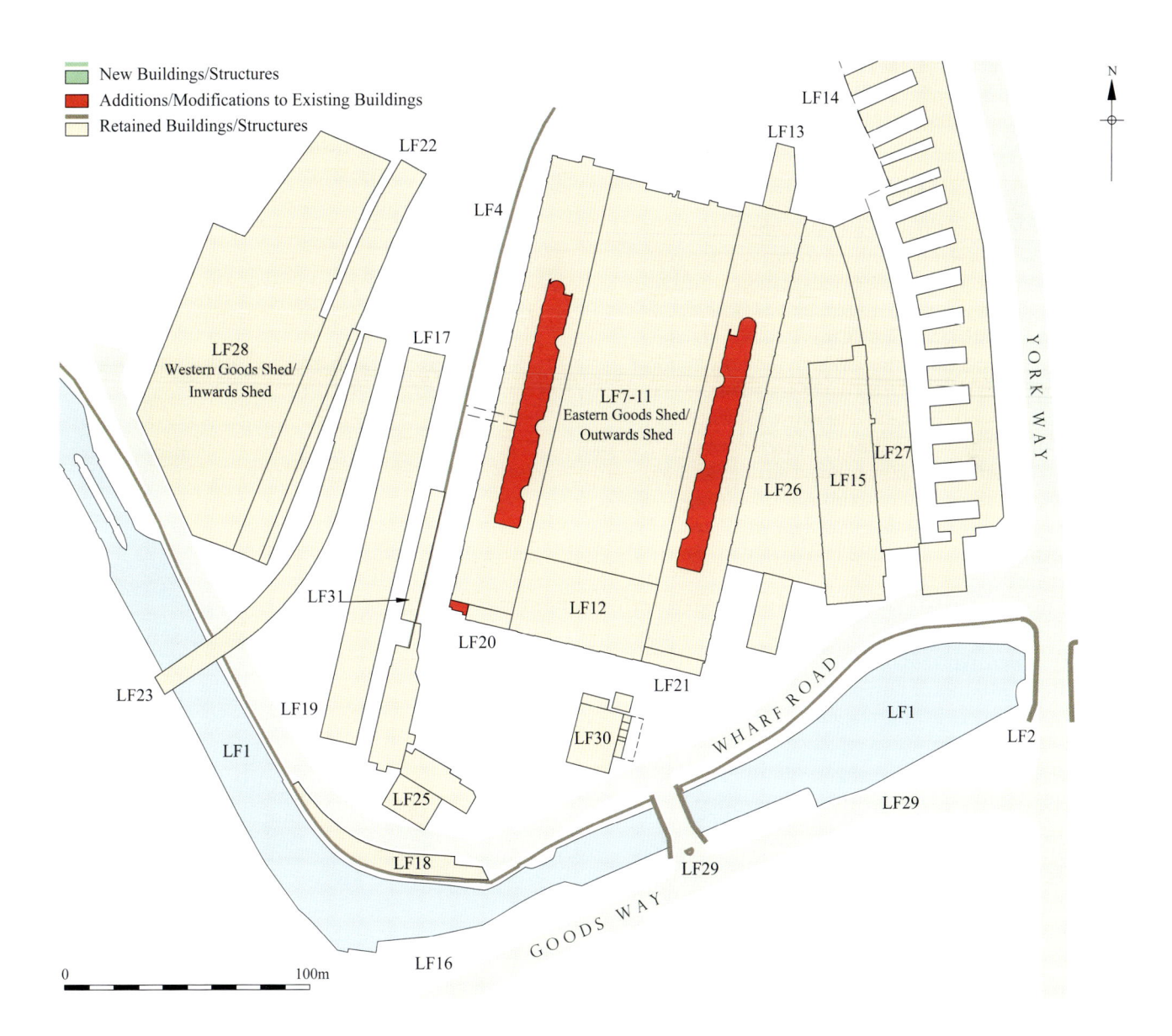

Fig. 10.2 The locations of the Landscape Features and ARP structures that were present in the Goods Yard from 1939–45, scale 1:2,500

to a greater degree than any other carrier by the increase in goods shipment that the war brought (Bonavia 1983: 15–6). This was exacerbated by the cessation of coastal shipping along the east coast, the bulk of which was transferred to the railways. The disparity between the fixed income received from the REC and the increasing volume of freight traffic meant that the company began to suffer from a shortfall of readily available funds as the war progressed. The increase in the number of freight trains took place at the immediate expense of other forms of railway traffic, bringing the 'Golden Age of Steam' to an abrupt conclusion as slower and less frequent passenger services became the norm. In addition to running a greater number of longer, heavier goods trains, a great deal of improvisation and innovation was needed in order to cope with an unprecedented growth in unscheduled freight specials. Such departures had to be arranged on a day-by-day or even an hour-by-hour basis rather than to a timetable, obliging the railway companies to factor in an extraordinary degree of flexibility into their systems.

The concentration of the LNER's main lines in the eastern half of Britain meant that a significant proportion of the company's infrastructure was vulnerable to attack from German aircraft based in France, the Low Countries and Norway. Given the strategic importance of London, the company's facilities in the capital were especially vulnerable to attack from the air. All of the company's principal London stations suffered to varying degrees, most notably the Marylebone Goods Station, which was gutted by fire on the night of April 16th 1941 (*ibid.*: 23). Less than a month later King's Cross Passenger Station was severely damaged during a raid in the early hours of 11th May, which put platforms out of action for several weeks (Brooksbank 2007: 89). The entire King's Cross District of the LNER, including Holloway, Finsbury Park and Barnet, was heavily bombed during the Blitz. The LNER managed to keep trains running throughout the period, the district's superintendent W.E. Green subsequently receiving an MBE in recognition of his department's efforts.

The majority of the LNER's pre-war Board committees were suspended at the outbreak of the war, when they were replaced by a standing Emergency Committee. Minutes of the committee's meetings are the primary source of information regarding wartime operations, expenditure and repairs in the Goods Station at King's Cross. In comparison with pre-war company records, the minutes of the Emergency are brief and somewhat terse, making it difficult to determine exactly when certain repairs and alterations took place. Post-war documentary evidence suggests that the finances necessary to implement all but the most essential repairs was not generally made available until after the war ended.

This chapter considers the impact of the war upon the yard's infrastructure, as well as exploring how effectively the LNER coped with the challenges that confronted the station during wartime.

10.1 ARP Structures in the Outwards Goods Shed: 1939–1945

Landscape Features 7–11

A plan of the Outwards Shed prepared in October 1942 represented the most detailed cartographic record of the various Air Raid Precaution (ARP) measures that were undertaken by the LNER in 1939/40 (Fig. 9.8).

The plan showed that the former stables under the platforms of the Eastern and Western Transits Sheds (since renamed 'A' and 'G' Banks respectively) were converted into shelters for staff of the Outwards Goods Shed and a surface air raid shelter was erected adjacent to the west end of the Western Goods Offices (Landscape Feature 20) for the use of employees (TNA RAIL 390/1165; NRRG 95-K-42: 19/10/1942). Few special arrangements appear to have been made concerning the conversion of the stables to shelters before 1942. It was, however, planned to improve the level of protection offered by infilling the brick arches that supported the platforms above and by providing a secondary entrance and exit from each stable shelter.

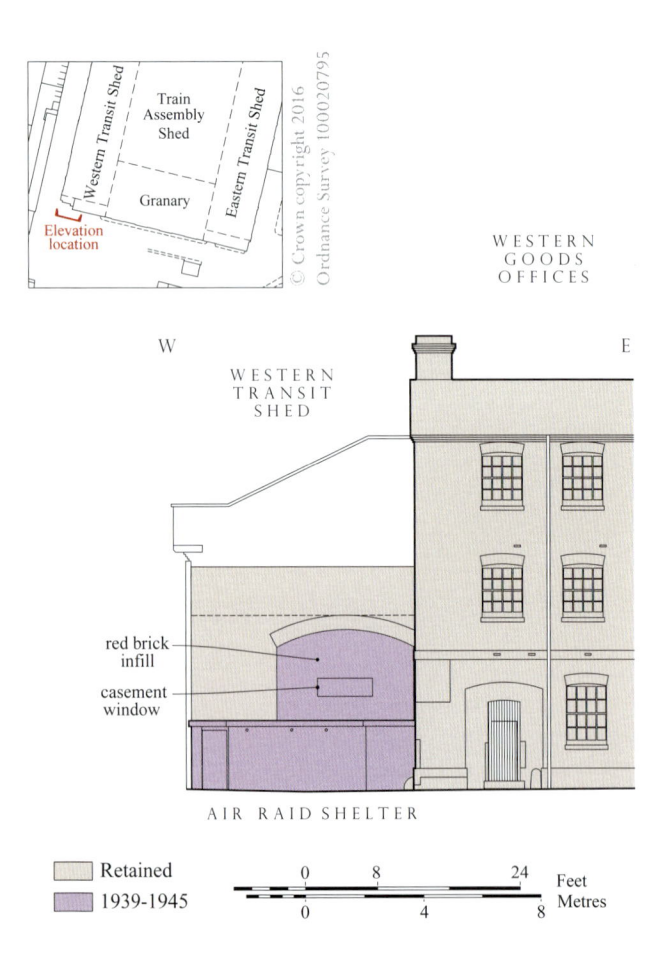

Fig. 10.3 External elevation of the south wall of the Western Transit Shed (Landscape Feature 7) showing the changes of 1939–42 that included the construction of an air raid shelter, scale 1:250

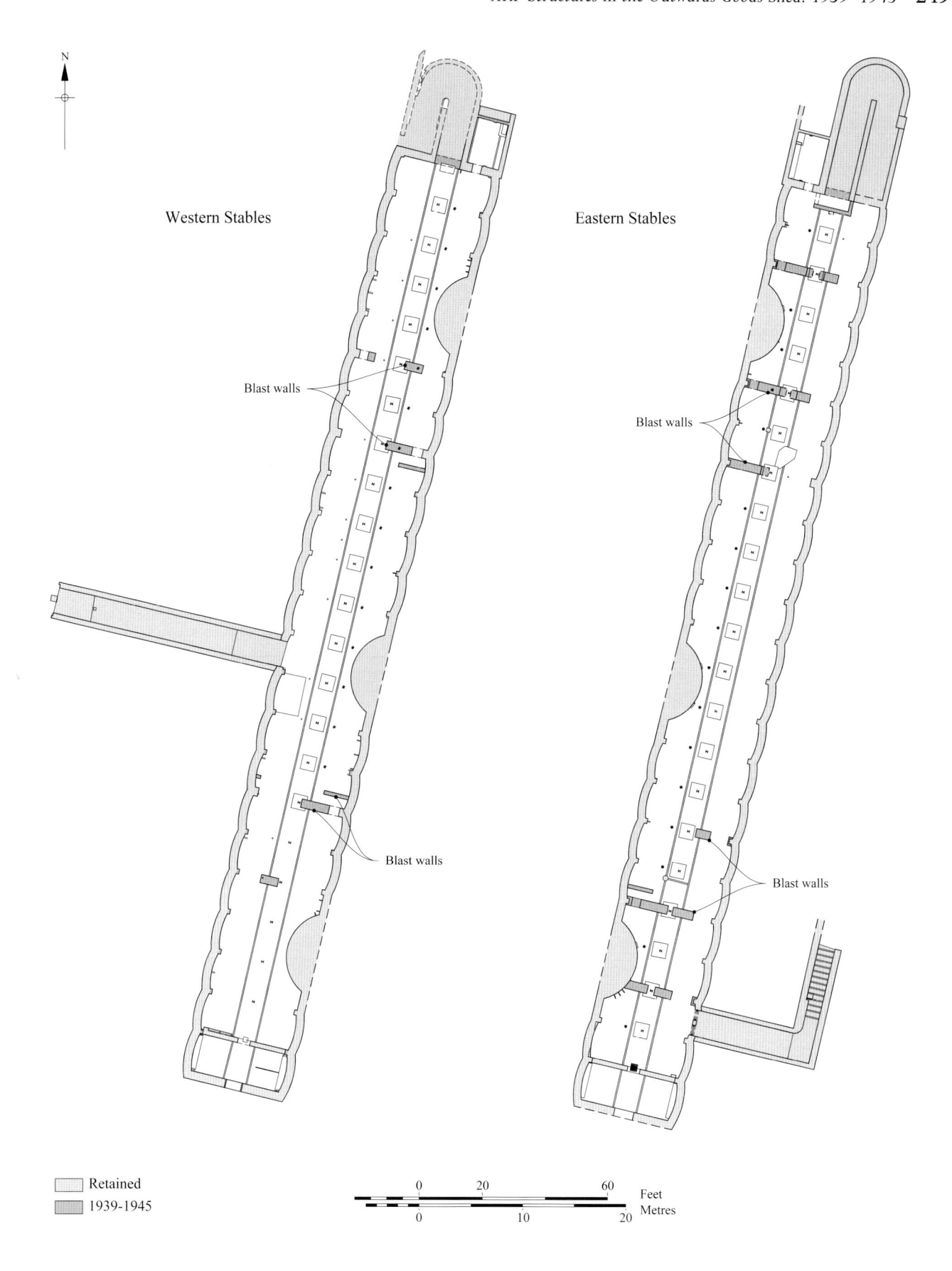

Fig. 10.4 The Western and Eastern Transit Shed Shelters (Landscape Features 9 and 10) in plan *c*.1942, reconstructed from built heritage and archaeological evidence, scale 1:500

The plan also indicated that platform shelters had been built in the Train Assembly Shed on 'F' Bank beside the east wall of the Western Transit Shed and a Fire Watcher's Shelter added at the western end of the 1936 traverser at the northern end of the Outwards Shed. The occupant of the shelter would have had a clear line of sight along the traverser into the Train Assembly Shed, ensuring that fires started by incendiary bombs that fell there could be spotted and promptly extinguished.

A plan thought to date to a couple of years after nationalisation showed a single cell rectangular northwest–southeast oriented 'shelter' on the site of the infilled Canal Basin, a short distance to the northwest of the electrical sub-station (Fig. 10.2). It was not clear whether the structure was a surface- or trench-shelter.

Built heritage evidence relating to ARP structures in the Eastern Goods Shed, 1939–1945

Landscape Features 7–11

A Second World War air raid shelter survived to the immediate south of the Western Transit Shed (Landscape Feature 7), where it abutted the south wall of that structure and the west wall of the Western Goods Offices (Landscape Feature 20). It obstructed the Western Transit Shed's southern cart road access point, which was probably bricked up when the shelter was constructed (Fig. 10.3). This was confirmed during the built heritage survey, which demonstrated that the bricks that were used in the blocking were identical to those that had been used to build the air raid shelter itself. This shelter was constructed during or prior to 1942 (Fig. 9.8).

Numerous alterations were made to the former stables below the Eastern and Western Transit Sheds (Landscape Features 9 and 10) in order to convert them to air raid shelters for shed staff (Fig. 10.4). The ramped entrance that once connected the former Western Stables (Landscape Feature 9) with the Western Transit Shed's cart way was blocked with bricks of mid-20th-century appearance. It may have been too large to offer protection from flying shrapnel and other debris, hence the decision to close it (Fig. 10.4). In the Eastern Transit Shed, the equivalent opening was also blocked with a partition wall, presumably for the same reason. That example, which was better preserved, demonstrated that a doorway had been incorporated in the blocking indicating that the ramps continued to provide access to the shelters throughout the war. Further protection from flying debris was provided inside by a number of blast walls, some of which were fully or partially removed in the wake of the conflict (Fig. 10.4).

The roofs of what had been the eastern and western 'tack rooms' (located at the northern ends of the former stable blocks) were lowered and ventilation slots were inserted in order to improve the quality of the air for those sheltering inside.

10.2 The Vulnerability of the Hydraulic Station: 1939–1945

Landscape Feature 16

The Hydraulic Station (Landscape Feature 16) was the power house that drove most of the non-electrical equipment in the Eastern (Outwards) and Western (Inwards) Goods Yards. If it was put of action by enemy action, goods handling operations at King's Cross would have been seriously impaired. The Hydraulic Station could therefore be described as the 'Achilles heel' of the yard.

The vulnerability of hydraulic power systems in railway depots to aerial attack was addressed by the Railway Executive Committee in the late autumn of 1940, when the 'Blitz' was already raging (Smith 2008: 20). It was not until February 1942 however, that the Emergency Committee of the LNER sanctioned the removal of a hydraulic pumping set from the Hydraulic Station to 'another part of the depot', where it was to be installed 'adjacent to an existing hydraulic accumulator' (TNA RAIL 390/1973: Minute 2971 19/02/1942). The contract was awarded to Pitchers Ltd in June of that year for £522.14s.4d (*ibid.*).

The latter comprised a 'back up' to the main hydraulic power system at King's Cross Goods Yard, which would have been capable of keeping the depot running on reduced power should the Hydraulic Station have suffered a direct hit.

The construction of the Pump House in the former Midland Goods Shed in 1942

Landscape Feature 15

A rectangular blast-proof enclosure was inserted in the northeast corner of the former Midland Goods Shed during this period. No internal access existed between it and the rest of the shed, although a doorway (now bricked up) did provide an internal connection with the accumulator tower to the immediate north. The windowless room could only be entered from the exterior via a pair of metal doors in the eastern elevation at ground floor level. A concrete platform was observed in the centre of the room, which probably functioned as a machine bed for an electric three-throw ram pump that was driven through reduction gear. The suction head would have been provided by a header tank that sat upon a series of rolled steel beams positioned above and regulated by the accumulator next door (Smith 2008: 3). A large slot (since infilled) on the eastern side of the base may have accommodated an engine flywheel or, more probably, a gear wheel (*ibid.*). The rectangular structure represents a blast proof pump house that contained relocated apparatus that had been moved from the Hydraulic Station during World War II.

10.3 New Messrooms for a Changing Workforce: 1939–1945

The first floor of the Western Goods Offices (Landscape Feature 20) had been converted to a mess facility for Granary staff after the end of the First World War, when the Way and Works Committee authorised expenditure on a number of new facilities of that nature (TNA RAIL 236/190: 39, 203; NRRG 66-ME-520: 02/1966). The second floor was converted for the same purpose at some point before the end of World War II (NRRG 46-LP-313: 31/12/1946).

More than 100,000 railwaymen were conscripted into the armed forces during the Second World War, creating many new employment opportunities for women on the railways. It is therefore conceivable that it was necessary to provide additional welfare facilities for the new arrivals, although official records reveal little detail of these works.

10.4 Air Raid Damage at King's Cross Goods Station: 1940–1945

The earliest recorded damage caused by enemy action at the King's Cross Goods Station occurred in the early morning of the second day of the Blitz. A bomb that fell on the Potato Market on 8th September 1940 caused fires and damage to points at No. 5 Arch (TNA AN 2/1104). During the night of 23rd October a large number of bombs fell to the east of the Goods Yard, destroying the Board School in York Way and a number of houses in the surrounding streets (TNA AN 2/1105). The blast that destroyed the school also damaged the roof and eastern boundary wall of the Potato Market, which was forced against the cast iron columns of the 1896/7 glazed canopy (TNA MT 6/2762; Figure 10.5).

At 10.30 in the morning of Saturday 9th November 1940 a number of high explosive bombs caused extensive damage to No. 1 Office (Regeneration House), the Midland Goods Shed, the Outwards Goods Shed and the RME Garage, a messroom and an electricity substation immediately to the south of the Granary (Figure 10.6). Photographs taken on the day of the raid show that the Eastern Transit Shed (Landscape Feature 8) was severely damaged, the blast blowing off part of its roof and damaging its loading platforms. The southern wall of the Eastern Goods Offices was also badly damaged, while the blast also blew out windows in the south and east walls of the Granary and Western Goods Offices, spraying the eastern wall of the Granary with shrapnel, creating a distinctive scar which has survived to the present (Figs. 10.8 and 10.9). Early reports indicated that ten people had been killed and a further 25 injured, however as the debris was cleared over the days that followed, several more bodies were discovered in the rubble (TNA AN 2/1106; TNA MT 6/2762). The number of confirmed deaths had risen to 23 by Monday 11th November, with the final toll declared as 28 dead (24 staff and four others) and 104 injured (Brooksbank 2007: 41). The garage was partially restored the following year, while the buildings were repaired and rebuilt towards the end of the war (TNA RAIL 390/1973).

Fig. 10.5 The damaged walls of the Potato Market following the air raid in Delhi Street, taken October 1940 (TNA ZR.5/79/5/3 © The National Archives)

Fig. 10.6 The Goods Yard in the immediate aftermath of the air raid that took place on 9th November 1940 (TNA ZR.5/79/5/3 © The National Archives)

Fig. 10.7 The remains of the Eastern Goods Offices (Landscape Feature 8) in the wake of the air raid of 9th November 1940 (TNA ZR.5/79/5/3 © The National Archives)

Fig. 10.8 The damage that was inflicted upon No. 1 Office on 9th November, 1940 (TNA ZR.5/79/5/3 © The National Archives)

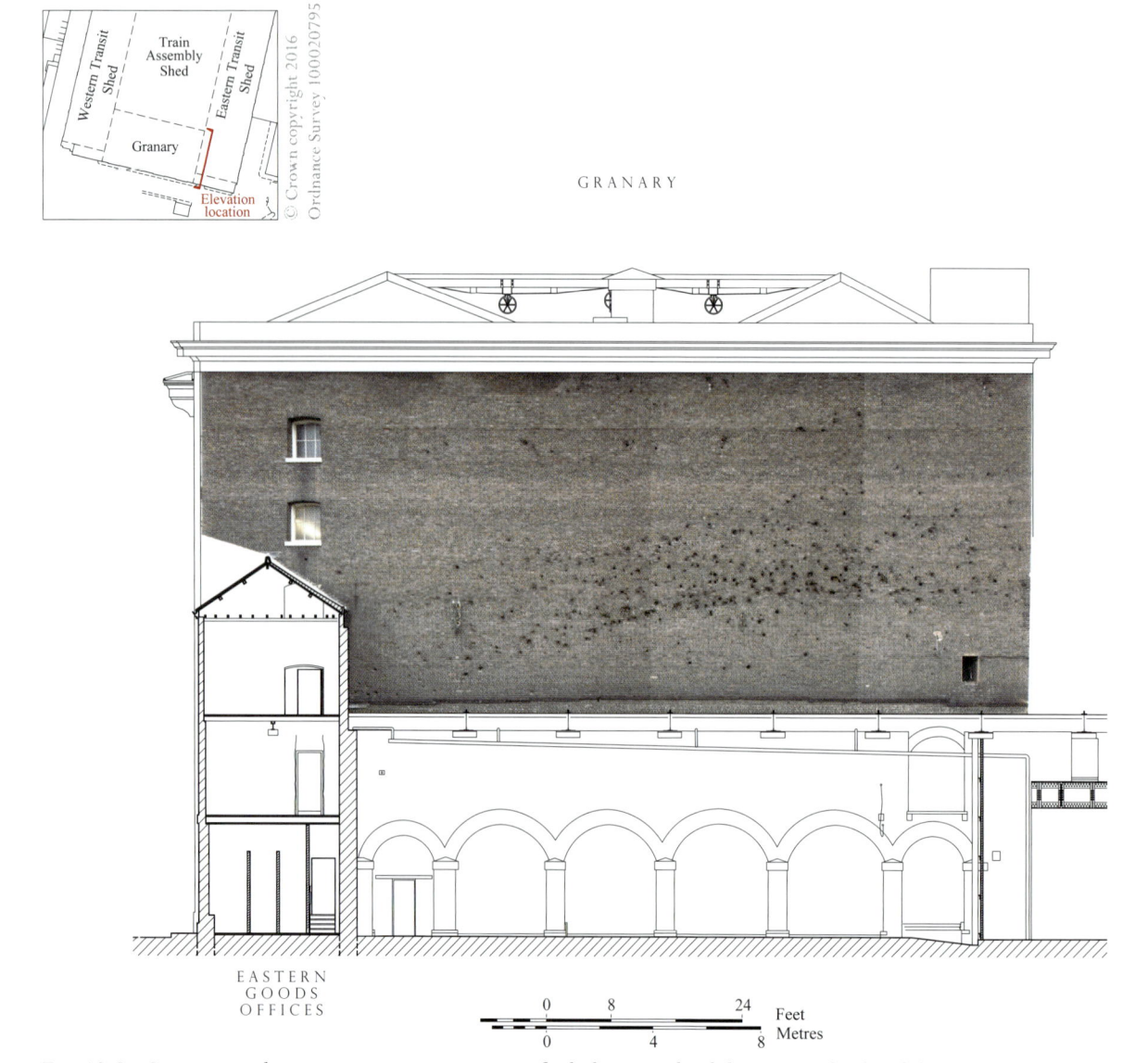

Fig. 10.9 Composite elevation incorporating a rectified photograph of the external side of the east wall of the Granary (Landscape Feature 12) showing shrapnel damage from the 1940 bomb blast, scale 1:250

The station was bombed again on 30th January 1941, when two HE bombs hit the Goods Yard. One of the bombs landed in a crater made during the raid of 9th November, near the bomb-damaged messroom, whilst the other hit the car park, damaging nine cars. Whilst the resulting damage was slight, five railwaymen were injured; of whom one later died of his wounds (TNA RAIL 390/1192). Authorisation to rebuild the much-damaged messroom was given in July 1943 (TNA RAIL 390/1974).

With Germany preoccupied by the war against the Soviet Union after June 1941, London was largely spared the attentions of the Luftwaffe over the next two and-a-half years. After the failure of the 'Little Blitz' of early 1944, it appeared that London's Anti-Aircraft (AA) defences were a greater threat to property than the Luftwaffe. In the early morning of 1st March 1944 an AA shell exploded at the Potato Market, destroying an office there (Brooksbank 2007: 107).

On 13th June 1944 the first V1 flying bomb to strike Britain crashed in a field near Dartford in Kent. The V1 offensive quickly gained momentum, and the King's Cross district was first struck by one of the new weapons on 26th June 1944. A V1 was reported to have fallen on a property in Tiber Street on the east side of York Way, causing heavy Category 'A' damage in Tiber Street and Category 'B' damage in Dennis Street, Tiber Street, Carlsbad Court and Copenhagen Court (TNA HO 198/82, 26/06/1944). The explosion caused extensive damage to windows in King's Cross Goods Yard and brought down ceilings in No. 1 Block of the General Offices, recently rebuilt after having been almost destroyed three and-a-half years earlier (TNA RAIL 390/1192).

In the early morning of 17th July 1944, a V1 fell on houses in Treaty Street on the east side of York Way (TNA HO 198/88, 17/07/1944). The resulting explosion destroyed three properties and caused widespread destruction, with blast damage extending to a radius of 200 yards (183m). The blast was reported to have caused considerable damage to the Potato Market and the General Enquiries and Outwards Offices in the Goods Yard (Brooksbank 2007: 116). The final recorded bomb damage incident in the vicinity took place nearly a month later at 07.00 in the morning of 14th August 1944, when a V1 hit the raised timber decking of the Camley Street Coal Drops (TNA HO 198/92, 14/08/1944). The blast damaged the coal drops and smashed windows in the Goods Offices, Locomotive Shed and Inwards Goods Shed in the Goods Yard (TNA RAIL 390/1192; Brooksbank 2007: 120).

Fig. 10.10 Bomb damage to the RME Garage (Landscape Feature 30) and the Granary Basin car park incurred on 9th November 1940 (TNA ZR.5/79/5/3 © The National Archives)

10.5 Repairing Bomb Damage: 1940–1945

Whilst strenuous efforts were made to clear bomb damage and carry out essential repairs in the immediate aftermath of air raids, budgetary constraints meant that there was often a considerable delay before permanent repairs could be undertaken.

Plans to repair the damage inflicted during the air raid of 9th November 1940 were drawn up in October 1942. In addition to the permanent repair of the damaged buildings, it was proposed to make a series of alterations to both Transit Sheds and the Train Assembly Shed (NRRG 95-K-42: 19/10/1942) The railway siding known as 'No. 11 Road' was to be lengthened so that it would extend the full length of 'G' Bank (i.e. Platform 1) and the existing Cart Road in the Western Transit Shed. The plan also suggested that 'G' Bank itself was to be done away with, to be replaced by a second Cart Road, and the Fire Watchers' Shelter at the west end of the traverser was to be removed. It was also proposed to build a new roof at the southern end of the Eastern Transit Shed, to replace the damaged roof covering and to construct a new loading bank in the southernmost section to replace the bomb damaged platforms. Other works proposed under the scheme included plans to take down the overhead platform at the north end of the Shed (built in 1881) and the removal of the columns that had supported it, cutting back the bank under the east side and reconstructing and relocating the Fish Office of 1886 (*ibid.*).

It was also proposed to make certain alterations in the Train Assembly Shed, where it was decided to create an additional entrance to the air raid shelter in the Western Transit Shed and to relocate the fire escape from the overhead offices. The 1942 plan also indicated that the eight sidings and five platforms were to be extended to the northern edge of the Cart Road.

It is not entirely clear to what extent the alterations that were proposed in October 1942 took place during

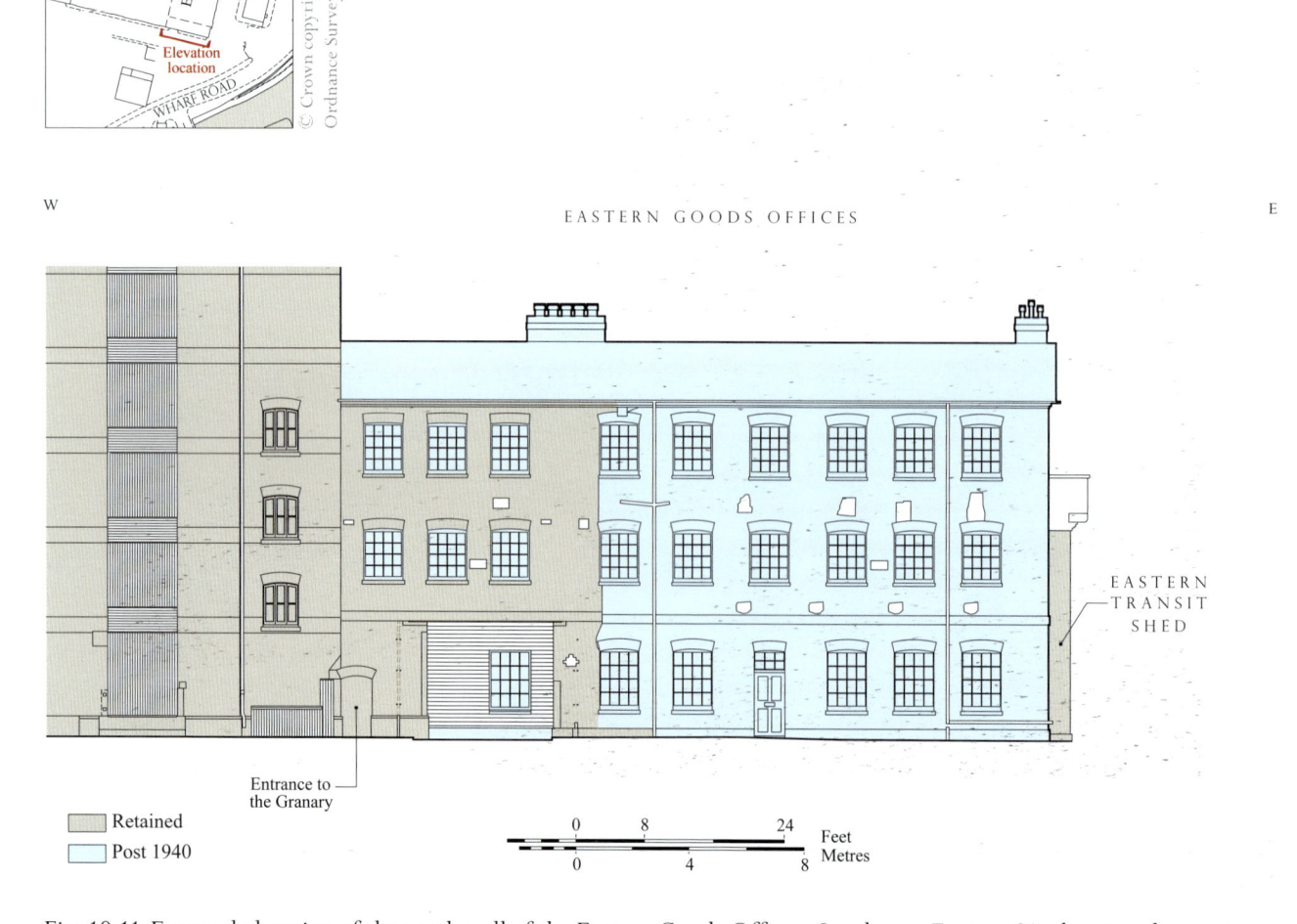

Fig. 10.11 External elevation of the south wall of the Eastern Goods Offices (Landscape Feature 21) showing the repair work that was carried out in the wake of the bomb blast of 9th November 1940 (TNA ZR.5/79/5/3 © The National Archives)

the Second World War. A poor quality reproduction of an unidentified and untitled plan of the Goods Yard, probably compiled during the early post-war years, demonstrates that a number of the proposals, such as the truncation of the railway lines at the southern end of the Eastern Transit Shed, did not take place during wartime (NRRG 50-LKC-105: 1950). Documentary evidence suggests that the finance necessary to undertake anything more than the most urgently required repairs did not become generally available until after the war ended, so it seems likely that the majority of the repair works were actually carried out in peacetime. This was certainly the case for the roof replacements schemes, which were not implemented until the 1950s. Unless a more specific date is provided by LNER records, sections of structures that were rebuilt because of bomb damage are discussed in this chapter for clarity.

Built heritage evidence of repairs to the Eastern Goods Offices and the Eastern Transit Shed after the air raid of 9th November 1940

Landscape Features 21 & 8

The entire east and central sections of the southern elevation of the Eastern Goods Offices (Landscape Feature 21), the southern side of its eastern wall and the entire roof of the building had been rebuilt with 20th-century materials (Fig. 10.11; Fig. 10.12). These presumably represent repairs to damage caused by the air raid of 9th November 1940. All the window frames and panes in the southern wall had been replaced with mid-20th-century examples, suggesting that all or most of the originals were blown out by the blast (Fig. 10.11).

It is highly likely that the remaining vehicular access point in the southern façade of the Eastern Goods Offices (Landscape Feature 21) and the corresponding access point in the south wall of the Eastern Transit Shed (Landscape Feature 8) were filled in during this phase of repair work (Fig. 10.11; Fig. 10.12). Indeed, they disappear from all cartographic depictions of the complex post-1942 when road communication with the southern end of the Eastern Transit Shed was severed for the first time. This dating was confirmed by the inclusion of a mid-20th-century window in the cladding that had been used to block the entrance in the south wall of the Eastern Goods Offices. The window in question was identical to the others that were inserted to replace those damaged during the air raid of November 1940 (Fig. 10.11).

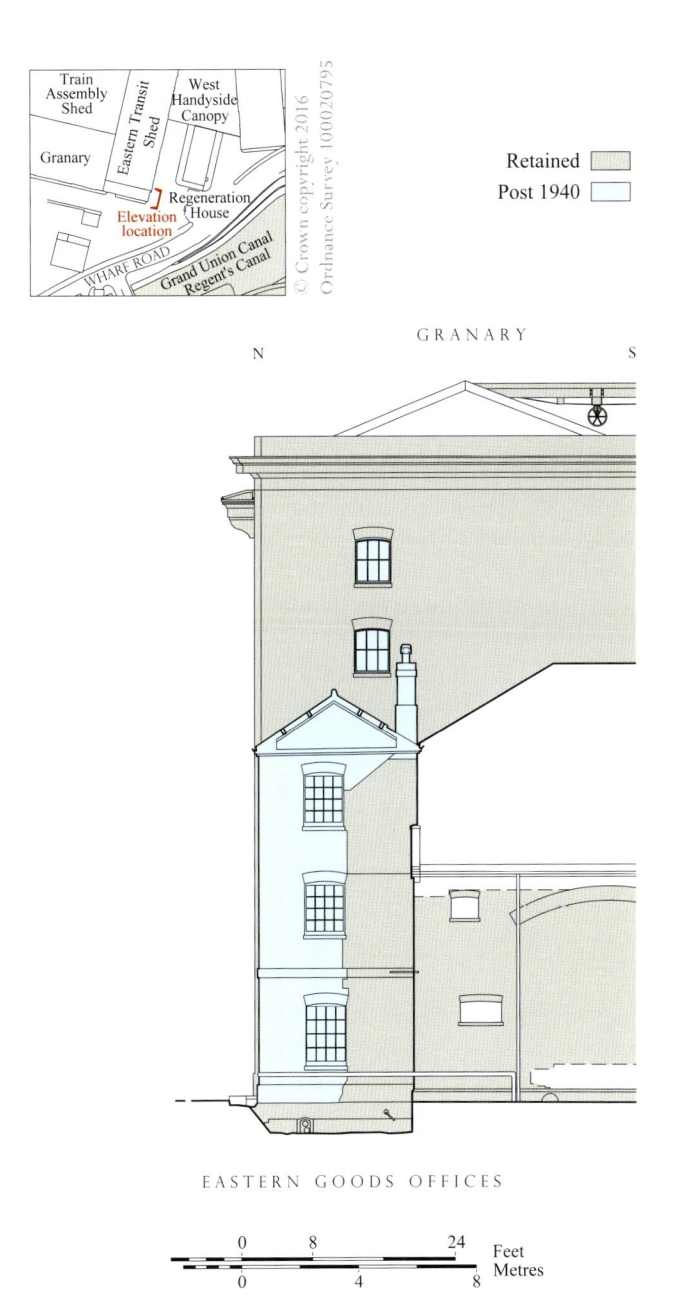

Fig. 10.12 External elevation of the east wall of the Eastern Goods Offices (Landscape Feature 21) showing the repair work that was carried out in the wake of the bomb blast of November 1940 (TNA ZR.5/79/5/3 © The National Archives)

10.6 Synopsis

The Second World War had a huge impact on operations at King's Cross Goods Station. That the station managed to function throughout the conflict without significant interruption to services is a testament to the success of the ARP measures undertaken in the months before the war broke out, the leadership of the LNER and the Railway Executive and above all the fortitude and resilience of the workforce.

The ARP measures undertaken by the LNER went beyond the construction of air raid and fire watchers' shelters, encompassing new working practices and traffic arrangements, as well as the stockpiling and distribution of materials necessary for spares and repairs. The records of the LNER Emergency Committee and the Railway Executive suggest that the command structures established in 1938–9 worked effectively in wartime, even at the height of the Blitz in 1940–1.

Both the Railway Executive and the LNER were obliged to exercise considerable financial restraint for the duration of the conflict. With the exception of essential expenditure on repairs and the necessary modifications that were needed to comply with wartime operating conditions, very little money was spent on new or existing infrastructure at King's Cross, despite the increase in freight shipment that the conflict brought about. Although the LNER's reliance upon freight revenues meant that it ought to have benefitted from this, the disparity between the fixed income it received from the Railway Executive and the increasing value of this traffic meant that it suffered more than fellow members of the Big Four from the financial consequences of state control. The implications for the company's infrastructure were predictable, and by the end of the war yards like King's Cross were suffering from years of underinvestment.

The restrictions imposed by the blackout represented one of the greatest challenges to operations at King's Cross during wartime. Work in the marshalling and goods yards was particularly difficult, not least owing to the increased volume of goods traffic at night as a result of the need to keep lines open to faster passenger traffic during the day. This did not mean that the yard was idle by day, as the company's daily reports to the Railway Executive make clear. The yard also had to cope with a greater number of longer, heavier goods trains, as well as accommodating unscheduled freight specials in a cramped and increasingly crowded yard.

The King's Cross Goods Yard was bombed on at least ten separate occasions during the Second World War, with further bombs landing nearby in the Locomotive Depot, the Camley Street coal drops and the streets east of York Way. The most serious incident occurred on the morning of Saturday 9th November 1940, when several of Lewis Cubitt's buildings were damaged by high explosive bombs. Despite the damage, disruption and comparatively heavy loss of life, the staff of the Outwards Shed returned to work in the immediate aftermath of the disaster. Their response provides the clearest example of how the yard's employees soldiered on in the face of aerial bombardment.

That the yard managed to keep functioning throughout the war is also a testament to the improvements that were made in the inter-war period, most notably the Transposition Programme of 1935–8. The transformation to the station's operating practices wrought by Transposition were sufficient to enable it to handle the additional freight that wartime brought. Although the financial contribution of the Goods Station to the effort is unquantifiable, there is no doubt that it was an essential cog in Britain's war machine.

CHAPTER 11

Nationalisation and Austerity: 1946 to the mid-1950s

'There are basic industries ripe and over-ripe for public ownership and management in the direct service of the nation. In the light of these considerations, the Labour Party submits to the nation the following industrial programme: Public ownership of inland transport. Co-ordination of transport services by rail, road, air and canal cannot be achieved without unification'

An extract from the Labour Party's 1945 General Election manifesto entitled 'Let us Face the Future', which led to the formation of British Railways

The General Election of July 1945 returned Britain's first majority Labour Government. Having made the nationalisation of the railways a campaign commitment, Clement Attlee's election as Prime Minister brought with it the realisation that the Big Four railway companies' days were numbered (Bonavia 1983: 43).

The resumption of peace was accompanied by post-war reconstruction and rebuilding schemes, which ensured that demand for goods and raw materials remained high. Continued petrol rationing meant that long distance road transport of manufactured goods remained unaffordable to many producers, whilst the roads themselves were suffering the consequences of long-term underinvestment during the interwar period and the Second World War. These factors perpetuated, in the short term at least, the boom in rail freight that began during the war years. In order to stimulate business more generally and aid the rebuilding process, the Government continued to fix prices across the railway network, which had the unintended consequence that the LNER was unable to benefit from the profits that the continued high levels of freight shipment might otherwise have brought. Despite the bleak outlook, the railway companies continued to plan for the future, in anticipation of the lifting of Government restrictions and the distribution of money from the Arrears of Maintenance Trust Fund (*ibid.*: 44, 47). Unfortunately for the railway companies, the country's post-war coffers were virtually empty, meaning that the State had little choice but to ration the limited sums available for even the most essential railway repairs.

Britain's transport infrastructure, including the railways, the waterways, the major ports, road haulage firms and bus companies was nationalised by the Transport Act 1947. Overall responsibility for the management of these disparate enterprises was vested in the British Transport Commission (BTC), which oversaw and coordinated the respective transport divisions, including the Railway Executive, the Docks and Inland Waterways Executive, the Road Haulage and Road Passenger Executives and so on. At midnight on 31st December 1947 the LNER ceased to exist, when it was absorbed into the newly founded British Railways, the trading arm of the Railway Executive. The old company's lines were divided between the Eastern, North Eastern and Scottish Regions of the new organisation, the King's Cross District becoming part of the Eastern Region.

Nationalisation coincided with the preparation of a comprehensive Government-sponsored road building campaign, which would ultimately make long distance road transport an attractive option to manufacturers. This took place at a time when Britain's industrial output had fallen owing to the return to a peacetime economy. International economic factors also meant that the country was no longer a preeminent exporter, a problem that had first begun to bite during the early 20th century, but which had been put on hold by the World Wars (Jackson 2013: 26).

Conditions at King's Cross Goods Yard were somewhat austere during the twilight years of the LNER and the dawn of British Railways. Continuing shortages of manpower, materials and money meant that normal working did not resume until some months after the war had ended, while authorisation for the removal of blackout conditions and the restoration of lighting at the Outwards shed was not granted until September 1946 (TNA RAIL 390/74: Minute 4550a 26/09/1946). Shortly afterwards the LNER set about demolishing or selling off redundant or surplus ARP assets. Whilst the majority of the temporary structures were removed, some wartime relics were put to alternate use, ensuring their survival into the early 21st century as a result. At least one brick-built air raid shelter was retained after the end of the war for the use of the crane foreman (TNA RAIL 390/1148: 'LNER Disposal of ARP Assets to 31/05/1946'). Claims for reimbursement of the costs of repairing war damages were generally made under the War Damage Acts 1941–3, although authorisation for substantial capital expenditure on permanent repairs or reconstructions could be slow to materialise. Consequently much of the work that was required to rectify the damage that had been incurred during the conflict was undertaken gradually and in a piecemeal fashion.

Shortly before nationalisation, the LNER authorised the expenditure of £47,600 for the renewal of the roofing of the Outwards Shed (see 11.1 below). The depletion of the company's finances during the war years ensured that there was no way that the LNER could afford to undertake such works on its own account, and it was only after the railways had been nationalised that it was practicable to implement them. Even then it was necessary for the Railway Executive to demonstrate that the programme fulfilled the BTC's criterion that it could be shown 'to improve the net revenue position, or of an essential renewal or maintenance character' (TNA RAIL 8/85: 67).

Although the quality and quantity of staff welfare facilities at King's Cross, such as canteens and wash rooms, had improved during the interwar period, the recruitment of female labour during the war years revealed a glaring shortage of purpose-built facilities. Matters were not helped by the loss of messroom facilities to enemy bombing during the war. Whilst many members of the temporary female workforce were replaced by men in the aftermath of the conflict, large numbers of female office staff retained their positions in the post-war period. Some of these women would have found themselves obliged to use the overcrowded female facilities in the Western Goods Yard, or even forced to share welfare facilities with the men. Consequently, extensive efforts were made during the years of post-war austerity to improve the provision of welfare facilities for the yard's female workforce.

Fig. 11.1 The newly introduced 'door to door' express services for modern containerised traffic, one of the few improvements made to freight transportation services in the 1950s (Gourvish and Blake 2011: 292). One of these was the 'Tees-Tyne Freighter' which provided a direct connection between Darlington and King's Cross. Here it can be seen departing from King's Cross pulled by an A1 Pacific Class locomotive, No. 60140 Balmoral, on 1st July 1960 (SSPL 1995-7233_LIVST_TF_245 © National Railway Museum and SSPL)

11.1 The Eastern Goods Shed and the Granary: 1946 to the 1950s

Landscape Features 7–12

One of the final acts of the LNER Board before nationalisation was to authorise expenditure totalling £47,600 for the renewal of the roofing over 'A' and 'G' Banks of the Outwards (i.e. Eastern) Shed (TNA RAIL 290/74: Minute 4830; TNA AN 13/198: Railway Executive Memorandum to British Transport Commission March 1948). Although a set of drawings showing the proposed reconstruction of the Eastern and Western Transit Shed roofs was prepared towards the end of 1946 when the yard was still owned by the LNER, the works themselves did not take place until after nationalisation in January 1948 (NRRG 46-LP-313: 20/12/1946, 31/12/1946; Fig. 11.3; Fig. 11.4). Owing to their advanced age and the damage sustained during the war it was decided to replace the existing timber trusses and slates with steel framed trusses covered with reinforced asbestos cement sheeting (TNA RAIL 390/74: Minute 4830; TNA AN 13/198: Railway Executive Memorandum to British Transport Commission March 1948). The Board's decision was approved by its successor, the British Transport Commission (BTC) at the beginning of April 1948 and the contract to complete the works was awarded to George Simpson

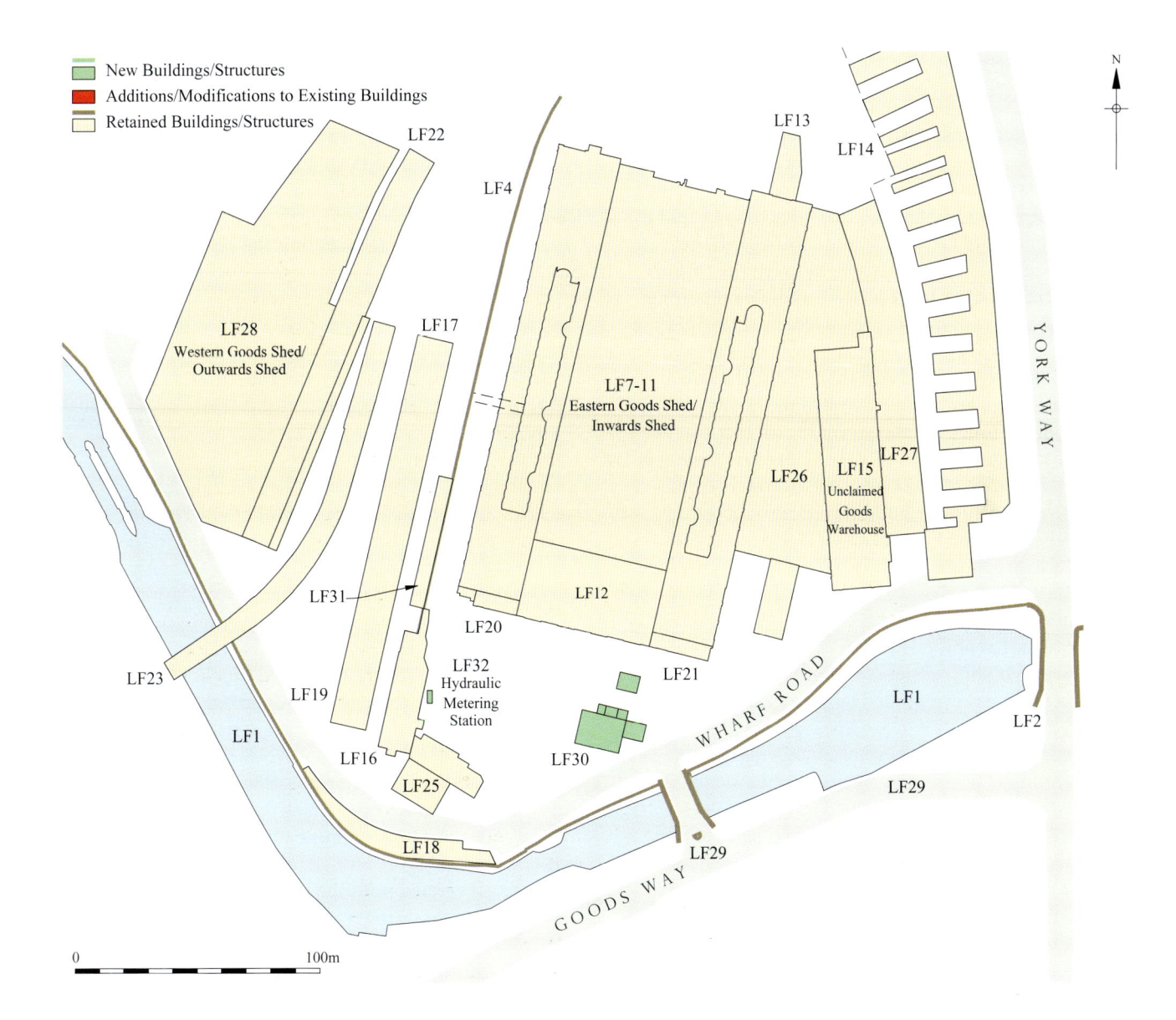

Fig. 11.2 The locations of the landscape features present in the Goods Yard from 1946 to the mid 1950s, scale 1:2,500

& Co. Ltd for £41,322 (*ibid.*: 02/04/1948 BTC minute). The specification called for the replacement of the timber trusses of Cubitt's original roof by steel trusses fabricated from rolled steel angles joined by 3/8" thick gusset plates, while the mixed slate and glazed roof covering was to be replaced with a combination of 'Trofsec' corrugated asbestos cement sheets and patent glazing (Fig. 11.3). Engineering drawings indicated that the works also necessitated a number of minor alterations to the south gable wall of the Western Transit Shed. In order to accommodate the hipped south end of the new roof, the brick gable wall of the Western Transit Shed was to be partially rebuilt (to a slightly different pattern to that of the Eastern Goods Shed) and surmounted with a new concrete coping. A concrete corbel was to be inserted beneath. In order to support the new roof three new brick piers topped by padstones were instated on the north side of the gable wall to support the two hip and one central channel hip rafters of the replacement roof (Fig. 11.3). The Ladies' Lavatory was to be retained, while the supporting girder was to be partially supported by the new roof trusses (Fig. 11.4). A contract for the replacement of the glass and what was described as 'maintenance work' in the roofs of the Outwards Shed was awarded to Pitchers Ltd early in January 1949 (*ibid.*: Minute 4616f 09/01/1947). The value of the contract was split between the cost of the replacement glass itself (£1,200) and the remaining £2,300, which presumably covered labour costs (*ibid.*).

In October 1952 plans were approved for the installation of a new goods lift shaft beside the north wall in the third bay from the west of the Granary, presumably in order to supplement the lift that had been installed during the 1920s (Fig. 11.5). Because the upper levels of the Granary had been used for warehousing since the mid-1920s, the building was also subject to the provisions of the Factories Acts of 1937 and 1948, both of which specified minimum standards for lifting machinery (TNA AN 8/136: 5f, 16a-h). The brickwork of the new installation was 13½" thick, considerably thicker than that of the 1920s lift shaft. The base was composed of reinforced concrete floor slabbing, whilst the structure was surmounted by a corrugated steel roof (NRRG 52-AR-MN-111: 1952). The rigorous building standards required by the Factories Acts may also have been behind the decision to reconstruct the 1920s 30cwt lift, drawings of which were issued in 1955 (NRRG 55-ME-BM-22: 1955). These plans were similar to those produced in 1952 and specified the widening of the brickwork of the shaft and the reconstruction of the rooftop machine room in the style of the earlier lift.

By the summer of 1953 an area of the roof of the Train Assembly Shed was in such a poor condition that it was considered in need of immediate attention in order to prevent it from collapsing altogether. A temporary solution was found by removing the slate roof covering, whilst options for the permanent repair or replacement of the structure were considered by the Railway Executive (TNA AN 13/198: Railway Executive Memorandum to BTC June 1953). It was estimated that if it was decided to retain the existing column structure the cost of the roof repairs would come to £75,000, whereas complete replacement with a modern single span design might amount to twice that figure (*ibid.*). Mindful of the fact that both proposals represented a substantial spending commitment during a period of austerity, the Executive instead proposed an alternative scheme whereby the Train Assembly Shed would be reconfigured so as to receive inward-bound loads, which would be transferred directly from the wagons to waiting road vehicles. This in turn would permit the demolition of the defective portion of the roof without the need for a replacement (*ibid.*).

The scheme proposed by the Railway Executive comprised three main elements:

1. The complete demolition of the defective portion of the Outward Shed [i.e. Train Assembly Shed] roof: £7,500
2. 'Necessary alterations' to the Outward Shed following demolition of the roof for a total cost £50,000, comprising:
   a. The removal of certain internal Banks
   b. The provision of flush paved roadways inside the shed to provide accommodation with road access for full load traffic
   c. Track alterations, including the provision of three new tracks
   d. The installation of a wagon traverser and associated capstans
   e. Lighting
3. 'Necessary alterations' to the Inward Shed [i.e. Western Goods Shed] for a total cost of £11,500, comprising:
   a. The extension of two lines
   b. The provision of two new connections to facilitate transfers of wagons and the removal of fixed cranes

The scheme necessitated the transposition of the function of the Inwards and Outwards Sheds, only fifteen years after this had last taken place. Despite the extremely ambitious nature of the proposal it was approved by a committee of the BTC at the end of June, on the basis that the estimated cost of £69,000 offered 'a good return on outlay' when considered against the estimated cost of the complete renewal of the roof (*ibid.*: 30/06/1953; TNA AN 8/85: 72, 201). It is evident, however, that the scheme did not proceed as was originally planned and a new double pitched steel framed roof comprised of four bays of rolled I-section east–west principal rafters tied together by circular tie rods was not erected until the second half of the 1950s.

Whilst proposals to renew the roofs of the Transit and Train Assembly Sheds had been authorised in 1947 and 1953, the programme was not extended to the Handyside canopies until 1956, when contract drawings for the partial rebuilding of the East Handyside Canopy (Landscape Feature 27) were issued (NRRG 138796: LNE. 56-ME-BM-12 1956; not illustrated). Although no comparable drawings relating to the reconstruction of the West Handyside Canopy (Landscape Feature 26)

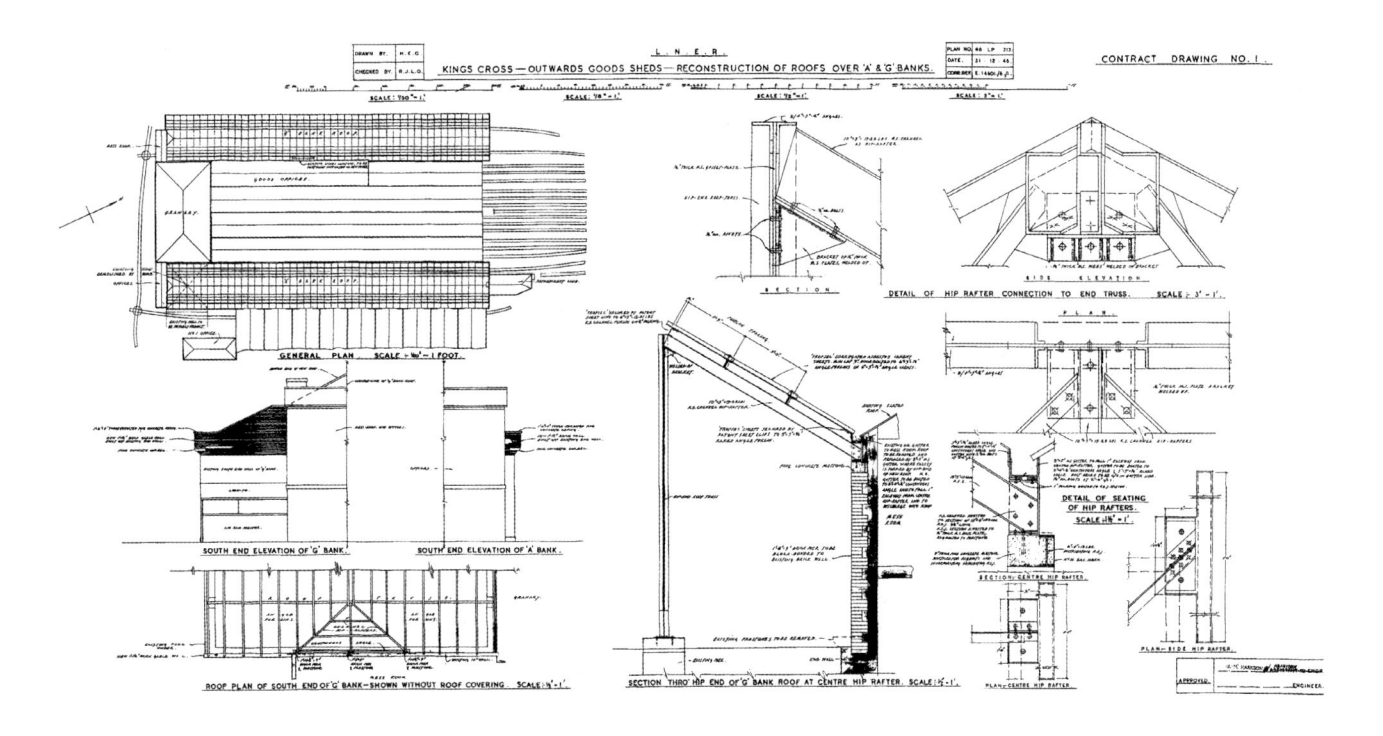

Fig. 11.3 A collection of LNER plans and sections from 1946 entitled 'proposed reconstruction of the Outwards Goods Shed Roofs (A and G Banks)', various scales (NRRG DMFP 00026357 LNER 46-LP-313)

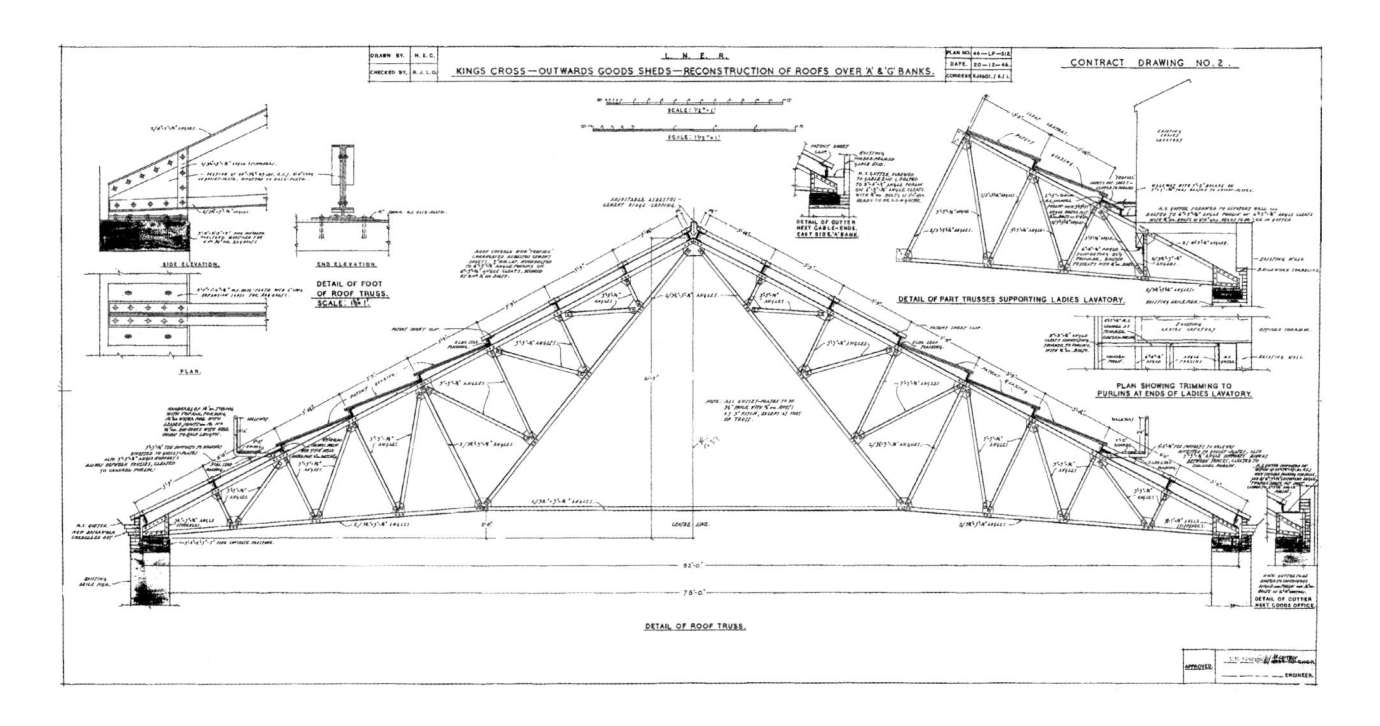

Fig. 11.4 Detail of a roof truss taken from a collection of LNER plans and sections from 1946 entitled 'proposed reconstruction of the Outwards Goods Shed Roofs (A and G Banks)', various scales (NRRG DMFP 00026357 LNER 46-LP-313)

262 *Nationalisation and Austerity: 1946 to the mid-1950s*

have been found during the course of this research, it is likely that it was replaced at approximately the same time given the fact that certain structural elements are shared by the two roofs.

Drawings of the East Handyside Canopy indicate that the riveted lattice girders that supported the east side of the roof to the north of the Goods Shed were to be retained, as were the supporting cast iron columns of the 1850 temporary passenger terminus. The three riveted compound I-section stanchions that supported the girders were to be encased in concrete up to 5'4" from ground level, presumably in order to protect them from vehicle impact. The 1888 wrought iron trusses of both canopies were also to be retained, whilst the roof cladding of the East Handyside was to be entirely replaced with 'Trofsec' corrugated asbestos cement sheets and 'Mellowes' wired patent glazing over a 6" x 8" timber ridge and 5" x 2½" timber purlins. The same materials appear to have been used for the re-cladding of the West Handyside Canopy.

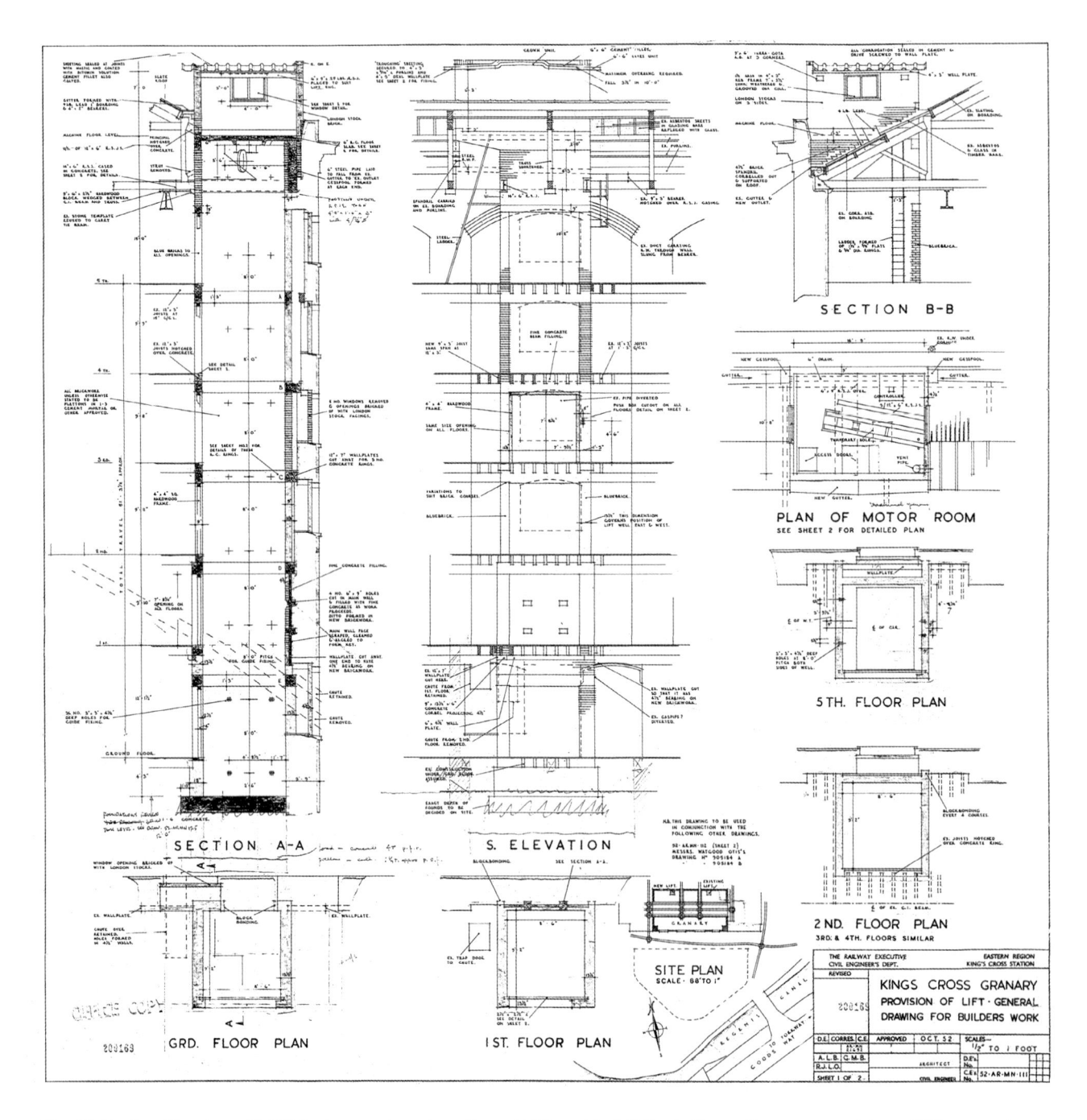

Fig. 11.5 A collection of LNER plans and sections from 1952 entitled 'King's Cross Granary provision of lift general drawing for builders work', not to scale (NRRG 52-AR-MN-111)

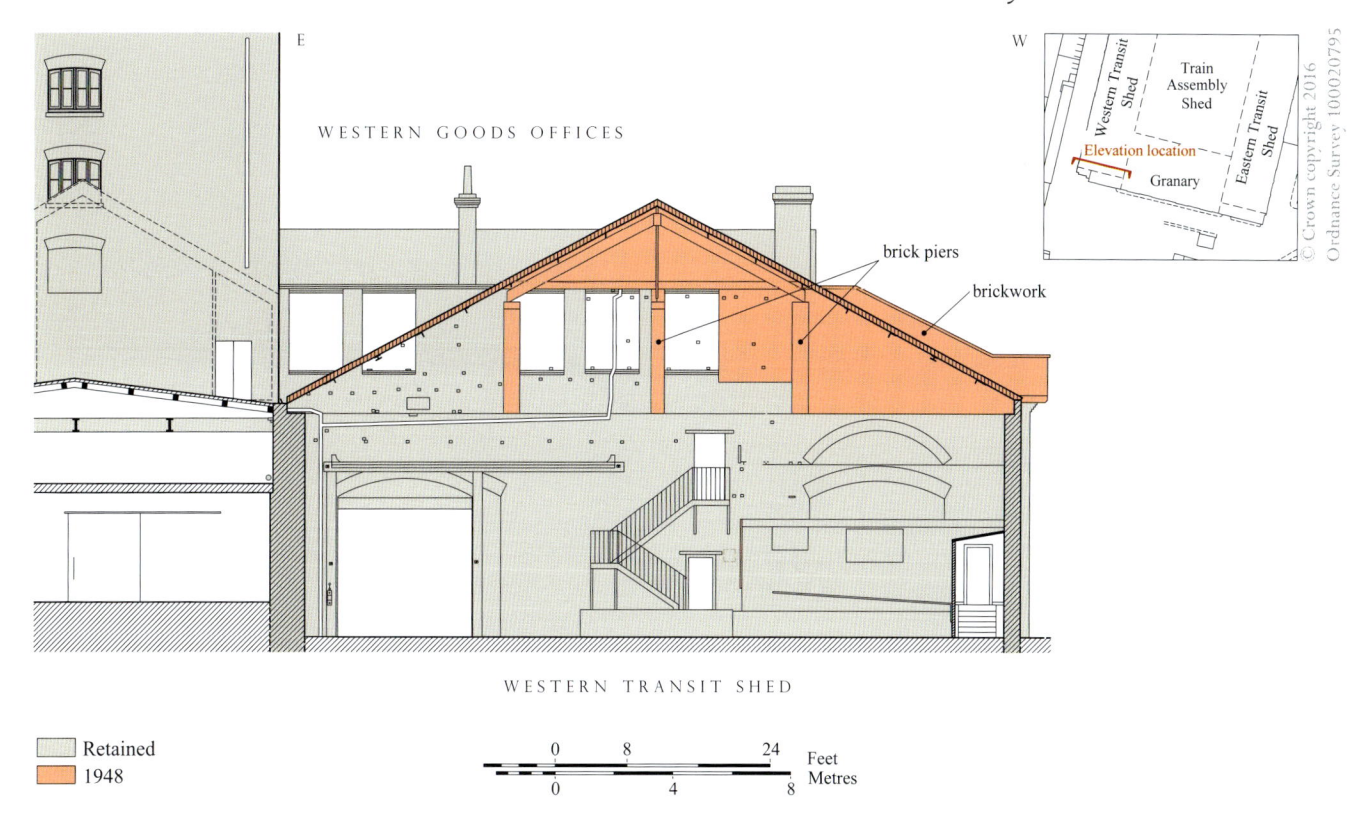

Fig. 11.6 Cross-sectional elevation through the Western Transit Shed (Landscape Feature 7) showing the internal face of the south wall and the changes of 1948, scale 1:250

Built heritage and archaeological evidence relating to the renewal of the roofs of the Eastern Goods Station and the Granary; 1948–1953

Landscape Features 7–12

When the roof of the Western Transit Shed (Landscape Feature 7) was replaced in 1948, the southern end was altered from fully hipped to half hipped through the inclusion of part of the north wall of the upper floors of the Western Goods Offices (Landscape Feature 20; Fig. 11.6). In order to achieve this, three red brick piers were inserted on top of the original south wall of the Western Transit Shed against the face of the Western Goods Offices to support three new, steel hip rafters (Fig. 11.6). In addition, a tall parapet wall was built at the west end of the south wall of the Western Transit Shed above eaves height. It was integral to an identically coloured rebuild that ran across the top of the southern external elevation, which was capped by a concrete corbel and coping slabs (Fig. 11.7). In the western elevation, the cornice of the building clearly sat upon a few courses of modern brickwork, suggesting that it too was added or reinstated in 1948. The Eastern Transit Shed roof was modified at the same time in much the same way. Both were reclad with corrugated asbestos sheeting, whilst the former glazed sections were covered with transparent panels and asphalt.

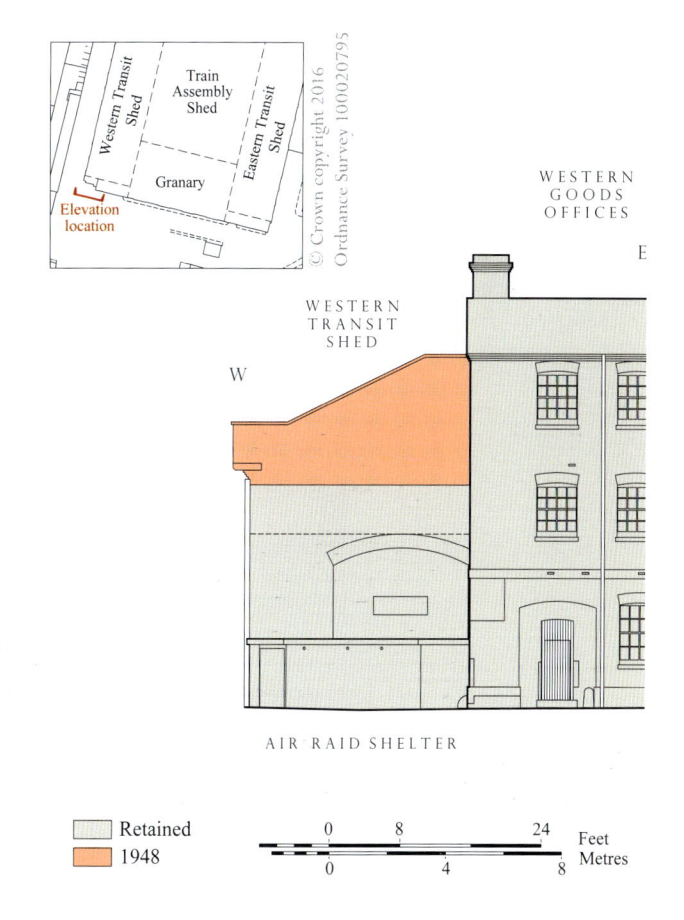

Fig. 11.7 External elevation of the south wall of the Western Transit Shed (Landscape Feature 7) showing the changes of 1948, scale 1:250

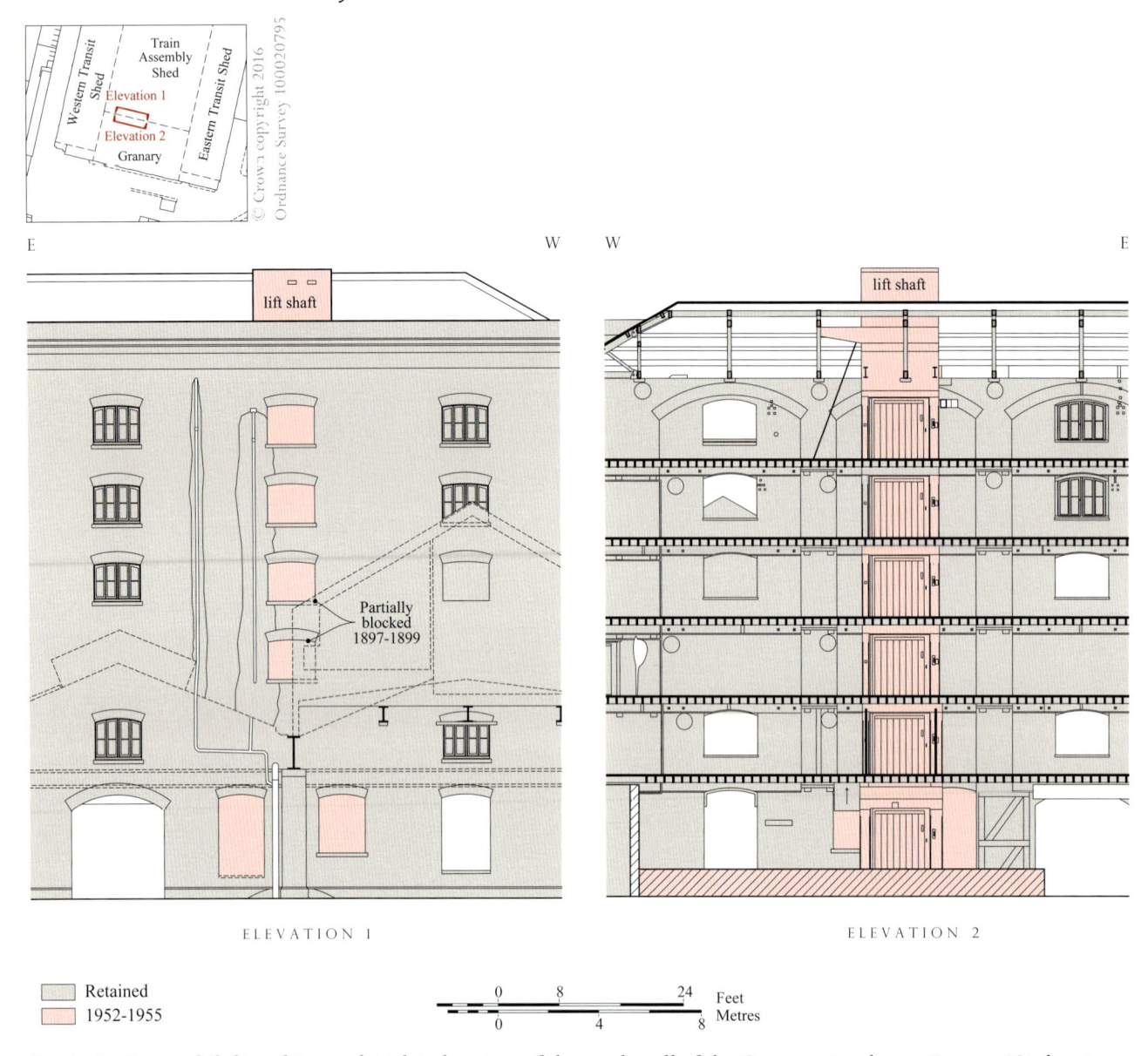

264 *Nationalisation and Austerity: 1946 to the mid-1950s*

Fig. 11.8 External (left) and internal (right) elevations of the north wall of the Granary (Landscape Feature 12) showing the changes of 1952, scale 1:200

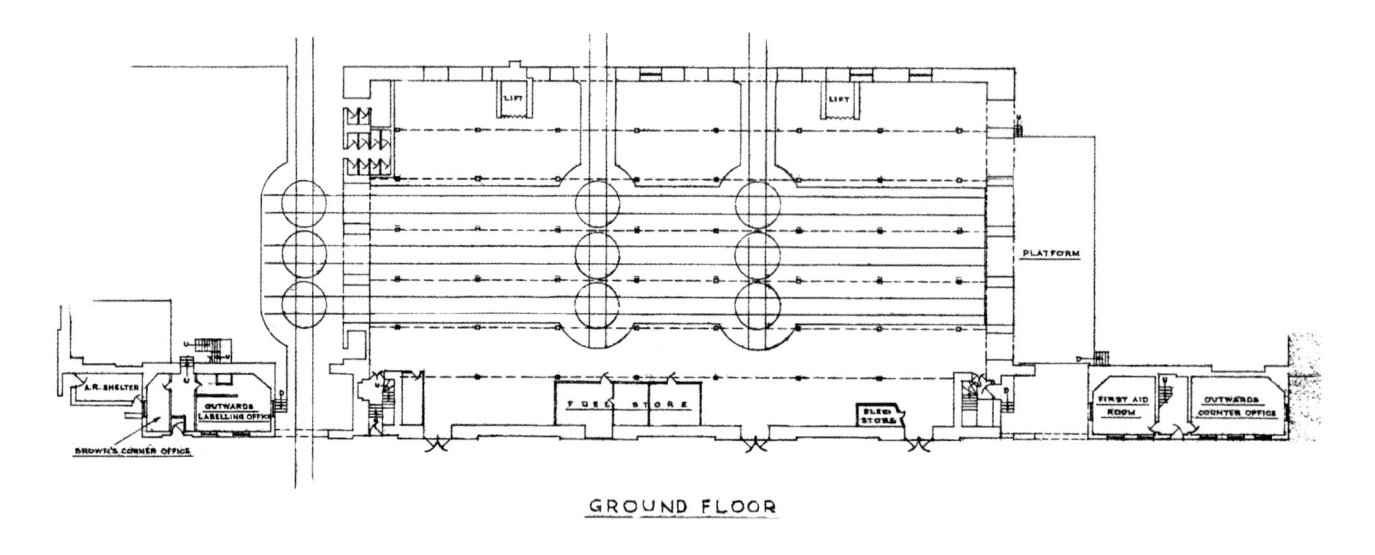

Fig. 11.9 LNER plan of *c.*1955 detailing the ground floor of the Granary (Landscape Feature 12) and the lift shafts that were inserted in 1952 and 1955, not to scale (NRRG 52-AR-MN-111)

The re-roofing programme that commenced shortly after the end of the Second World War spanned several years, presumably due to the financial constraints that were imposed upon British Railways by the BTC during a period of austerity. Consequently work on the canopies of the Train Assembly Shed (Landscape Feature 11) did not commence until the latter half of the 1950s. It was at this time that the original 45' span, composite timber roof trusses were replaced with light steel versions. Most of the shed's columns were also removed and steel stanchions were installed in their stead. The only columns that survived the process were those that supported the floor of the shed's Overhead Offices. Like the roofs of the Transit Sheds, the new roof cladding consisted of corrugated asbestos sheeting and strips of glazing, whilst the valley gutters were also made of asbestos. No clear-cut archaeological or historical evidence revealing the precise nature of the alterations that were undertaken inside the Eastern Goods Shed in 1953 were found during the course of this project, despite the fact that some of the changes that were called for were substantial. For example no trace was found of the new traverser that was proposed by the Railway Executive. It does appear that the Train Assembley Shed's banks (i.e. Platforms 6 to 10) were probably removed at this point in time in order to improve road vehicle access and enable a direct transfer system from wagon to lorry. Taken together, the archaeological record therefore suggests that certain cheaper elements of the 1953 scheme were implemented in the late 1950s, whilst the more expensive elements, such as the acquisition of an additional traverser, were postponed or cancelled altogether. It is likely that the scheme was revised on grounds of cost.

The Installation of new lifts in the Granary in 1952–1955

Landscape Feature 12

In October 1952, a new lift shaft was inserted against the north wall of the Granary (Fig. 11.8; Fig. 11.9). In order to accommodate it, a number of windows had to be infilled with London stock bricks (Fig. 11.8). The machine housing for the mechanism was erected directly above the shaft on the north side of the roof.

The lift installed in the 1920s was replaced in 1955 in order to comply with the Factories Acts of 1937 and 1948 (TNA AN 8/136). The work involved the widening of the original lift shaft and the insertion of a new mechanism at roof level. Although the specification stated that the machine room should be built in the style of its earlier predecessor (NRRG 55-ME-BM-22: 1955), it is worth noting that it was also visually identical to its 1952 counterpart. These machine rooms could be seen from ground level protruding above the stone parapet of the north elevation.

11.2 The Unclaimed Goods Warehouse: 1946 to the 1950s

Landscape Feature 15

Contract drawings for the partial rebuilding of the East Handyside Canopy issued by the British Railway Eastern Region Chief Engineer's Department in 1956 (NRRG 138796: LNE. 56-ME-BM-12 1956) demonstrated that the former Midland Goods Shed (Landscape Feature 18) was by then used as an 'Unclaimed Goods Warehouse'. Though the relevant documentary evidence has not been traced, it is understood that the leaking and dilapidated roof of the shed was replaced with clear-span steel roof trusses manufactured by Colville's Steel the following year (Bussell and Tucker 2004c: 5). Like many other sections of the complex, it was then clad with asbestos sheeting and wired glass panels.

The Unclaimed Goods Warehouse in the latter half of the 20th century

Landscape Feature 15

A goods lift communicating with the first floor of the Unclaimed Goods Warehouse (Landscape Feature 15) was installed in the eastern side of the north-central section of the building. Whilst no documentary evidence relating to the installation of this feature has been found, the brick fabrics that were used in its construction suggested that it was inserted in the second half of the 20th century, presumably before the shed fell into decline or disuse in the early 1980s.

Lines discernible on the floor of the shed in the south-central section marked the location of a series of former partitions. The presence of truncated pipes protruding from the floor suggested that a toilet facility was once housed beside a kitchen area. The facilities cannot be original since they blocked a former hatch that was used to transfer goods between floors during an earlier period. The lavatories had also been divided into ladies' and mens' sections, suggesting that they post-date the beginning of the Second World War, during which many more women were recruited to work in the yard. In all probability the block was added in the latter half of the 20th century, a period when the yard's welfare facilities were modernised.

It may have been around this time that some modern offices were erected in the southeast corner of the warehouse. They extended over two levels and were connected to one another via a staircase that ran up one side of the structure. It is likely that they were associated with an east–west timber partition that traversed the entire width of the southern end of the warehouse.

11.3 The Western and Eastern Goods Offices: 1946 to the 1950s

Landscape Features 20 & 21

The Goods Offices continued to be an integral part of the Outwards Goods Depot after the Big Four railway companies were nationalised in 1948. Permanent repairs may have been carried out to the bomb damaged Eastern Goods Offices during the years immediately following the Second World War, although the extent of the works appears not to have been documented and it remains possible that the undertaking occurred prior to 1945.

Specifications issued in December 1946 for the rebuilding of the roofs of the Eastern and Western Transit Sheds called for a number of minor alterations to be made to the north wall of the Western Goods Offices. In order to accommodate the hipped south end of the new roof, the existing gutter to the roof of the second floor messroom was to be removed and replaced with a new gutter, raised above the level of the rolled steel channel hip rafter of the new canopy (Fig. 11.3). Three new Fletton brick piers were to be inserted on the north side of the gable wall of the Western Transit Shed in order to support the hip rafters of the raised replacement roof.

11.4 Staff welfare in the Outwards Goods and Granary Complex and the Goods Offices: 1946 to the 1950s

Landscape Features 7–12 & 20–21

Following the end of the war the various messrooms at King's Cross Goods were extensively reorganised. In late 1947 it was decided to provide new equipment in the Outward Shed Male Staff Messroom and to relieve overcrowding by allocating the former Inwards Women's Messroom for use by the Outwards Checkers (TNA RAIL 390/65: Minute 4411 30/10/1947). The structure known since the 1990s as the 'Laser Building' (Landscape Feature 31) remained in use as a messroom during the decades that followed nationalisation in 1948.

A Ladies' Lavatory was erected c.1946 in a somewhat precarious position between the Train Assembly Shed Goods Offices and the roof of the Western Transit Shed (NRRG DMFP 00026357 46-LP-313: 31/12/1946). It was probably provided in order to rectify the lack of female toilet facilities in the Train Assembly Shed Goods Offices, which was highlighted by Peter Erwood in his reminiscences of working at the Goods Yard during the late 1930s (Erwood 1988: 135). It is possible that the toilet was one of a number of conveniences referenced in a minute of a meeting of the LNER Works Committee in March 1946, at which expenditure of £542 was authorised in order to improve lavatory accommodation for male and female staff (TNA RAIL 390/74: Minute 44490f 28/09/1946).

Owing to the loss of a number of messroom facilities to enemy action during the war it was decided at the end of October 1947 to reallocate the surviving facilities in accordance with the need of those departments worst affected (TNA RAIL 390/65: Minute 4411 30/10/1947). The former Inwards Women's Messroom was handed over to the Outwards Checkers, while members of the fish-handling staff were ordered to vacate the air raid shelter adjacent to the Western Goods Office (which they seem to have occupied since the end of the war) and instructed to move to the former Outwards Women's Messroom (one or more of the Outwards Messrooms were located on the upper floors of the Western Goods Offices). This latter arrangement was only temporary and the fish handlers were provided with a purpose built clothes drying room in the Western Transit Shed in 1952 (NRRG 52-LKC-28: 1952). The new room was inserted into the southwest corner of the Western Transit Shed (due north of the air raid shelter) and was heated by a gas fired boiler situated beneath the steel staircase to the Western Goods Office. In order to build the room it was necessary to brick up the 1850s arch in the southern wall of the shed.

11.5 The London Hydraulic Power Company Metering Station: 1946 to the 1950s

Landscape Feature 32

In 1954 representatives of the Midland Region of British Railways entered into negotiations with the London Hydraulic Power Company (LHP) in order to secure a supply of hydraulic power for a number of railway goods depots in the capital (Smith 2008: 20). Although part of the Eastern Region's estate, King's Cross Goods Yard was subsequently selected along with Broad Street and Somers Town (both Midland Region) to be connected to the LHP mains supply, in accordance with the terms of a three year agreement that was eventually signed and sealed in March 1957 (Smith 2008: 20; LMA B/GH/GP/06/003: 20/03/1957). In October of that year the BTC accepted an offer from the General Hydraulic Power company (the parent company of LHP) to sell the three-throw electric ram pumps then operating at King's Cross Goods for £1,6250 (LMA B/GH/GP/06/003: 16/10/1957). Following the connection of the depot to the LHP supply the pumps were to be dismantled and overhauled in advance of being installed at the GHP Athol Street Pumping Station in Liverpool (*ibid*.:18/09/1957).

The preliminary work of connecting trunk mains

in the City Road for connection to the King's Cross and Somers Town depots began in April 1957. By January 1958 some 2,900 yards had been laid, although the connection to the existing internal mains at King's Cross was not completed until later that spring (*ibid.*: 21/05/1958). Everything was in place by May 1958 and the Hydraulic Station at King's Cross Goods was switched off for the final time.

Archaeological evidence relating to the London Hydraulic Power Company (LHP) Metering Station of 1957

Landscape Feature 32

A six-sided, yellow brick structure was unearthed to the immediate east of the southeast corner of the Hydraulic Station's Engine House (Fig. 11.10). It was aligned north–south being 4.20m long, 1.10m wide and 0.79m deep. The walls were 0.20m wide and had been constructed with frogged, yellow bricks that were bonded together with indurated mortar. A poured concrete floor was present in the base, the top of which was found to be at a level of 23.21m OD. Remnants of pipework survived within the structure, along with a circular metal fixing that had been driven through the concrete floor at the southern end. The floor had also been truncated by seven rectangular and two circular holes, some of which still contained brick or timber supports. Together, these probably supported upstanding sections of hydraulic pipework and associated apparatus (Tim Smith pers. comm.).

It has been suggested that the six-sided structure was a metering chamber that was installed by the London Hydraulic Power Company (LHP) in order to monitor the yard's power consumption for billing purposes (Tim Smith pers. comm.). As such it is reasonable to assume that it was created in 1957, the year in which the hydraulic power supply was switched from the yard's Hydraulic Station (Landscape Feature 16) to the LHP.

11.6 The RME Garage and Associated Messroom: 1946 to the 1950s

Landscape Feature 30

The RME Garage had been seriously damaged during the air raid of 9th November 1940. Short term fixes had been made in the immediate aftermath of the attack, which enabled refuelling and minor repair work to continue on the premises, however the main repair shop had to be transferred to the garage in Blundell Street (TNA RAIL 390/1973: Minute 2873 24/07/1941). Permanent repair of the garage and the adjacent messroom had been postponed until after the war and it was not until 1950 that plans were approved for their reconstruction (NRRG 50-LKC-105B: 1950). It was envisaged that the existing mess block would be demolished and the ground made good and concreted over, whilst a new single storey range comprising a storeroom, an office, a kitchen and messroom and a toilet and washroom would be built along the north elevation of the RME Garage. These collective changes have been reconstructed from cartographic evidence and are illustrated in Fig. 11.2.

Fig. 11.10 The London Hydraulic Power Company Metering Station (Landscape Feature 32) exposed by excavation, photograph faces north

11.7 The Goods Department Offices: 1946 to the 1950s

Landscape Feature 18

The former Coal and Fish building (Landscape Feature 18) continued to be used as office accommodation by British Railways during the 1950s. Unfortunately, it has not been possible to identify which departments of the organisation were based in the building during this period.

11.8 Synopsis

By the summer of 1945 the King's Cross Goods Yard literally bore the scars of six years of war; bomb-damaged, over-worked and in desperate need of repair and renewal. Although the two years after the end of the war brought no let up in the volume of traffic handled by the yard, the parlous finances of the LNER ensured that there was no real prospect of obtaining the money necessary to conduct anything more than essential maintenance works. As if things couldn't get any worse, the winter of 1947 brought further misery in the form of freezing temperatures and coal and power shortages. Conditions scarcely improved at nationalisation in 1948, when the Big Four railway companies were added to the balance sheet of a near-bankrupt State encumbered by a huge national debt. Although the nationalised railways had an operating surplus of £19 million in 1948, austerity demanded that spending on the railways was restricted to schemes proven 'to improve the net revenue position, or of an essential renewal or maintenance character' over the years that followed. It says much about the poor condition of the Goods Yard that the replacement of the roofs of the Transit Sheds and the Train Assembly Shed fulfilled these criteria within five years of nationalisation. For the sake of spending £6,000 on roofing materials, the BTC was willing to countenance the immense upheaval of reversing the transposition scheme of 1935–8 a mere fifteen years later. In the event, the roof of the Train Assembly Shed was not replaced until the late 1950s, when the age of post-war austerity had come to an end.

In conclusion, the years of wartime overworking, underinvestment and post-war austerity appears to have had a significant negative impact upon operations at King's Cross Goods Yard. Yet British Railways was placed in an invidious position from the outset, as the post-war period saw costs rising steeply and competition from the road haulage sector increasing. The lifting of fuel rationing coupled with new road building projects meant that the road transport was becoming an increasingly attractive option for businesses previously obliged to rely upon the railways. The BTC only launched its own freight traffic policy as part of its modernisation programme in 1955, by which date the organisation was already in deep trouble, having amassed an annual operating deficit of £17 million (Wragg 2012: 183).

As the road transport option became ever more attractive, the multitude of railway goods depots in London inherited by British Railways came to be seen increasingly as a burden rather than an asset, at a time when rail freight levels were predicted to fall. There can be little doubt that this factor further discouraged significant investment in the infrastructure of King's Cross Goods Station.

King's Cross Goods Yard And The Bottle Trade

Chris Jarrett

Archaeological excavations in the Eastern Goods Yard unearthed an interesting assemblage of bottles that amounted to 178 individual vessels. The following places them in context by looking at their geographical origins, their distributions within the Goods Station itself and the growing reach of the railway connections that fed King's Cross.

Wares from the Eastern Transit Shed and the eastern side of the Train Assembly Shed

The Eastern Transit Shed and the eastern side of the Train Assembly area was used in a complementary fashion throughout most of their lives. With the exception of the milk bottles (explored subsequently) the vessels that were retrieved from them are therefore discussed together. Since the roles of these areas changed on three occasions between 1850 and 1954 the glass assemblage has been further divided along chronological lines.

The Eastern Transit Shed as an inbound facility: 1850–1935

A substantial number of complete glass vessels was discovered below Platform 2 in the Eastern Transit Shed, the date ranges of which suggest that they were deposited between the late 19th and the early 20th centuries at a time when the Eastern Transit Shed was functioning as an arrivals area.

A highly decorated beer bottle, manufactured between c.1880 and 1921, represents the most unusual vessel that was found, thanks to the presence of an intricate motorised lorry motif (National Telephone Company 1910b: Section 49 p.11; 1921: Section 31 p.13). The wording 'NORTH EASTERN BOTTLING CO LTD' surmounted the image, whilst 'THORNABY ON TEES' could be discerned below. Hailing from the area of present day Middlesbrough, this vessel could have been brought to King's Cross by the Great Northern Railway with the help of their east coast allies, the North Eastern Railway (Simmons and Biddle *et al.* 1997: 191).

Another example from a Lincolnshire source, embossed with the words 'IMPERIAL half pint/NEW TRENT BREWERY/CROWLE/NEAR/DONCASTER', was unearthed in a similar context. Operating under that name from 1878 to 1916, the owners of the brewery chose to position their business beside a railway and connect it to that transport artery via private sidings (Townley 2013; Kelly & Co. 1900: 154). Doncaster Station had formed part of the GNR mainline ever since it opened in the early 1850s (Nock 1974: 18), so New Trent products could have been conveyed to King's Cross by train as soon as that brewery came into being in 1878.

North Eastern Bottling Company Ltd beer bottle heralding from Thornaby on Tees, showing a motorised lorry and raised hand motif

In association with the Teeside and Lincolnshire vessels, two 'WALKERS WARRINGTON & BURTON LTD' bottles were found. This brewery has a long history dating back to 1817; however the Warrington branch did not open until c.1866, whilst the Burton-on-Trent arm only operated between 1877 and 1923 (TNA M380PWK/4). Indeed, the 'champagne' shape of the bottles also suggested an early 20th-century date (Orser 2002: 69). Whilst Burton-on-Trent lay beyond Great Northern territory, the GNR had been able to access Warrington ever since they became a joint partner in the 'Cheshire Lines Committee' during the 1860s (Simmons and Biddle et al. 1997: 82, 238). This bottle could therefore have been imported to the Eastern Transit Shed after that time.

Three oval-sectioned bottles for alcoholic spirits were found, one of which was made for the blended scotch whisky manufacturer James Buchanan & Co. This business was founded in 1884, however the black label on the bottle refined its dating to the 1890s. Three other Scottish whisky bottles accompanied the vessel, namely a post-1913 'Bells' miniature, a 'JOHN H[AIG &] CO GOLD LABEL' bottle, produced from 1825 onwards and another brandy type example labelled '[O]LD SCOTCH', made by an unknown producer. The Great Northern had strong links with Scotland thanks to their long established relationships with the North Eastern and North British Railways so they could have transported these products to their London Goods Station from as early as 1860 after access to Scotland was consolidated (*ibid.*: 191).

A soft drinks bottle with a pointed base, variously called Hamilton, torpedo or egg-soda types, embossed 'W. EKINS' and 'HUNTINGDON' was found in association with the beer and spirit bottles. Made for a medicinal company trading under that name between 1786 and 1877, the bottle would have contained one of their many carbonated beverages (Kelly & Co. 1869: 256; TNA HINCH 11/204). Huntingdon was positioned on the GNR's main-line into the capital so it would have been possible for W. Ekins to convey their products into King's Cross by rail from 1850 (Nock 1974: 18) until the demise of the company in 1877.

A 'CALEY' of 'NORWICH' bottle accompanied the Ekins example. Founded by A. J. Caley in 1857, the company that bore his name began bottling mineral water six years later (Gurney-Read 1987). The GNR did not establish a link with Norwich until it acquired a share of the Lynn & Fakenham Railway in 1882 (Simmons and Biddle et al. 1997: 355) after which King's Cross could receive wares from that area of the country.

New Trent Brewery beer bottle from the Doncaster area

Two 'Walkers' beer bottles imported from Warrington

'Caley of Norwich' and 'Schweppes' soft drinks late flat-bottomed torpedo-type bottles

'Kendall Bros' Codd-type soft drinks bottle from the Chesterfield area and a Champagne-type bottle made by 'Sherwood Morris' of Wolverhampton

From left to right, a 'Shaker Cocktails' vessel; mid 20th-century soft drinks bottles manufactured for Fling Soda, Pepsi-Cola and Coca-Cola and a bell-shaped miniature 'Bells Whisky' bottle

A 'Codd' bottle, embossed with the words 'KENDALL BROS STAVELEY & SHIREBROOK' most probably heralded from the Chesterfield area (National Telephone Company 1910a: Section 9 p.41; Kelly & Co. 1912: 619). Great Northern Railway access to Chesterfield was improved at the turn of the century thanks to the creation of the Lancashire, Derbyshire and East Coast Railway, which connected the city with a Great Northern and Great Eastern joint line west of Lincoln (Hill 1974: 210). In 1907 that railway passed into the hands of the Great Central, who remained allied with the GNR despite the creation of their own London Goods Depot at Marylebone (ibid.). Any one of these working relationships could account for the arrival of Kendall Brothers merchandise at King's Cross (Simmons and Biddle et al. 1997: 189).

Two 'Schweppes' bottles of a shape produced after c.1799 (Talbot 1974: 37) were found in the Eastern Transit Shed, whilst another 'egg-shaped' 'Schweppes' bottle was retrieved from an unstratified context (ibid.: 37, 44). Although predominantly London based, Schweppes also bottled mineral water elsewhere, opening factories in Derby and Glasgow in 1812 and 1865 (Simmons 1983: 36, 45). Great Northern trains gained access to the latter city in the 1860s, whilst the former could be reached by 1878 (Simmons and Biddle et al. 1997: 127, 191). Although these bottles may well have been lost after their contents were consumed by shed employees, they could alternatively represent northern imports to the capital that either arrived empty for bottling in the London Schweppes factory or ready filled with mineral water from Derby or Glasgow.

A 'Codd' type bottle, embossed with the words 'PURE TABLE WA[TER] AS SUPPLIED TO THE QUEEN/H D RAWLINGS LTD' was also discovered. Founded in 1754, this company traded from Nassau Street, Soho between 1870 and 1891 (Lake n.d.). Unlike Schweppes, H.D. Rawlings did not have a northern bottling arm so this vessel either represents an imported commission from a northern glass maker or an accidental loss by a member of station staff.

A flat lid from a jar produced by the Yorkshire firm Kilner Brothers was discovered in the southeastern side of the Train Assembly area. The importation of Kilner wares to London over Great Northern rails was a long-standing arrangement. Indeed, the relationship was so strong that the company chose to establish their London depot in the yard itself, renting the Midland Goods Shed from 1869 before moving to new premises in the Eastern Coal Drops during World War I (TNA RAIL 236/144: 376; TNA RAIL 236/189: 213; Goad 1921). Rail facilities for the reception of their goods existed in both locations, so the presence of a Kilner product in the Train Assembly Shed was not surprising. It was found in an area that did not function as an inbound facility until 1872, which suggests that it was not deposited before the company occupied the Midland Shed in 1869. More plausibly, Kilner goods might have briefly been brought into the Main Shed whilst their warehouse was being relocated in 1915.

A selection of bottles made for the soft drinks manufacturer Batey of London

The Eastern Transit Shed as an outbound facility: 1935–54

The Eastern Goods Shed, of which the Eastern Transit Shed formed a part, became an outbound facility in 1935.

Four glass bottles made for the London based soda manufacturer Batey probably date to the first half of the 20th century. Batey's produced ginger beer, lemonade, cream soda and kola, which may account for the bottles' different colours and shapes. Although it is possible that these vessels post-date the Transposition Programme of 1935 thus representing London exports bound for the north, they could equally represent *ad hoc* discards by Shed staff.

The Eastern Transit Shed and the Train Assembly Shed as an inbound facility: 1954–66

A number of mid 20th-century soda bottles were discovered in the Eastern Goods Shed, which was converted back into an 'arrivals' area in 1954. In the eastern half of the complex, single examples of 'Coca-Cola' and 'Pepsi-Cola' bottles were found along with two 'Fling' soda bottles. Manufactured in Brentford, Middlesex, the latter was popular in the 1960s until Fling Soda went into receivership in 1969 (Anon 1969: 1028). The Eastern Goods Shed handled ever-decreasing volumes of incoming general freight between 1954 and 1965, before being converted into a rail parcels hub in 1966. Some or all of these bottles could therefore represent pre-bottled or empty commissions that were imported to the Eastern Transit Shed and the Train Assembly area before 1966. However, given that volumes of incoming rail freight had started to fall drastically by the 1950s, deliberate or accidental discard by employees working within the Goods Station itself or the later parcels depot might more plausibly account for their deposition.

Wares from the Western Transit Shed

The role of the Western Transit Shed changed on four occasions at pivotal points in its lifetime so the glass and stoneware assemblage that was recovered from it must again be divided chronologically.

The Western Transit Shed as an outbound facility: 1850–99

From 1850 to 1899, the Western Transit Shed handled outgoing commodities that were destined for the north. An assumption has therefore been made that much of the shed's pre-1899 glass bottle assemblage represents southern goods destined for export by rail to the north of the country. This section will explore the validity of that assertion.

An intact light green 'Hennessy' brandy bottle was discovered in an unstratified context in the Western Transit Shed. Richard Hennessy, an Irish officer serving in the army of Louis XV, established this cognac distillery in 1765 on the Charent River, southwest France and soon afterwards he began exporting his product internationally (Anon n.d.a). It is therefore possible that this cognac bottle was brought to King's Cross after arriving at a southern port for onward transshipment by rail to the north of Britain.

A beer bottle stopper was recovered from the lower reaches of Canal Dock 1 in the Western Transit Shed. Recessed upon it was the name 'KEYS' and 'N.E.', printed around a circular medallion with a polygonal symbol in relief. This attested to the Cleveland City Brewery, North East England, which operated for one decade between 1895 and 1905 and was owned by Daniel H. Keys (http://www.oldbreweries.com/breweries-by-state/ohio/cleveland-oh-61-breweries/daniel-h-keys-brewery-oh-126c/). This time window pre-dates the infilling of the Granary Basin and the canal docks and coincides with the Western Transit Shed's last few years as an outbound facility before its conversion into part of the Eastern Goods Shed in 1899.

The Western Transit Shed as an inbound facility: 1899–1935

The Western Transit Shed was converted into an inbound reception facility in 1899 when the Western Goods Station was created, a role that it fulfilled until 1935. Consequently, it is reasonable to assume that some or all of the artefacts that were found in the shed dating to that time period were imported into London from the north by rail.

A bottle stopper was recovered from the lower reaches of Dock 1 in the Western Transit Shed. It originated from the Hewitt Brothers' brewery in Grimsby, an operation that commenced in 1874 and was still going in 1921 (Anon 2014; Hughes 1921: 133). Great Northern links with that fishing port were established in 1866 thus making the importation of Hewitt's products to the Western Transit Shed possible (Simmons and Biddle et al. 1997: 191).

A total of 22 vessels were found below Platform 1 adjacent to the Western Stables access ramp. These included two food storage jars, a cache of seven identical clear glass sauce bottles, six other similar-shaped, squatter flat bottles, three oval-sectioned bottles and another with a distinctive shape that must represent a special commission for a brand. Although it was impossible to deduce the precise nature of the products that these vessels contained, it is reasonable to assume that most if not all held condiments, table sauces, preserves and so forth. Since the assemblage dates to the early 20th century, by which time the Western Transit Shed was functioning as an inbound goods reception area, it is reasonable to assume that they represent imports from northern factories. The food containers were accompanied by two blue glass bottles for poisons or toxic substances and a hexagonal section phial. This unusual group of glassware also dates to the first quarter of the 20th century and most probably represents the importation of specialised containers to the south.

A selection of food, poison and medicinal vessels that were found together in the Western Transit Shed (Landscape Feature 7)

A stoneware ginger beer bottle marked 'BATEY' was also found in the Western Transit Shed, the fabric of which suggested that it was manufactured in London. The printing techniques that were used to stamp the bottle were indicative of a post-1890 date, being more typical of the early 20th century. Although Batey was a London based bottler, this vessel probably does not represent a north-bound export given its likely early 20th century date since the Western Transit Shed received inbound commodities at that time. More probably it was lost or discarded by a shed employee after its contents were consumed.

The Western Transit Shed as an outbound facility: 1935–54

The Western Transit Shed once again reverted to an 'inbound' area in 1954. The following vessels might have been lost during that time window.

Six wine bottles were discovered below Platform 1 in the Western Transit Shed. They were not sufficiently diagnostic to provide refined date ranges, although what can be said is that five of them are thought to date to the 20th century. They presumably represent foreign imports from continental wine producing areas, so it is reasonable to assume that they were brought to England by boat before being forwarded to King's Cross for onward transshipment. If they pre-date the Transposition Programme of 1935 then they could have been sent to the Western Transit Shed from a port on the eastern coast for onward dispatch to the metropolis. In contrast if they post-date that event then they were presumably forwarded from the southern ports for onward dispatch to northern markets. The latter explanation is deemed to be far more probable since Britain presumably received far more shipments of French and Mediterranean products via the English Channel rather than the North Sea due to the proximity of the former to those geographical regions.

The Western Transit Shed: 1935–66

One other mid 20th-century bottle in the shape of a cocktail shaker was found in the Western Transit Shed, which held an alcoholic drink. Embossed with the brand name 'SHAKER COCKTAILS', this bottle presumably represents another instance of goods lost in transit. Whether this product was made in London or elsewhere remains unknown, as does its exact date of production. Whilst it probably post-dates Transposition in 1935 it is therefore impossible to state with confidence whether it pre-dates or post-dates the shed's conversion back to inwards use in 1954 thus making its direction of travel undeterminable.

Bottles from other areas

Fourteen milk bottles were found during excavations in and around the Main Shed, all of which came from London dairies. Eight were unstratified, three had found their way into the wells of Turntables A and B and another three were discovered in the eastern side of the Train Assembly Shed. The export of milk away from a huge consumer base like London seems highly improbable so most of these vessels must represent the consumption of milk by Goods Station staff, during tea breaks for example. Alternatively some or all of the examples that were found in the Train Assembly Shed could represent empty vessels that were imported by train for bottling in London, since that part of the station functioned as an 'arrivals' area for a substantial part of its lifespan.

Two intact 'Gordons' gin bottles made for the famous London distillery of that name were found in the modern backfill that was dumped inside the Eastern Stables when the Eastern Goods Shed was converted into a road parcels depot in the 1980s. Although they could have been shipped to the Goods Station by rail from a northern glass works before eventually finding their way into the backfill of the Eastern Stables, more probably they were brought to the site from elsewhere along with the rest of the modern infill that characterised that mixed deposit.

An early 'champagne' type soda bottle, discovered outside the Main Shed, was stamped with the words 'RAY & SONS EFFERVESCING LEMONADE 23 ARTILLERY ROW WESTMINSTER'. Established in 1816, this company traded from the address given on this bottle between 1839 and 1882 (Pigot 1839: 201; Kelly & Co. 1882: 195; 1895: 1338). A small early 20th-century cylindrical 'pop' bottle with an external screw thread was also discovered in an external context, as was another 'champagne' example, made for Sherwood & Morris of Wolverhampton (Anon n.d.c). The city of Wolverhampton was firmly situated in London & North Western Railway territory, later passing into the hands of the London, Midland and Scottish. Even after unification under British Railways the pre-existing track network meant that it would have made little logistical sense to take West Midlands produce to King's Cross. The local origin of the first bottler, the fact that the third was situated beyond the yard's reach coupled with the fact that all three were discovered outside the Main Goods Shed suggests that these bottles were brought to the yard by station employees before being consumed and discarded.

Discussion

The evidence strongly supports the notion that the vast majority of the alcoholic drinks bottles, many of the soda bottles and in all probability all of the food containers, sauce bottles and specialised glasswares that were discovered within the Goods Station represent losses of goods in transit through the facility. The spread of these items within different usage areas of the Main Goods Shed correlates extraordinarily well with what would be anticipated given the nature of the products that they contained and the geographical locations of the companies that produced them. Luxury foreign goods brought to King's Cross from the southern ports such as wine, spirits and exotic beers consistently occurred in outgoing sections of the station, whilst vessels from London bottlers were skewed towards outbound areas. In contrast, items produced in northern areas that could be reached from King's Cross were always found in incoming areas. This distribution shifted as time moved on but always tallied with the Goods Station's changing working practices.

This conclusion can be asserted with a high degree of confidence with one vital caveat, which must be applied to the soft drinks vessels and milk bottles. They were deposited in a variety of ways that included loss of goods in transit as well as loss after consumption by shed staff. All of the milk bottles came from London dairies, suggesting that the latter agent was primarily responsible for their deposition. In contrast, the geographical bias that defined the distribution of soft drinks containers heralding from southern and northern areas suggests that some but certainly not all of them represent goods that were lost in transit.

CHAPTER 12

The Decline of the Former Eastern Goods Yard in the late 20th Century

'The site of King's Cross Goods Station, together with the adjacent loco depot and a group of sidings known in my day as Four Arch, now presents a picture of the Prophet Daniel's abomination of desolation. Its 120-odd acres, bounded by York Road, Islington on the east, the North London line to the north, the former Midland Railway main line from St Pancras to the west, and Battle Bridge Road and the Regent's Canal on the south, are now for the most part derelict and are awaiting a profitable sale'

Reminiscences of a former LNER employee taken from 'Memories of King's Cross Goods' (Erwood 1988)

The end of austerity in the mid-1950s heralded an era of rapid social, technological and economic change in Britain. Despite increasing GDP, higher wages and consumer spending, British manufacturing and heavy industry would continue to decline as the country's producers competed on a more even footing with those of other nations (Jackson 2013: 26). This, coupled with the construction of better road networks and the availability of larger and more fuel efficient road vehicles, would lead to a rapid decline in railway passenger numbers and volumes of railway freight throughout the period (Goslin 2002: iv). These factors impacted greatly upon British Railways, which only managed to maintain an operating profit until 1952 (British Railways Board 1963: 3). After that date the organisation's revenues began to decline rapidly to the extent that by 1960 it was running at an annual loss of £67.7 million, a figure that rose to £86.9 million just one year later (*ibid.*). In order to address the spiralling deficit, the Government passed the Transport Act 1962, which broke up the British Transport Commission (BTC) and established separate boards to manage the railways, the waterways, the docks and the state-owned road hauliers. The new British Railways Board (BRB) was charged with ensuring that its operating profits met its running costs. It was decided that those sectors of the railways that were profitable should no longer be used to subsidise the rest of the network, meaning that loss-making lines would face the axe. The practical consequences of this policy was described by Dr Richard Beeching in his report 'The Reshaping of British Railways' (better known as the 'Beeching Report') of 1963, the implementation of which would result in the loss of 31% of the railway passenger network in just one decade. Dr Beeching also considered the future of freight operations, concluding that while the bulk carriage of coal was in decline and the transport of 'freight sundries' was currently uneconomical, considerable potential lay in the 'liner' concept (British Railways Board, 1963: 42). These services (known as 'Freightliner' by British Railways) comprised fast, through running trains made up of low flat wagons which carried large road-portable containers, which were loaded or unloaded at purpose-built depots in centres of industry and population. The contents of the containers would be distributed locally by road.

In 1955 the British Transport Commission (BTC) had unveiled a fifteen year plan to modernise and re-equip British Railways (TNA AN 8/4; 8/9). Though primarily concerned with the elimination of steam traction, the plan also proposed to transform freight traffic policy by concentrating wagon loads in fewer and more efficient goods terminals and new 'Freight Transfer Depots' which would streamline transshipment of goods from rail to road and vice-versa (TNA AN 8/4: 30; TNA AN 8/136). The BTC's plans were unsuccessful and goods handling at King's Cross continued to operate in much the same way as it had since the late 1930s. Following the dissolution of the BTC in 1962, the British Railways Board (BRB) sought to rationalise freight terminal provision across the national network and to roll out the Freightliner concept (TNA AN 183/113). In 1965 British Railways launched its inaugural Freightliner service from a newly completed terminal at York Way (TNA AN 115/237). Over the next three years, a further three Freightliner terminals were built in the capital at King's Cross, Stratford and Willesden. The King's Cross terminal stood in the northwest corner of the Goods Yard on the site of the former Locomotive Depot. Although modest in comparison with its sister depots, the King's Cross Freightliner terminal handled many of the bulk goods that had previously arrived at the Goods Station, including bricks from the London Brick Company's works at Stewartby in Bedfordshire and beer brewed by Scottish & Newcastle (TNA AN 156/613; TNA AN 209/119: 18/10/1984).

At the local level it was accepted that the provision of railway goods handling facilities in the London Borough of Camden was excessive, and both the local authority and the BRB was keen to see surplus land released for development (TNA AN 169/631). By the mid-1960s duplication of functions was widespread among the borough's railway goods depots; three of which (King's Cross Goods, Somers Town Goods and Camden Town Goods) handled freight sundries traffic alone. A separate working group established in 1966 to consider the feasibility of closing one or other of King's Cross or St Pancras passenger stations also proposed that freight sundries facilities in the area be rationalised and replaced by a single "large modern freight terminal" to be built on the site of St Pancras Goods Station (TNA AN 183/113).

Fortunately for King's Cross, this scheme would come to nothing, although the other goods yards in the borough were not spared. As the former depots of the LNWR and the Midland were closed and demolished, the former Eastern Goods Yard would continue to function as a transport depot in one guise or another until the 1990s. This chapter considers the history of the yard during the latter half of the 20th century, describing the impact of the dramatic changes that took place during the period. The gradual demise of the complex is described in order to better understand the chain of events that led to its eventual abandonment by rail and road hauliers and charts its fate after their departure.

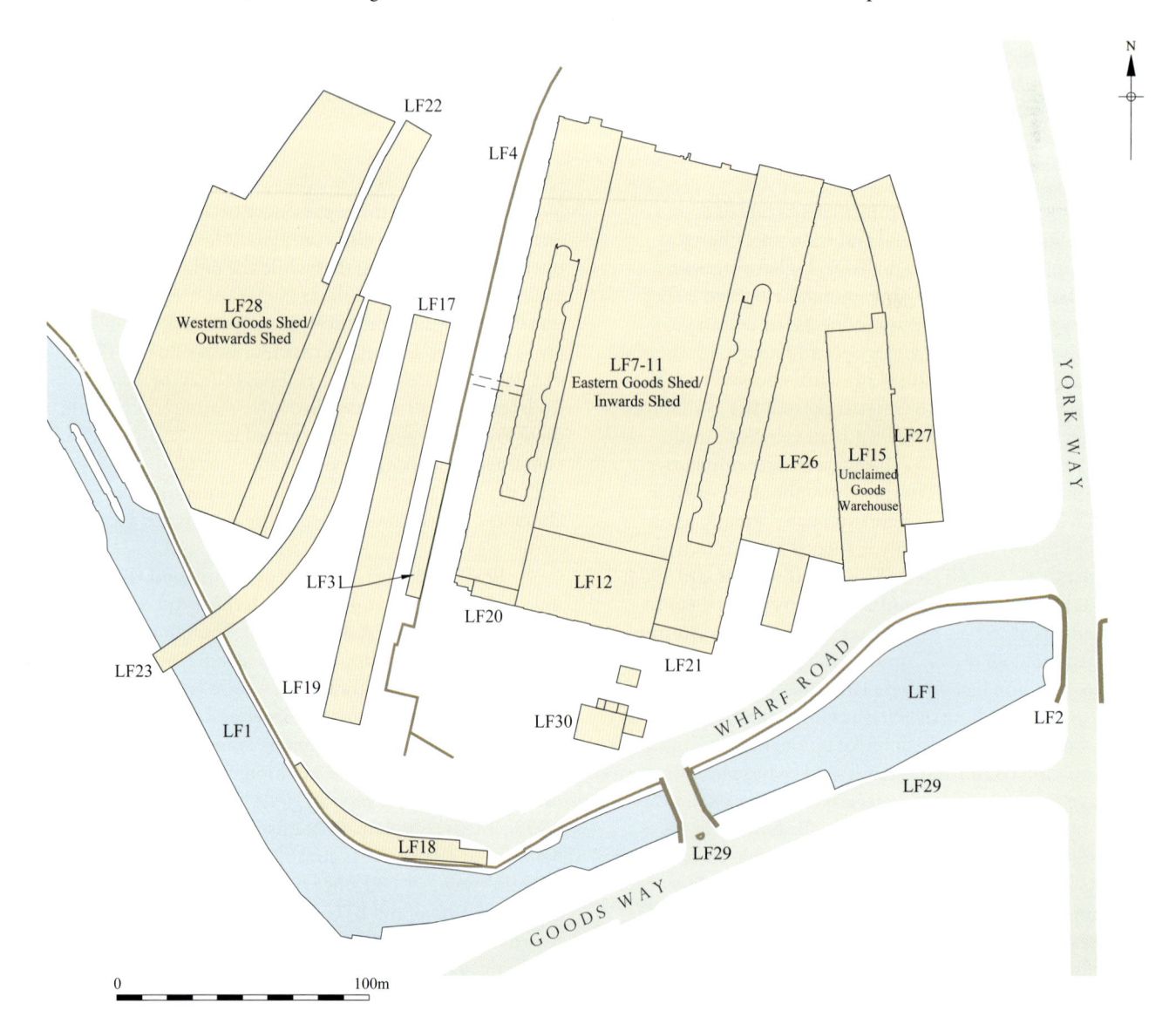

Fig. 12.1 The locations of the Landscape Features that were present in the Goods Yard in the latter half of the 20th century, scale 1:2,500

12.1 The Eastern Goods Shed and Granary Complex as a British Rail Parcels Depot: 1966–1981

Landscape Features 7–12, 20 & 21

While the BRB was debating the future of railway freight services at King's Cross and St Pancras, British Railways Eastern Region (ER) proposed to concentrate all parcels and post previously handled by King's Cross passenger station and King's Cross Goods at the latter terminal (TNA AN 183/113). This decision was fortuitous for the survival of King's Cross Goods Yard since it meant that its railhead would be spared from the BRB's wide-reaching reforms of the mid-1960s. The new Rail Parcels Depot was scheduled to open as early as June 1966, although the decision to abandon plans to establish a new freight terminal on the site of St Pancras Goods Station meant that it continued to receive some rail freight into the early 1970s (*ibid*.; TNA AN 169/631: 21/12/1970). Photographs taken in the middle of that decade indicate that the Train Assembly Shed formed the main hub of the Rail Parcels Depot.

Whilst the former Outwards Goods Shed was converted into a rail parcels depot, the Midland Goods Shed (Landscape Feature 15) became part of the holdings of Freightliners Ltd, though it is not clear as to what use the building was put.

In October 1966 the Chief Civil Engineer's Department at King's Cross proposed to erect an external steel staircase that would have provided a means of escape from the second floor of the Western Goods Offices in the event of fire (NRRG 66-ME-520: 02/1966). It was intended to build the stairs over the Air Raid Shelter, which was to be retained, even though it had long fallen out of full time use. It is possible that the plans were prepared in response to the 1961 Factories Act, which gave local authorities greater powers to ensure that premises were fitted with fire escapes, and the Public Health Act of the same year. This led to the issue of the first set of Building Regulations in 1965, which introduced statutory building standards across England and Wales (Factories Act 1961 Section II: 40, 41; Public Health Act: 1961). Although the fire escape does not seem to have been built, the plans that accompanied the proposal confirmed that the railway tracks and turntables in the ground floor of the Granary remained *in situ* in the mid-1960s. The ground floor of that building was shared by fuel and electrical stores, whilst toilets had been built in the northeast corner of the second and fourth floors along with a messroom in the southeast corner of the fourth floor.

In the mid-1960s the Eastern Goods Offices (Landscape Feature 21) were divided somewhat unequally between a First Aid Room to the west and the Outwards Counter Office to the east (NRRG 66-ME-520: 02/1966). Although the vehicle arch remained open, it led to an extended loading dock at the south end of the Eastern Transit Shed. On the first floor of the offices, a messroom was situated to the west, whilst the eastern room was used as the Tobacco Office (*ibid*.). The entirety of the second floor was given over to the Unclaimed Goods Office (*ibid*.).

The first and second floors of the Western Goods Offices appear to have continued to be used as messrooms by shed staff into the 1960s, whilst the ground floor was divided between 'Brown's Corner Office (west room) and the larger Outwards Labelling Office in the east room (*ibid*.).

The Granary meanwhile appears to have been used by the road haulage company, Pickfords, which was then part of the nationalised conglomerate British Road Services. The state owned road haulage sector was reorganised in January 1969 by the Transport Act 1968, which compelled British Railways to surrender its road transport arm (including vehicles, buildings and depots) to the newly established state owned National Freight Corporation (NFC), of which BRS became a subsidiary (Transport Act 1968: Part 1 Section 1). The ownership of the entire Granary Group was ceded to British Road Services (BRS), in whose hands it remained until the late 1980s (LBC London Borough of Camden 1985a).

12.2 The Road Distribution Terminal: the 1980s and 1990s

Landscape Features 7–12, 20 & 21

In 1981 British Rail announced the cessation of its rail parcels services. The King's Cross Parcels Depot was closed and the railway tracks lifted after 130 years' service. The buildings however continued to be used as a freight interchange terminal, which was exclusively served by road hauliers such as BRS and other NFC subsidiaries including National Carriers Ltd (NCL), Pickfords and Roadline UK Ltd (LBC London Borough of Camden 1994).

A series of modifications were required to convert the former Eastern Goods Shed (Landscape Features 7–11) from a road and rail interchange to a depot that was entirely devoted to road transport. Major alterations included the installation of new truck loading bays in the Transit and Train Assembly Sheds and the creation of new road vehicle and pedestrian access points (*ibid.*).

Pickfords used the Granary (Landscape Feature 12) for document storage and occupied the west ground floor and first floor rooms of the Eastern Goods Offices (Landscape Feature 21), which were used for the administration of the adjacent warehouse. It is also possible that the company used the messrooms on the upper floors of the Western Goods Offices as a welfare facility for their staff (Thompson *et al.* 2011).

A fleet of NFC vehicles had been stationed at the Goods Depot since the late 1960s. The Handyside Canopies appear to have been utilised for truck parking by various NFC subsidiaries. and a photograph taken in the mid 1970s suggests that the Western Transit Shed and the adjacent roadway were used by the NFC subsidiary National Carriers Limited (NCL) for truck trailer parking (Fig. 12.2).

In 1982 the National Freight Corporation was sold to its employees in one of the earliest privatisations carried out by Margaret Thatcher's Conservative Government. Thereafter the company traded as the National Freight Consortium (also known as the NFC). The Goods Depot was subsequently used by a multitude of NFC subsidiaries including BRS and Lynx Express Ltd (formerly Roadline UK Ltd, the BRS parcels division). The Train Assembly Shed was used by the Lynx Express road parcels service until BRS vacated the buildings in the late 1980s (Thompson *et al.* 2011). Photographs of the former Goods Yard taken at the turn of the 1990s (from Tim Smith's personal collection; not illustrated) indicate that BRS continued to use the car park on the site of the former Canal Basin for its truck rental fleet into the following decade.

The remaining rail tracks that entered the Midland Shed (Landscape Feature 18) were also removed during this period, though it is not known whether this was associated with an intended change of use, or whether it represented a step towards the decommissioning of the shed.

Fig. 12.2   NCL Trailers parked by the Western Transit Shed (Landscape Feature 7) in 1975, photograph faces north (© Tim Smith)

The conversion of the Parcels Depot to a road distribution terminal in 1981

Landscape Features 7–12

Cartographic evidence revealed little that could be used to reconstruct the internal configuration road parcels depot of the 1980s and 1990s. The archaeological investigations similarly shed no light on this developmental stage since most of the surfaces that were removed in order to enable the excavations to take place were instated at that time. Because of that fact, however, the built heritage survey (much of which was undertaken prior to the main excavation and watching brief phases of work), provided a great deal of information concerning the conversion of the yard from a rail parcels depot into a road freight terminal.

The Western Transit Shed's (Landscape Feature 7) vehicular openings were extensively modified during this period. Four of the six original arched examples that were present in the western elevation were altered in a relatively minor way via the attachment of roller shutters and a fifth was partially infilled with red bricks during the insertion of a new pedestrian doorway. A sixth fell out of use and was bricked up in its entirety whilst the large entrance that accommodated the former cart road near the southern end of the shed was retained in an unmodified state. Two new, larger entrances with roller shutters were then cut into the central and southern sections of the elevation, bringing the total number of vehicular access points in the western façade up to seven (Fig. 12.3). In contrast, all the surviving openings in the eastern wall that formerly communicated with the Train Assembly area were infilled with bricks or concrete blocks, whilst all but two of the arches that permitted access to the Granary were permanently sealed off. In the northern elevation, two pre-existing vehicular access points were altered through the insertion of riveted compound girders set in concrete and the attachment of roller shutter doors (Fig. 12.4). As previously discussed, the former road vehicle entrance point in the southern elevation had already fallen out of use during World War II when an air raid shelter was erected in front of it. Since the shelter was retained until the early 21st century, the entrance was not reinstated. The equivalent exit situated immediately to the east (that formerly accommodated the railway siding that ran through the Western Transit Shed and the Western Goods Offices) fell out of use and was blocked (Fig. 12.5).

Changes were made to the Western Transit Shed's pedestrian as well as vehicular access and egress points. Two of the four original surviving pedestrian thresholds in the western wall of the Western Transit Shed were destroyed when two new vehicular entrances were inserted (Fig. 12.3). At around the same time, the remaining two were narrowed through the use of red brick infilling in order to insert smaller door frames with concrete lintels (Fig. 12.3). Entirely new entrances

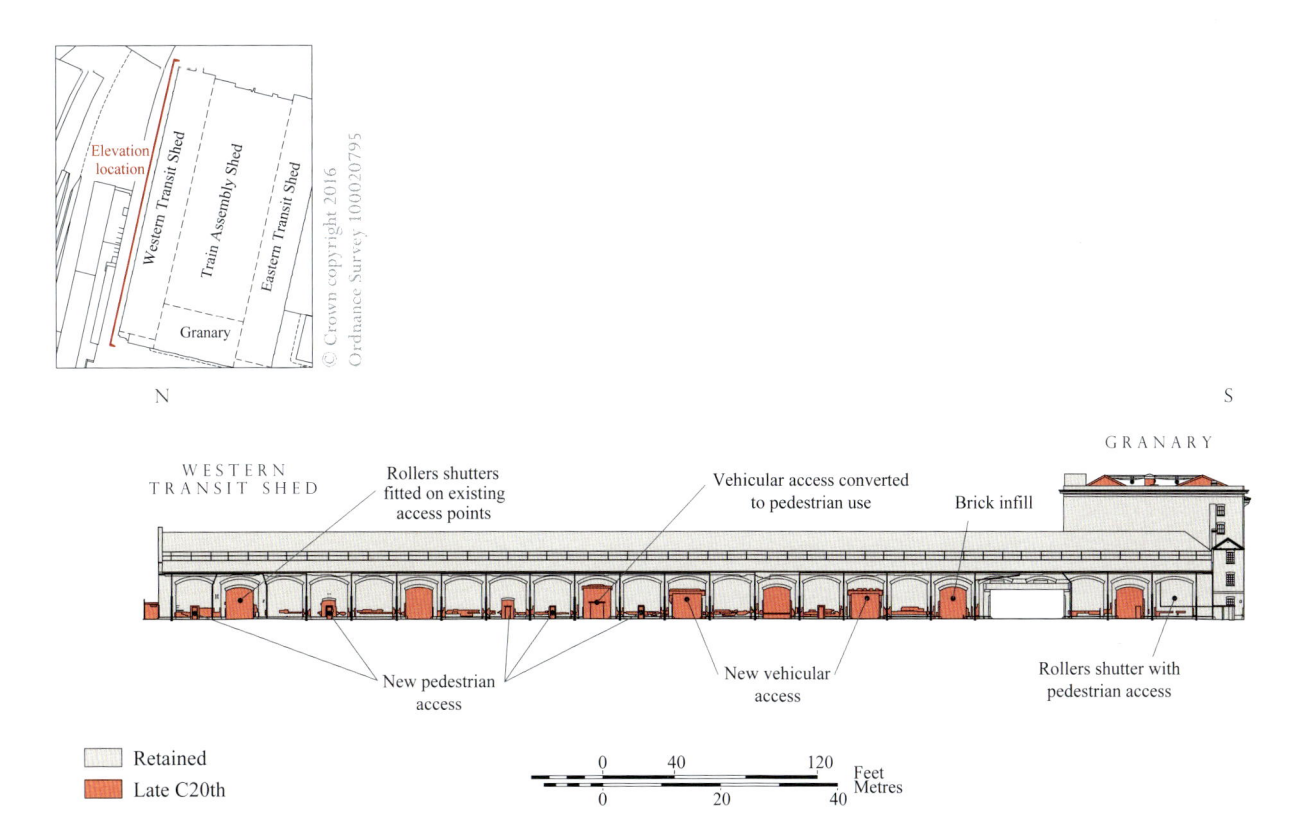

Fig. 12.3 External elevation of the west wall of the Western Transit Shed (Landscape Feature 7) showing the modifications that were made during the late 20th century, scale 1:1,250

282   *The Decline of the Former Eastern Goods Yard in the late 20th Century*

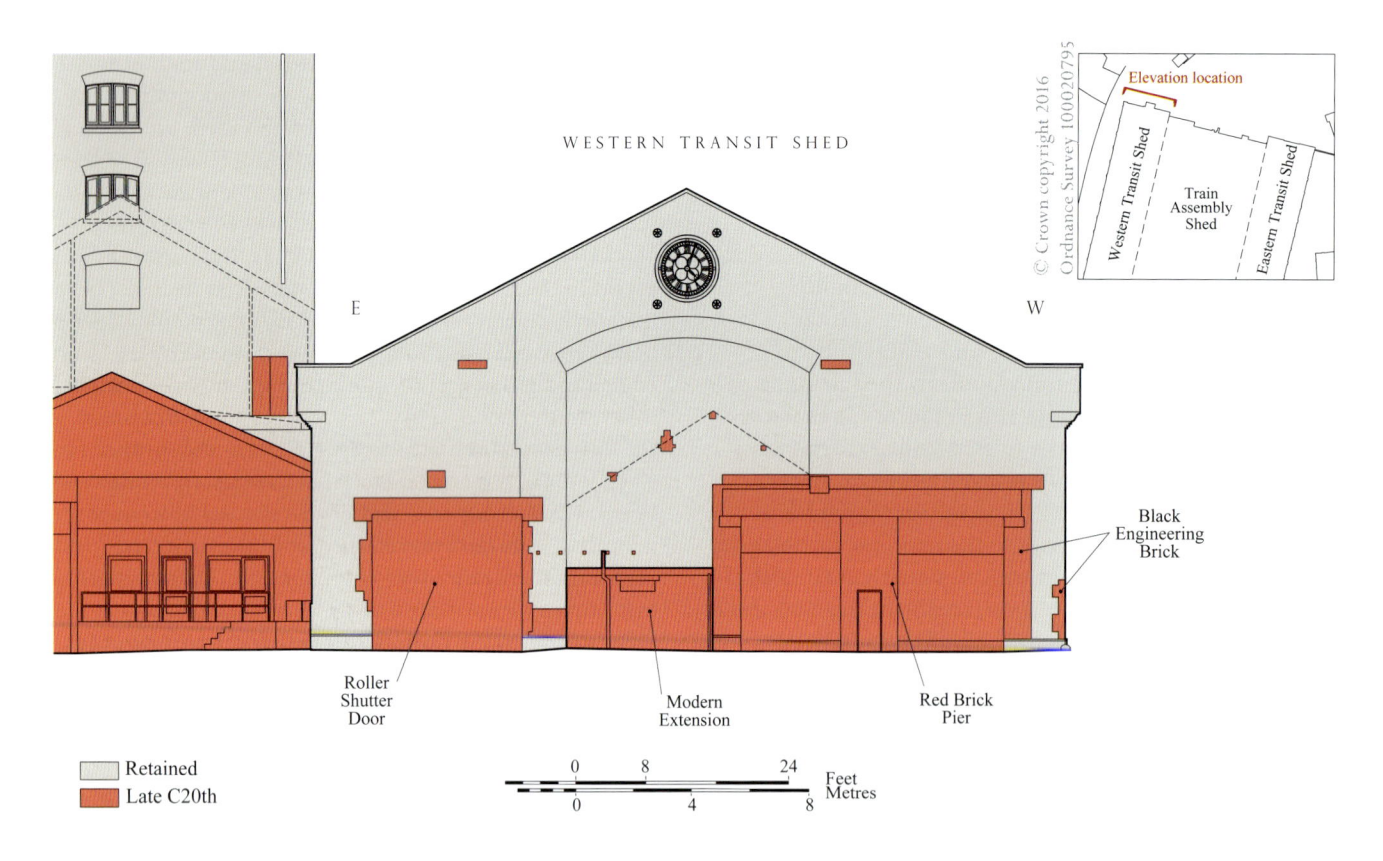

Fig. 12.4   External elevation of the north wall of the Western Transit Shed (Landscape Feature 7) showing the modifications that were made during the late 20th century, scale 1:250

Fig. 12.5   External elevation of the south walls of the Western Transit Shed (Landscape Feature 7) and the Western Goods Offices (Landscape Feature 20) showing the modifications that were made during the late 20th century, scale 1:250

for pedestrian use were then cut into the façade, bringing the total number to seven. This would later be increased to eight after the roller shutter and corrugated cladding that covered the most southerly vehicular access point in the western elevation was modified in order to allow pedestrian access to the temporary, pre-fabricated structures that had been installed on the internal side of the Western Transit Shed (Fig. 12.3; Fig. 12.6). Like the vehicular openings, all the pedestrian doorways in the eastern wall of the Western Transit Shed were infilled. In the centre of the northern façade another doorway was inserted, which led from the shed into a new toilet block that had been erected against the external wall (Fig. 12.4).

At the start of this period, no vehicular openings remained in the southern face of the Eastern Transit Shed (Landscape Feature 8) and none were reinstated. Instead, efforts were concentrated on modifying the other three walls of the building. The six original arched openings and the later, larger road exit in the eastern façade were retained and fitted with roller shutters. They were supplemented by three new access points, bringing the total number of vehicular openings in the eastern façade up to ten (Fig. 12.8). In the northern elevation, the two original arched openings were also retained. Since both had suffered from subsidence, it was necessary to insert rolled steel joists and brick up the space between the new lintels and the original arches (Fig. 12.9). An internal raised platform was installed to the immediate south of the west opening, demonstrating that this access point was used for the loading of road vehicles that were parked immediately outside (Fig. 12.9). With the exception of the large exits that accommodated the roadway that had been inserted by 1938, all the vehicular access points were then fitted with roller shutter doors. In the Eastern Transit Shed

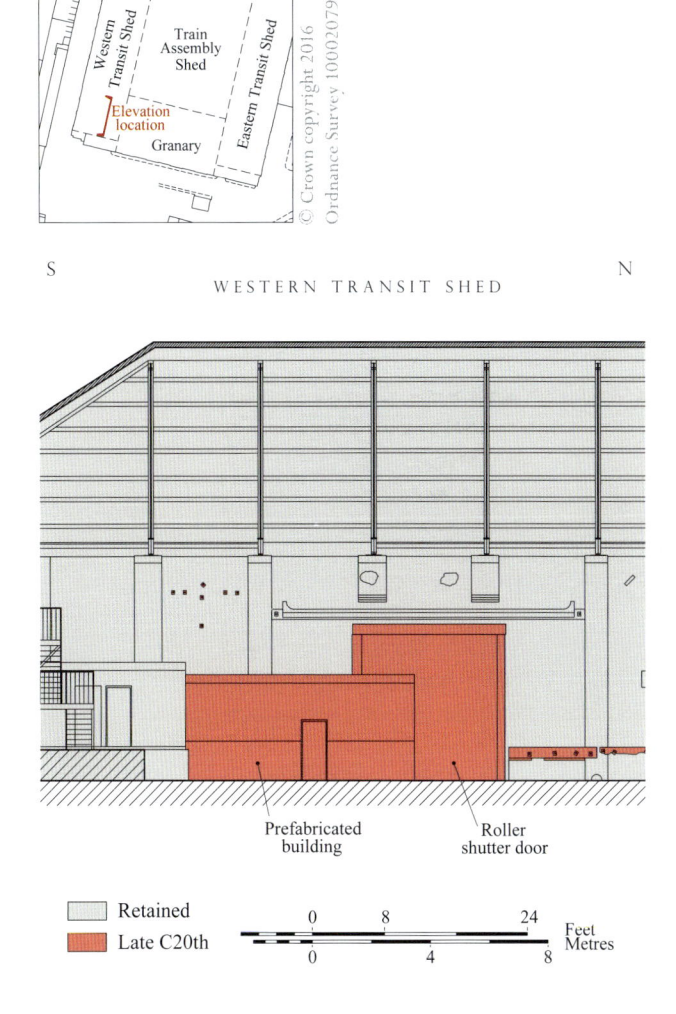

Fig. 12.6 Cross-sectional elevation through the southern end of the Western Transit Shed (Landscape Feature 7) showing the internal face of the west wall and the changes of the late 20th century, scale 1:250

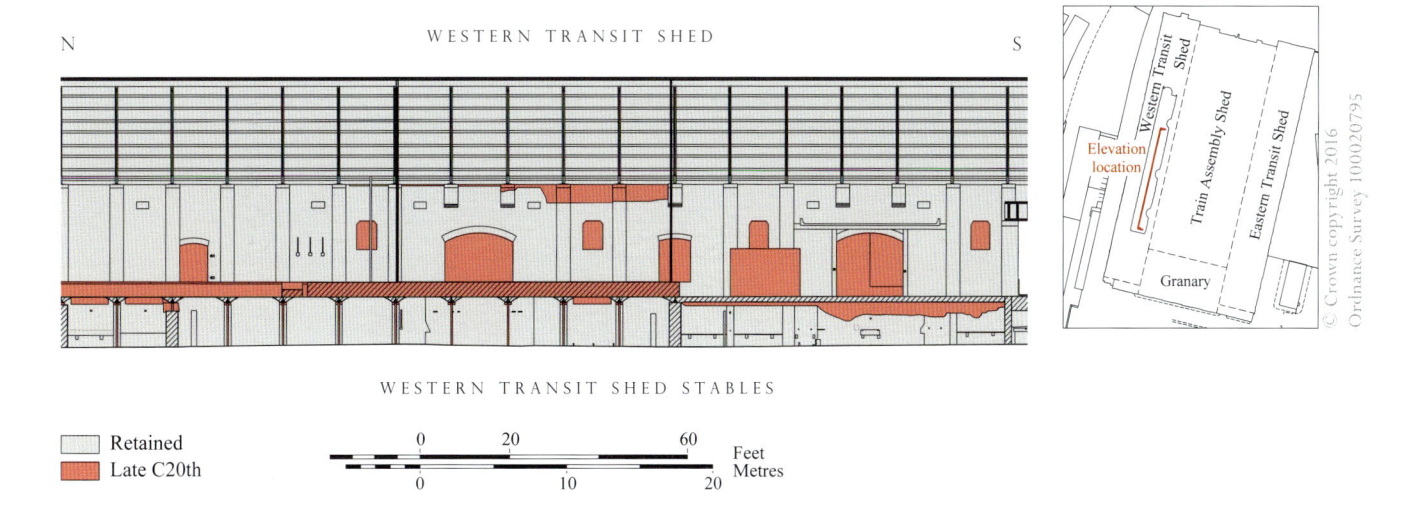

Fig. 12.7 Cross-sectional elevation through the Western Transit Shed (Landscape Feature 7) and the Western Transit Shed Stables (Landscape Feature 9) showing the internal face of their east walls; note the modifications that were necessary when the replacement ceiling was installed along the length of the eastern half of the stables after Platform 1 was narrowed and lowered in the late 20th century, scale 1:500

all former vehicular and pedestrian access points in the western wall that formerly communicated with the Train Assembly Shed and the Granary were blocked or infilled. Direct internal access to the Train Assembly Shed therefore no longer existed, whilst internal access to the Granary was limited to one pedestrian doorway.

Like the Western Transit Shed, the Eastern Transit Shed's pedestrian access points were also modified. Four of the five pedestrian doorways that had been incorporated in the eastern wall of the shed had survived until the start of this period. Three were then destroyed in their entirety by the insertion of new

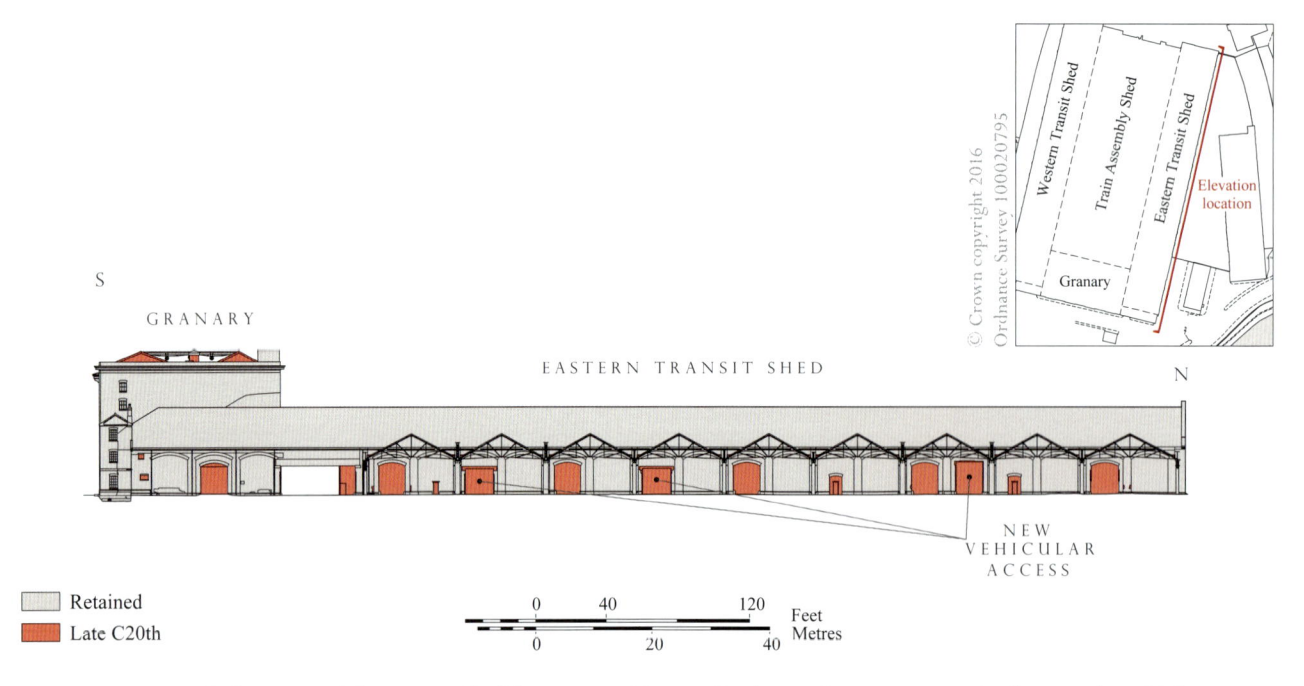

Fig. 12.8 External elevation of the east wall of the Eastern Transit Shed (Landscape Feature 8) showing late 20th century modifications, scale 1:1,250

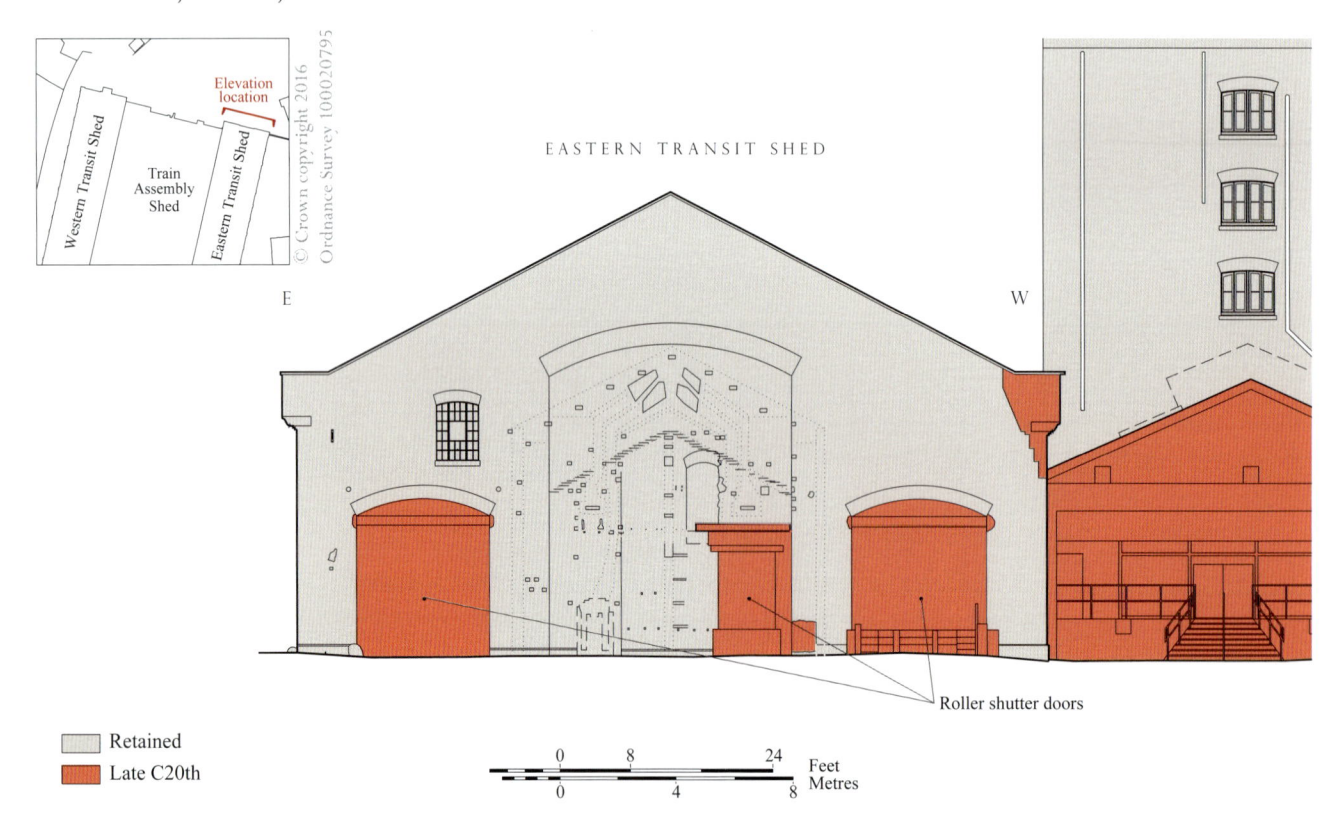

Fig. 12.9 External elevation of the north wall of the Eastern Transit Shed (Landscape Feature 8) showing later 20th century modifications, scale 1:250

vehicular openings whilst two others were bricked up and a replacement with a concrete lintel was then created. In the northern wall, just west of centre, a new pedestrian opening was inserted (Fig. 12.9). It accommodated a loading stage that extended through the opening where it was covered by a small canopy (Fig. 12.9). Together, the doorway and the stage were used to manually load haulage vehicles parked immediately outside.

The larger openings that were used by articulated vehicles during this period were typically flanked by guard stones designed to protect their sides from strikes. Striations along the length of the eastern and western exteriors of the both Transit Sheds (Landscape Features 7 and 8) at the approximate height of a modern lorry wheel and around the edges of the many road vehicle exits suggest that vehicle strikes were something of an occupational hazard. Modern bullnosed engineering bricks were typically used to repair the damage, strongly suggesting that it was incurred after the complex was converted to a road haulage depot.

Platforms 1 and 2 in the Eastern and Western Transit Sheds (Landscape Features 7 and 8) were narrowed and lowered in order to make them more suitable for use with road haulage vehicles. Since the former Western and Eastern Stables (Landscape Features 9 and 10) were integral to the overlying platforms, it was necessary to modify their ceilings by lowering the eastern half in the case of the former and the western side in the case of the latter. This in turn demanded that the former ventilation slots that were present in the affected walls be filled in (Fig. 12.7). The brick arch jack vaulting in the affected sections of each stable was replaced with a framework of rolled steel joists (RSJs) supported by a north–south line of upright RSJ stanchions set on concrete pads (Fig. 12.7). The disused stable blocks were then filled with rubble and soil, presumably in order to permit their ceilings to carry a greater load, which in turn increased the amount of floor space that could be used by road vehicles.

In the Train Assembly Shed (Landscape Feature 11), numerous truck loading bays and loading stages were created.

The cumulative effects of the various alterations to the Eastern Goods Shed can be summarised as follows: the Train Assembly Shed was now isolated from the Transit Sheds, whilst direct internal communication between the Transit Sheds and the Granary was limited to pedestrian access only. This meant that, for the first time in their history, these buildings henceforth became independent of one another. Instead of performing related roles they could now be used by different road hauliers. The extra vehicular openings that had been inserted in the walls of the Transit Sheds meant that they could be accessed by a greater number of road vehicles at any one time. Lorries and vans presumably backed into the sheds up to the edges of the narrowed platforms, where they were loaded.

12.3 The Former Coal and Fish Offices: the 1960s to the 1990s

Landscape Feature 18

As a consequence of the Transport Act of 1968, responsibility for the former Coal and Fish Offices was transferred from British Railways to National Carriers Limited (NCL), the road transport arm of the newly established National Freight Corporation (Transport Act 1968: Part 1 Section 1). They appear to have been used as offices in the first instance, but by the 1980s they had been vacated and had fallen into a state of disrepair (Senatore 1982: 4).

12.4 Re-use, Neglect and Abandonment: King's Cross Goods Yard: the late 1990s to the early 21st Century

During the late 1980s certain vacant areas of the Goods Yard became the venue for warehouse parties, reflecting an emergent countercultural boom driven by electronic dance music and recreational drugs. By the middle of the 1990s the former Coal Drops had become the permanent home of three influential nightclubs: 'Bagleys'[15] (later renamed 'Canvas') 'The Cross' and 'The Key'. Thanks to the effort of their owner, Billy Reilly, the clubs became central elements of the London dance music scene until their eventual closure in 2008 (Dodson 2007; Swindells 2008). During their latter years the clubs hosted several festivals and performances by a number of famous bands and artists, the most well-known of whom included Madonna, Prince and the Rolling Stones (Dodson 2007).

The former goods shed also attracted a number of small businesses, including several associated with the trade and repair of motor vehicles. For example, the larger of the two vehicular access points in the northern elevation of the Western Transit Shed (Landscape Feature 7) was transformed into a pair of display windows associated with an outlet that specialised in the installation and supply of spare parts for TVR sports cars (Fig. 12.4).

After the departure of Lynx Express in the early 1990s the Train Assembly Shed was used intermittently for storage whilst sections of the two former Transit Sheds were occupied by small businesses. In December 1994 Camden Council granted Listed Building Consent to Cobra Freeways Ltd to convert part of the Train

---

15   The club was named after the bottle merchant who tenanted the Coal Drops from 1879.

Assembly Shed into a go-karting circuit named the 'Raceway', which was licensed to operate for a period of five years (LBC London Borough of Camden 1994). The works necessary to convert the premises involved filling the truck loading bays with hardcore and rubble, which was overlain with a concrete slab to create one large elevated platform, in addition to the construction of a new wall that partially infilled the north elevation and the insertion of three new external doors. Internal offices and hospitality facilities were created, certain internal arches were infilled and an internal wall made from profile metal cladding was erected. At around the same time, the platforms in the southern part of the ground floor of the Granary (Landscape Feature 12) were removed and the roof was re-covered in corrugated zinc sheeting (Sabel and Tucker 2004: 4).

Despite the efforts of these tenants, a shortage of permanent occupiers meant that much of the yard became increasingly vulnerable to accident, vandalism and arson. A series of fires caused considerable damage, with one conflagration laying waste to the entire western corner of the complex. The Train Assembly Shed's overhead offices, built by Kirk Knight in 1898, survived beyond their centenary year but were severely damaged by fire in 2001 and were demolished as a result.

More than three decades earlier the empty Hydraulic Station's Boiler House (Landscape Feature 16) was also gutted by fire, which led to its demolition at some point before 1970. Its redundant Accumulator Tower and Engine House survived until the late 1970s, although they too had been demolished by the early 1980s. Despite their demolition, sections of their eastern sides remained extant as a continuation of the Retaining Wall (Landscape Feature 4).

The West and East Handyside Canopies (Landscape Feature 26 and 27) also fell into disrepair during this period. Flimsy and cheap materials such as corrugated metal and plastic sheeting replaced much of the 1950s glazing. Many of the connections between the guttering and downpipes that were installed back in 1888 failed in the intervening years, causing the corrosion and decay of numerous roof trusses. Nevertheless, the Handyside Canopies fared better than the adjoining Potato Market (Landscape Feature 14), which was entirely demolished during this period. In contrast, the 'Laser Building' (Landscape Feature 31) remained in existence until the first decade of the 21st century. It was used as a storage area and as offices by a number of different tenants.

Although damaged by fire in the early 1980s, the former Coal and Fish Offices (Landscape Feature 18) survived into the 21st century, with the exception of the triangular extension that abutted the eastern side of the complex (demolished between the publication of the Ordnance Survey Maps of 1975 and 1982; not illustrated; Senatore 1982: 4). The offices were almost lost in 1983 when National Carriers Limited applied to the London Borough of Camden (LBC) for demolition consent (LBC London Borough of Camden 1985b). After receiving numerous representations from interested parties and an associated press campaign over the following year, the Council rejected the NCL's application and their subsequent appeal of February 1985 (LBC London Borough of Camden 1985b). Following the 1983 fire that gutted the interior of one of the buildings and partially destroyed another, several proposals to refurbish them were devised (e.g. LBC CLAWS 1988). None of them came to fruition and the buildings were not made weatherproof until the late 1990s.

The Eastern and Western Goods Offices (Landscape Features 21 and 20) and sections of the Granary (Landscape Feature 12) continued to be used during the late 20th century when various internal stud and plaster walls were erected and new toilet blocks were inserted. Extensive refurbishment work, undertaken during the 1990s, included an internal 'soft strip' of the Goods Offices. This led to the removal of superficial modern partitions along with most of the doors and doorframes and any other remaining original features. Only the main entranceway and the three doors that led to the southeast Granary stairwell survived.

12.5 Synopsis

The decision of the Eastern Region of British Railways to concentrate its rail parcel operation at King's Cross enabled the complex to survive the 1960s intact, at a time when the British Railways Board was making strenuous efforts to divest itself of surplus infrastructure across the capital. By the second half of the 1970s rail traffic at the depot had dwindled to a trickle, whilst the cessation of British Rail's parcels services in 1981 led to the closure of the yard's rail head. The earlier conversion of the depot into a parcels hub ensured that it survived into the 1980s as a road parcels transport facility.

The privatisation of the National Freight Corporation in 1982 initially had little impact on the long-term viability of the depot, which was extensively modernised for road vehicle use during the years that followed. The change in use necessitated numerous changes to the fabric of the former Eastern Goods Shed. The infilling of access points during this episode broke the long lasting physical connections that had existed between the Train Assembly Shed and the Transit Sheds and severely restricted internal communication with the Granary. However, these breaks were functional as well as physical since those structures, which had previously worked together, were subsequently put to different uses by the various arms of the NFC, a division that was continued after privatisation.

The reprieve that was offered by road haulage sector was a temporary one. A general lack of investment throughout the period meant that the former Goods Station was slowly crumbling. From the early 1990s onwards, the road haulage firms therefore began to move out of the complex, relocating to better maintained, better situated and more modern premises elsewhere. Sections of the site were subsequently occupied with varying degrees of success by nightclubs, a go-karting circuit and a number of small businesses however large areas of the complex were left empty, its historically important buildings falling prey to vandalism, accident and disrepair as a result of acute neglect.

Hydraulic Power At King's Cross Goods Yard

Tim Smith

When the Great Northern Railway began building King's Cross Goods Depot hydraulic power was in its infancy. The term 'hydraulic power' was coined by the Victorians for the transmission of power using high pressure water distributed to scattered machines in cast iron pipes.

Hydrostatic machines (i.e. those powered by the pressure of water) were first used in mining in the middle of the 18th century (Reynolds 1983: 331–5). Joseph Bramah patented his hydraulic press in 1795 and, by 1802, had used it to power a crane at his Pimlico works. In 1812 he described all the elements of a hydraulic system in a patent on water supply. He died in 1814 and his dream of a hydraulic power system died with him, but his press found widespread use in industry (McNeil 1968).

Cranes and Sack Hoists

In 1846, William Armstrong, a solicitor, demonstrated a hydraulic crane on the Town Quay in Newcastle-upon-Tyne. He took water from the town's water supply. He patented the crane and, with others, founded the firm of William Armstrong & Co. to manufacture cranes and other hydraulic machinery with a works at Elswick, Newcastle (Armstrong 1877: 11319).

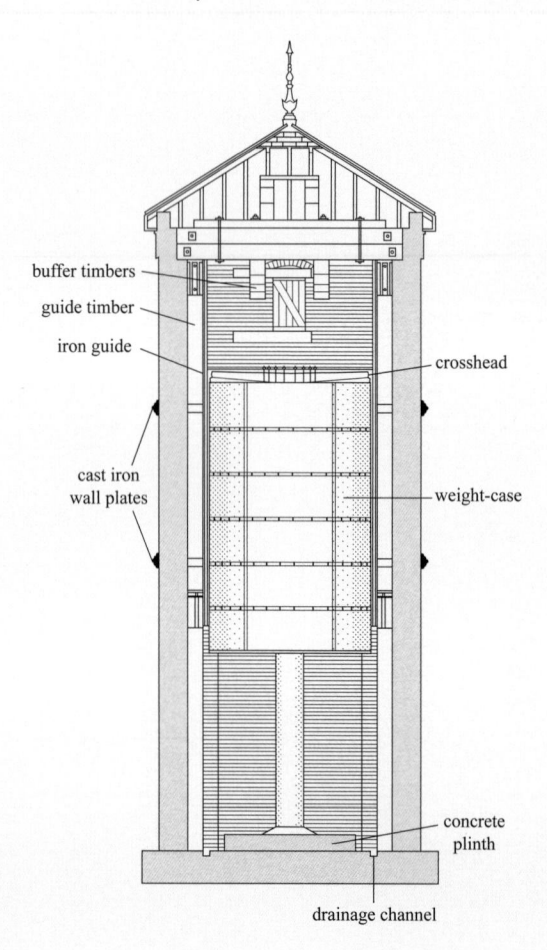

Schematic cross-section through a typical mid 19th-century accumulator, not to scale

Armstrong had had mixed success with his low pressure hydraulic cranes. Those used in Newcastle gave long, trouble free service (Ellington 1888: 77–8). They took their water from the local town water supply which was of sufficient pressure and was constant and reliable. But cranes supplied to Liverpool and Glasgow, which also took pressure water from local town water supplies, suffered from variable pressure reducing their reliability and usefulness (Smith 1991: 86–8). Many towns and cities, including London, had intermittent water supplies. A high pressure system promised not only reliability of supply but a reduction in running costs, since these cranes needed less water to raise the same load as did low pressure cranes.

In 1850, Armstrong received an order from the Manchester Sheffield & Lincolnshire Railway for hydraulic cranes at their new dock at New Holland, the southern terminus of the ferry across the Humber from Hull. Prior to this, all Armstrong's hydraulic machines had worked at low pressure, no more than about 100lbs/in^2 (about 7bar). At New Holland the pressure was raised to 600lbs/in^2 (about 41bar) by using steam powered pumps and a weight loaded hydraulic accumulator. This was successful and Armstrong soon received orders for other high pressure hydraulic systems (Dougan 1970; McNeil 1972; Heald 2010).

The high pressure system needed one or more boilers and at least one steam pumping engine at a central pumping station. Weight loaded accumulators were used to provide an artificial head of water and a reserve of energy (left). A network of pipes distributed the pressure water to the various cranes and other machinery. Thus the high pressure hydraulic system anticipated the introduction of electricity later in the century. Both used a central power plant, the pumping station in the case of hydraulic power and the generating station (or power station) in the case of electricity. Both could provide power to widely scattered machines through cast iron pipes (hydraulic power) or copper wires (electricity).

The weight loaded accumulator comprised a vertically mounted cast iron cylinder, within which a ram was raised and lowered on a column of water. The ram supported a cross-head above. Armstrong's first accumulators had cast iron weights suspended from the cross-head (Robinson 1887: 34). By the end of 1850 he had introduced the wrought iron weight case, again suspended from the cross-head and filled with gravel (TWCA MS: 1975/1). A branch from the high pressure main fed into the bottom of the cylinder and the resulting column of water supported the ram, cross-head and weight. The accumulator was housed in a tower. The cross-head ran in guides on each side of the tower. In operation, the accumulator rose and fell slowly according to the demand for power and supply from the pumps. The accumulator was a large and very heavy device. The weight bearing on the column of water could be anything from 80 to 120 ton, depending on the pressure in the system and the diameter of the ram.

The New Holland system was ordered in early May 1850 (*ibid.*; TNA RAIL 463/1). Armstrong read a paper to the Institution of Civil Engineers, on hydraulic cranes and other machinery, on 7th May 1850 (Armstrong 1850). A footnote detailed the impending introduction of the weight loaded accumulator. At the end of May, Joseph Cubitt recommended that Armstrong hydraulic cranes should be used at King's Cross (TNA RAIL 236/70: 275 28/05/1850). Cubitt was probably at the meeting because he was clearly au fait with the developments. The Great Northern Railway company ordered twenty 2-ton cranes from Armstrong, to be hand powered initially but capable of conversion to hydraulic power (TWCA 175/1). These were delivered and eighteen had been erected by the end of November 1850 (TNA RAIL 236/933: 30/11/1850). The order for the elements of the hydraulic system at King's Cross goods depot was amongst the first for high pressure hydraulics. Armstrong was to supply hydraulic machinery for the hand powered cranes, sixteen additional 1-ton cranes, a steam pumping engine, two locomotive type boilers, three accumulators and all the necessary pipes and valves. An order was also placed for eleven sack hoists for the Granary (TNA RAIL 236/71: 295, 29 01/1851; TWCA 1975/1).

To house the power plant a pumping station was built on the west side of the canal basin in front of the Granary. With the boiler house to the south of the engine house and the accumulator tower to the north it was a typical arrangement for an early hydraulic pumping station. The tower housed the largest of the three accumulators. The smaller ones were placed at the far end of the mains; i.e. one at the north end of each of the transit sheds. Experience at New Holland had shown the need for such remote accumulators to avoid loss of pressure when working cranes (Armstrong 1877: 71).

Other London orders placed that year included coal derricks for Poplar Dock, cranes and other machinery for the Great Western Railway at Paddington, and coal cranes at the Regent's Canal Dock. Outside London, Armstrong had received orders for hydraulic machinery at a Sheffield goods depot and for Birkenhead Docks (TWCA 1975/1).

The most pressing need was for the Granary hoists, since the Great Northern Railway hoped to attract custom from the 1851 grain harvest in Lincolnshire. But it soon became clear that Armstrong could not supply these hoists in time (TNA RAIL 236/16: 274 29/07/1851). From the historical perspective it seems likely that Armstrong's lead times for delivery of machinery were hopelessly optimistic given the number of orders he had received since the introduction of the high pressure system at New Holland. In the three years from 1847–9, Armstrong & Co. received orders for 41 low pressure cranes (TWCA 1975/1). In 1850 33 cranes, both high pressure and low pressure, were ordered. But in 1851 orders were received for no less than 94 cranes, together with steam pumping engines, boilers, accumulators and other hydraulic machines such as hoists (Saunders and Smith 1970: 8). This would require a step change in output capacity at their Elswick Works in Newcastle (Dougan 1970: 48–50).

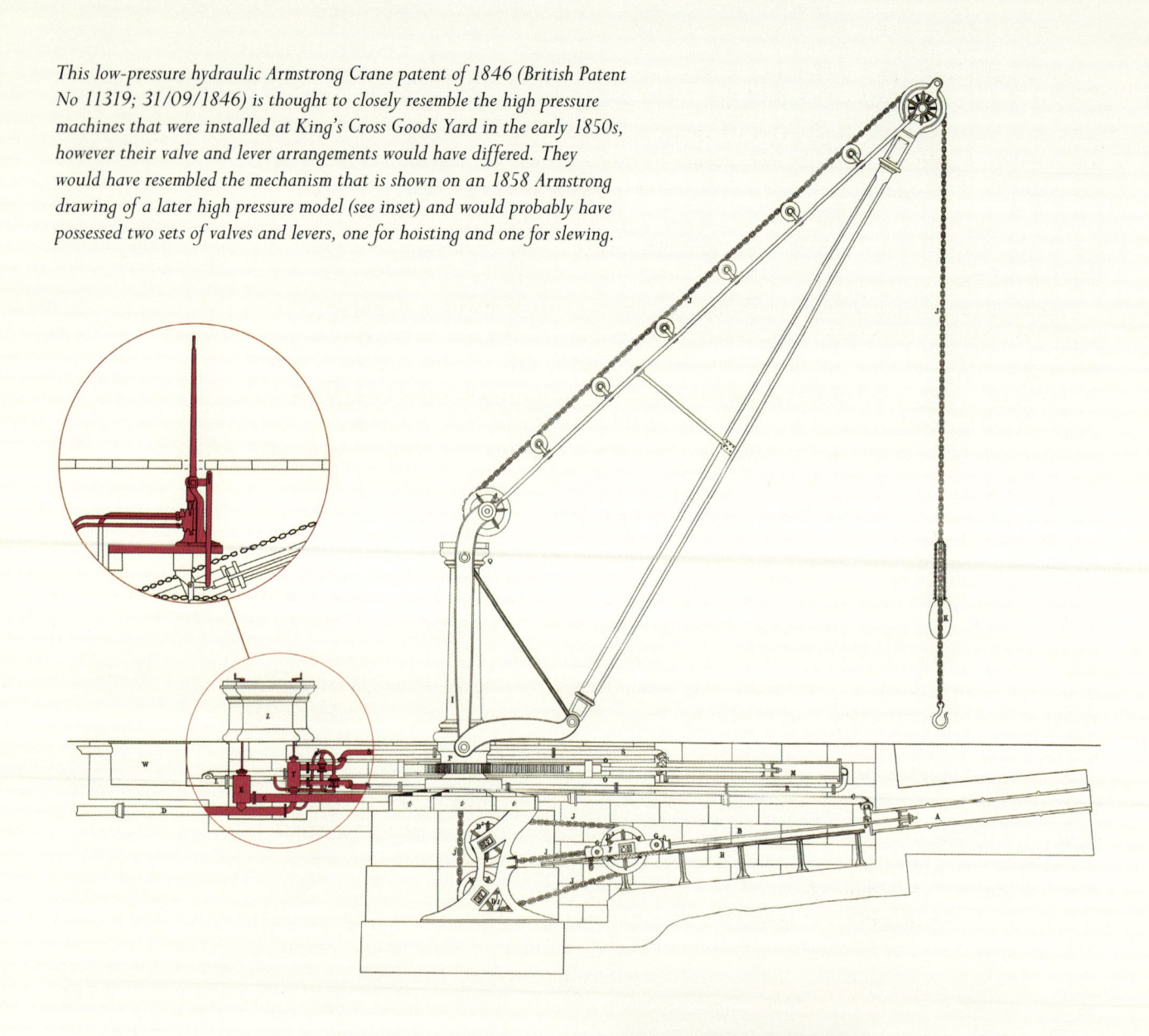

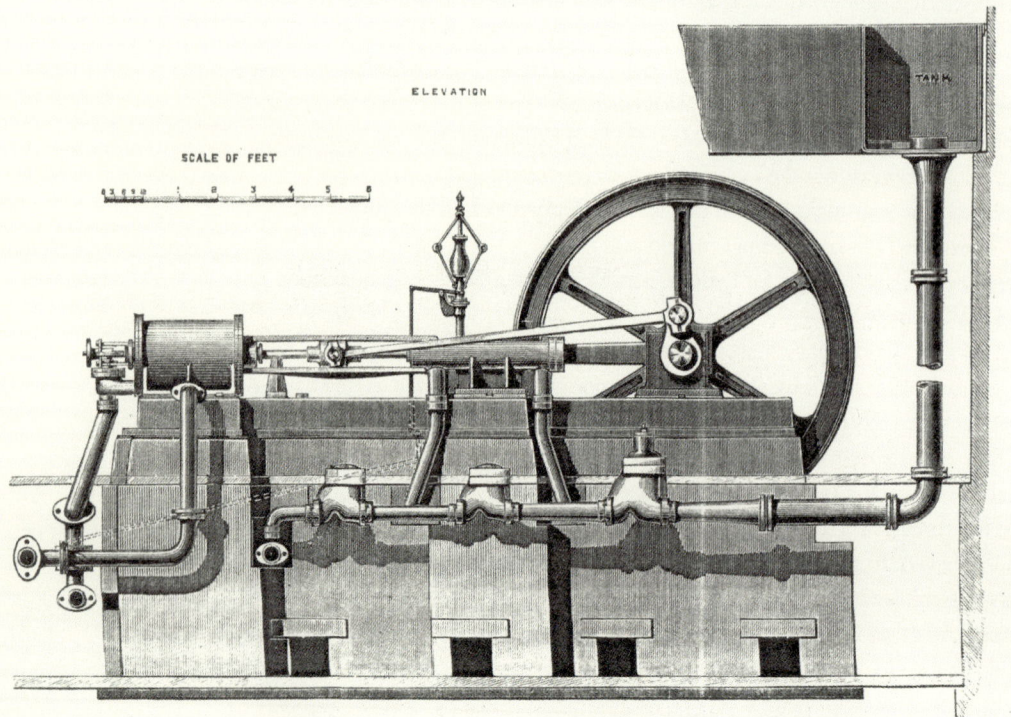

This low-pressure hydraulic Armstrong Crane patent of 1846 (British Patent No 11319; 31/09/1846) is thought to closely resemble the high pressure machines that were installed at King's Cross Goods Yard in the early 1850s, however their valve and lever arrangements would have differed. They would have resembled the mechanism that is shown on an 1858 Armstrong drawing of a later high pressure model (see inset) and would probably have possessed two sets of valves and levers, one for hoisting and one for slewing.

William Armstrong's patent drawing reproduced from The Engineer of January 1885 showing an elevation of a steam pumping engine similar to those that were instated in the Hydraulic Station (Landscape Feature 16) in the 1850s

290

Armstrong probably knew that supplying high pressure sack hoists and the necessary pumping plant before the 1851 harvest was out of the question and he may have recommended low pressure sack hoists instead. This would account for the large tanks in the roof space of the Granary, said to hold water for the hydraulics and for fire fighting. Such high level tanks were not necessary for a high pressure system and to the best of the author's knowledge no other high pressure system had them. Of course, without further evidence, this is speculation. Only ten sack hoists of the eleven that were ordered were delivered (TNA RAIL 236/933: 01/04/1853).

The arrangement of the sack hoists can only be conjectural since little remains. There were traps through the floors and gantries on the roof, which could account for six of the ten hoists. There were also four sets of loops in the façade, which might account for the other four. In other granaries, where there was no hydraulic power, the usual arrangement was for two ropes or chains to be wound counterwise on the winch drum such that, as the drum turned, one rope ascended with one or more loaded sacks, the other descended with no load. As the full sacks were unattached at the top more sacks could be attached at the bottom, thus doubling the work load. There are ways in which this could be done using hydraulic machinery. One way would be to have double jiggers for each hoist. While one raised a load the other would lower the rope. Another method would be to use a single cylinder but with two sets of fixed pulleys. It is also possible that the hoists used a hydraulic engine to turn a winch drum. Armstrong patented such an engine in 1848 (BL 1/1617-14359/1852: No. 12157 1848; Woodcroft 1857: 869).

Two paintings by the architect, Lewis Cubitt, show the interior of the transit sheds. Both were reproduced 50 years later in *The Engineer* (28 November 1913 supplement), where they show very different cranes. One view shows hand operated timber whip cranes, typical of the period but taller to reflect the height of the building. Whip cranes of this sort were commonly used in railway goods sheds. Interestingly, timber whip cranes were probably used in the Midland Shed at King's Cross goods depot in 1858 (TNA RAIL 236/934). They came from a known supplier of whip cranes and were much cheaper than their hydraulic equivalents. Later in the century, as a goods shed was built over the Western Coal Drops, there was evidence of whip cranes in the roof timbers of that building. Cubitt probably produced the painting before he knew exactly what the cranes would look like meaning that his image represents an impression of what the transit shed would be like rather than the finished article. The second view shows hand operated, cast iron platform cranes, again taller than usual. These probably represent the 20-ton hand cranes that were supplied by Armstrong and later converted to hydraulic power.

The standard machine for hydraulic cranes was the hydraulic jigger. This evolved from Armstrong's original cranes. The main differences were that the original jiggers used pistons whereas the later ones used rams. Originally the fixed pulleys were not attached to the jigger whereas later fixed pulleys were attached to the blind end of the jigger cylinder. The original arrangement was that a single moving pulley was attached to the outer end of the piston rod and hydraulic pressure was applied to the annular space around the piston rod, driving the piston into the cylinder and moving the pulley away from the fixed pulley, effectively lifting the load. The cylinder was inclined so that water could leave it by gravity. Besides the hoisting jigger, hydraulic cranes had slewing jiggers allowing them to be slewed left or right.

Control levers for both hoisting and slewing had three positions. The lever was kept in the central position, with the valve closed. Moving the hoisting lever in one direction would raise the load; the other direction would lower it. Similarly, moving the slewing lever one way would turn the crane to the right; the other would turn it to the left.

The great advantage of the hydraulic crane was in saving labour costs. One man could operate a hydraulic crane. On a visit to the Grimsby Docks in 1853, representatives of the East and West India Docks in London saw a young lad of about fifteen operate a hydraulic crane lifting around 12 to 15cwt. The superintendent said that for the same job six men were required at the cranes of the West India Dock (Smith 1991: 86).

There was a 10-ton hydraulic crane next to the canal basin, at the west end of the Granary. The crane base and jigger pit of this crane were revealed during archaeological investigations, together with an adjacent valve pit. The arrangement of these pits was identical to those shown on Armstrong's patent drawings. The only difference was that the valve pit was deeper.

Armstrong's low pressure cranes were controlled by slide valves, but he found that the higher pressures tended to lift these valves off their seats. So he introduced a weight loaded mitre valve for use at higher pressures. These needed a deeper pit and were opened and closed using levers.

Of the sixteen 1-ton cranes ordered only fourteen were delivered, though the reason for this is not known (TNA RAIL 236/933: 01/04/1853). A further two were ordered in 1852, bringing the total number of cranes for the goods sheds to 36 (TNA RAIL 236/74: 16/11/1852). The Humber plan of 1866 and the GNR plan, also of that year, show only 32 (Humber 1866; Fig. 3.8). Over the coming years other cranes were ordered, but it is not easy to identify them on plans.

To supply hydraulic power, water was taken from the canal basin in front of the Granary under an agreement of 1850 with the Regent's Canal Company (TNA RAIL 860/54: 446–7 08/07/1868). The 30hp steam pumping engine was a two cylinder, simple, non-condensing horizontal engine driving two force pumps (TWCA 1975/1). In order to even out the flow of water into the pressure mains, Armstrong designed a force pump in which the water capacity of the annular space around the piston rod was exactly half of that at the other side of the piston. Suction water was taken from a header tank, at a sufficient height to open the suction valve, allowing water into the larger side of the pump cylinder. The water was then forced through an intermediate valve with half of it passing into the other side of the pump cylinder and half through a delivery valve into the mains. Effectively, therefore, water was pumped into the mains four times for every revolution of the engine. The engine also drove two sets of lift pumps, presumably one set to pump water from the canal basin into the Granary tanks, the other to pump water into the station header tank. A second steam pumping engine with force pumps was installed in 1857 (TNA RAIL 236/933: 06/08/1857).

The station accumulator, with 10" diameter ram, was slightly larger than the two remote accumulators, which each had an eight and a ½" diameter ram (TWCA 1975/1). All three had a stroke of 15', giving a total height when raised of just under 30'. Each accumulator would be fitted with a relief valve, called a momentum valve by Armstrong, fitted to the connecting pipe at the bottom of the accumulator. This valve released water to waste if the accumulator rose too high, preventing the ram from coming out of the cylinder. A timber buffer at the top of the tower served the same function. The two remote accumulators were inside the transit sheds, and could have had timber guide frames without a brick tower.

The pressure main would be connected to the two force pumps of the engine and to the accumulator cylinder. There would probably then be two branches, one along the length of each transit shed, to the remote accumulators. These are likely to have run along the roof timbers since that is where the crane jiggers were situated. Cranes would be connected by a supply pipe of smaller diameter teed off the main. The accumulator connection and each supply pipe would almost certainly be fitted with a screw down stop valve to allow the accumulator and each crane to be isolated. There could also have been stop valves on each branch of the main so that, in the event of breakage of a pipe, the affected section could be isolated to prevent further loss of water, and for repair.

During operation of the hydraulic system, the accumulators would rise to the top of their strokes as the steam engines pumped water into the mains. Remote accumulators were always loaded slightly less than station accumulators. Once the station accumulator had reached the top of its stroke pumping would be stopped. This could be done either by using a trip on the accumulator, connected to the throttle valve of the steam engine, or manually by the engineman. In the latter case, a bell would be sounded to alert the engineman. There would probably also be an indicator board to show the position of the accumulator.

Using cranes would draw water from the mains giving a temporary reduction in pressure. This would, in turn, cause the accumulator to fall. As it did so, at a predetermined point the steam engine would be started again. Or if it were already pumping water into the system, the throttle valve would be opened further to increase the speed. If the engine were working at maximum capacity then the second engine would be started. If both were working to capacity pressure in the system would fall until cranes were stopped. In practice the accumulators would be rising and falling slowly according to demand throughout the day.

Shunting Capstans

As a general rule, the steam railway used steam locomotives for the movement of rolling stock. But there were exceptions: where there were physical constraints on the use of locomotives, where there was a risk of fire (as was the case in and around goods depots) and warehouses where steam locomotives were banned. The King's Cross goods depot was no exception. But some other railway companies were beginning to use capstans instead of horses.

As early as 1849 a capstan was being used in Pickford's shed at Camden (Head 1849). It was driven from line shafting by the steam engine that powered the cranes. In the 1860s similar steam driven capstans were in use elsewhere. From the mid 1850s there were Armstrong hydraulic capstans at Paddington (Armstrong 1877). In 1863, Armstrong's introduced a hydraulic engine which was more suitable for capstans (BL 1/1852-18225/1915: No. 1712 1863). Up to three capstan heads could be driven from one hydraulic engine by means of shafts and gearing. With steam capstans, the capstan heads revolved continuously, whether being used to shunt waggons or not. This design was the cause of many accidents (Royal Commission on Railway Accidents 1877: Minutes of Evidence 16767–828).

Similarly, with the early hydraulic capstans, once the hydraulic engine had been set in motion, all capstan heads driven by it would revolve.

In the 1870s two compact hydraulic motors suitable for driving single capstans were developed, by Percy Westmacott for Armstrong's, and by Peter Brotherhood (BL 1/1852-18225/1915: No. 60 1874, No. 2003 1873). In each case the hydraulic motor was placed beneath the capstan head, which was attached to the crankshaft. The Leeds firm of Tannett Walker took a different approach. They used a variation of the Armstrong hydraulic engine to drive each individual capstan head, using a horizontal crankshaft and bevel gearing (TNA RAIL 236/317).

In 1877, the Great Northern Railway decided to replace horse shunting in the Potato Market at King's Cross with capstan shunting (TNA RAIL 236/146: 15/03/1877, 31/05/1877; TNA RAIL 236/43: 01/06/1877, 13/07/1877). Six hydraulic capstans and associated snatch heads, were ordered from Tannett Walker & Co. of Leeds. The order included an accumulator and all the necessary pipes. The accumulator was installed in a tower at the north end of the Midland Shed. In 1881 a further order for capstans was placed with Tannett Walker (TNA RAIL 236/174: 10/08/1881, 06/10/1881). The pumping station had to be enlarged to provide enough capacity to drive the new capstans. A steam pumping engine and an accumulator were included in the Tannett Walker order. New engine and boiler houses were built adjoining the existing pumping station, and a new double accumulator tower was built at the north end. This tower housed the Tannett Walker accumulator and a new one from Armstrong's, ordered in 1882 (TNA RAIL 236/175: 30/03/1882).

Tannett Walker hydraulic capstan patent drawing showing the internal mechanism

An operative uses a hydraulic capstan to shunt a wagon at King's Cross Goods Yard in 1956 (SSPL 1995-7233_LIVST_DA_38 © National Railway Museum and SSPL)

Just how many shunting capstans were installed is not know as the quoted cost included the engine and accumulator. But the capstans were to replace 35 horses and eight men.

In 1882, Benjamin Walker, one of the partners in the firm of Tannett Walker & Co, took out a patent for a new hydraulic motor designed to drive a capstan (BL 1/1852-18225/1915: No. 1715 1882). It had four similar cylinders, arranged in pairs. Each pair had the cylinders face to face with a gap between them, and they shared a ram. As pressure water entered one cylinder it was exhausted from the other, and vice versa. Drive was taken from the centre of the ram, between the two cylinders, using a forked connecting rod which acted on a crank of a vertical shaft. The capstan head was fitted to the upper extension of this shaft. The shaft had two cranks, one above the other. The second pair of cylinders, positioned immediately below the first, acted on the lower crank.

The working valves had three ports. The two end ports were connected to each cylinder in the pair, connecting them to pressure water in turn. The middle port was connected to the waste water pipe. The valves were operated by a connecting rod and lever system with a spindle connecting with the slide valve. The motor was contained in a cast iron box. The pressure pipe and exhaust pipe entered the box on one side. There was a stop valve operated by foot pedal on the pressure pipe. This was used to control the capstan.

Hydraulic capstans could be used to move wagons along a railway track or to turn a wagon on one of the many small turntables around the depot. Hemp rope, or similar fibre rope, was kept coiled by the capstan when it was not in use. A capstan man and a runner, often a young lad, were needed to work the capstan. The runner would attach the rope to the wagon and take it to the capstan where it would be wound a couple of times round the head. If it were necessary to alter the direction of the rope one or more intermediate snatch heads would be used. These looked like smaller versions of the capstan, but were un-powered. The capstan man would hold the rope behind the capstan in his hands whilst operating the control valve with his foot. As the capstan drew the wagon towards him he would coil the rope. He would stop the capstan at an appropriate point, the runner unhooking the rope and pinning down the wagon brakes to bring it to a stop.

Expansion of the System

Towards the end of the 19th century a new goods shed, the Western Goods Shed, was approved. It was fitted out with hydraulic cranes from Armstrong's and hydraulic capstans from Tannett Walker (TNA RAIL 236/212: 17/11/1898). More hydraulic power was needed to supply the additional machinery. This seems to have been carefully planned. A duplicate pumping engine, from Tannett

Walker & Co, was installed in the western section of the engine house. This allowed the two old Armstrong engines to be scrapped, freeing the space in the eastern side of the engine house for a new pumping engine. The engine and boiler houses were subject to major rebuilding. Four new 8' diameter Lancashire boilers were purchased from Daniel Adamson, an old established firm in Hyde, now in Greater Manchester (ibid.: 02/05/1898).

In 1899 a compound steam pumping engine was ordered from Tannett Walker to increase the capacity of the pumping station (ibid.: 28/09/1898). Compounding (where the steam is used twice, first in a high pressure cylinder and then in a low pressure cylinder) had been used for many years to increase the efficiency of stationary steam engines, particularly in the textile industry. It had been used in hydraulic power stations from the 1870s. But there was, perhaps, reluctance on the part of railway companies to employ it. Compounding of steam locomotives had not proved a great success in Britain, though the French used it to advantage.

There were two basic arrangements for a horizontal compound stationary steam engine. The tandem-compound had its two cylinders in line whereas the cross compound had its cylinders side by side. An engineer's report described the Tannett Walker engine at King's Cross as a tandem-compound, yet the archaeology suggests that it was, in fact, a cross compound. It was also the only condensing engine used at King's Cross goods depot. With the other, non-condensing engines steam was exhausted to atmosphere. After use in the low pressure cylinder of the compound the steam passed to a condenser, where cold water drawn from the canal basin condensed the steam creating a vacuum, which improved the efficiency of the engine.

In early 1915, a steam pumping engine from Farringdon Goods Depot, where hydraulic power had been taken from the London Hydraulic Power Company since July 1914, was moved to King's Cross. This was another Tannett Walker engine (of 1893) and probably replaced the 1881 engine (TNA RAIL 236/216: 02/12/1914).

In 1915 there were plans to fill in the canal basin. Almost at the last minute the Great Northern Railway realised that they needed to make other arrangements for the water supply for the hydraulic system. A new agreement was reached with the canal company to take water directly from the canal. As before, it was returned to the canal after use (TNA RAIL 1189/1426).

Electrification of the Hydraulic Station

As early as 1907–8, pumping at the Underfall Yard hydraulic station in Bristol was electrified (Fisher and Powell 1979: 6). By the 1920s a number of firms were offering suitable electrically driven pumps. Electrification gave substantial savings in pumping costs, since boilers could be dispensed with and pumps could be automatically controlled from the accumulators. The Great Northern Railway had installed electric pumps at Derby Friargate in 1922 (Palmer and Neaverson 1992: 164). In the 1920s and 1930s the LNER continued the programme of electrification of hydraulic pumping stations which began in pre-Grouping days. Electric pumps for Forth Goods, Newcastle and York were ordered in 1924, and other orders followed (TNA RAIL 390/30: 26/06/1924, 09/10/1924).

One of the electrically driven pumps from the Hydraulic Station (Landscape Feature 16) at King's Cross photographed after its removal from the site and taken to a store at Princes Dock (© Tim Smith)

Three electrically driven, three-throw pumps from Hathorn Davey & Co. for King's Cross were ordered in 1925 (*ibid.*: 26/11/1825 06/01/1827). Each was capable of delivering 150 gallons per minute. Concrete machine bases to support these pumps were constructed on top of the engine beds. The boiler house became redundant and was converted into a repair shop for the company's electrical engineers.

During World War II there was concern that a direct hit on the pumping station would put much of the depot out of action (TNA AN 3/32: 07/11/1940; TNA AN 3/26: 03/10/1941, 21/11/1941). Therefore, one of the pumping sets was moved to a new engine room in the northeast corner of the Midland Shed, next to the remote accumulator. The accumulator was modified to control the set.

Power Purchased

The London Hydraulic Power Company (LHP) provided a supply of hydraulic power to customers throughout central London. The company began in a small way in September 1883, when its first pumping station opened (Ellington 1888). The network extended until there were nearly 200 miles of mains and around 4000 customers. The Great Northern Railway took advantage of LHP at its Royal Mint Street and Farringdon goods depots, closing its own pumping stations (TNA RAIL 783/220; TNA RAIL 236/216: 02/12/1914). But the LHP mains came nowhere near King's Cross. In 1954 negotiations began between LHP and British Railways to extend supplies to other goods depots. Eventually agreement was reached to supply Broad Street, King's Cross, Marylebone and Somers Town depots (LMA B/GH/LH/02/003: passim; LMA B/GH/LH/02/004: 20/03/1957). This involved laying 4300 yards of new main from the existing 6" main in Pentonville Road to King's Cross and Somers Town. The new main continued past Somers Town to join another main near Euston (LMA B/GH/LH/02/004: 14/03/1956; LMA B/GH/GP/06/003: 15/01/1958). As part of the deal the electric pumps at King's Cross were sold to the Liverpool Hydraulic Power Company for use at Athol Street pumping station in that city. LHP and the Liverpool company were both subsidiaries of the General Hydraulic Power Company. On 22nd May 1958 the LHP supply to King's Cross goods was turned on (LMA B/GH/GP/06/003). Pumping ceased at the last LHP pumping station (Wapping) on 1st July 1977 but the use of hydraulic machinery at King's Cross had ended sometime before that.

CHAPTER 13

Conclusions

'Our objective must be to lift the sights and raise the aspirations of those with a serious strategic interest in the future of the railway heritage...we need, between us all, to establish... [its]... credentials... The people for whom we are doing it are the public at large, whose heritage it is. We must not lose sight, through our interests, our passions and our obsessions for the railway and all it represents, of the fact that it is one of the most important dimensions in our cultural history, its heritage is noble and significant, and the future of that heritage should be the concern of us all'

Taken from 'An agenda for railway heritage' by Sir Neil Cossons (Cossons 1997: 16)

Having presented a detailed description of the developments that took place in King's Cross Goods Station, from its creation by the Great Northern Railway in 1850 to its gradual decline and eventual abandonment at the turn of the 21st century, this chapter collates that evidence to explain how, when and why the complex evolved in the way that it did. By combining archaeological and archival evidence with cartography it has been possible to recreate 'snapshots' that show how operational changes within the yard altered the allocation of space to different commodities over time and, where the data allows, has also enabled traffic flow through the complex to be approximated. These reconstructions show at a glance where and how the various categories of goods that passed through King's Cross were handled, which were given priority (in terms of space) at different times and when reception and dispatch facilities for different commodities were improved or down-graded.

In order to further determine whether events at King's Cross fit within wider trends in the history of London based railway goods handling, this chapter also goes on to compare the findings of this study with what is known about two of the yard's direct competitors. The results are then used to further explore the value of a combined archaeological and historical approach when investigating railway structures and industrial sites more generally. Closing remarks focus upon the importance of King's Cross Goods Yard to the nation at large and the significance of railway heritage to the history of modern Britain.

13.1 A Reconstruction of Goods Handling, Use of Space and Traffic Flow Over Time and an Exploration of the Forces that Drove Change at King's Cross

Insufficient data was gathered during this project to enable detailed traffic management and goods flow diagrams to be compiled for the entire yard. It was however possible to attempt to reconstruct the flow of traffic in the Main Goods Shed,--- later known as the Eastern Goods Shed, (Landscape Features 7–11) because it was recorded and excavated in considerable detail. In order to illustrate the myriad changes and their effects with greater clarity, they have been grouped into six fixed time periods (corresponding with the formation phase and the major episodes of operational change that occurred afterwards[16]) and presented in a series of diagrams, which enable the effects of these changes on traffic and goods flow over time to be more easily understood. Archival evidence and the arrangement of

---

[16] The six main periods of change that took place in the yard were as follows: the 1850s to the 1870s (Chapter 4); the introduction of capstan shunting in 1882 (Chapter 5); the construction of the Western Goods Shed in 1899 (Chapter 7); Transposition in 1935 (Chapter 9); the reversal of the roles of the Main Sheds in the 1950s (Chapter 11); and the transformation of the building into a road haulage depot post-1981 (Chapter 12).Chapters omitted from this inventory represent periods of relative stability.

the station's goods platforms (i.e. banks) reveal which activities dominated the different spaces. In turn the flow of goods and vehicles through the building can be approximated in terms of likely trends rather than definitive reconstructions. Colour coded plans indicate the tasks that dominated a particular area, although it is likely that sections of the shed were used more flexibly than these diagrams may suggest.

By placing the changes that they illustrate in a wider social, economic and political context, it is possible to discern how effective the yard's responses to these outside influences were. In doing so, an account of how and why the complex flourished, declined and eventually came to an end has been created. The following discussion naturally divides into three larger chronological periods that reflect wider socio-economic trends that affected not only King's Cross Goods Yard but the entirety of London, Britain and in some cases the entire Western world. Economic expansion and industrial growth typified the earliest phase (i.e. the mid 19th to the early 20th century; Chapters 3 to 7), instability characterised the intermediate period (i.e. the years that are bracketed by the two World Wars; Chapters 8 to 10), whilst the decline of manufacturing and the concomitant rise of the service sector during the latter half of the 20th century forms the final phase (i.e. 1946 to the early 21st century; Chapters 11 and 12).

Unless stated otherwise, the following text is based on the many strands of evidence that are presented in Chapters 3 to 12.

13.2 Britain's Industrial Boom. The Causes and Effects of Infrastructure and Working Practice Changes at King's Cross Goods Station: 1849–1913

Throughout the 19th and early 20th centuries, London was at the economic and political heart of an empire that extended from Canada to India. It is therefore unsurprising that the growth of the City continued unabated throughout this period. Sustained by goods and produce transported by sea, river and road, the coming of the railways further fuelled pace of expansion as the population of the metropolis and its hinterland boomed. Even before the first goods train pulled into King's Cross, it became apparent that the station would form a vital supply artery that would be instrumental to this process. Indeed, the expansion of London and the continuing development of its railway connections would form a loop of positive feedback that would cause both to grow until the process was disrupted in 1914 by the advent of war.

When the Great Northern Railway constructed its Permanent Goods and Temporary Passenger Stations at King's Cross between 1849 and 1852 the company realised that the revenue that would be generated by the London end of the line would be crucial to its survival. The Directors were well aware that most of their profits would come from the haulage of commodities to and from the metropolis so it is unsurprising that a great deal of thought, time, energy and money was invested in the London goods rail head and transshipment facilities. From the outset it was essential that the Goods Station could compete for business against the London & North Western's rail, road and canal interchange at Camden. It was therefore decided that King's Cross would have everything that its rival had and more thanks to its grand scale and its state-of-the-art hydraulic network. Whilst construction did not proceed without the occasional glitch, both the Permanent Goods and the Temporary Passenger Stations were completed more or less on time and on budget, enabling the Great Northern to function as a London carrier as early as the latter half of 1850.

Along with a solitary coal drops facility and a canal basin dedicated to the transshipment of bricks and stone (Fig. 13.1), the Main Goods Shed (Landscape Features 7–11) and the Granary (Landscape Feature 12) functioned as the architect Lewis Cubitt had intended during the first year or so of their existence (Fig. 13.7). Incoming wagons were unloaded in the Eastern Transit Shed (Landscape Feature 8) or the Granary (Landscape Feature 12) and were reloaded in the Western Transit Shed (Landscape Feature 7) before being reformed into trains for onward dispatch in the Train Assembly Shed (Landscape Feature 11). Delivery and onward shipment by water also took place thanks to the link with the Regent's Canal, whilst a huge company of horses was able to move goods to and from the complex via London's road network. The interchange rapidly proved to be a success as huge volumes of goods began to pour in. Weaknesses in Cubitt's design soon became apparent however, as the growing volume of rail and road traffic began to overwhelm the Goods Station within months of it entering service. Indeed for the duration of this period the pace of growth was so great that the development of rival depots at Agar Town in the 1860s, at Somers Town in the 1880s and at Marylebone in 1899 (all of which directly competed with King's Cross) did result in the loss of additional traffic but did not cause the numbers of vehicles frequenting the complex to significantly decrease. Instead, the continuing growth of goods traffic forced the yard's managers to find ways to combat the chronic rail and road congestion that plagued the yard for most of this period.

Inbound coal traffic, the coal handling infrastructure and adjacent, affiliated coal reception facilities: 1849–1913

It soon became apparent that inadequate reception facilities existed at King's Cross for many of the imports that London consumed in vast quantities, the most significant of which was coal. In 1850, facilities at the Goods Depot for the supply and administration of that commodity were few, comprising just one set of coal drops (Landscape Feature 17), a few dedicated sidings and two office blocks (Landscape Feature 18). As the Great Northern extended its reach into the northern coalfields and the demands of the metropolis grew, it became evident that the company had underestimated the amount of infrastructure that would be required to handle this trade. Consequently, just sixteen years after the yard opened for business, coal handling capacity had more than trebled. The original coal drops were

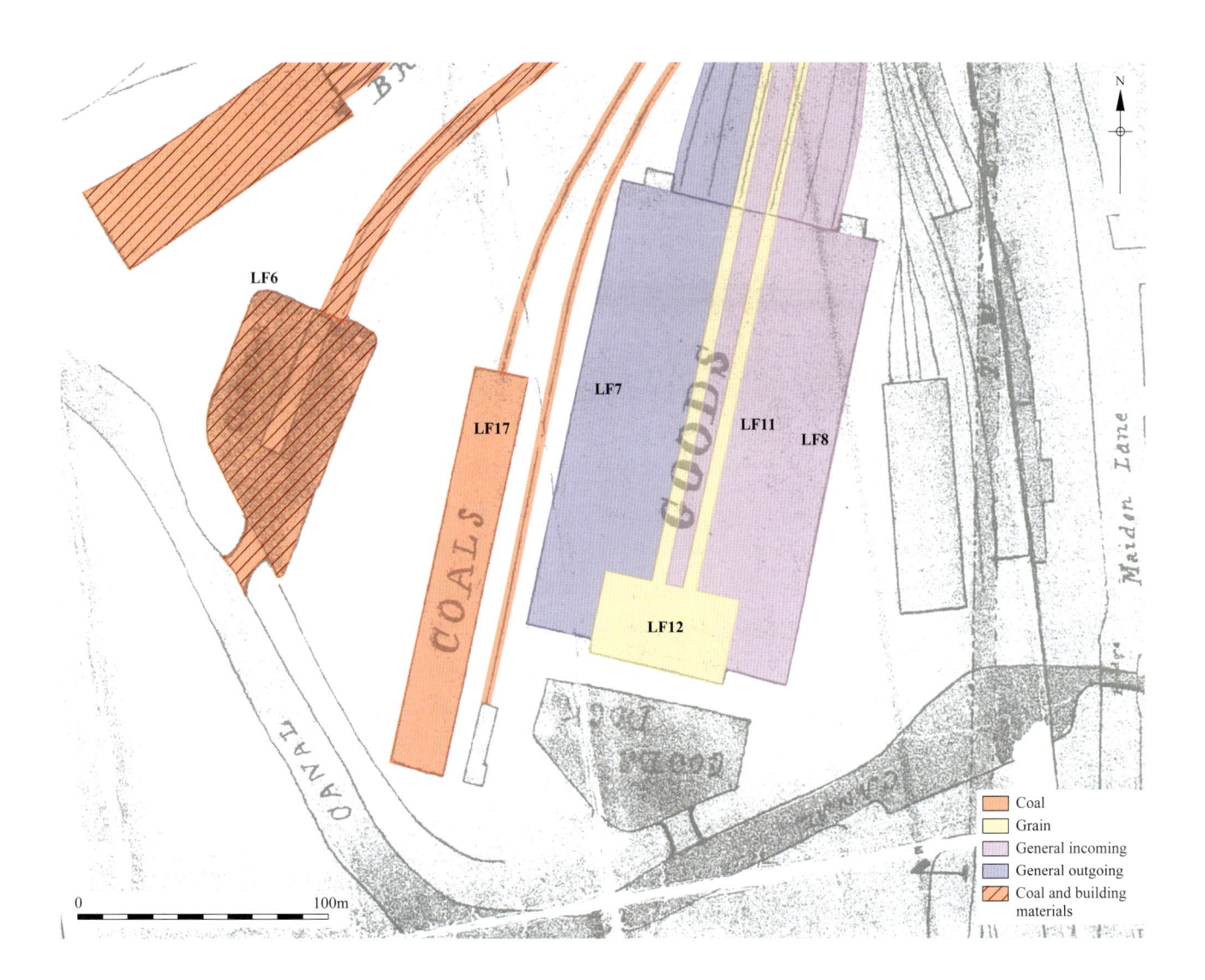

Fig. 13.1 Incoming and Outgoing Commodities in the Goods Yard *c*.1852 overlain on Captain Galton's Sketch Plan of October 1852 (TNA MT 6/10/38), scale *c*.1:2,500

supplemented by an additional facility to the west (Landscape Feature 22) and an enormous coal stacking ground was created at great expense in the area between the Goods Yard proper and the Great Northern's running sheds (Fig. 13.2). Then in 1867 a viaduct was constructed across the Regent's Canal (Landscape Feature 24) so that fuel could be delivered by the Great Northern to the Imperial Gas Light and Coke Company directly, an arrangement that continued until the St Pancras Gasworks closed in 1904 (Fig. 13.3). In addition to these infrastructural developments, the Coal Offices (Landscape Feature 18) were greatly enlarged in the early 1860s so that a greater number of independent merchants could be admitted after the Great Northern was stripped of its monopoly over the sale of coal carried over its rails (Fig. 13.2).

Between the 1860s and the 1880s, the Great Northern Railway continued to strengthen its hold over the coal producing areas in the Midlands and the north of England, at a time when demand in London continued to grow unabated. The fact that coal reception facilities at King's Cross actually started to shrink after 1871 may therefore seem curious, however the phenomenon may be easily explained. As London's local and suburban railway network improved, it became obvious that it would be more efficient to ship the commodity across the city by train to numerous coal yards dotted throughout the capital instead of bringing it all to one large depot at the end of the mainline. Decentralisation meant that the Great Northern could access a far larger share of the London market, both north and south of the river, whilst long and inefficient journeys by horse-drawn dray as a means of onward distribution could be minimised. It also meant that an increasingly large percentage of London-bound coal would bypass King's Cross, meaning that the yard's coal handling infrastructure did not need to be enlarged after 1871.

Meanwhile the success of the entrepreneur Samuel Plimsoll's technologically superior coaldrops at Cambridge Street (Landscape Feature 23) quickly drew the focus of the coal trade away from the Goods Yard proper to his facility after it opened in 1866. The Great

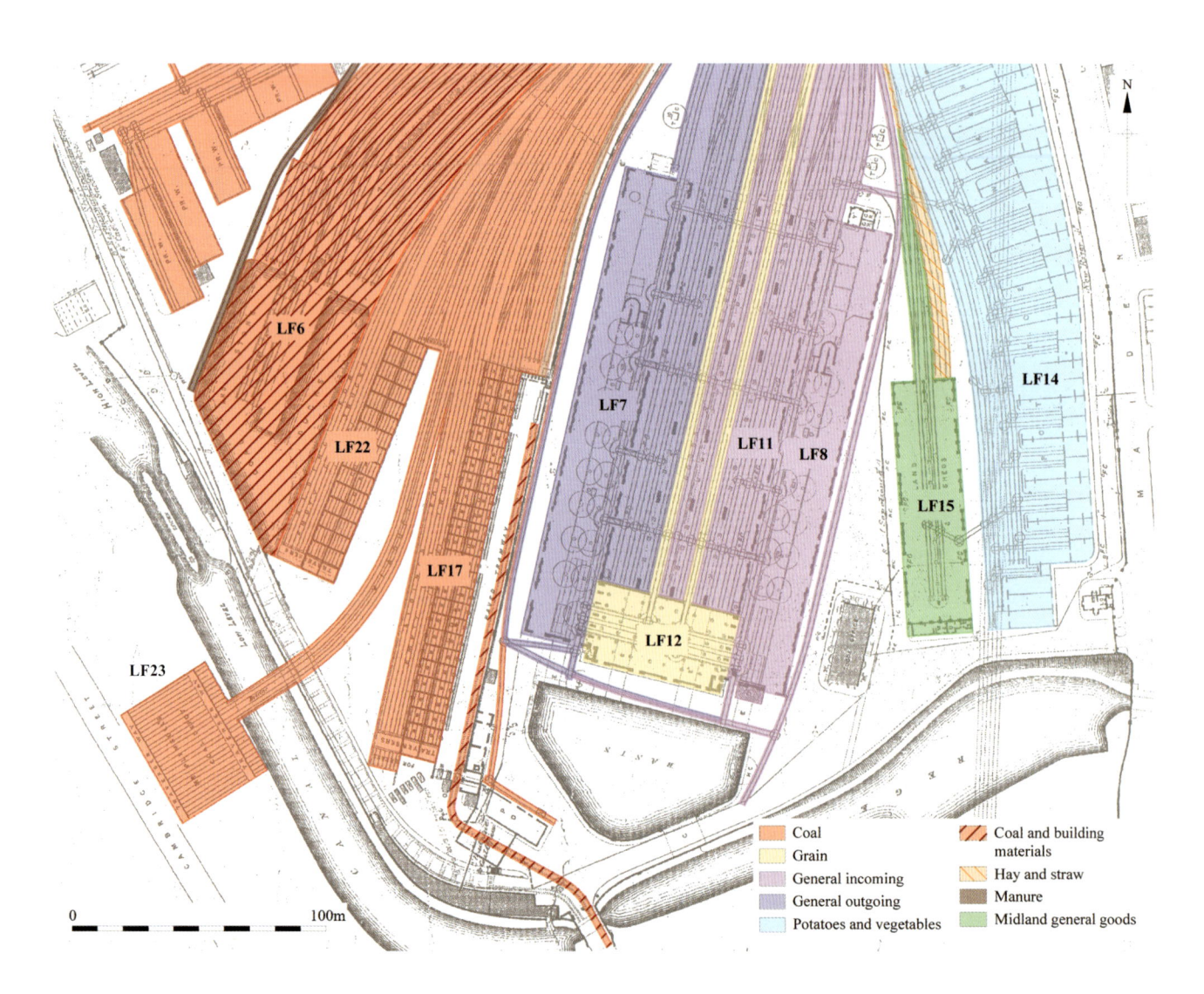

Fig. 13.2 Incoming and Outgoing Commodities in the Goods Yard *c*.1866 overlain on the Humber Plan of 1866 (NNRG DMFP 00026266), scale 1:2,500

Northern appears to have had neither the space nor the inclination to fight Plimsoll, instead using his success to its advantage. After all, the company was were still making a great deal of money from profits of carrying coal to his depot. After Cambridge Street opened, it became possible to free up much needed space in the Goods Yard and rent the newly-vacated sections of the former coal drops (Landscape Features 17 and 22) out as warehousing to new customers. Plimsoll would be left to his own devices until the Great Northern was given the opportunity to purchase Cambridge Street in 1891, which returned the reception of coal at King's Cross to the company.

The success of Cambridge Street also impacted upon the former Coal Offices (Landscape Feature 18), the original administrative heart of the coal trade. They became increasingly detached from the practicalities of coal dealing as Plimsoll's depot began to dominate, becoming increasingly unpopular with the merchants. From the 1880s to the 1900s the coal traders that once resided within them relocated to new premises around Cambridge Street and the old offices were occupied by hay and straw merchants, clerical staff of the Horse Department, a harness workshop and Great Northern Goods Department employees tasked with administering the fish trade at the yard.

Inbound vegetable traffic in the Potato Market and the East and West Handyside Canopies: 1849–1913

Landscape Features 14, 26 & 27

As with coal, the Great Northern appears to have somewhat underestimated the vast quantities of agricultural produce, especially potatoes, that would flow into King's Cross thanks to the relative proximity of the latter to Covent Garden. In the first instance, such traffic must have been received in the Eastern

Fig. 13.3 Incoming and Outgoing Commodities in the Goods Yard *c*.1871 overlain on the Ordnance Survey Plan of 1871, scale 1:2,500

Goods Shed (Landscape Feature 8) but within the space of a few months it must have become apparent that it would struggle to cope if separate facilities were not provided. It was therefore fortuitous that the old temporary passenger station (Landscape Feature 14) became available when it did, allowing its conversion into a dedicated potato market as early as 1852 (Fig. 13.2). Doubtless the new complex took considerable pressure off the Eastern Transit Shed, but its design was far from perfect. Whilst it provided a welcome separate space for the potato merchants to carry out their trade it was a poor working environment, not least because the offices provided by the railway company were small and cramped and the majority of the unloading of wagons took place in the open. The hasty conversion of the Temporary Passenger Station resulted in insufficient thought being given to the perishable nature of the goods that the Potato Market would receive and the working conditions of those who toiled there.

A series of improvements to the market's infrastructure were made over the years, however it was not until the mid 1880s when the Midland Railway opened a rival facility at Somers Town that the Great Northern responded to the concerns of the potato traders. After vociferous threats and protests from the merchants, the Great Northern constructed the East and West Handyside Canopies (Landscape Features 26 and 27) in 1888, thus creating a large covered area where vegetable traffic could be received without hindrance or inconvenience (Fig. 13.4).

Inbound fish traffic and the East Handyside Canopy: 1849–1913

Landscape Feature 26

As the decades rolled by, King's Cross started to handle a growing quantity of fish from the northern and eastern ports. The volume of fish traffic increased after the Great Northern consolidated its access to the Norfolk coast through the execution of a joint venture with the Midland in 1889. Once this deal had been struck,

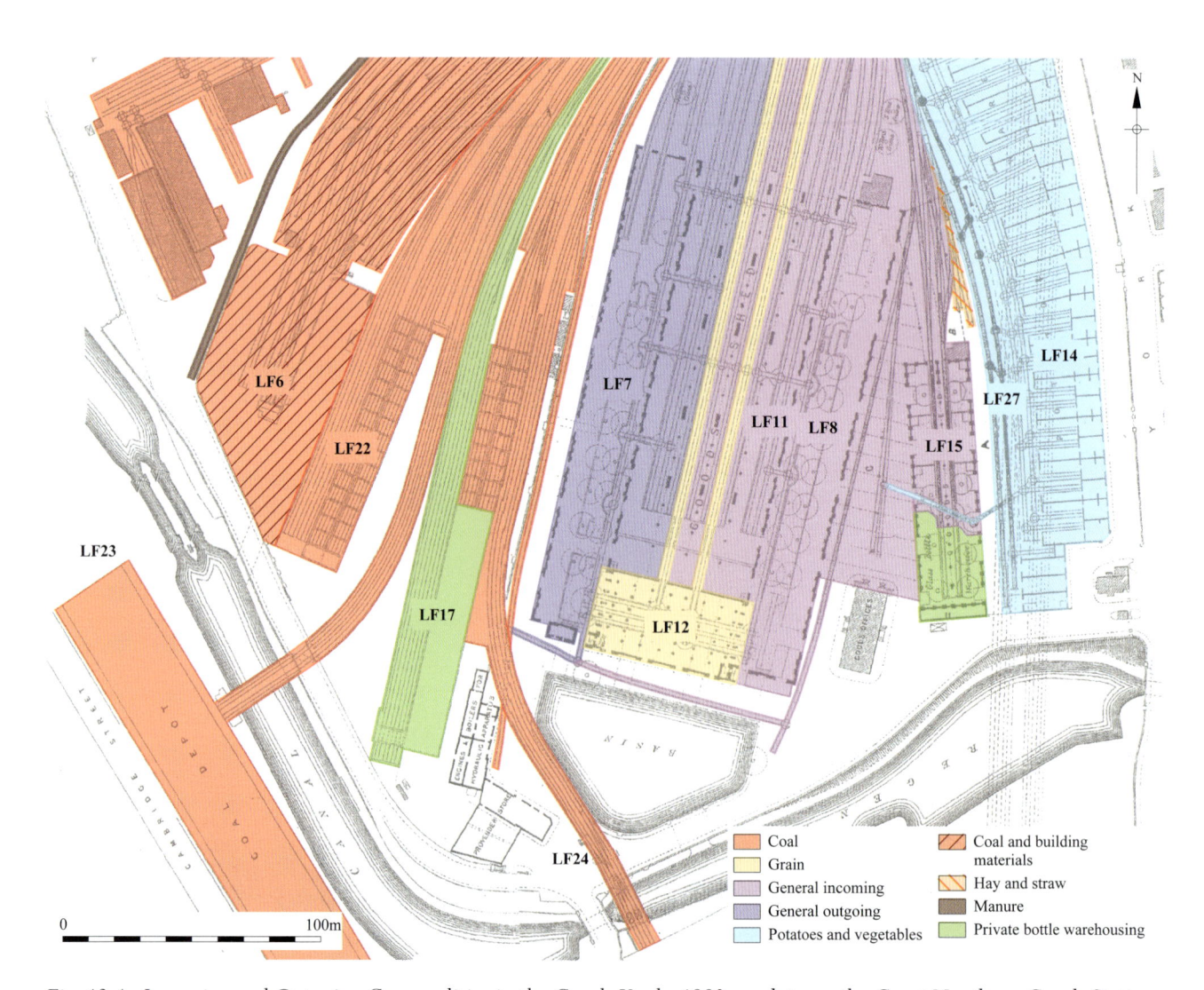

Fig. 13.4 Incoming and Outgoing Commodities in the Goods Yard c.1882 overlain on the Great Northern Goods Station Plan of 1882, scale 1:2,500

it appears that the majority of this traffic entering the goods yard moved away from the Eastern Transit Shed (Landscape Feature 8), towards the newly completed West Handyside Canopy (Landscape Feature 26), from where that perishable commodity could be dispatched onwards by road more quickly (Fig. 13.5). This may also have coincided with the migration of the clerical staff who handled the administration of the fish trade to the former Coal Offices (Landscape Feature 18), after the latter fell out of favour with the coal merchants of Cambridge Street.

Bricks and stone and the Coal and Stone Basin: 1849–1913

Landscape Feature 6

Dedicated facilities for the reception of the yard's stone and brick traffic were improved between 1850 and 1866 but developments were not as dramatic as they were for many other commodities. The Great Northern realised from an early stage that the growing city would need these materials in huge quantities and ensured that there was sufficient space to increase the number of railway sidings that surrounded the Coal and Stone Basin (Landscape Feature 6) with relative ease (Fig. 13.1; Fig. 13.2). Sandwiched between coal reception facilities to the east, west and south, it is likely that these lines were utilised flexibly by coal wagons as well as those carrying bricks and stones. Brick traffic was also received and handled at a number of sheds dotted around the goods yard that the Great Northern leased to independent brick manufacturers.

As the volume of canal traffic entering the yard fell over time, the Coal and Stone Basin gradually became underused. By the late 1890s it was a largely redundant feature, which is why the Great Northern chose to infill it and use it as the site of a second general goods station (Landscape Feature 28) which they opened in 1899. From that point onwards brick and stone traffic continued to be

Fig. 13.5 Incoming and Outgoing Commodities in the Goods Yard c.1894–6 overlain on the Ordnance Survey Map of 1894–6, scale 1:2,500

received by other 'inwards' areas within the Yard, a feat made possible by the construction of the new 'general' shed (Landscape Feature 28), which doubled handling capacity at King's Cross after it opened (Fig. 13.6).

Effects upon incoming and outgoing 'general' goods traffic in the Main Goods Station; 1849–1913

Landscape Features 7–11

It appears that in the first instance incoming and outgoing wagons were kept separate from one another in the Main Shed (Landscape Features 7–11) to minimise collisions and rationalise traffic flow (see Section 3.12). Assuming that this assumption is correct then the arrangement of railway lines in the Main Goods Station, particularly those that were used to access the unloading areas in the Granary (Landscape Feature 12), strongly suggest that just over half of the lines and banks in the Train Assembly Area (Landscape Feature 11) were used for breaking up and unloading trains, whilst the remaining space was set aside for disassembly and loading (Fig. 13.7). This arrangement, which seems to have been in existence from the outset, suggests that the Great Northern Railway always anticipated that the flow of general goods into London from the north would be greater than the flow in the opposite direction.

Infrastructure for the transhipment and temporary storage of general goods in and around the Main Goods Shed was soon found wanting. Too much space had been devoted to train reassembly and disassembly, whilst too little had been allocated to loading and unloading (Fig. 13.7). However, in contrast to developments elsewhere, the size of the building could not be increased since it was hemmed in on all sides by other facilities (Fig. 13.1). Until the construction of the Western Goods Shed (Landscape Feature 28) in 1899, the only recourse was to keep it as technologically up to date as possible and use the existing space within it more efficiently.

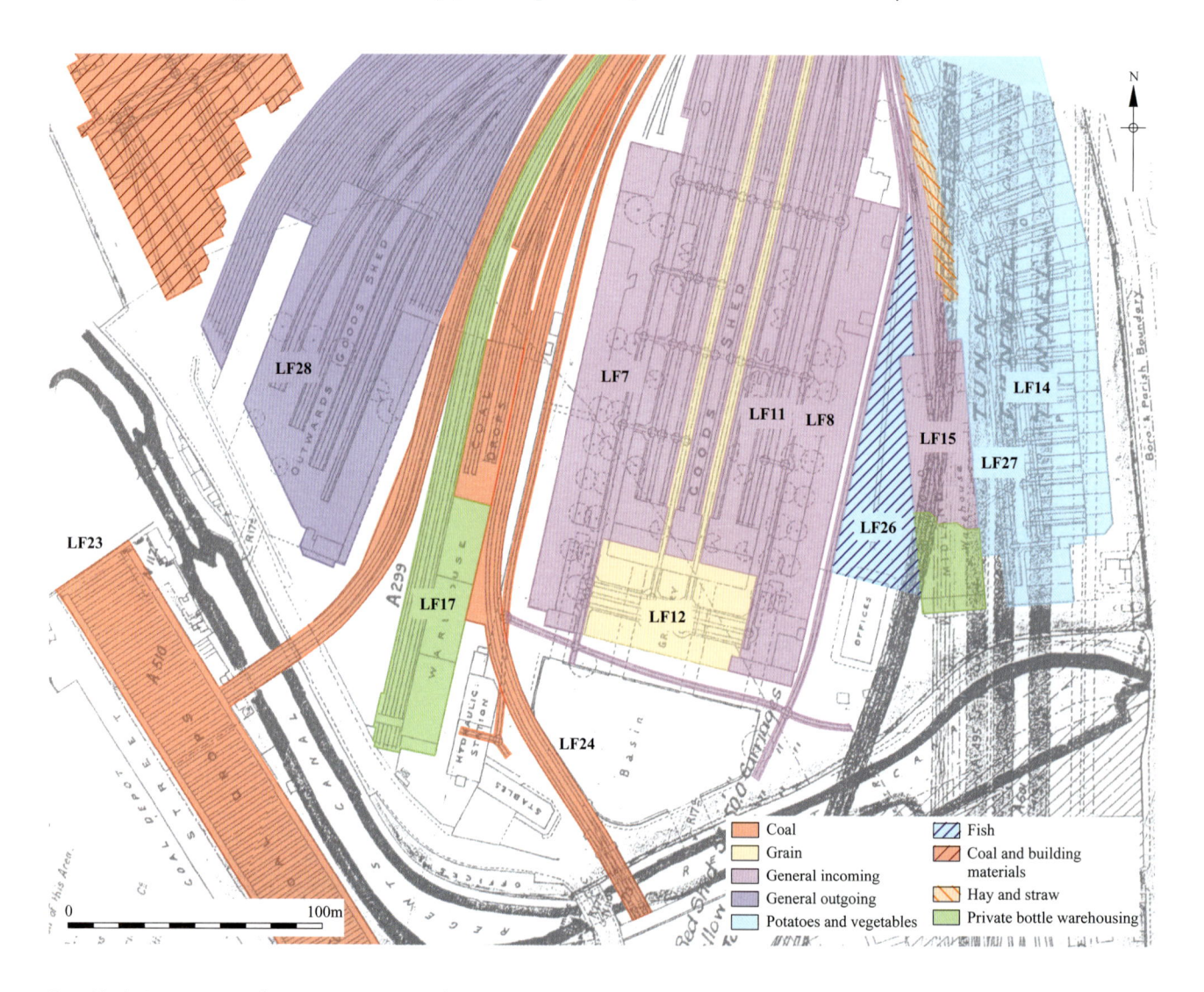

Fig. 13.6 Incoming and Outgoing Commodities in the Goods Yard c.1905 overlain on the Great Northern Railway Station Plan of 1905, scale 1:2,500

The period between 1850 and 1879 was characterised by a series of alterations to the 'general' arrivals area (i.e. the Eastern Transit Shed, Landscape Feature 9, and its affiliated outdoor infrastructure). All were designed to increase capacity and streamline operational procedures and to reduce road and rail congestion. In the 1850s additional platform accommodation and unloading space in the Eastern Transit Shed was created. Capacity was increased on at least three occasions between 1855 and 1857, which together boosted by approximately one quarter the amount of space that was available for unloading and temporary storage (Fig. 13.8). Although these measures must have offered a modest improvement, the station reached its maximum operational capacity just eight years later. The southern end of the shed was therefore altered once again in 1871–2 through the enhancement of the ground floor platform accommodation and the addition of an upper storey, after which little more could be done to increase capacity without seriously disrupting the flow of road and rail traffic. The only option was to utilise part of the Train Assembly Shed (Landscape Feature 11) for the purpose of unloading and in 1872 a square unloading platform was erected in the southeast corner. That decision represented the first major alteration to the allocation of space in the Main Goods Shed (i.e. to the four different tasks that were carried out within it: train disassembly, unloading, loading and train reassembly; Fig. 13.8). Similar alterations were made to the departures area (i.e. the Western Transit Shed, Landscape Feature 7) during the 1850s and 1870s for much the same reasons but the changes were less drastic. A modest amount of additional space was created next to the stables ramp in 1857, whilst a more substantial platform enlargement scheme was implemented in 1870–1 in the northern section, which involved the insertion of two additional sidings (Fig. 13.8). This suggests that, as anticipated by the Great Northern Railway, the flow of general goods from the north into London was greater and was increasing at a somewhat faster rate than the flow in the opposite direction.

Together, these alterations necessitated that the road and railway infrastructure in and around the Main Goods Shed be modified, thus forcing its operational procedures to change in terms of the use of space, the flow of traffic and the transit of goods through the complex (Fig. 13.8). The changes were modest in comparison with what would come later, but they do represent the first significant departure from the station's design as envisaged by Lewis Cubitt and his Great Northern associates.

The adaptations of the 1850s, 1860s and 1870s were not sufficient to eliminate the congestion that had plagued the Main Goods Shed since it entered service. In 1882 a substantial modernisation drive was launched in order to accelerate goods handling in the shed and to resolve the problem of congestion. This involved the introduction of hydraulic capstan shunting to the shed, thereby ensuring that the shed was equipped with the latest technology. The installation of a substantial quantity of additional platform accommodation in the Train Assembly Shed (Landscape Feature 11) also formed part of the scheme, and that area was used thereafter for loading and unloading as well as train assembly and disassembly. This represented a much more efficient use of the space that was available, but resulted in a series of major operational changes to the flow of traffic and goods through the building (Fig. 13.9). The removal of the southernmost transverse track meant that longer sections of the shed's north–south lines became bidirectional, whilst better access to the banks would have been possible providing that the remaining 'long' east–west lines were used more flexibly (Fig. 13.9).

Although the introduction of capstan shunting necessitated a substantial financial outlay on the part of the Great Northern, the money was well spent as the scheme does appear to have achieved its objectives in the Main Goods Shed, for the time being at least.

Rail traffic to and from the capital continued to increase throughout the remainder of the 19th century, which meant that congestion problems would soon return to the Main Goods Shed. The catalyst for the company's decision to upgrade facilities at King's Cross in the late 1890s may have been the imminent arrival of the Great Central Railway (the creators of the last main line that would run into London) and its state-of-the-art goods handling facilities at Marylebone. In order to retain as much traffic as possible, the Great Northern took decisive action and implemented an enlargement programme that would double capacity at King's Cross Goods Yard by the end of the decade. This involved the construction of a sister station, the centrepiece of which was the Western Goods Shed (Landscape Feature 28). It would handle all outgoing traffic, whilst the original Main Goods Shed, henceforth known as the Eastern Goods Shed (Landscape Features 7–11), would receive incoming commodities only (Fig. 13.6). This represented the most significant operational change that had been made to the yard since it opened in 1850. The result was that the congestion issues were largely resolved whilst the Great Central was deprived of any technological advantage that it might otherwise have had.

With the exception of some minor modifications to the arrangement of turntables, the Eastern Goods Shed's existing railway infrastructure was not significantly altered in 1899, although cartographic evidence does suggest that the central bay of the building was being used as an unloading and storage area by that time. With that exception, the changes were largely operational rather than physical (Fig. 13.10). The shed must have functioned more flexibly than it previously had done in order to maximise access to all the unloading banks (for example by running wagons in two directions along all of the transverse tracks), whilst all the former 'down' lines henceforth accommodated 'up' traffic (Fig. 13.10).

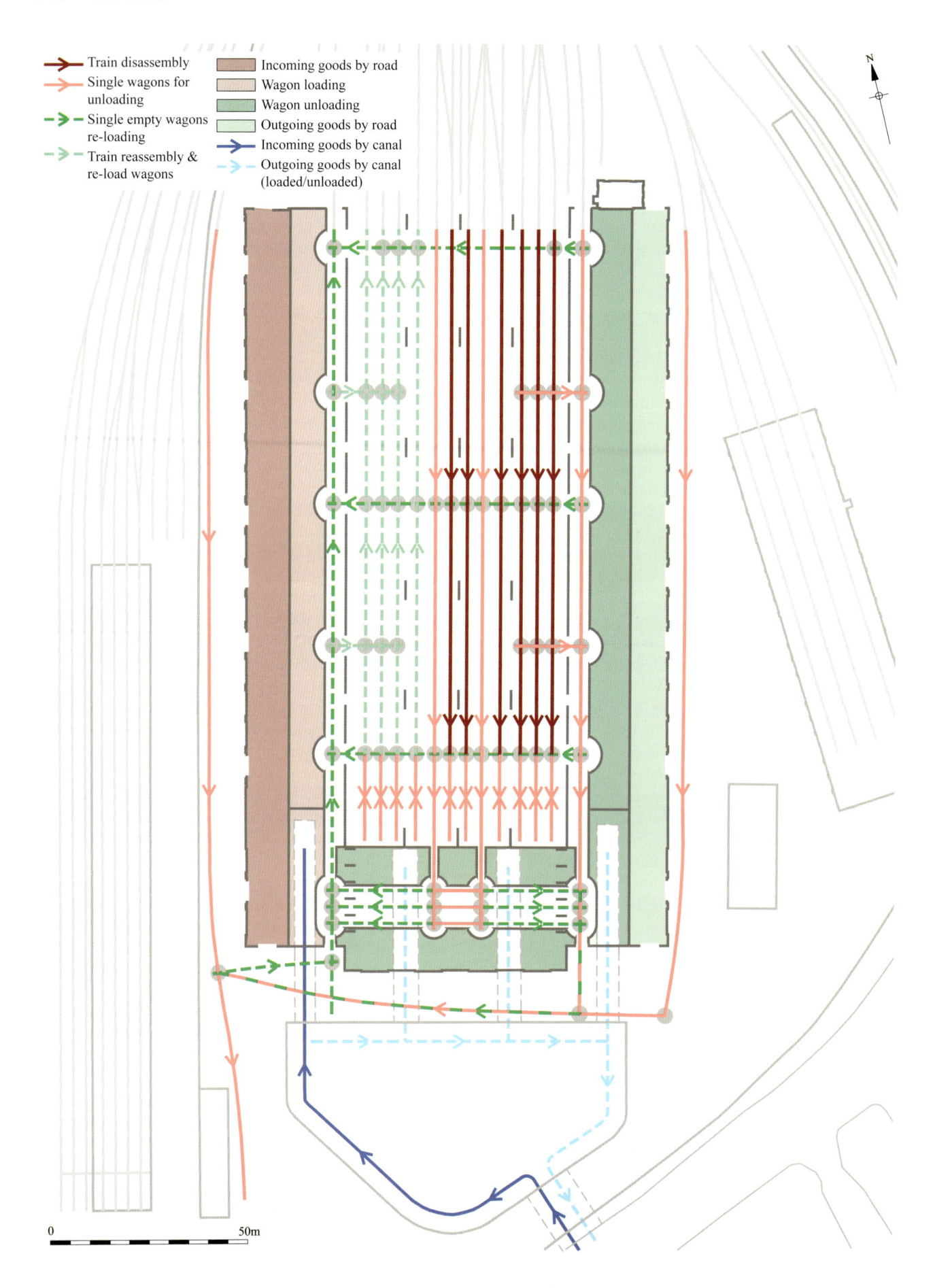

Fig. 13.7 Reconstruction of the probable goods management procedures and goods flow through the Main Goods Shed (Landscape Feature 7–11) and the Granary (Landscape Feature 12) c.1852, scale 1:2,000

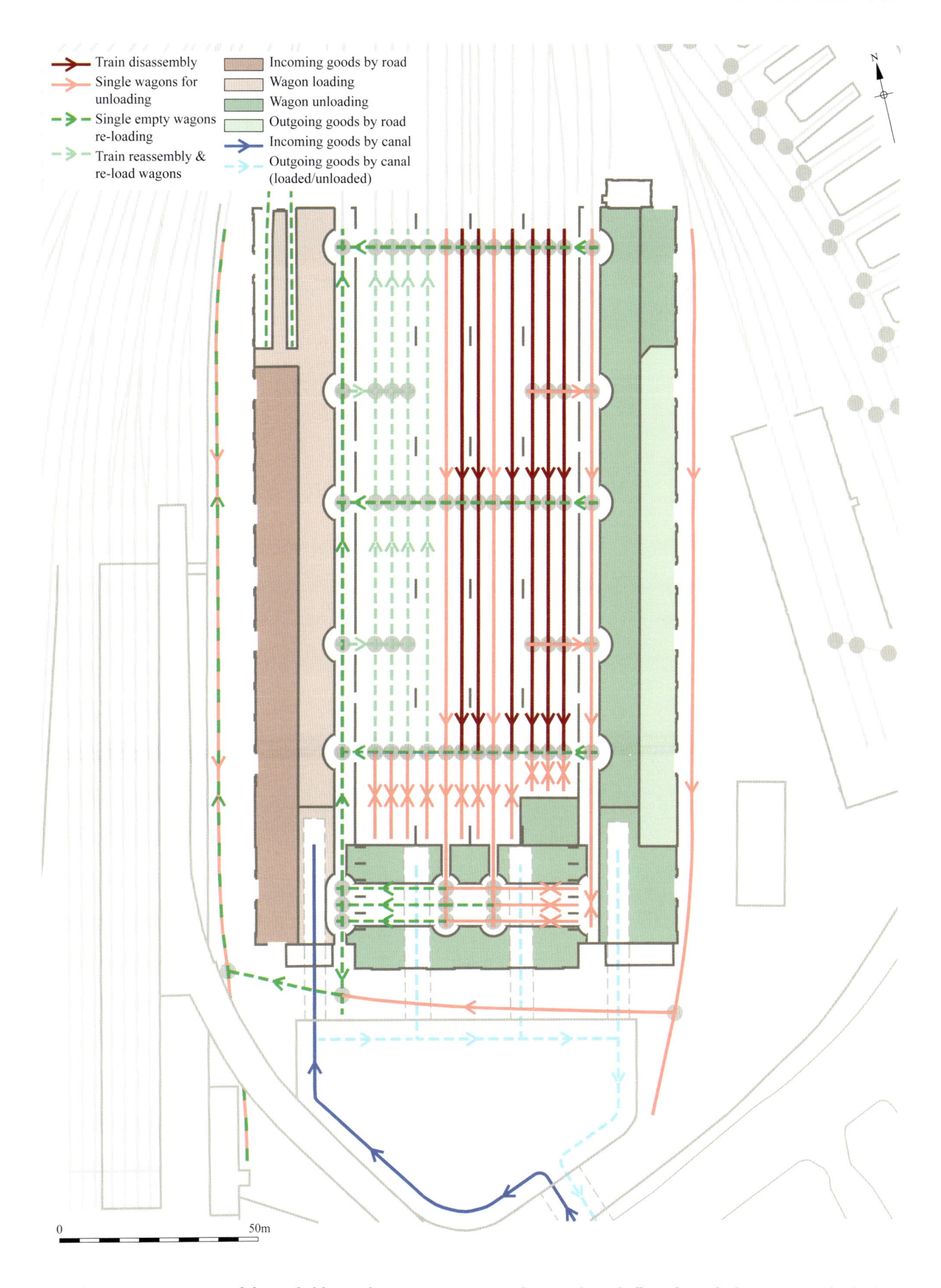

Fig. 13.8 Reconstruction of the probable goods management procedures and goods flow through the Main Goods Shed (Landscape Features 7–11) and the Granary (Landscape Feature 12) c.1879, scale 1:2,000

308 Conclusions

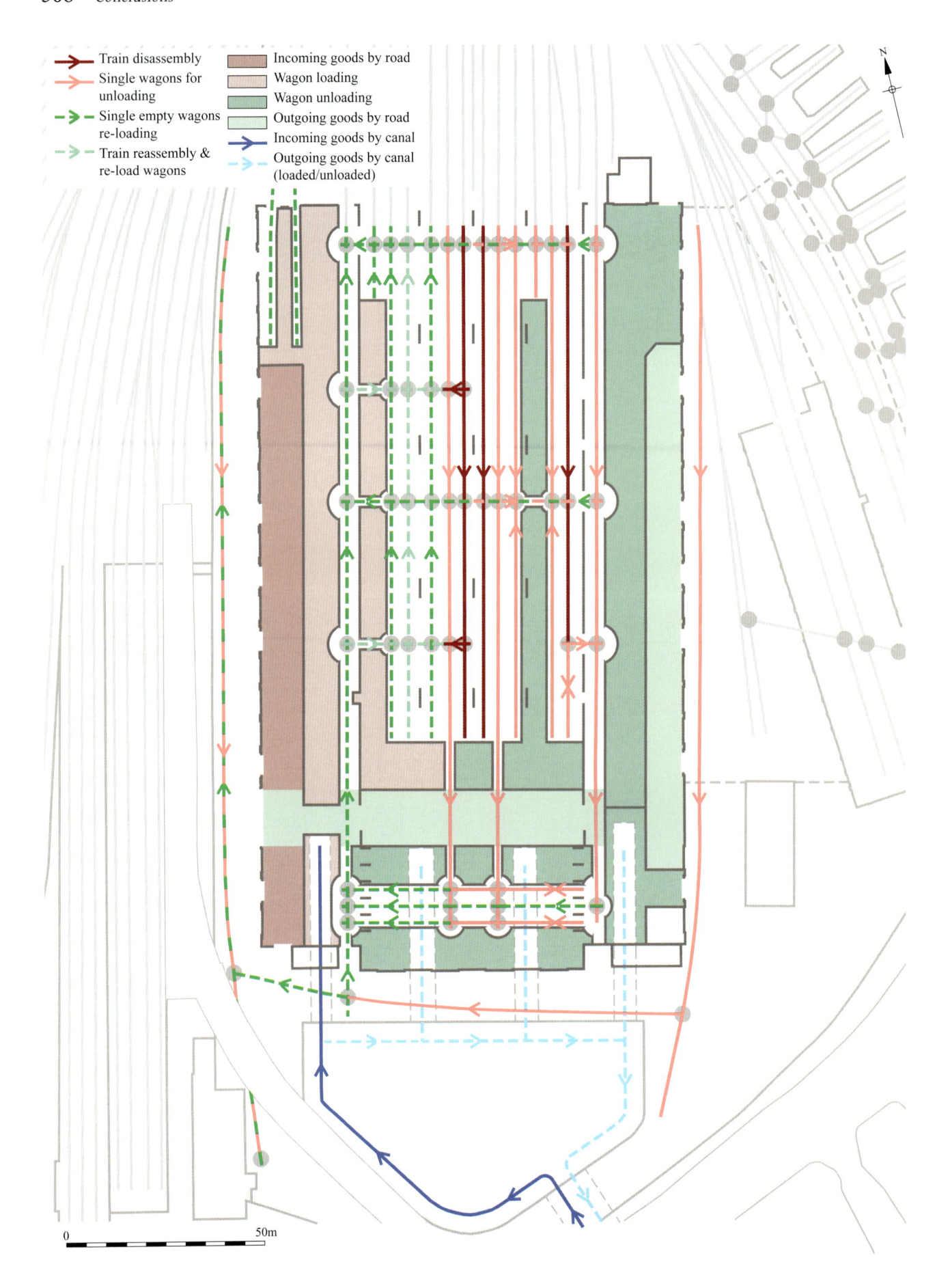

Fig. 13.9 Reconstruction of the probable goods management procedures and goods flow through the Main Goods Shed (Landscape Features 7–11) and the Granary (Landscape Feature 12) c.1882, scale 1:2,000

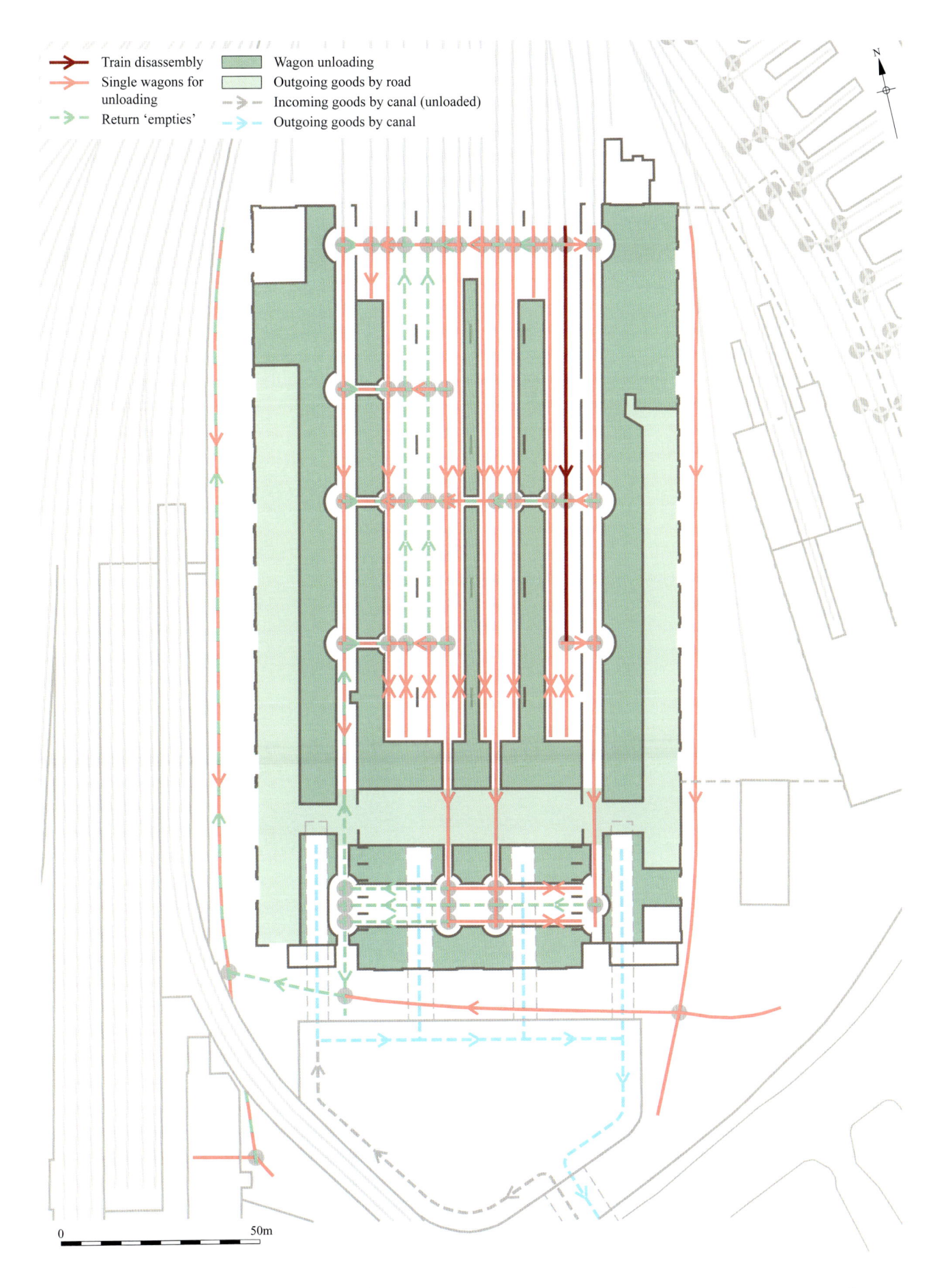

Fig. 13.10 Reconstruction of the probable goods management procedures and goods flow through the Eastern Goods Shed (Landscape Features 7–11) and the Granary (Landscape Feature 12) *c*.1899, scale 1:2,000

Effects upon incoming grain traffic and the Granary: 1849–1913

Landscape Feature 12

The fact that relatively few alterations were made to the Granary (Landscape Feature 12) during the first decade of its existence suggests that it functioned reasonably effectively at first. However, by the mid-1860s, Seymour Clarke had declared that the structure had become 'quite full', thus triggering the construction of additional railway sidings for the reception of grain and potatoes, presumably in the vicinity of the Potato Market (Landscape Feature 14). Inside the building itself, improvements were limited to the addition of a few extra gravity driven chutes and trapdoors and the rationalisation of traffic flow through the building after the southern tip of the Eastern Transit Shed's (Landscape Feature 8) railway siding was lifted prior to 1871 (Fig. 13.8). The removal of that part of the siding prevented the return of empty grain wagons to the Western Transit Shed (Landscape Feature 7) via the eastern side of the Granary, meaning that they would henceforth have to exit via the western side only (Fig. 13.8). This was a positive change, since the journey back to the Western Transit Shed via the lost eastern route was a longer and therefore less efficient pathway than the one that could be found to the west. As a consequence it was probably under-used meaning that it is unlikely that the shed workers mourned its passing.

When capstan shunting was introduced to the Main Goods Shed (Landscape Features 7–11) in the early

Fig. 13.11 Reconstruction of the Granary showing internal arrangement for movement of grain around the building. By Chris Mitchell (c.mitchell@btinternet.com)

1880s, an access point was inserted across the southern end of the complex (Fig. 13.9). This must have ameliorated road congestion in that area to a degree, however it was the road vehicle access that it afforded to the north face of the Granary that made the greatest difference. That building's northern loading bay doors and windows at ground floor level would thereafter be used for the onward shipment of grain by road, a development that boosted its efficiency further.

At a similar time, two turntables were removed from the eastern side of the Granary, which had earlier been used to return empty wagons from the Eastern Goods Shed (Landscape Feature 8) to the Western Goods Shed (Landscape Feature 7) and the Train Assembly Area (Landscape Feature 11) via the Granary's transverse tracks (Fig. 13.8; Fig. 13.9). Evidently, just one track was sufficient to achieve this purpose and it is presumed that the central example was chosen because, unlike the other two east–west tracks, it did not connect with any of the Granary's hydraulic winches (which were used for direct unloading from train wagons; see Section 3.11). This meant that the central track was henceforth reserved for 'return empties' whilst the outer tracks were exclusively used by wagons laden with grain (Fig. 13.9). In all probability, this arrangement was a long standing one since it would have been the most efficient traffic management option in the Granary. However, the removal of the two turntables would have formalised it and precluded any alternative choices that might on occasions in the past have disrupted the unloading of grain and the flow of traffic through the building.

The fact that the Granary functioned relatively effectively in comparison with the adjoining Main Goods Shed is highlighted by the fact that no other major modifications were made to it during this time, even when the Western Goods Shed was constructed in 1899. The original design of the Granary was well thought through, but that alone cannot account for its long term success since Great Northern records suggest that it was full to bursting as early as 1860. Perhaps a more important factor was that after the repeal of the Corn Laws the grain trade was becoming an increasingly international affair. It therefore seems likely that it was London's dockland warehouses that bore the brunt of the growth in demand rather than King's Cross.

The growth of rented accommodation: 1849–1913

When it was established in the mid-1840s, the Great Northern Railway could not have anticipated that several of its customers, most significantly a number of bottle makers, would soon carry such vast quantities of their products into King's Cross that they would press for rented warehousing within the yard itself. Other than building temporary sheds in the yard (as it did for some of the brick manufacturers), the only recourse available to the Great Northern was to utilise redundant buildings as and when they became available. Consequently the Yorkshire firm Kilner Brothers moved into the southern section of the former Midland Goods Shed (Landscape Feature 15) in 1869 whilst sections of the Eastern and Western Coal Drops (Landscape Features 17 and 22) were rented out as bottle warehousing after the focus of the coal trade shifted away from the Goods Station to Cambridge Street and the wider London area.

The relationship between the road, the railway and waterborne transport: 1849–1913

The heyday of railway transport coincided with the Great Northern Railway's tenure at King's Cross. Although transport by sea and inland waterway barge remained viable options for the long distance movement of bulk goods for most of the 19th century, waterborne transport was slow in comparison to the railway and as a consequence the train would erode the volume of waterborne goods as time marched on. The relatively high speeds at which goods could be transported by rail was one of the selling points of the London & York railway scheme as early as 1844, by which date producers had already begun moving their goods by rail in preference to using the canals.

The interplay between the shipping of commodities by sea and by rail was much more complicated. The coming of the railways did poach some domestic business away from the docks, for example through the disruption of the trade in sea coal, however the situation was not straightforward in terms of international commerce; after all, Britain's 19th-century economy was already a global one. Much trade with the Empire as well as unaffiliated nations was taking place and certain commodities could only be obtained from abroad whilst others could be produced more cheaply in other countries. London's docks received vast quantities of goods throughout the 19th century, sending the imports on via the canal and, during the latter half of the century, the railway in a similar way to the Great Northern's interchange at King's Cross. However, virtually all of those commodities presumably bypassed the southern end of the Great Northern's mainline into London, whilst in contrast some exotic items arriving at ports on the west coast of England may well have been carried all the way into the capital on Great Northern rails[17].

The national road network was poor in comparison to the swift connections that were offered by the train. Until the internal combustion engine became dependable in the early 20th century, the roads could

---

17   This certainly happened from the 1890s onwards (e.g. transshipment of bananas from the port of Liverpool to King's Cross).

not compete with the railways in terms of speed and efficiency. Consequently, the primacy of long distance goods haulage by train could not be challenged by the road for the duration of the 19th and early 20th centuries. The same cannot be said for short-haul goods movement, for which road cartage was indispensible. Although London's local and suburban railway network was improved considerably during the Victorian and Edwardian periods (and was used very effectively by the Great Northern and other railway companies to distribute goods throughout the metropolis), road vehicles were still needed for door to door deliveries. Consequently, the railway and the road co-existed symbiotically for the duration of the Great Northern's existence as they slowly usurped both short and long-haul canal transport.

The evolution of the King's Cross Goods Yard reflects these wider economic trends, infrastructural developments and technological changes. Canal infrastructure was not enlarged at any point during the yard's history. In contrast to the stagnation and eventual decline of the Regent's Canal, reception facilities for long distance incoming and outgoing trains grew rapidly whilst the station's road network and stabling facilities were improved and enlarged to keep pace with the growing demand for short-haul pick-ups and deliveries by road. The only exception was the Granary (Landscape Feature 12), the handling capacity of which was only moderately increased between 1852 and 1914. It is likely that this building was not tested in the same way as the rest of the Goods Yard's Inwards facilities since an ever increasing percentage of the capital's grain was arriving from abroad via the London docks.

As time marched on, long term weaknesses inherent to a design that allocated such a large amount of much needed space to canal infrastructure were exposed. Although the Western Goods Shed (Landscape Feature 28) had replaced the Coal and Stone Basin (Landscape Feature 6) by 1899, the Granary Basin (Landscape Feature 5) survived into the 20th century, by which time it was barely used at all.

13.3 Conflict and Crisis. The Causes and Effects of Infrastructure and Working Practice Changes at King's Cross Goods Station: 1914–1945

The rising prosperity of Victorian Britain was both a cause and an effect of the steadily increasing proportion of freight carried by rail as the 19th century progressed. However, the outbreak of war in 1914 disrupted that process and heralded the start of a period of political, social and economic instability that affected Britain and the wider Western world until after the end of the Second World War. Bookended by two World Wars, this tumultuous period also saw the most severe economic depression of the 20th century. In contrast to what had gone before, the early to mid-20th century was therefore characterised by huge variations in industrial output as the nation struggled to meet the demands of the two wars and battle through the economic collapse in between. Developments at King's Cross reflect these wider economic trends.

The impact of the First World War upon the Goods Yard and its operational procedures: 1914–1918

The First World War saw the volume of goods entering London by rail double in the space of four years as the Government prioritised the transport of freight above civilian passenger traffic. State control of the railways ensured that while the cost of the enlargement of the Western Goods Shed and the conversion of the former Midland Goods Shed (Landscape Feature 15) to goods use was initially met from the Great Northern's own funds, the company would have been compensated by the Government for works considered to be of benefit to the war effort (Fig. 13.12). Yet the state's interest in the railways was strictly confined to ensuring that they effectively served the national interest for the duration of the war. This disparity between the short-term priorities of the state and the long-term financial interests of the railway companies meant that weaknesses in the system exposed by the war were usually resolved by short-term fixes rather than by investment in infrastructural improvements. Consequently, few other changes were made to the fabric of the Goods Station during the war. The fact that the dramatic rise in rail freight traffic that took place during the First World War does not appear to have overwhelmed the Great Northern's operation at King's Cross suggests that the enlargement programme of 1899 was largely successful, enabling the yard to handle the ensuing influx tolerably well.

The economic crises of the interwar period and their effect upon the Goods Yard and its operational procedures: 1919–1938

At the end of the First World War Britain's railway companies were impoverished. Four years of limited funding, rising wage bills and maintenance arrears had drained the companies' coffers and left the country's railway infrastructure in a parlous state. The Ministry of Transport's solution to its railway problem was intended to encourage greater coordination and less competition by amalgamating companies into four large regional groups. From the outset however, the London and North Eastern Railway inherited a range of structural problems from its constituent parts which were to leave it uniquely exposed to the economic turbulence that characterised the 1920s and 1930s. During the 19th and early 20th centuries, the Great Northern's fortunes had flourished on the receipts of freight haulage, most notably that of coal from the company's northern heartlands. Given that as much as two thirds of the LNER's profits came from freight haulage, the company suffered disproportionately from the collapse in British coal exports after 1925 and the general decline in the volume of goods transported across its network during the interwar period.

Despite the company's limited finances, some improvements had to be made in order to enable the King's Cross Goods Yard to function in an increasingly modern world and compete for a decent share of the falling levels of rail freight traffic. These included the infilling of the Granary Basin (Landscape Feature 5), the improvement of the road network (both of which had been initiated by the Great Northern Railway) and the introduction of the latest lifts and cranes thus allowing the depot to handle an influx of containerised goods from the Continent (Fig. 13.12). The Hydraulic Station (Landscape Feature 16) was also electrified in order to boost its efficiency and a better communications system was installed in the telegraph office. Motor vehicles were steadily introduced at the expense of the cart horse and a garage facility (Landscape Feature 30) was built on the recently infilled Granary Basin (Fig. 13.13).

Fig. 13.12 Incoming and outgoing commodities in the Goods Yard c.1921 overlain on the Goad Fire Insurance Plan of 1921, scale 1:2,500

These changes must have enhanced the operational efficiency of the complex. Most notably, road vehicle access was greatly improved and the waste of space that the redundant Granary Basin represented was finally resolved. Meanwhile, the Eastern and Western Goods Sheds (Landscape Features 7–11 and 28) continued to operate using what was essentially Victorian infrastructure until the latter half of the 1930s. The limited improvements of the 1920s were however overshadowed by the Transposition Programme of the mid-1930s. In response to the Great Depression at the beginning of the decade, the Government introduced a number of measures to stimulate the economy including the Guarantees and Loans Act 1934, which was designed to encourage private capital investment in major infrastructural works. Financed by highly competitive Government loans, the 'Big Four' railway companies submitted 'wish lists' of schemes 'of proven public utility' which also promised to be of benefit to the finances of the companies and their shareholders. Having chosen to use the majority of the available finance to fund improvements to its freight handling infrastructure, the LNER used a proportion of the money to switch the roles of the Eastern and Western Goods Sheds so that the former handled outgoing commodities only whilst the latter dealt with all outgoing goods (Fig. 13.13). By 1939 the LNER was in possession of a fully modernised yard at King's Cross, which was a considerable achievement given the hurdles that had stood in the company's way.

Effects upon incoming and outgoing 'general' goods traffic: 1919–1938

Owing to the economic recessions and the ensuing depression of the interwar period, the overall volume of goods entering the yard probably fell during this period. Consequently the years between 1919 and 1939 were the first to show a small decrease in general goods handling space in the depot through the loss of the Midland Goods Shed (Landscape Feature 15) after it was omitted from the Transposition Scheme (Fig. 13.13).

Amongst the plethora of improvements that were made to the Eastern Goods Shed (Landscape Features 7–11) during Transposition the most significant by far was the replacement of the outmoded hydraulic capstans and turntables with a modern system of traversers and

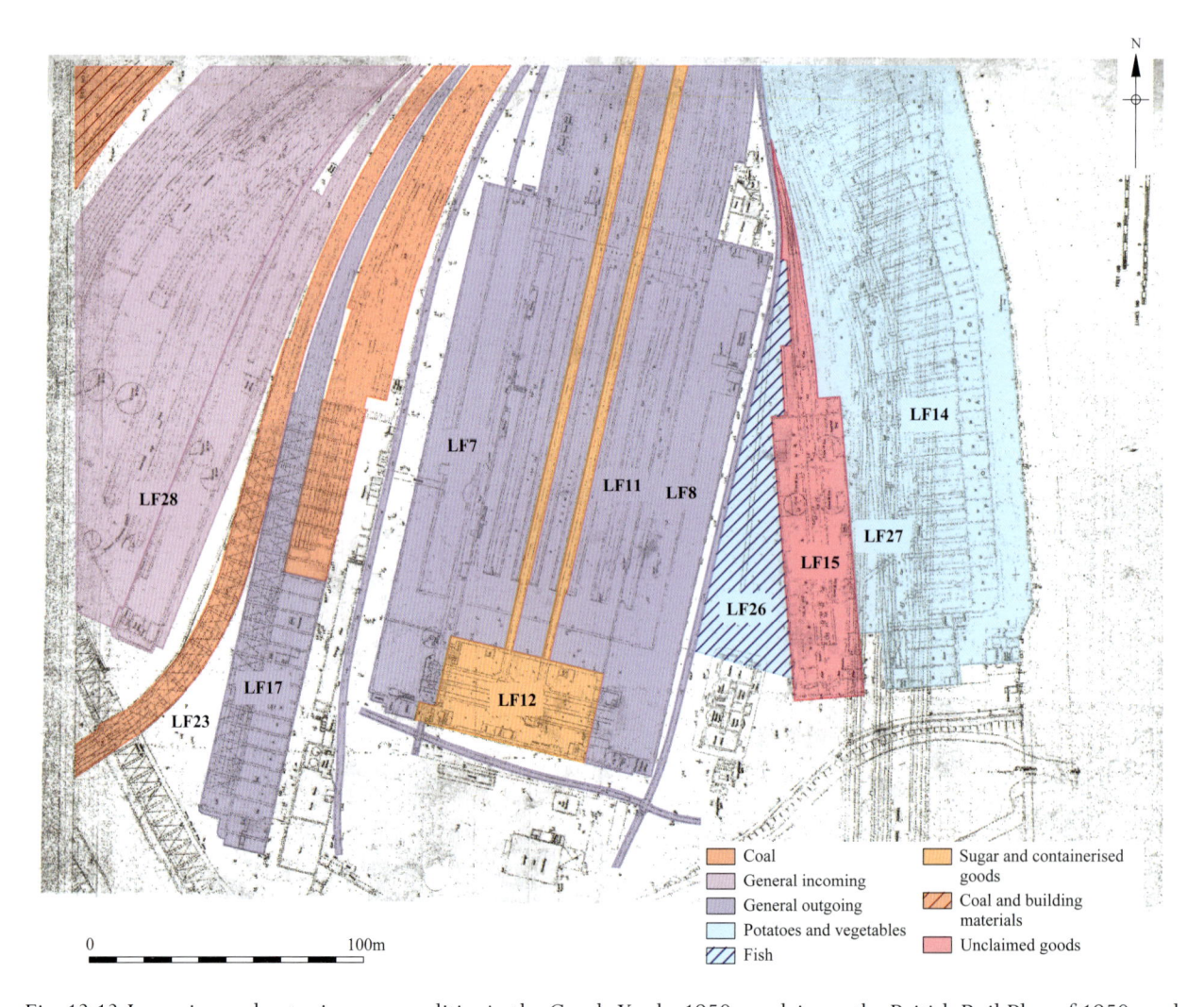

Fig. 13.13 Incoming and outgoing commodities in the Goods Yard c.1950 overlain on the British Rail Plan of 1950, scale 1:2,500

Conflict and Crisis: 1914–1945 315

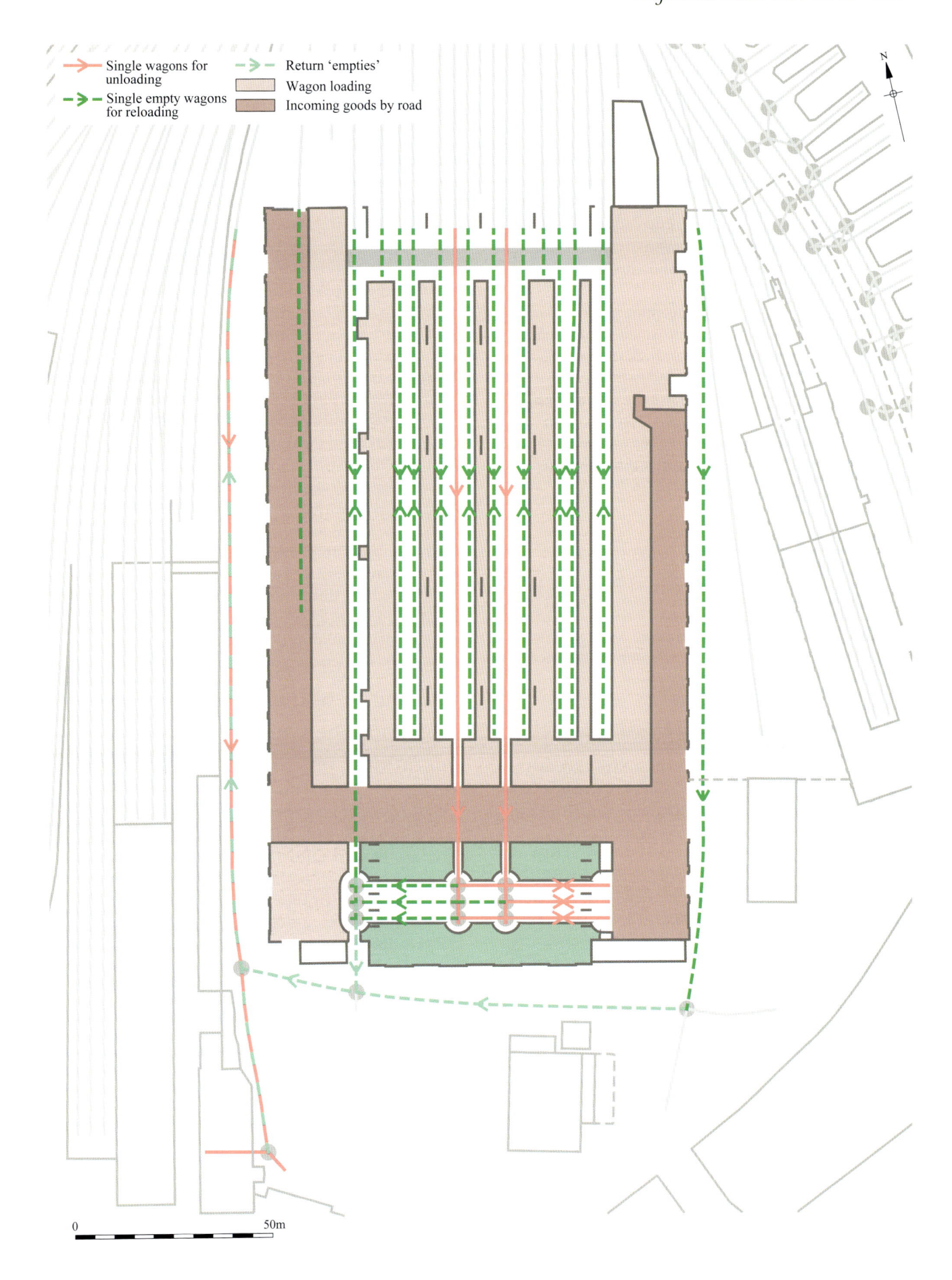

Fig. 13.14 Reconstruction of the probable goods management procedures and goods flow through the Eastern Goods Shed (Landscape Features 7–11) and the Granary (Landscape Feature 12) *c.*1935, scale 1:2,000

the installation of a substantial quantity of additional bank space. With the possible exception of the sidings that ran into the Granary (Landscape Feature 12), the layout of the redesigned Eastern Goods Shed suggests that virtually all of its railway lines became bidirectional after Transposition. The close arrangement of the banks indicates that loading and train assembly must have been carried out in the same space thus further decreasing the amount of manoeuvring that was necessary (Fig. 13.14). Together these changes would have had the effect of greatly reducing shunting times and improving the shed's efficiency.

It is telling that after 1935, incoming general goods traffic was housed in a fractionally smaller space than outgoing general goods traffic for the first time (Fig. 13.13). This no doubt reflects the industrial decline that was taking place in the north of Britain, which would have caused incoming 'general' goods levels to fall whilst the economy of London and the output of the city weathered the economic storm better than any other area of the country.

Effects upon inbound 'staples' i.e. coal, vegetables and fish: 1919–1938

Taking into account the station's coal, vegetable and fish reception facilities, the overall amount of space dedicated to the reception of incoming commodities after Transposition still exceeded that which was dedicated to goods travelling in the opposite direction (Fig. 13.13). Traditionally the Great Northern had supplied the capital with vast quantities of coal, fish and vegetables. While receipts from the carriage of coal fell dramatically after 1925, the volume of agricultural produce and fish carried by the company actually grew as the LNER captured a majority share of that traffic. The economic resilience of London and the South-East, characterised by their rapid recovery from the depression of the early 1930s, fuelled a period of sustained growth during the remainder of that decade. The extraordinary growth of London's suburbs fuelled demand for many of the 'staples' in which King's Cross Goods specialised, ensuring that the station's staff remained busy throughout the period.

Effects upon inbound grain and the Granary: 1919–1938

Landscape Feature 12

The import of grain from the agricultural heartlands of the north and East Anglia to King's Cross was the only obvious example of an imported 'staple' commodity that bucked the trend described above. London's population grew dramatically during the interwar period, so demand for grain must have continued to increase. Therefore the answer to the falling volume of grain received by King's Cross lies in the globalisation of the grain trade. During the late 19th and early 20th centuries consumers increasingly turned away from more expensive British cereals in favour of cheaper foreign imports from the 'bread baskets' of Europe, Argentina, Australia and in particular North America (Atkin 1995: 16–22). The impact of this development on the railways was a reduction in the volumes of grain carried by 'traditional' routes in favour of the growth of routes that carried international imports from the nation's docks. It is therefore reasonable to assume that the Granary at King's Cross had to some extent been effectively been bypassed by the mid 1920s.

Fortunately two new factors ensured that the Granary gained a new purpose during the interwar period. Firstly, thanks to the Great Eastern's monopoly of the carriage of containers from the Continent, the LNER was able to use the facility to store these goods after the Spitalfields Goods Station burnt down in 1925 (Fig. 13.12). Secondly, the dramatic growth of domestic sugar production enabled the LNER to use the upper floors of the building for the storage of sugar beet from 1926. The Granary may also have continued to receive some flour alongside the sugar during this period, since the words 'Flour and Sugar 1st Floor' had been painted onto the ground floor entrance of the building's eastern stairwell.

The impact of the Second World War upon the Goods Station and its operational procedures: 1939–1945

As was the case in the First World War, the Government assumed control of the railways in the national interest when war broke out in September 1939. The demands of the wartime economy and the transfer of bulk goods previously carried by the east coast shipping trade ensured that the volume of goods traffic carried by the railways dramatically increased. Once again, King's Cross Goods Station became an important cog in Britain's war machine. The development of aerial bombardment as an element of total warfare meant however that it was a strategically important target for enemy aircraft, and it was bombed on at least ten separate occasions during the Second World War. Although the Outwards Goods Shed had to be closed on a number of occasions as a result of bombing, the ensuing disruption rarely lasted for more than a few hours. Even after the devastating raid of Saturday 9th November 1940, staff returned to work the same day. King's Cross therefore managed to overcome the hardships of war in large part thanks to the successful implementation of the pre-war Transposition Programme, which had greatly improved efficiency, thus enabling the station to handle the increased volume of traffic irrespective of any disruption caused by the Luftwaffe's bombing campaign.

The relationship between the road, the railway and waterborne transport: 1914–1945

Fluctuations in the volume of rail freight handled by King's Cross during this period were predominantly a consequence of the economic factors rather than technological changes. Whereas the reliability of motor vehicles and the state of the nation's roads were being improved during the interwar period, these developments were insufficient to challenge long distance freight haulage by rail. Consequently rail would continue to dominate from 1914 until the end of the Second World War, whilst lorries and vans took over from horse-drawn vehicles on short-haul journeys. The canals meanwhile had been in long-term decline for decades and had long fallen out of favour with manufacturers for the carriage of goods. Even before the start of the Second World War, Britain's waterways were falling into disuse.

In keeping with these trends, the Granary Basin and the Goods Station's surviving canal infrastructure were infilled over a number of years from 1915 onwards, which enabled its road network to be greatly improved. During the interwar period, the prior removal of the Basin provided sought after car parking space and a much needed garage facility for the maintenance of the LNER's growing fleet of short-haul motor vehicles.

13.4 The Decline of Industry in Post-War Britain. The Causes and Effects of Infrastructural and Working Practice Changes at King's Cross Goods Station: 1945–1990

Two and a half years after Clement Attlee's Labour Government assumed office in July 1945, the railways were nationalised and the LNER was absorbed into British Railways. The new concern inherited a bomb-damaged Goods Station at King's Cross, along with numerous other worn-out facilities spread across the capital. The nation's war-depleted finances meant that little money was available to be spent on the railways, despite the continuing growth of rail freight in the immediate post-war period. The ten years after 1945 would therefore be characterised by continued neglect at King's Cross thanks to post-war austerity and a continuing shortage of money. Only the most vital repairs would be carried out, which were generally done as cheaply as possible. By the time that money did became available in the mid-1950s, the future of railway freight operations was under threat from the fast expanding road haulage sector. Although the Goods Station escaped closure thanks to its conversion into a parcels hub in 1966, this was a temporary reprieve.

The 1960s and 1970s would see rail freight services continue to decline as road haulage became increasingly practical and popular. This led to the eventual closure of the Goods Station's rail head in 1981. Road delivery companies continued to make use of it in the short term (Fig. 13.16), but by the early 1990s it was gradually being vacated in favour of more modern and accessible premises situated beyond inner London.

Effects upon goods reception facilities in the Eastern Goods Yard: 1945–1990

It is not clear whether the 'reverse transposition' scheme of 1953 took place in the form envisaged by the British Transport Commission (BTC). The precise effects of the alterations carried out during the 1950s upon traffic and goods flow in the Eastern Goods Shed (Landscape Features 7–11) and the Granary (Landscape Feature 12) are therefore hard to model. The available archaeological and historical evidence suggests that the proposals originally approved by the BTC were scaled back in order to save money. The reconstruction that is shown in Fig. 13.17 is therefore a rough approximation of some likely changes, since little clear evidence pertaining to the internal layout of the shed post-1953 has come to light. It is likely that a major aim of the scheme was to enable the direct transfer of goods from train wagons to road vehicles, so it is probable that the proposal to demolish some of the banks within the Goods Station did come to pass. Direct road access to the Transit Shed platforms (i.e. in Landscape Features 7 and 8) was a long standing arrangement, which explains why they were not removed. However, the same cannot be said for the Train Assembly Shed banks, some or all of which might have been demolished as shown (Fig. 13.17). It is also telling that such a large amount of storage space in the form of the banks could be done away with at this time. Had such a change been implemented at an earlier stage in the life of the Eastern Goods Shed then it is likely that the building would have been overwhelmed by the volume of incoming goods, indicating that rail freight levels had already fallen drastically.

The Eastern Goods Shed survived the streamlining and rationalisation schemes that the British Railways Board (BRB) implemented during the 1960s due to its fortuitous conversion into a parcels hub by the Eastern Region. However, serious money would not be invested in it again. Traffic flowing into the Handyside Canopies and the Potato Market also fell sharply during this period, whilst coal traffic, already in steep decline owing to the Government's decision to end production of 'town' gas had largely been redirected elsewhere and was to cease altogether by the end of the decade (Fig. 13.15). Although the exact date went unrecorded, the Potato Market fell out of use and was a redundant feature by 1970, after which it was demolished in stages.

After the closure of the railhead in 1981, the Eastern Goods Shed was converted into a parcels depot for road hauliers, an event that resulted in the loss of the connections that formerly linked the Eastern Transit

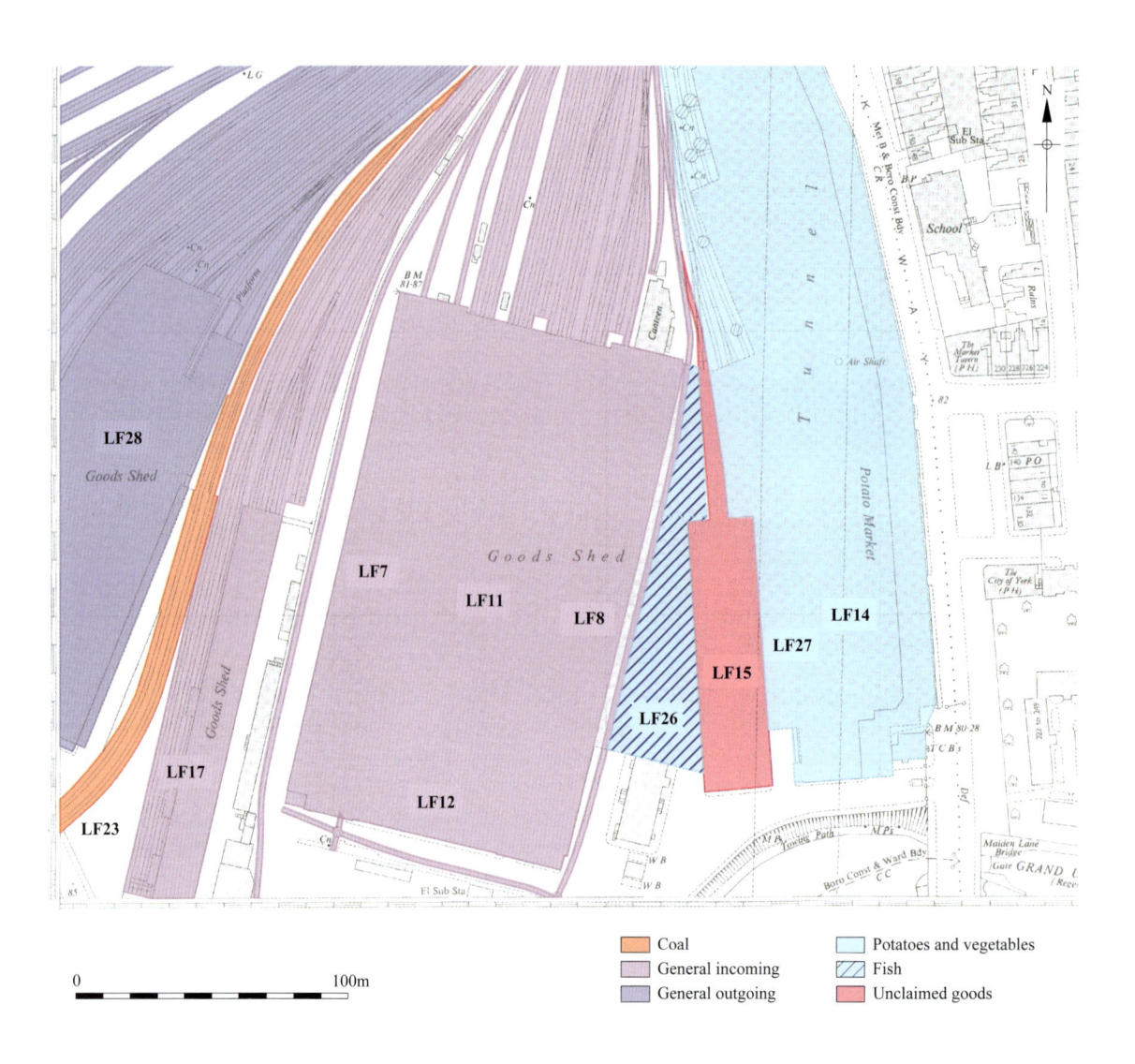

Fig. 13.15 Incoming and outgoing commodities in the Goods Yard c.1953 overlain on the Ordnance Survey Map of 1953, scale 1:2,500

Shed (Landscape Feature 8), the Western Transit Shed (Landscape Feature 7), the Train Assembly Shed (Landscape Feature 11) and the Granary (Landscape Feature 12). As a result, those four buildings, which had been interlinked for their entire history prior to that point, functioned as separate structures from then on (Fig. 13.18). In keeping with what had gone before it is probable that separate areas were set aside for incoming and outgoing parcels. Unfortunately no historical or physical evidence in support of this assertion was found, whilst matters were complicated post-privatisation by the arrival of several different delivery companies, all of whom must have had their own working practice arrangements. Along with the main Goods Station, other structures such as the Handyside Canopies were used by the haulage operators. Many more were abandoned or were used by short term tenants who invested little in them (Fig. 13.16).

The relationship between the road and the railway: 1945–1990

The decline in rail traffic that made the direct transfer of goods between train wagons and road vehicles possible in the Eastern Goods Shed post-1953 was due to the falling share of goods transported by rail and the corresponding rise in the proportion carried by road. The nation's arterial roads and motorway networks were improving rapidly, a development that was being actively encouraged and subsidised by the governments of the day. Efforts to ensure the continued viability of rail freight, most notably the introduction of the Freightliner depots in the second half of the 1960s, meant that the old railway goods depots were being superseded by new facilities. The combination of these factors and the cessation of railway parcel services in 1981 eventually led to the closure of the station's rail head. Consequently from 1981 until its virtual abandonment in the early 1990s, the Station was fed exclusively by road (Fig. 13.16; Fig. 13.18).

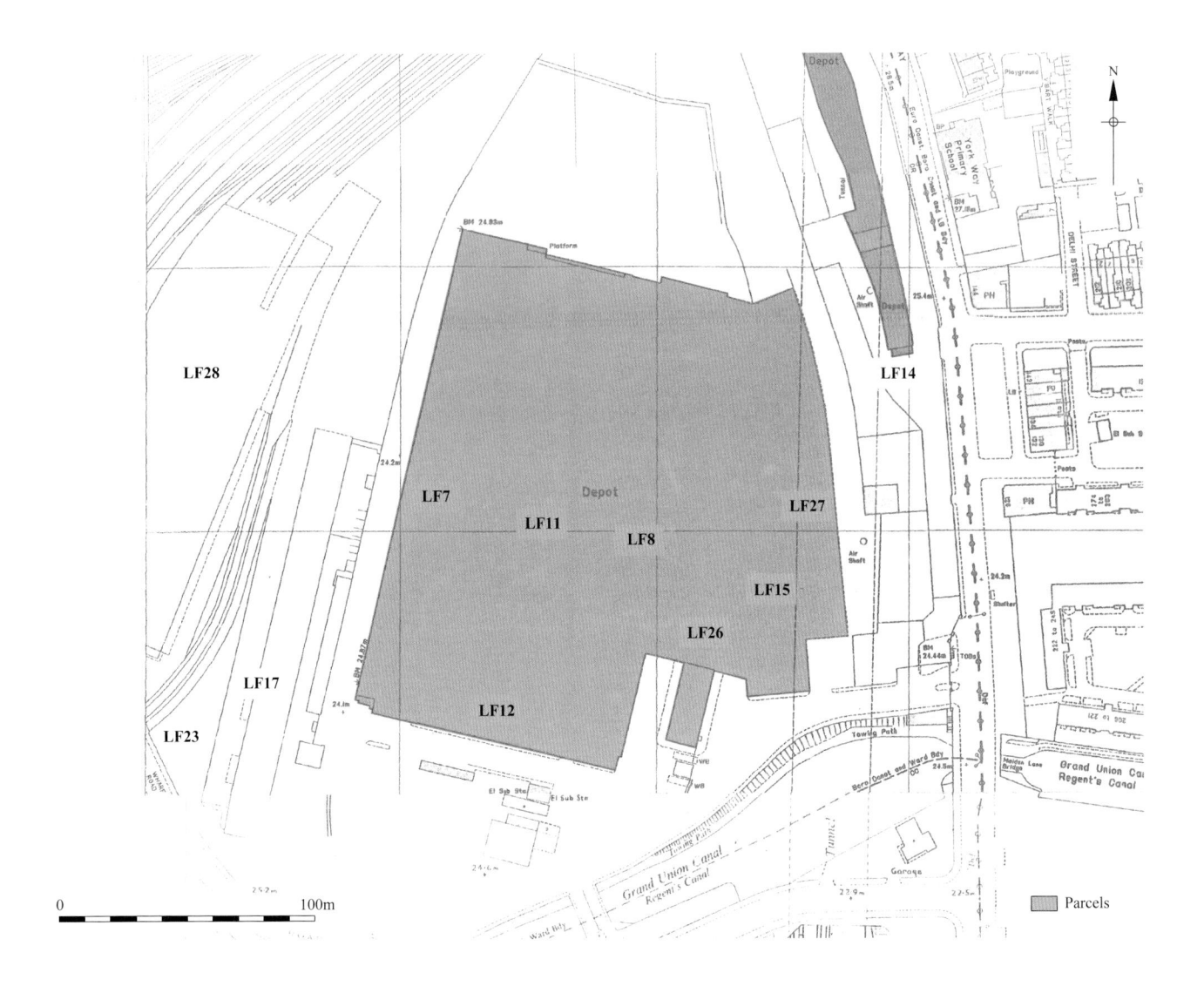

Fig. 13.16 Incoming and outgoing commodities in the Goods Yard c. 1982–3 overlain on the Ordnance Survey Map of 1982–3, scale 1:2,500

320 Conclusions

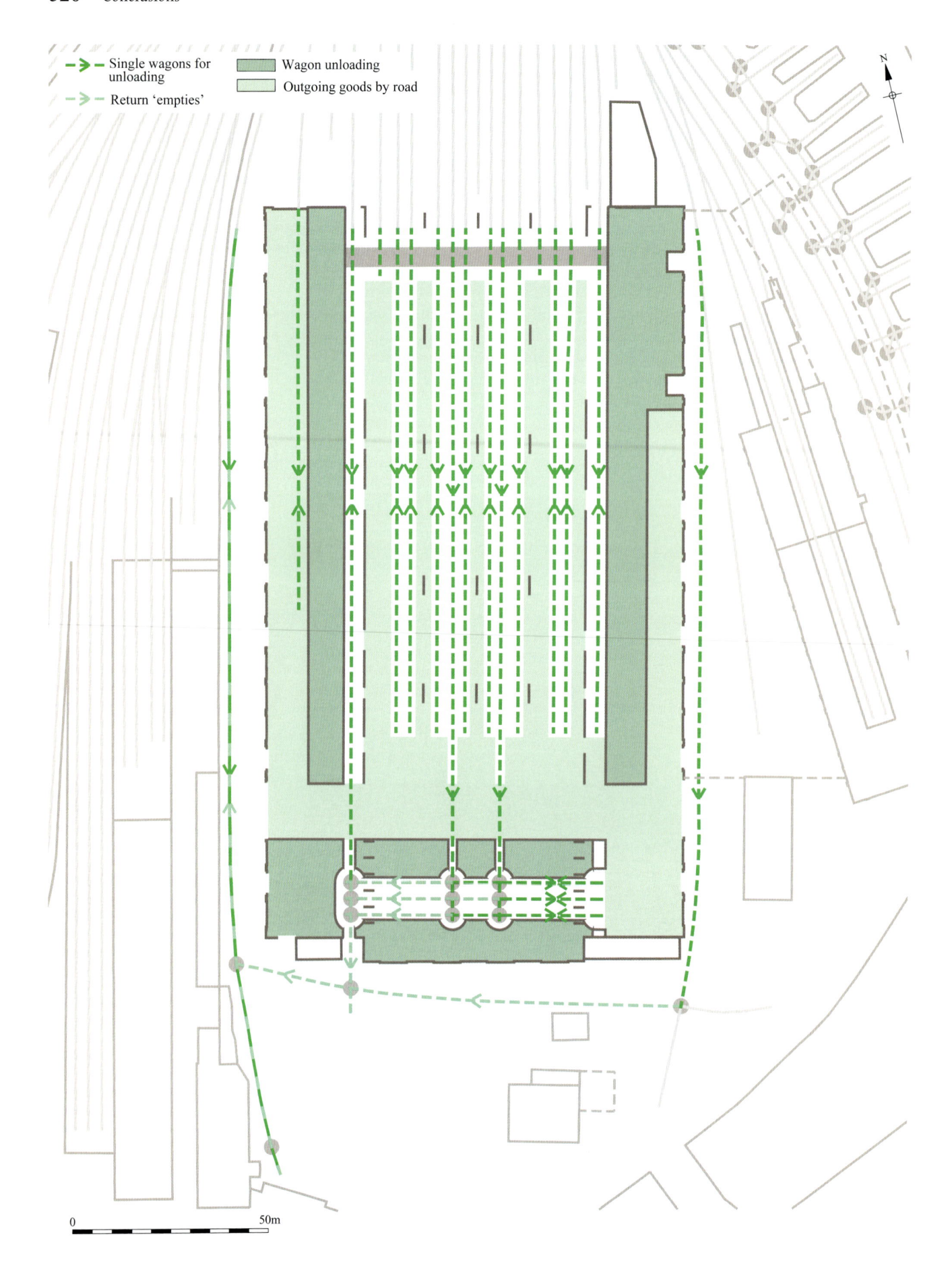

Fig. 13.17 Reconstruction of the probable Goods Management procedures and goods flow through the Eastern Goods Shed (Landscape Features 7–11) and the Granary (Landscape Feature 12) c.1953, scale 1:2,000

13.5 Developments at King's Cross Goods Yard Compared and Contrasted with Two Major Rivals in the Capital

In order to determine whether or not wider trends in railway goods handling and goods station development existed in the capital, the historical development of the Goods Yard at King's Cross is compared below with two major London based railway goods termini: the Midland railway's Goods Station at Somers Town and the London & North Western's yard at Camden. These facilities ought to represent good analogies for King's Cross because they imported and exported similar sorts of commodities between overlapping geographical regions and were thus driven by similar technological, social, economic and political developments. Theoretically these factors should have triggered similar adaptive responses in all three yards.

An excellent synopsis of the history of the Camden yard has been compiled by Peter Darley of the Camden Railway Heritage Trust (see Darley 2013). However, Darley's methodology did not extend to archaeological excavation or high level building recording and was not designed to reconstruct the depot's history in minute detail. Information on Somers Town is primarily drawn from a publication detailing the results of an excavation that was recently undertaken within the northern section of the depot, the objectives of which were to present the findings of the fieldwork rather than to present the history of the entire yard in great depth (see Lewis 2013/2014). The following comparisons are therefore only illustrative since they are drawn against the findings of two studies that were not directly comparable with the present investigation of King's Cross.

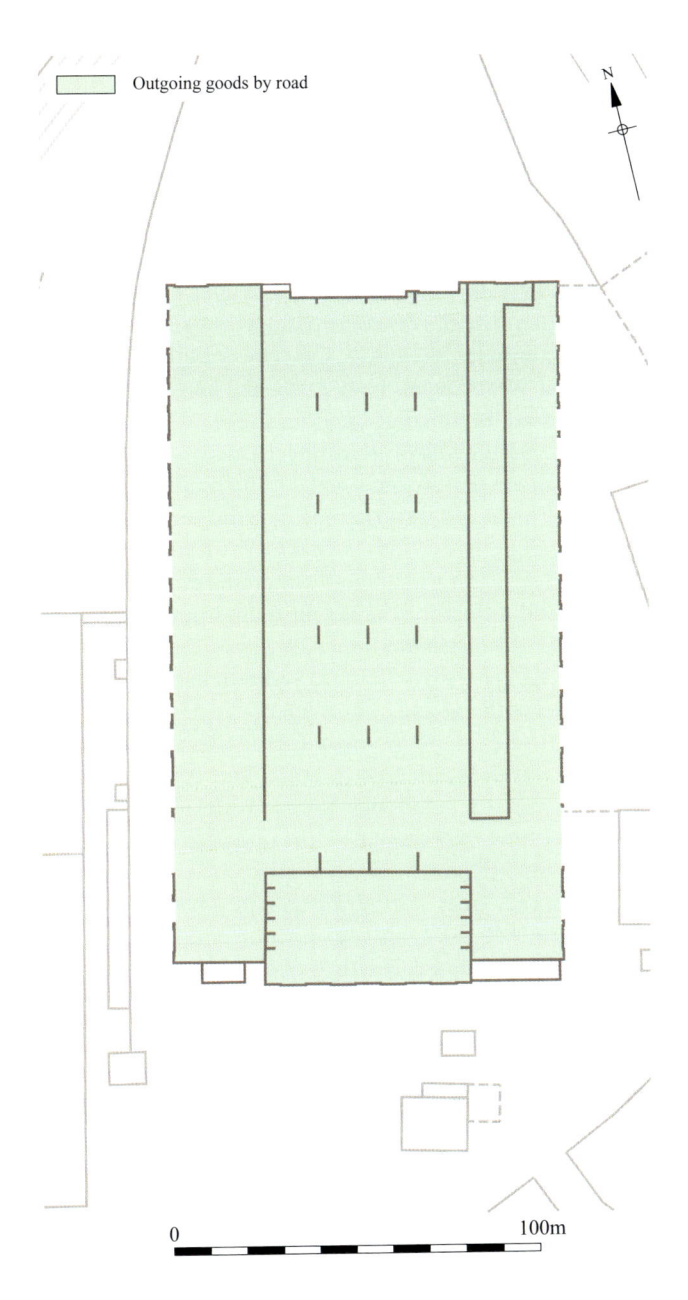

Fig. 13.18 Reconstruction of the probable Goods Management procedures and goods flow through the Parcels Depot (Landscape Features 7–12) *c*.1983, scale 1:4,000

King's Cross Goods Yard and the London & North Western Railway's goods depot in Camden

The earliest incarnation of Camden Goods Yard was opened by the London & Birmingham railway in 1837, the same year that they company began running passenger services from Euston. It was inherited by the London & North Western after that company absorbed the London & Birmingham at its creation in 1846 (Darley 2013: 3, 33).

In a similar way to the invisible line that prevented the extension of railway passenger services beyond the New Road, when Camden Goods Yard was created goods trains were barred from travelling beyond the Regent's Canal. The London & Birmingham therefore built its goods station on a ten hectare plot on the north bank of that waterway. Like King's Cross, the site of the Camden Goods Yard was raised and levelled through the redeposition of London Clay so that a flat surface suitable for railway use could be created. A large stretch of the retaining wall, affectionately known as 'The Great Wall of Camden', still survives (ibid.: 25).

Early locomotives departing from Euston were not powerful enough to overcome the station's steep approaches so a stationary steam engine in the Camden Goods Yard winched trains up the incline on a 'continuous rope' (ibid.: 3, 19, 24)[18]. After arriving at the depot, they were coupled to a locomotive that carried them on. When travelling in the opposite direction, passenger and goods services were uncoupled from their locomotives within the goods yard and passenger trains were transported onwards into Euston under the influence of gravity. This is why both Camden and King's Cross goods stations were surrounded by engine stables, repair shops and other railway infrastructure despite the former being much further removed from its sister station than the latter. That factor would have long term effects upon the ways in which both depots could be enlarged to meet the demands of the future; however issues of space would prove to be less acute at Camden.

Like the Great Northern's London goods terminus, the London & Birmingham's earliest railway goods reception facilities in the capital included two goods sheds, stores, administrative blocks and stabling for horses. However, unlike King's Cross, Camden initially possessed its own cattle pens for the reception of livestock since the creation of the Metropolitan Cattle Market on Copenhagen Fields (which the Great Northern relied upon) had not yet been proposed. Camden's unusual 'pinnate' goods sidings were also quite different in form to those that would be created by the Great Northern, however they were presumably used for very similar purposes.

It quickly became apparent that the size and balance of the original goods handling infrastructure at the Camden yard was imperfect so the next few decades witnessed a drive to improve matters (ibid.: 25, 29, 30, 33, 42). This is directly comparable with general trends at King's Cross. An 'arms race' also prompted change as the two railway companies tried to out-compete each other. For example hydraulic power at King's Cross was switched on in 1852, an event that spurred the London & North Western to do the same just one year later. Although it was the Midland that first brought coal to London by rail, it was the Great Northern's pioneering decision to carry vast amounts of the commodity into King's Cross that drove its competitors to enlarge Camden in 1854 so that a coal drops and a coal yard could be included (ibid.: 42, 54; Brandon 2010: 83).

As was the case at King's Cross, accommodation for independent merchants and road hauliers existed in the Camden yard. In both instances that capacity grew over time through the reuse of redundant railway structures, however the fact that the latter relied to a far greater extent upon sub-contractors for short-haul road transshipment in comparison with the formative years of the Great Northern meant that more space and freedom was made available to its tenants (Darley 2013: 30, 47, 49: May 2003: 21). Many therefore occupied private, purpose built goods sheds, the most influential of which was commissioned by the road haulage company Pickford's. Described as being twice the size of Westminster Hall, it was the first canal, rail and road interchange depot in the world (Darley 2013: 29). Designed by the architect Lewis Cubitt, the infrastructure that it contained will no doubt sound familiar. It possessed subterranean canal tunnels that enabled barges to sail into it, roadways for horse-drawn carts, railway sidings, manually operated cranes for loading and unloading, banks for temporary storage and stabling in the basement (ibid.: 29, 31). Mirroring early events at King's Cross, Pickford's warehouse was quickly inundated with custom after it opened in 1841, however in that instance room for an extension was available and the shed was duly enlarged in 1845 (ibid.: 30). Its success would inspire the London & North Western to create its own three-way interchange on the opposite side of the canal in 1848 (ibid.: 48, 85).

When the Great Northern began its advance on London in 1849 the company aimed to emulate and outstrip the London & North Western. The success of Pickford's warehouse and the London & North Western's own interchange thus encouraged the Directors of the Great Northern to hire Lewis Cubitt to design a rival facility. Unlike the Main Shed at King's Cross and the earliest incarnation of the London & North Western's interchange at Camden, Pickford's warehouse took the form of a large, multi-storey building with an unlucky history. It suffered two destructive conflagrations in 1857 and 1867, the last of which caused it to be vacated (ibid.: 31). It was duly reconstructed in a modified form and was henceforth used for the reception of incoming grain and potatoes. This was not a success so it was allocated to one of Camden's most important tenants, the alcohol merchants Gilbey, who used it for vatting, blending, bottling and packing their wares (ibid.: 47, 50). This history suggests that the London & North Western viewed the building as superfluous to their needs when they inherited it in 1867. This is unsurprising since by then, in addition to the company's own three-way interchange, the Camden yard sported the largest general goods shed in Britain (ibid.: 48, 81). In 1905 it would break another record when its interchange facility was enlarged, improved and transformed into the most 'sophisticated example of storage and three-way transfer' in the country (ibid.: 85).

When designing the Great Northern's Main Goods Shed complex in the early 1850s, Lewis Cubitt took the canal infrastructure that he had instated in Pickford's multi-storey warehouse and incorporated it within a road and railway transshipment facility

---

18 This continued until 1844 when locomotives became powerful enough to make the climb with the aid of a banking engine.

that was at that time the largest in London. His design was sensible because there was not enough room on the Great Northern's chosen site to create separate three-way transshipment and general goods sheds. The design also discouraged long term storage, which was helpful given the high volume of custom that was expected. The complex therefore represents a bespoke commission rather than a direct copy of an existing shed, a factor that must have facilitated its longevity. However, only modest room for future expansion existed in the immediate vicinity of the main shed, in part because the Great Northern had initially chosen to better the London & North Western by devoting more space to the reception of barges through the creation of two large mooring basins. Until one of these was sacrificed in the late 1890s the only way of improving goods reception capacity was to upgrade, modify and maintain the existing complex. Unlike Pickford's Shed, this ensured that the Main Goods Shed at King's Cross received general goods until the late 20th century but it also meant that the London & North Western's sheds on the Camden site would soon outstrip it in scale and complexity. It also meant that Camden was able to retain and upgrade its interchange with the waterway for longer, whereas the King's Cross canal connection had to be severed in the early 20th century in order to lessen road and rail congestion.

Post-grouping, the Camden yard passed into the hands of the London Midland and Scottish (LMS) Railway. Like the London & North Eastern Railway the LMS was slow to replace its horse stock with motorised vehicles and the last working horses at Camden retired in the 1950s (ibid.: 65).

After the yard became the property of British Railways in the late 1940s, general goods trains continued to frequent it until its closure in 1980 (ibid.: 89). Unlike the core of King's Cross Goods Station, which survived into the 1990s as a road parcels depot, the Camden facility was parcelled up and sold to different developers for a wide variety of uses (ibid.: 89). This caused some areas of the former site to be completely cleared of railway structures whilst in other areas buildings were adaptively re-used with great success.

King's Cross Goods Yard and the Midland Railway's goods depot in Somers Town

Somers Town Goods Yard opened in 1887 to relieve the goods facilities at Agar Town, which had been recognised as being insufficient as early as 1874 (Lewis 2013/2014: 289). Like the Great Northern, the Midland had therefore been inundated with custom since arriving in the capital in the 1860s and the company was similarly struggling to cope.

The Midland's new depot at Somers Town was centrally situated beside the New Road. The advantages of the location were obvious: the position facilitated onward transportation by horse-drawn dray and enabled cross London goods traffic to be transferred to the Metropolitan Railway, which ran below that thoroughfare. The long-term decline of the inland waterways meant that the Midland saw no need to include canal infrastructure in the design of the new depot.

Although built on a smaller tract of land than King's Cross Goods Yard, the extensive construction programme that resulted in the creation of Somers Town took as many as five years to complete. In contrast, King's Cross began receiving goods trains after less than two years' worth of construction work. These differences were primarily caused by the fact that Somers Town was a central London terminus that was constructed at a comparatively late stage in the development of the capital's railway network. Dramatic differences between its layout and that of King's Cross existed as it had to be crammed into a smaller and less topographically favourable position because large tracts of land in this area of the city were no longer readily available.

When building its passenger station at St Pancras, the Midland chose to overcome the topographic obstacles that made life difficult for steam locomotives departing from this area of north London by elevating their railway upon a viaduct (see Section 1.2 for further details)[19]. This is why trains departing and arriving at St Pancras Passenger Station do so above street level even today. An identical solution was deployed at Somers Town Goods Yard which meant that the Midland's goods sheds in that depot were multi-storey. Trains arrived, were disassembled, reassembled and departed from above whilst storage and goods interchange with the road carried on below. Communication between storeys was provided by large hydraulic lifts that were capable of transporting single loaded railway wagons and their contents between floors thus maximising the space that was available for goods handling.

Like King's Cross the earliest incarnation of Somers Town Goods Depot was home to a main goods shed, a hydraulic power station, a coal drops and a potato market though unlike its competitor it possessed a dedicated 'Milk and Fish Depot' (ibid.: 290–1). Recent archaeological excavations undertaken on its former site revealed the partial remains of the Hydraulic Station, with an accumulator base that was remarkably similar to the examples that were unearthed at King's Cross. Within the former 'Milk and Fish Depot', railway sidings where single wagons were unloaded were discovered alongside cobbled cart ways and banks for the transfer of goods to the road.

The railway and goods handling infrastructure at Somers Town therefore shared many similarities with King's Cross, although a plethora of operational differences must also have existed. Unlike the Main

---

[19] In contrast the Great Northern chose the alternative solution: the construction of the Gasworks and Copenhagen Tunnels, which took their trains under rather than over the Regent's Canal, the Imperial Gasworks and the high ground that could be found to the north.

Goods Shed at the Great Northern depot, which dealt with the reception and dispatch of a wide array of commodities, the role of the general goods shed at Somers Town was more specific. Its focus was on the delivery of perishable foodstuffs such as fruit and vegetables, meaning that in terms of the kinds of goods that it was handling and their direction of travel it must have had more in common with the East and West Handyside Canopies and the Potato Market at the Great Northern's yard than it did with the Main Goods Shed.

In contrast to the dearth of provision at King's Cross, dedicated warehousing space for the temporary storage of commodities like bananas was provided in the Main Shed at Somers Town along with office accommodation for the various private fruit and vegetable merchants that traded out of it. The Potato Market was also far superior to the example at King's Cross, which prompted the Great Northern to improve their facilities through the construction of the Handyside Canopies in 1888.

The Great Northern Railway's Main Goods Shed was of a comparable size to the Midland's however the same cannot be said of their coal reception facilities. The coal drops at Somers Town were smaller than the two examples at King's Cross, despite the fact that the Midland also had excellent access to major coal producing areas in the north of England, eventually becoming the largest single importer of that commodity to the capital as a result (Simmons *et al.* 1997: 93; Brandon 2010: 83). Like the Great Northern, the Midland had access to Samuel Plimsoll's coal drops at Cambridge Street and owned another set of drops to the immediate north of their passenger station (Christopher 2013: 75), as well as its own network of suburban coal yards (Williams 1988: 146). It is therefore unsurprising that the company did not devote a larger amount of space to the reception of the commodity at Somers Town.

By the time that Somers Town was built, it would have been obvious where the greatest need and the greatest potential profit lay in terms of goods transshipment in this area of the city. The Midland evidently used that prior knowledge in order to craft a depot that focused upon the importation and transshipment of the most profitable kinds of goods by the most efficient methods whilst simultaneously resolving deficiencies in the pre-existing goods reception facilities that existed at King's Cross. The company was also able to get the balance right in terms of the scale of the depot's coal reception infrastructure and wisely chose to omit a grain warehouse from their scheme given the declining domestic production of wheat. Armed with such knowledge and experience, the company was therefore able to maximise the profitability and usefulness of their smaller Somers Town site.

When the Great Northern Railway built King's Cross the company had comparatively few experiences to draw upon since railway transshipment to large central sites was in its infancy. The decades that followed the construction of the goods yard therefore saw the company make a flurry of changes, all of which were designed to redress the imbalances that emerged after it opened its doors. Despite the challenges it faced, King's Cross had many advantages. It was able to offer importation and exportation of a much wider range of goods despite retaining the bulk of its outmoded canal infrastructure until the early 20th century. Its more favourable position and its larger size also meant that in the longer term it was able to expand and modernise more flexibly, most strikingly in 1899 when the Western Goods Shed was constructed. That luxury was impossible at Somers Town, the footprint of which barely changed from the year that it opened in 1887 until its closure in the late 1950s. Any future improvements at that yard must therefore have been limited to changes to working practice and enhancements to goods handling and rail reception infrastructure. Unfortunately the nature of those improvements, if they did indeed take place, cannot yet be compared with those that occurred at King's Cross because the history of Somers Town Goods Yard has not yet been researched in comparable detail and a smaller proportion of the site has been excavated.

The above evidence suggests that King's Cross and Somers Town were able to thrive into the early 20th century despite their proximity to one another and the overlap in the territory that they served because they each had different strengths and weaknesses. Although they did compete for certain commodities their operational differences appear to have been somewhat complimentary, enabling both to cope with the growing volumes of rail freight that characterised the 1890s.

After passing into the hands of the London Midland and Scottish, Somers Town Goods depot suffered bomb damage during the Second World War and it was never restored to its former glory after the conflict (Lewis 2013/2014: 290). In the post-war period it became the property of British Railways but it received none of the lucky breaks that enabled King's Cross to survive into the 21st century, perhaps because its small, multi-storey layout made it less compatible with modern modes of working. Consequently it fell during the rationalisation schemes and budgetary cuts of the 1960s. Somers Town was not spared from the wrecking ball and the British Library and the Francis Crick Institute now occupy its former site.

Similarities, differences and wider trends

The preceding discussion demonstrates that more differences than might be expected characterise the histories of these three neighbouring yards. Given that the territories that they served overlapped, it was assumed that they would react in similar ways to shared socio-economic and political trends. Evidence suggests however, that their responses were instead complicated by chronological differences in their construction dates and variations in their relative positions within London's changing cityscape[20].

In the 19th and early 20th century developments at King's Cross and Camden Goods Yards had much in common because they were direct rivals competing for the import and export of the same sorts of commodities by identical means, to and from similar areas of the country. As time passed however, Camden was able to adapt more flexibly and rapidly because much less space had been dedicated to canal infrastructure at the outset, which meant that the pre-existing site of the main goods sheds could be modified with relative ease. At King's Cross, substantial enlargement of general goods reception facilities could not take place until the Coal and Stone Basin was infilled, which represented a major undertaking. In contrast any superficial similarities between King's Cross and Somers Town were outweighed by major functional variations that were caused by topographical and chronological variables. Given its small size and its later inception, a specialised depot lacking a canal interchange was the only practical option at Somers Town, whereas King's Cross was large enough to survive as a general facility that incorporated canal infrastructure for longer.

Although the Great Northern Railway, the London & North Western and the Midland would probably never have admitted it, the sheer volume of rail freight that passed through London during the 19th and early 20th centuries transformed all three of these 'rival' yards into relatively compatible companions that together coped acceptably well with that immense burden. In different ways, all three were therefore able to thrive for the duration of that time period.

As the 20th century progressed and far reaching socio-economic, political and technological developments began to bite their histories converged. Post-grouping, Somers Town and Camden would join forces under the London Midland & Scottish, whilst all three would become the property of British Railways after the Second World War. Their fates would then diverge once more after a series of business decisions and planning judgements caused Somers Town to be demolished, Camden to survive in a piecemeal fashion and King's Cross to be saved in its virtual entirety.

To bolster the conclusions that are presented above a more detailed comparison of the flow of traffic and goods through Camden, Somers Town and King's Cross over time would have been advantageous. Similarly, a more thorough discussion of the comparative effects of the two world wars, the impact of the Grouping Act and the economic troubles of the interwar period would have been desirable. Unfortunately such comparisons cannot be drawn as yet because sufficiently detailed historic information relating to such issues only exists in a published format for King's Cross. It is therefore hoped that future work will help to produce a better and more thorough account of the history of goods handling in London in due course.

13.6 The Value of a Multi-disciplinary Approach When Investigating Industrial Sites: Case Studies from King's Cross Goods Yard

The number of instances that have emerged from this investigation that demonstrate the advantages of a multidisciplinary approach are too numerous to repeat here. Instead three case studies, each of which illustrates a different aspect of the benefits of such a method, have been identified for illustrative purposes.

Ground preparation and raising in the early 1850s: testing an historical narrative through an exploration of the archaeological record

As detailed in Sections 3.1 and 3.3 of this volume, Joseph Cubitt's reports to the Great Northern's Board of Directors concerning the earliest phases of the Goods Station construction project indicate that ground levelling work commenced in May 1849, continuing until 'formation level' was reached in the spring of 1850. Shortly afterwards a temporary construction surface in the form of a layer of burnt clay ballast was deposited across the entire site, after which work began on the construction of the foundations of buildings and services. Cubitt's correspondence therefore appears to provide a detailed and precise inventory of developments on site; however the extant remains that were revealed by excavation demonstrated that it was simplified or idealised.

The made ground and the temporary surfaces that Cubitt described had been dumped against the sides of the earliest footings, which strongly suggests that they pre-dated at least some of that ground raising activity. Given what else is known about the programme of works, these findings prove that landscaping and ground raising continued well beyond the spring of 1850.

---

20 It would be interesting to evaluate the interplay between such factors and the socioeconomics that affected goods shipment more generally for a wider geographical area (both within the capital itself and the nation at large). Unfortunately, the amount of research that would be required to address that question in a meaningful way is so considerable that it must be set aside as a topic for a different book.

The evolution of the Hydraulic Station: aiding archaeological interpretation through the use of historical information and vice versa

As was the case in so many other areas of the yard, the archaeological and historical information relating to the evolution of the hydraulic station proved to be, for the most part, complementary[21]. As is described in detail elsewhere in this volume (see Sections 3.15, 4.16, 5.2 and 7.4 for further details), the changing footprint of the building could not have been deduced from the historic cartography or the minutes of the Great Northern Railway alone. Likewise, the changing distribution of machine supports inside the building over time would have remained largely unknown. Instead these aspects of the building's history were primarily reconstructed through stratigraphic analysis of the below ground remains via excavation in combination with typological dating via brick fabric analysis. Without access to the archives of the Great Northern Railway and the LNER it would have been impossible to state with any degree of confidence what kinds of machines were installed upon the various engine beds and boiler supports that were found inside the Hydraulic Station. This was only achieved because direct and indirect references to the contents of the building at various points in its history were discovered in railway archives. Likewise, without access to GNR and LNER minutes, the names of the contractors that built the Hydraulic Station, the financial implications for the railway and, crucially, the reasons behind the expansion of the hydraulic network would have remained unknown. This clearly illustrates how archaeology and history can work together to build a comprehensive story.

Archaeological 'stability' in the early 20th century: the importance of the wider historical context

From a purely archaeological and built heritage standpoint, it is tempting to see the early 20th century at King's Cross Goods Yard as a comparatively uneventful time that was characterised by comparatively little change within the depot. In isolation, this could lead a commentator to conclude that this period of archaeological 'stability' was due to the fact that the complex was running smoothly and that few large changes were required. Of course, this could not have been further from the truth. The early 20th century was a time of great political and financial uncertainty generated in the first instance by the outbreak of war in 1914 that was quickly followed by a period of immense socio-economic and political instability in the 1920s. In very different ways these events put great strain upon the nation's railway goods handling infrastructure, causing huge variations in demand and manpower whilst limiting the amount of money that was available for change. This wider historical context clearly demonstrates that this period of 'stability' in terms of a comparative dearth of major building programmes therefore owed more to the uncertainties that pervaded Britain in the early 20th century than it did to the success of the depot itself. The wider historical context is therefore vital when interpreting a site of this age and type.

The importance of a multidisciplinary approach

In different ways each of these case studies demonstrates that, when investigating an industrial site, archaeology and history are far more powerful when used together. The evidence discussed above confirms the premise that unless a direct contradiction between the various lines of evidence exists, in which case the physical must always take precedent over the written, they complement and aid each other (Morris 2003: 11; Palmer et al. 2012: 15). Only when they are compared, contrasted and placed in a wider context can questions concerning how, when and why an industrial site functioned and evolved be comprehensively addressed.

At the time of writing King's Cross is the only example of a railway goods yard where all of these datasets have been collated in detail, combined and presented in a wider technological, socioeconomic and political context. Without access to all of those lines of evidence its story would be incomplete.

---

21   Only one major contradiction arose between the archaeological and historical records pertaining to the engine house: the former demonstrated that, in contrast to the historic evidence which suggested that a tandem-compound engine was instated in 1899, a cross-compound example was actually installed instead.

13.7 The Importance of King's Cross Goods Yard Both Past and Present

This archaeological and historical study has, we hope, not only demonstrated the value of a combined archaeological, built heritage and historical approach when investigating a site of this nature but has also highlighted the importance of King's Cross Goods Yard itself.

In the past, the yard was pivotal to the success of the Great Northern Railway and the London & North Eastern Railway in their turn, however its historical significance was far greater than this. For over a century the British economy was driven by the railways, so the depot was a crucial cog in a machine that sustained the country, her capital and the Empire. It played an important role in maintaining the wealth of the nation throughout the 19th and the first half of the 20th centuries and was of considerable strategic significance during both World Wars. Fundamentally, this was why the various owners of the complex did everything that they could to ensure that it remained able to fulfil its vital and therefore profitable purpose for the duration, which was why the depot was modified and adapted on so many occasions. That long standing trend typifies most of the Goods Station's existence however it cannot be applied to the post-war years. Instead, they witnessed the slow decline of the complex, an event that is of equal historical significance since it fits within a wide and important socioeconomic trend that saw the completion of Britain's transformation from a pre-eminent industrial powerhouse to a country that became increasingly reliant on global production. On top of that, the period would see a plethora of 'A' roads and motorways be constructed that would enable road hauliers to gain a competitive advantage over the railways. In summary, technological advances and economic changes from 1945 onwards turned the Goods Station from a vital piece of infrastructure into a relic from a bygone era. Nevertheless, the magnificent buildings that survive within the confines of the yard remain imposing reminders of the former primacy of the railways and the might of Britain's industrial past.

Despite their historical importance, major as well as minor railway warehouses and freight yards have generally been neglected by railway commentators, architectural historians and archaeologists (Stratton 1997: 41). What is more, few large urban goods sheds survive and many of those that do are amongst the most endangered railway structures in Britain today (Palmer *et al.* 2012: 257). In a modern context the station's tremendous heritage value is therefore elevated because it represents one of the largest, best preserved and best understood examples of its kind. Its significance is further elevated because the Great Northern decided to install pioneering hydraulic machinery from the very beginning, whilst all elements of the main shed merit an increased significance because of the emphasis that the company initially gave to interchange with the canal. Although the London & North Western's interchange at Camden would eventually become the largest and most sophisticated example of a three-way interchange in the world, no London carrier would outstrip King's Cross in terms of canal capacity.

Cubitt's elegant Transit Sheds represent transshipment facilities that were capable of the rapid transfer of goods between the railway, the road and the canal; they also contained banks for temporary storage. Consequently their key elements and plan forms define them as 'combined internal and external loading single storey goods sheds' (*ibid.*: 277). Similar large sheds were commonly constructed beside major docks, whilst modest examples lacking canal infrastructure characterised smaller towns and rural locations[22]. In the Victorian period this design was comparatively rare in urban areas, where railway goods depots more often took the form of large multi-storey warehouses because a greater emphasis was commonly placed on longer term storage.

The transshipment facilities at King's Cross therefore represent early examples of a comparatively unusual form of urban railway goods handling architecture. In terms of extant examples, they have more in common with the single storey depot that survives in York (now home to the National Railway Museum) than the famous and more typical multi-storey example that can be found at Liverpool Road in the major urban centre of Manchester, which is now occupied by the Museum of Science and Industry (Palmer *et al.* 2012: 280). Exactly why the Great Northern opted for rapid urban transshipment over longer term storage in their first London goods yard was not recorded in great detail, however it is reasonable to assume that the company adopted an innovative concept that the Directors and senior officers of the company would encourage a constant stream of goods to pass through what they anticipated would be their busiest yard. That in turn would result in the constant flow of money into the company's coffers. Although imperfect in execution, the popularity of the depot for the duration of the 19th century demonstrates that their thinking was fundamentally sound.

Being a multi-storey storage facility, the Granary had more in common with standard urban Victorian railway goods stations than the Transit Sheds and Train Assembly area, however the layout of its canal inlets, gravity-driven shoots and mechanical hoists suggests

---

22   Rural examples and some dockland sheds did not possess elevated platforms thus removing storage from the equation completely (they are defined as 'transshipment sheds'; see Palmer *et al.* 2012: 279).

that inspiration was also drawn from canalside grain reception facilities and dockland warehouses (*ibid.*: 277–8). Despite these obvious influences, its plan is unique because it was tailor made to interact with the Main Goods Shed so that it could form an integral part of the three-way interchange.

Together these innovations remind us of the fact that the Great Northern was striving to do things differently when it arrived at King's Cross in the early 1850s. The company intended to make its mark upon the railway scene by building what was at that time the largest, most technologically advanced and innovative goods yard that London had ever seen.

The importance of the Goods Yard as a heritage asset to King's Cross, London and the nation at large cannot be overstated. It is hoped that the results and conclusions that are presented in this volume will prompt new interest in the study of these important industrial facilities and offer guidance regarding their successful recording, preservation and interpretation in the future.

Neil Cossons has argued that all too frequently railway-related subjects are erroneously viewed or portrayed as of interest only to the train enthusiast; however the reality of the situation could not be more different (Cossons 1997: 16). Ever since Stephenson's Rocket captured the public imagination in 1829, the railways and their associated infrastructure have been inextricably bound to the political, social and economic development of Britain and its people. As demonstrated by the content of this study, a thorough understanding of our railway heritage via archaeology, built heritage and history is thus vital to a deeper and wider understanding of our nation's past.

Fig. 13.19 The granary after regeneration. Water features in front of the renovated Granary building mirror the location of the former granary basin

CHAPTER 14

Epilogue

The adaptations that were made to the Goods Yard at King's Cross enabled it to function as a goods handling facility from the 1850s to the 1990s, a phenomenal lifespan for a complex that retained many of its original Victorian features. Although the last few decades of its existence were characterised by gradual abandonment and neglect, this may actually have been a positive event from a heritage perspective. In their book *Change at King's Cross,* Robert Thorne and his co-writers make a salient point. Serious investment during this period may well have resulted in the complete demolition of these historically important buildings that were at that point in time much maligned. The improvements would have done nothing to stem the continued development of the national road network so it is likely that the haemorrhaging of haulage contracts would have continued unabated. Consequently, whilst these actions may well have maintained the viability of the station's rail head for a few more years, the loss of its historic buildings would have been tragic given the temporary nature of the reprieve (Thorne *et al.* 1990: 109).

The long term survival of the Goods Yard's remaining buildings has been assured thanks to a sympathetic and well considered programme of redevelopment that has placed heritage at its core. It was that project that generously funded this archaeological and historical study, meaning that the Goods Station at King's Cross is now not only one of the most historically important and well preserved examples in the country but is also one of the best understood. Most importantly, this handsome redevelopment has preserved this important complex for the benefit of future generations, breathing new life into those beautiful yet long neglected buildings in a dynamic and practical way. Because of the success of this tasteful restoration project, it is once again possible to appreciate the grandeur of Lewis Cubitt's original designs whilst recalling the achievements of the Great Northern Railway and its successors.

Fig. 14.1   The King's Cross regeneration area by night, seen from the south, with the Regent's Canal in the foreground and the Granary building at the centre

Appendix 1:

The archaeological sites and built heritage surveys that are included in this document and an inventory of the relevant interventions

Name of Archaeological Site or Built Heritage Survey	Site Code	Central National Grid Reference	Terminology of the Relevant Archaeological Interventions	Dimensions of the Archaeological Interventions	Nature of the Work
The Western Stables and the	KXC06	TQ 3017 8361	Western Stables	10.00m E–W by 100.00m N–S	Archaeological Watching Brief
The Granary Complex		TQ 3010 8349	Test Pits 1, 2A, 2B, 3, 4, 4A, 5, 6, 9, 10, 13, 15	6.48m to 1.16m E–W by 5.00m to 1.20m N–S	Archaeological Watching Brief
The Excel Bridge and Regent's Canal Wall	KXD07	TQ 3017 8345	N/A	N/A	Historic Building Recording
Coal and Fish Offices	KXE08	TQ 3005 8344	N/A	N/A	Historic Building Recording
The Laser Building and Retaining Walls	KXF07	TQ 3015 8360	N/A	N/A	Historic Building Recording
The Eastern Goods Yard: The Granary Group of Buildings	KXF07	TQ 30068 83560	N/A	N/A	Historic Building Recording
King's Cross Central	KXI07	TQ 3010 8349	Area A1	9.79m E–W by 11.36m N–S	Archaeological Excavation
			Area A2	10.24m E–W by 12.21m N–S	Archaeological Excavation
			Area A3	10.06m E–W by 15.80m N–S	Archaeological Excavation
			Area A4	54.38m E–W by 17.30m N–S	Archaeological Excavation
			Area A5	4.40m E–W by 15.66m N–S	Archaeological Excavation
			Area A6	9.42m E–W by 12.71m N–S	Archaeological Excavation
			Area A7	12.71m E–W by 10.24m N–S	Archaeological Excavation
			Area B1	11.67m E–W by 11.65m N–S	Archaeological Excavation
			Area B2	9.30m E–W by 11.65m N–S	Archaeological Excavation
			Area B3	7.26m E–W by 26.40m N–S	Archaeological Excavation
			Area B4	55.15m E–W by 11.00m N–S	Archaeological Excavation
			Area B5	7.42m E–W by 16.00m N–S	Archaeological Excavation
			Area B6	10.03m E–W by 12.63m N–S	Archaeological Excavation

Name of Archaeological Site or Built Heritage Survey	Site Code	Central National Grid Reference	Terminology of the Relevant Archaeological Interventions	Dimensions of the Archaeological Interventions	Nature of the Work
King's Cross Central	KXI07	TQ 3010 8349	Area B7	8.32m E–W by 12.63m N–S	Archaeological Excavation
			Area C1	26.74m E–W by 15.79m N–S	Archaeological Excavation
			Area C2	55.75m E–W by 15.00m N–S	Archaeological Excavation
			Area C3	23.75m E–W by 15.35m N–S	Archaeological Excavation
			Area D	9.32m E–W by 14.94m N–S	Archaeological Excavation
			Area E	23.75m E–W by 22.50m N–S	Archaeological Excavation
			The Granary	53.00m E–W by 28.80m N–S	Archaeological Watching Brief
			East Stables	9.30m E–W by 88.60m N–S	Archaeological Watching Brief
			West Stables	9.10m E–W by 92.80m N–S	Archaeological Watching Brief
			East Platform North	6.00m E–W by 5.60m N–S	Archaeological Watching Brief
			East Platform South	5.20m E–W by 31.00m N–S	Archaeological Watching Brief
			West Platform	5.00m E–W by 25.00m N–S	Archaeological Watching Brief
Hard Landscaping, Wharf Road and York Way	KXK08	TQ 30140 83571	N/A	N/A	Historic Building Recording
Midland Goods Shed	KXM08	TQ 3024 8356	N/A	N/A	Historic Building Recording
			Test Pits 1 to 5	3.40m to 2.00m E–W by 3.75m to 1.70m N–S	Archaeological Watching Brief
The Engine House	KXO08	TQ 30067 83496	Area A	16.00m E–W by 28.00m N–S	Archaeological Excavation
			Area B	12.80m E–W by 22.00m N–S	Archaeological Excavation
			Area C	8.00m E–W by 16.00m N–S	Archaeological Watching Brief
			Area D	5.00m E–W by 7.00m N–S	Archaeological Watching Brief
			Trench 1	1.70m E–W by 25.60m N–S	Archaeological Watching Brief
			Trench 2	12.20m E–W by 1.60m N–S	Archaeological Watching Brief
Turntable A and Turntable B	KXP08	TQ 30182 83499	Turntable A	7.63m E–W by 7.28m N–S	Archaeological Excavation
		TQ 30103 83516	Turntable B	30.40m E–W by 15.94m N–S	Archaeological Excavation

Appendix 2:

An inventory of features (buildings, Landscape Features and infrastructure networks) discussed in this publication (arranged by landscape feature number)

Aliases of Building, Landscape Feature or Infrastructure Network	Landscape Feature No.	Site Code(s)	Intervention Type
Regent's Canal, Towpath and Canal Walls	LF1	KXD07	HBR
Maiden Lane Bridge	LF2	KXD07	HBR
Gasworks Basin	LF3	KXD07	HBR
Retaining Wall	LF4	KXF07, KXI07, KXO08	HBR
Granary Basin and Canal Tunnels	LF5	KXI07, KXO08	AEX, AWB
Coal and Stone Basin	LF6	KXD08	HBR
Western Transit Shed (Part of the original Goods Shed complex, later known as the Eastern Goods Yard)	LF7	KXF07, KXI07	HBR, AEX, AWB
East Transit Shed (Part of the original Goods Shed complex, later known as the Eastern Goods Yard)	LF8	KXF07, KXI07	HBR, AEX, AWB
Western Stables (Part of the original Goods Shed complex, later known as the Eastern Goods Yard)	LF9	KXC06, KXI07	HBR, AWB
Eastern Stables (Part of the original Goods Shed complex, later known as the Eastern Goods Yard)	LF10	KXI07	HBR, AWB
Train Assembly Shed (Part of the original Goods Shed complex, later known as the Eastern Goods Yard)	LF11	KXF07, KXI07	HBR, AEX, AWB
Granary (Part of the original Goods Shed complex, later known as the Eastern Goods Yard)	LF12	KXF07, KXI07	HBR, AWB
The Lamp Room and Coffee Shop / The Refreshment Club	LF13	KXI07	AWB
The Temporary Passenger Station / The Potato Market	LF14	N/A	N/A
The Carriage Shed / The Midland Goods Shed	LF15	KXM08	HBR
Hydraulic Station	LF16	KXO08	AEX, AWB
The Coal Drops (later the Eastern Coal Drops)	LF17	N/A	N/A
Coal and Fish Offices	LF18	KXE08	HBR
Wharf Road Viaduct	LF19	KXK08	HBR
Western Goods Offices (Part of Granary Group of Buildings)	LF20	KXF07, KXI07	HBR
Eastern Goods Offices	LF21	KXF07, KXI07	HBR
Western Coal Drops	LF22	N/A	N/A
The Plimsoll Viaduct and the Cambridge Street Coal Drops	LF23	KXK07	HBR
Gasworks Viaduct	LF24	KXK07	HBR
Horse Provender Store / Stables	LF25	KXO08	AWB
West Handyside Canopy	LF26	KXF07, KXI07	HBR, AWB
East Handyside Canopy	LF27	KXF07	HBR, AWB
Western Goods Shed	LF28	N/A	N/A
Goods Way and Excel Bridge	LF29	KXD07	HBR
RME Garage	LF30	N/A	N/A
Laser Building	LF31	KXF07	HBR
Hydraulic Metering Station	LF32	KXO08	AEX

HBR: Historic Building Recording, AEX: Archaeological Excavation, AWB: Archaeological Watching Brief

Appendix 3:

Bricks used at King's Cross, their fabrics and origins

by Kevin Hayward

The composition of a brick, its 'fabric', including texture and colour, is the product of different types of clay and other materials used in its manufacture. The use of these clay sources changes over time, so it is possible to date a brick merely by its colour, texture and inclusions. Different coloured bricks were frequently used to add decorative details to a building.

Most of the fabrics listed below, used in construction at King's Cross, were obtained from local clays. Specialist bricks, on the other hand, used for the purpose of withstanding high temperatures and loads were acquired from much further afield.

Each brick fabric mentioned in the text is designated by a separate Museum of London four digit code as detailed below (MoLA 2007).

3032; 3034 (1664–1900) Post-Great Fire Brick. Maroon-yellow-red coloured brick with abundant inclusions of clinker, sometimes even clay tobacco pipes. Those from King's Cross have a frog (a shallow indent) that dates their period of manufacture to between 1850–1900 and were manufactured locally for construction purposes.

3033 (1800–1950) Blaes brick. Very dense orange construction brick made from waste shale or Blaes from the Coal Measures throughout the British Isles.

3032R; 3034R (1664–1900) Post-Great Fire Brick. As above but with more brickearth used in its manufacture giving the brick a distinctive red colour.

3035 (1780–1940) Yellow Stock Brick. Fine bright yellow fabric with occasional inclusions of clinker, manufactured from the estuary clays around the Medway. Those from King's Cross have a deep frog which dates their period of manufacture to between 1850 and 1940.

3261 (1850–1950) Kiln Brick. A very dense yellow brick fabric usually with a maker's mark. These bricks are manufactured from Coal Measure clays, which are high in alumina. They were manufactured for the purpose of withstanding high temperatures and are also called refractory bricks. At King's Cross they are associated with the Engine House but also cobbling from the stable areas. Those stamped 'CRAVEN', named after John Craven the inventor of the stiff plastic process in brick production, were manufactured from the Roundwood brickworks at Alverthrope, Wakefield. Other Yorkshire kiln brick manufacturers identified from their stamp at King's Cross are 'JOSEPH CLIFF & SONS WORTLEY', 'INGHAM WORTLEY', 'LEEDS FIRECLAY COMPANY', 'CINDERHILLS' from Halifax and 'HOPKINSONS PATENT HUDDERSFIELD'.

3220 (1890–Present Day) Fletton Brick. Construction bricks manufactured from the Upper Jurassic Oxford Clays in the Bedfordshire/Cambridgeshire area. Those characterised by the stamp 'LBC PHORPRES' from Bedfordshire are so named because they had been pressed twice in each direction so that they are literally 'four pressed'.

3498 (1890–Present Day) Staffordshire Blue. Black very dense bricks manufactured to withstand high temperatures and pressures acquired from the Etruria Marls (Upper Coal Measures) from Staffordshire.

Glossary

Archivolts	Ornamental curve or moulding on the underside of an arch.
Capstan	A broad revolving vertical cylinder for winding rope or cable.
Fairlead	Device to guide a line, rope or cable around an object, typically a ring or hook.
Intrados	The inner curve of an arch.
Jigger	A machine component that oscillates to and fro.
Pinnate	An arrangement of sidings that radiate from both sides of a central axis. When viewed from above the layout resembles the fronds of a fern.
Rowlock bedding	Bricks laid on their side.
Scrip	Substitute for legal tender, a form of credit or 'chit'.
Snatch head	A device that can be opened on one side to insert a rope, so called because the rope is inserted quickly (i.e. 'snatched').
Soldier bedding	Bricks laid vertically, on end.
Solebar	A longitudinal beam that runs either side of a piece of rolling stock upon which the bodywork is mounted.
Whip crane	A basic crane design that relies upon the principle of the wheel and the axle. Typically it possesses a rope that runs over a small drum upon which a load is suspended, raised and lowered. The device is turned by another rope that is wound around a larger drum on the same axle.

Bibliography

Archives consulted

BL: The British Library
CCRO: Cambridgeshire County Record Office, Huntingdon
COL: Corporation of London Guildhall Library
ICE: The Institution of Civil Engineers Archive
LBC: London Borough of Camden Metropolitan Archives Centre
LMA: London Metropolitan Archive
LSE: London School of Economics
NRRG: The Network Rail Group York
TNA: The National Archive
TWCA: Tyne and Wear County Archives

Primary Sources

BL Cubitt, L. 1850a. Schedule of Prices and Terms of Contract Intended to be applied to the execution of the several works which may be directed by the Architect to be performed in the Buildings or other constructions forming the London Permanent Goods and Temporary Passenger Station for the Great Northern Railway on their Land, North Side of the Regent's Canal, Maiden Lane, King's Cross, London. British Library BL 8244f5.

CCRO 148/2/839 Articles of Agreement: George Thornhill of Diddington, Hunts. Esq. and John Smith of Battle Bridge, parish of St Pancras, Middx. Brick maker, 17/02/1812.

CCRO 148/2/844 Lease for 14 years: George Thornhill of Diddington, Hunts. Esq. and William Ford Hickman, Henry Hickman, both Bricklayers of Battle Bridge, St Pancras, Middx, 22/12/1821.

COL MS 14943/13 London Wharf & Warehouse Committee Surveyors Reports, 1891–1892.

ICE Cubitt, L. 1850. Conditions of Contract and Specification of Works (with 10 accompanying drawings) prepared for the works to be done in the construction of iron roofing to adjoin the buildings of the Permanent Goods Shed and Temporary Passenger Station, Maiden Lane, near King's Cross, March 1850.

LBC CLAWS (Community Land and Workspace Services Ltd) 1988. Proposals for the Coal Offices, King's Cross. London Wildlife Trust.

LBC (London Borough of Camden) 1985a. LB Camden Planning and Communications Committee, King's Cross Goods Depot: Report of the Director of Planning and Communications, 16/04/1985. Catalogue 48.6626.

LBC (London Borough of Camden) 1985b. Department of Planning & Communication. King's Cross Goods Yard. Report of the Director of Planning and Communication, 05/02/1985.

LBC (London Borough of Camden) 1994. LB Camden Environment (Development Control) Sub-Committee, Change of use and works of convenience to provide Go Kart track Central Transit Shed, King's Cross Goods Yard, Wharf Road, York Way N1, Application No. PL/9400151, 08/12/1994. Catalogue 48.6626.

LMA B/GH/GP/06/003 General Hydraulic Power Co. Ltd Engineers' Reports, 1957–1962.

LMA B/GH/LH/02/003 General Hydraulic Power Co. Ltd. Minute Book, 1940–55.

LMA B/GH/LH/02/004 General Hydraulic Power Company Ltd. Minute Book, 1955–68.

LMA B/IMP/GLC/1 Imperial Gas Light & Coke Company Director's Minutes and Orders No. 1, 1821–1823.

LMA B/IMP/GLC/2 Imperial Gas Light & Coke Company Director's Minutes and Orders No. 2, 1823–1825.

LSE Booth B356 Walk with Inspector Bowles, of the Somers' Town Police sub-division, District 18 [Somers Town and Camden Town], 14 November 1898.

NRRG DMFP 00026218. LNER Southern Area-King's Cross Locomotive Depot, 'King's Cross Goods', 50'=1", c.1933.

NRRG DMFP 00026298 GNR 15761 London Goods Yard Alterations to Old Midland Goods Shed, 08/04/1915.

NRRG DMFP 00026308 55-ME-BM-22, Provision of Lift, 1955.

NRRG DMFP 00026357 LNER 46-LP-313 LNER King's Cross Outwards Goods Sheds, reconstruction of roofs over A & G Banks, Contract Drawing No. 1, 31/12/1946.

NRRG DMFP 00026358 LNER Remodelling of Goods Station; detail of traverser for present Inwards Goods Shed, 14/09/1936.

NRRG GNR [LNER] King's Cross Goods Station, c.1935.

NRRG LNER King's Cross Goods Shed Alterations to lay-out, 1935.

NRRG LNER 95-K-42 Proposed Alterations to Outwards Goods Shed, 19/10/1942.

NRRG RAILTRACK 138796 LNE. 56-ME-BM-12. King's Cross Goods Depot, Renewal of Roof Covering, 10 o'clock Road, 1956.

NRRG 138812 LNE GNR New Road at King's Cross on north side of Regent's Canal, 08/12/1920.

NRRG 50-LKC-105B KXG – RME Depot: Staff Accommodation, 1950.

NRRG 52-AR-MN-111 King's Cross Granary. Provision of Lift, General Drawing for Builder's Work, October 1952.

NRRG 66-ME-520 King's Cross Granary, Plan as Existing, February 1966.

St Mary Islington burial records 1850. Online at http://interactive.ancestry.co.uk/1559/91280_197215-00038 [Accessed 14/04/2016].

St Mary Islington burial records 1851. Online at http://interactive.ancestry.co.uk/1559/91280_197215-00040 [Accessed 14/04/2016].

TNA MT 6/10/38 Captain Gatton's inspection, Kings Cross Station, 1852.

TNA AN 2/1104 Railway Executive Committee Daily Reports Part 10, 27/08/1940–01/10/1940.

TNA AN 2/1105 Railway Executive Committee Daily Reports Part 11, 01/10/1940–07/11/1940.

TNA AN 2/1106 Railway Executive Committee Daily Reports Part 11, 07/11/1940–31/12/1940.

TNA AN 8/4 Modernisation and Re-equipment of British Railways. First Report of Traffic Survey Group, March 1956.

TNA AN 8/9 Modernisation and Re-equipment of British Railways. Report outlining the anticipated effect on the main locomotive, carriage and wagon works and carriage and wagon outdoor depots of the modernisation and re-equipment plan. Mechanical Engineering Committee, November 1956.

TNA AN 8/85 British Transport Commission. Modernisation and Re-equipment of British Railways. Reports of the Sub-Committees of the Planning Committee, 1954.

TNA AN 8/136 British Transport Commission. Modernisation and Re-equipment of British Railways. Handbook of Freight and Parcels Terminals Planning, March 1957.

TNA AN 13/198 Kings Cross Goods Depot: renewal of roofing of A and G banks of Outwards Shed and proposed method of working, 1948–1953.

TNA AN 115/237 Freightliner Terminal York Way, 1968–1971.

TNA AN 156/613 British Railways Board/ National Freight Corporation Liaison Committee: Road-Rail Transfer Schemes, 1973–1974.

TNA AN 169/631 British Rail Board King's Cross: proposed lease or sale of land to Post Office, 1970–1972.

TNA AN 183/113 King's Cross/St Pancras Rationalisation – Feasibility Report (Draft), 24/02/1966.

TNA AN 209/119: 18/10/1984.

TNA HINCH 11/204: Notices in the matter of proceedings for liquidation. Hinchingbrooke Collection http://discovery.nationalarchives.gov.uk/details/rd/126ec158-f024-4f37-9367-1572feb2189d>>. [Accessed 26th November 2014].

TNA HO 45/17610 Civil Defence: Air Raid Precautions Act, 1937: interpretation, 1938.

TNA HO 198/92 Ministry of Home Security Pilotless Aircraft Reports, Region 5, 07–22/08/1944.

TNA M380PWK/4 WALTER CAIN http://discovery.nationalarchives.gov.uk/details/rd/8b55c521-9f9c-455e-ad8a-4273bdef2173 [Accessed 26th November 2014].

TNA MT 6/2762 LNER Air Raid Damage, 1940–1942.

TNA RAIL 236/15 GNR Board of Directors Minute Book No. 3, 1849–50.

TNA RAIL 236/16 GNR Board of Directors Minute Book No. 4, 1850–1.

TNA RAIL 236/17 GNR Board of Directors Minute Book No. 5, 1851–2.

TNA RAIL 236/18 GNR Board of Directors Minute Book No. 6, 1852–3.

TNA RAIL 236/19 GNR Board of Directors Minute Book No. 7, 1853–4.

TNA RAIL 236/20 GNR Board of Directors Minute Book No. 8 1854–5.

TNA RAIL 236/21 GNR Board of Directors Minute Book No. 9, 1855–6.

TNA RAIL 236/22 GNR Board of Directors Minute Book No. 10, 1856.

TNA RAIL 236/24 GNR Board of Directors Minute Book No. 12, 1857–8.

TNA RAIL 236/25 GNR Board of Directors Minute Book No. 13, 1858.

TNA RAIL 236/26 GNR Board of Directors Minute Book No. 14, 1858–9.

TNA RAIL 236/27 GNR Board of Directors Minute Book No. 15, 1859–60.

TNA RAIL 236/28 GNR Board of Directors Minute Book No. 16, 1860–1.

TNA RAIL 236/29 GNR Board of Directors Minute Book No. 17, 1861–2.

TNA RAIL 236/30 GNR Board of Directors Minute Book No. 18, 1862–3.

TNA RAIL 236/31 GNR Board of Directors Minute Book No. 19, 1862–3.

TNA RAIL 234/32 GNR Board of Directors Minute Book No. 20, 1864–5.

TNA RAIL 236/33 GNR Board of Directors Minute Book No. 21, 1865.

TNA RAIL 236/34 GNR Board of Directors Minute Book No. 22, 1865–6.

TNA RAIL 236/35 GNR Board of Directors Minute Book No. 23, 1866–7.

TNA RAIL 236/38 GNR Board of Directors Minute Book No. 26, 1871–2.

TNA RAIL 236/39 GNR Board of Directors Minute Book No. 27, 1872–3.

TNA RAIL 236/43 GNR Board of Directors Minute Book No. 31, 1877–8.

TNA RAIL 236/49 GNR Board of Directors Minute Book No. 37, 1884–6.

TNA RAIL 236/50 GNR Board of Directors Minute Book No. 38, 1886–7.

TNA RAIL 236/51 GNR Board of Directors Minute Book No. 39, 1887–9.

TNA RAIL 236/52 GNR Board of Directors Minute Book No. 40, 1889–90.

TNA RAIL 236/53 GNR Board of Directors Minute Book No. 41, 1891–2.

TNA RAIL 236/54 GNR Board of Directors Minute Book No. 42, 1892–3.

TNA RAIL 236/70 GNR Executive Committee Minute Book No. 2, 1849–50.

TNA RAIL 236/71 GNR Executive Committee Minute Book No. 3, 1850–1.

TNA RAIL 236/74 GNR Executive Committee Minute Book No. 5, 1852–3.

TNA RAIL 236/85 GNR Executive and Traffic Committee Minute Book No. 17, 1859–60.

TNA RAIL 236/90 GNR Executive & Traffic Committee Minute Book No. 22, 1864.

TNA RAIL 236/91 GNR Executive & Traffic Committee Minute Book No. 23, 1864–5.

TNA RAIL 236/92 GNR Executive & Traffic Committee Minute Book No. 24, 1865–6.

TNA RAIL 236/144 GNR Traffic Committee Minute Book No. 2, 1870–2.

TNA RAIL 236/145 GNR Traffic Committee Minute Book No. 3, 1873–6.

TNA RAIL 236/146 GNR Traffic Committee Minute Book No. 4, 1877–9.

TNA RAIL 236/147 GNR Traffic Committee Minute Book No. 5, 1879–81.

TNA RAIL 236/148 GNR Traffic Committee Minute Book No. 6, 1881–3.

TNA RAIL 236/149 GNR Traffic Committee Minute Book No. 7, 1883–4.

TNA RAIL 236/150 GNR Traffic Committee Minute Book No. 8, 1884–6.

TNA RAIL 236/151 GNR Traffic Committee Minute Book No. 9, 1886–8.

TNA RAIL 236/152 GNR Traffic Committee Minute Book No. 10, 1888–9.

TNA RAIL 236/158 GNR Traffic Committee Minute Book No. 16, 1896–7.

TNA RAIL 236/160 GNR Traffic Committee Minute Book No. 18, 1899–1900.

TNA RAIL 236/170 GNR Way & Works Committee Minute Book No. 1, 1867–9.

TNA RAIL 236/171 GNR Way & Works Committee Minute Book No. 2, 1869–73.

TNA RAIL 236/172 GNR Way & Works Committee Minute Book No. 3, 1873–6.

TNA RAIL 236/173 GNR Way & Works Committee Minute Book No. 4, 1876–9.

TNA RAIL 236/174 GNR Way & Works Committee Minute Book No. 5, 1879–81.

TNA RAIL 236/175 GNR Way and Works Committee Minute Book No. 6, 1881–3.

TNA RAIL 236/178 GNR Way and Works Committee Minute Book No. 8 1888–90.

TNA RAIL 236/179. GNR Way & Works Committee Minute Book No. 10, 1890–2.

TNA RAIL 236/181. GNR Way & Works Committee Minute Book No. 12, 1893–5.

TNA RAIL 236/182 GNR Way & Works Committee Minute Book No. 13, 1895–7.

TNA RAIL 236/183 GNR Way & Works Committee Minute Book No. 14, 1897–9.

TNA RAIL 236/185 GNR Way & Works Committee Minute Book No. 16, 1901–3.

TNA RAIL 236/187 GNR Way & Works Committee Minute Book No. 18, 1905–9.

TNA RAIL 236/189. GNR Way & Works Committee Minute Book No. 20, 1913–7.

TNA RAIL 236/190 GNR Way & Works Committee Minute Book No. 21, 1917–21.

TNA RAIL 236/191. GNR Way & Works Committee Minute Book No. 22, 1922.

TNA RAIL 236/211 GNR Locomotive Department Reports to Directors, 1895–8.

TNA RAIL 236/216 GNR Locomotive Department Reports to Directors, 1912–5.

TNA RAIL 236/212 GNR Locomotive Department Reports to Directors, 1898–91.

TNA RAIL 236/227 Stores Committee Minute Book No. 1, 1856–8.

TNA RAIL 236/235 GNR Horse Committee Minute Book No. 1, 1890–1900.

TNA RAIL 236/236 Horse Committee Minute Book No. 2, 1900–15.

TNA RAIL 236/239 GNR Minutes of Occasional Committees No. 1, 1846–51.

TNA RAIL 236/273 GNR Reports to Board, 1847–51.

TNA RAIL 236/275 GNR Reports to Board, 1850–2.

TNA RAIL 236/275/17 GNR Fish traffic from the north into King's Cross 1852.

TNA RAIL 236/275/23 GNR Reports to Board, 1850–2.

TNA RAIL 236/276/6 GNR Reports on Works 1852–3.

TNA RAIL 235/276/23 GNR Reports to Board 1853–4.

TNA RAIL 236/280 GNR Board Reports & Papers, 1857–69.

TNA RAIL 236/283 GNR Reports to Board 1853–62.

TNA RAIL 236/283/4 Cessation of Coal Business by the Company and letting of Coal Offices to outside tenders, 1860.

TNA RAIL 236/284 GNR Reports to Board, 1860–71.

TNA RAIL 236/285 GNR Board Reports & Papers, 1860–2.

TNA RAIL 236/294/6. GNR Engineer's Report on the State of Way & Progress of Works, 11/01/1864.

TNA RAIL 236/303 GNR Reports to Board, 1863–70.

TNA RAIL 236/305/20 GNR Reports to the Board: Construction of Experimental Means of Unloading Coal at King's Cross Goods by Mr Plimsoll, 17/08/1871.

TNA RAIL 236/306/19: Engineer's Report on Way & Works, 29/11/1871.

TNA RAIL 236/316, GNR Engineer's Reports, 1874.

TNA RAIL 236/317 GNR Reports to the Board, 1875–6.

TNA RAIL 236/317/3. GNR Goods Manager's Report on London, 03/07/1875.

TNA RAIL 236/350 GNR Reports to the Board, 1883–1900.

TNA RAIL 236/361 GNR Reports to the Board, 1888–94.

TNA RAIL 236/362 GNR Reports to the Board, 1883–91.

TNA RAIL 236/364/3 Engineer's Reports, New Tunnel at King's Cross, 1889–92.

TNA RAIL 236/370 GNR Reports to the Board, 1891.

TNA RAIL 236/449 GNR Chimney & Alterations to Hydraulic Station at London Goods Yard, Specification, Bills of Quantities & Tender, June 1898.

TNA RAIL 236/469 Regent's Canal Company and GNR: Agreement of Works, 14/05/1850.

TNA RAIL 236/530 Specification, Bills of Quantities & Tender for the several works required to be done in the erection of Additional Offices over a portion of the London Goods Yard (Outwards Shed) at King's Cross Station on the GNR, 12/1897.

TNA RAIL 236/600GNR Horse Department Book: Record of Staff, 1894–1922.

TNA RAIL 236/933 GNR Accountant's Records Capital Journal No. 2.

TNA RAIL 236/934 GNR Accountant's Records Capital Journal No. 3.

TNA RAIL 250/58 GWR Minutes of the Board of Directors No. 55, 1938–41.

TNA RAIL 390/30 LNER Locomotive Committee Minute Book, 1923–6.

TNA RAIL 390/31 LNER Locomotive Committee Minute Book, 1926–30.

TNA RAIL 390/58 LNER Traffic Committee Minute Book No. 1, 1923–5.

TNA RAIL 390/59 LNER Traffic Committee Minute Book No. 2, 1925–7.

TNA RAIL 390/61 LNER Traffic Committee Minute Book No. 4, 1929–30.

TNA RAIL 390/63 LNER Traffic Committee Minute Book No. 6, 1934–6.

TNA RAIL 390/64 LNER Traffic Committee Minute Book No. 7, 1936–8.

TNA RAIL 390/65 LNER Traffic Committee Minute Book No. 8, 1938–47.

TNA RAIL 390/67 LNER Works Committee Minute Book No. 1, 1923–4.

TNA RAIL 390/68 LNER Works Committee Minute Book No. 2, 1924–7.

TNA RAIL 390/71 LNER Works Committee Minute Book No. 5, 1932–5.

TNA RAIL 390/72 LNER Works Committee Minute Book No. 6, 1935–7.

TNA RAIL 390/74 LNER Works Committee Minute Book No. 8, 1939–47.

TNA RAIL 390/296 LNER Traffic Committee proposed new works at King's Cross: pneumatic tube between Goods Depot and Passenger Telegraph Office, 1923.

TNA RAIL 390/547 LNER Partial Conversion of Granary into a Warehouse for Case Goods from the Continent via Harwich, 1925–6.

TNA RAIL 390/704 LNER King's Cross Goods Yard: New Repair Shop and Garage, 1928–34.

TNA RAIL 390/1148 LNER Construction of Air Raid Shelters in Southern Area, 1939–46.

TNA RAIL 390/1165 LNER Air Raid Precautions, Civil Defence Act 1939, 1938–46.

TNA RAIL 390/1192 LNER Damage by enemy action (through bombs and rockets), 1940–1945.

TNA RAIL 390/1973 LNER Emergency Board Minutes, 1940–2.

TNA RAIL 390/1974 LNER Emergency Committee Minutes, 1943.

TNA RAIL 393/321 LNER newsletter, May 13th 1926.

TNA RAIL 436/1 MS&L Minutes and reports; meetings of shareholders and directors, 1847–51.

TNA RAIL 783/110 GNR Correspondence File: Closure of Gas & Coke Works, purchase of site by GNR, 1904–12.

TNA RAIL 783/112 GNR Correspondence File: Gas Light & Coke Co. Works St Pancras, 1904–6.

TNA RAIL 783/220 Correspondence File: GNR correspondence files: London Hydraulic Power Company correspondence and guide to the London Hydrualic Power Company, 1906–7.

TNA RAIL 783/519 Three counties sidings; Arlesey Brick Co (Bearts) Ltd, 1920–2.

TNA RAIL 860/1 Regent's Canal Company, General Meetings of Proprietors Minutes, 1812–9.

TNA RAIL 860/14 Regent's Canal Company General Committee Minutes, 1818.

TNA RAIL 860/16 Regent's Canal Company General Committee Minutes, 1819.

TNA RAIL 860/18 Regent's Canal Company General Committee Minutes, 1820.

TNA RAIL 860/42 Regent's Canal Company General Committee Minutes, 1848–50.

TNA RAIL 860/44 Regent's Canal Company General Committee Minutes, 1851–3.

TNA RAIL 860/54 Regent's Canal Company General Committee Minutes 1865–8.

TNA RAIL 1124/127 Railway Companies Returns of Fish rates, 1888.

TNA RAIL 1189/1423 New road at King's Cross (Goods Way), J.L. Davies (Coal Merchants) Ltd, 1919–20.

TNA RAIL 1189/1424 GNR Land at Pancras Dock leased to the GNR, Regent's Canal & Dock Company, 1914–21.

TNA RAIL 1189/1425/3 GNR New Road at King's Cross Working Plan drawing No. 2, n.d. c. 12/1920.

TNA RAIL 1189/1426 GNR New road at King's Cross (Goods Way), water from Regent's Canal, 1915–6.

TNA RAIL 1189/1426/2 Pipes for supplying water for Hydraulic Machinery from the Regent's Canal, 16/08/1915.

TNA RAIL 1189/1428 GNR Land easement associated with new road (Goods Way), Regent's Canal and Dock Company, 1917–21.

TNA RAIL 1189/5 Kings Cross: claim by Thomas Owen for compensation for remaining twenty-one years of lease of Freemasons Arms, Suffolk Street West; includes plans of Kings Cross area with all contemporary public houses in half mile radius of Freemasons Arms marked, 1871–1886 TWCA MS: 1975/1 Elswick Works Machinery Shop Work-in-Hand Book, 1847–52.

Published Sources

Anon 1845. "Removal of King's Cross" *The Illustrated London News,* 15th February 1845.

Anon 1850. "The Great Northern Railway" *London Evening Standard*, 17th December 1850.

Anon 1851a. "The Accident at the Great Northern Railway" *Lloyd's Weekly London Newspaper,* 12th January 1851.

Anon 1851b. "Dreadful Accident and Loss of Life at King's Cross Station" *Jackson's Oxford Journal*, Saturday 11th January 1851.

Anon 1851c. "Fatal Occurrence at the Terminus, King's Cross" *The Kentish Independent*, 11th January 1851.

Anon 1851d. "Strike at the Great Northern Railway" *Morning Chronicle*, 20th October 1851.

Anon 1951e. "The Strike at the Great Northern Railway" *Morning Advertiser*, 21st October 1851.

Anon 1852. *London Gazette,* 19th November 1852.

Anon 1853. "Great Northern Railway Terminus" *The Illustrated London News*, 28th May 1853.

Anon 1855a. *London Gazette,* 9th February 1855.

Anon 1855b. *London Gazette,* 11th May 1855.

Anon 1856. *London Gazette,* 4th July 1856.

Anon 1859. "A visit to the Great Northern Railway Company's Stables and Goods Department" *The Illustrated London News*, 10th September 1859.

Anon 1860. *London Gazette,* 6th July 1860.

Anon 1899. "New Goods Terminal Accommodation at King's Cross (GNR)", reprinted from *Transport: the Weekly Railway Review* of 7th July 1899.

Anon 1969. *The London Gazette*, 28th January 1969.

Anon 2014. *The Grimsby Telegraph,* July 2014. Online at http://www.grimsbytelegraph.co.uk/Brewing-success/story-21445569-detail/story.html [Accessed 26/11/2014].

Anon n.d. a http://www.hennessy.com/en-gb/maison-hennessy/history [Accessed 26/11/ 2014].

Anon n.d. b http://www.oldbreweries.com/breweries-by-state/ohio/cleveland-oh-61-breweries/daniel-h-keys-brewery-oh-126c/ [Accessed 26/11/ 2014].

Anon n.d. c http://blackcountryhistory.org/collections/getrecord/GB149_DB-63 [Accessed 26/11/2014].

Archer, R. 2009. An Archaeological Watching Brief at the Midland Goods Shed, King's Cross Central, London Borough of Camden. Pre-Construct Archaeology Ltd. Unpublished Report.

Armstrong, W.G. 1850. "On the Application of Water Pressure, as a Motive Power, for working Cranes and other descriptions of Machinery" *Minutes of Proceedings of the Institution of Civil Engineers* Volume 9, 375–383.

Armstrong, W.G. 1877. "The History of the Modern Development of Water-pressure Machinery" *Minutes of Proceedings of the Institution of Civil Engineers* Volume 50, 64–88.

Atkin, M. 1995. *The International Grain Trade Second Edition* Cambridge: Woodhead Publishing Limited.

Baker, T.F.T. and Elrington, C.R. 1985. *A History of the County of Middlesex: Volume 8.* Online at *http://www.british-history.ac.uk/source.aspx?pubid=30* [Accessed 10/06/2015].

Bardwell, W. 1854. *Healthy Homes and How to Make Them* London: Deane and Son.

Barker, F., Jackson, P. and Saunders, A. 2008. *The Pleasures of London* London: London Topographical Society Publication 167.

Biddle, G. 1990. "The Making of the Passenger Termini" in M. Hunter and R. Thorne (eds) *Change at Kings Cross* London: Historical Publications Ltd., 59–74.

Bonavia, M.R. 1982. *A History of the LNER Volume 1. The First Years, 1922–33,* London: George Allen & Unwin.

Bonavia, M.R. 1983. *A History of the LNER Volume 3. The Last Years, 1939–48.* London: George Allen & Unwin.

Brandon, D. 2010. *London and the Victorian Railway* Stroud: Amberley Publishing.

Braybrook, T. 2012. King's Cross Central Building J London N1 London Borough of Camden post-excavation assessment report London: MOLA. Online at: http://archaeologydataservice.ac.uk/archiveDS/archiveDownload?t=arch-702-1/dissemination/pdf/molas1-118122_1.pdf [Accessed 10/06/2015].

British Railways Board 1963. *The Reshaping of British Railways.* London: Her Majesty's Stationery Office.

Brooksbank, B.W.L. 2007. *London Main Line War Damage,* London: Capital Transport.

Bussell, M. and Tucker, M. 2004a. King's Cross Heritage Study Part 1: Historic Building Baseline Report: Train Assembly Shed. ICHM Unpublished Report.

Bussell, M. and Tucker, M. 2004b. King's Cross Heritage Study Part 1 Historic Building Baseline Report: West Handyside Canopy. ICHM Unpublished Report.

Bussell, M. and Tucker, M. 2004c. King's Cross Heritage Study Part 1 Historic Building Baseline Report: Midland Goods Shed. ICHM Unpublished Report.

Campbell, G. and Turner, J. 2010. "The Greatest Bubble in History": Stock Prices during the British Railway Mania. Queen's University Belfast MPRA Paper No. 21820. Online at: http://mpra.ub.uni-muenchen.de/21820/1/MPRA_paper_21820.pdf [Accessed 10/06/2015].

Casson, M. 2009. The World's First Railway System. Enterprise, Competition, and Regulation on the Railway Network in Victorian Britain. Oxford: Oxford University Press.

Christopher, J. 2012. King's Cross Station through time Stroud: Amberly Publishing.

Christopher, J. 2013. St Pancras Station through time Stroud: Amberly Publishing.

Cossons, N. 1997. "An agenda for the railway heritage" in P. Burman and M. Stratton (eds) Conserving the railway heritage Abingdon: Routledge, 3–17.

Darley, P. 2013. Camden Goods Station through time London: Amberly Publishing.

DEGW Architecture Planning Design. c.1987. Office Group North of Canal (extract from report).

Denford, S.L.J. 1995. Agar Town: The Life and Death of a Victorian "Slum" London: Occasional Paper of the Camden History Society.

Denny, M. 1977. London's Waterways London: Batsford.

Dodson, S. 2007. "End of the line for King's Cross clubland?" The Guardian, 23rd November 2007. Online at: http://www.theguardian.com/music/musicblog/2007/nov/23/endofthelineforkingscros [Accessed 10/06/2015].

Dougan, D. 1970. The Great Gun-maker Warksworth: Sandhill Press.

Duckworth, S. and Jones, B. 1990. "The English Heritage Inventory of the King's Cross site" in M. Hunter and R. Thorne (eds) Change at King's Cross From 1800 to the Present. London: Historical Publications Ltd., 141–55.

Earnshaw, A. 1990. Britain's Railways at War 1914-1918 Penryn: Atlantic Transport Publishers.

Ellington, E.B. 1888. "The Distribution of Hydraulic Power in London" Minutes of Proceedings of the Institution of Civil Engineers. Volume 94, 1–85.

English Heritage 2006. Understanding Historic Buildings: A guide to good recording practice. Online at https://www.english-heritage.org.uk/publications/understanding-historic-buildings/ [Accessed 28/08/2014].

Erwood, P. 1988. "Memories of King's Cross Goods, 1937/38" Railways South East, Volume 1 (No. 3), Winter 1988–9, 130.

Factories Act 1961. London: Her Majesty's Stationery Office.

Fairman, A. 2007. An Archaeological Watching Brief at the Granary Complex, King's Cross Central, London Borough of Camden. Pre-Construct Archaeology Ltd. Unpublished Report.

Faulkner, A. 1990. "The Regent's Canal" in M. Hunter and R. Thorne (eds) Change at Kings Cross, 40–57.

Fisher, T. and Powell, J. 1979. "The hydraulic system in Bristol City Docks" Bristol Industrial Archaeological Society Volume 12.

Gilbert, B. c.1985. The King's Cross Cut: A City Canal and its Community London: Lasso Ltd.

Godfrey W.H. and Marcham, W.M. 1952. Survey of London: Volume 24, the Parish of St Pancras Part 4: King's Cross Neighbourhood. Originally published by London County Council. URL: http://www.british-history.ac.uk/survey-london/vol24/pt4 [Accessed 10/06/2015].

Gordon, W.J. 1893. The Horse World of London. Online at: http://www.victorianlondon.org/index-2012.htm [Accessed 10/06/2015].

Goslin, G. 2002. Goods Traffic of the LNER. Didcot: Wild Swan Publications.

Greaves, J. 2007. "The First World War and its Aftermath" in F. Carnevali and J. Strange (eds) 20th Century Britain. Economic, Cultural and Social Change Harlow: Pearson Longman, 127–144.

Grinling, C.H. 1898. The History of the Great Northern Railway, 1845–1895. London: Methuen & Co.

Grinling, C.H. 1905. "Railway Companies as Road Carriers" reproduced in the Journal of the Railway & Canal Historical Society, Vol. 36, Part 1 No. 201, March 2008, 20–30.

Gurney-Read, J. 1987. A.J. Caley Limited online at http://www.heritagecity.org/research-centre/industrial-innovation/caleys.htm [Accessed 26/11/2014].

Harrison, M. 1965. London Growing London: Hutchinson & Co.

Haslam, R., Thompson, G. and Maher, S. 2011. Archaeological Excavation at "The Eastern Goods Yard" Kings Cross. Part 1: The Hydraulic Station, Engine House, Turntables A & B and an Interim Report on Kings Cross Central. Pre-Construct Archaeology Ltd. Unpublished Report.

Haslam, R. 2013. "The Goods Handling Facilities at the Kings Cross Railhead". *Institute for Archaeologists Yearbook and Directory 2013*, 29–32.

Hawkins, H. 2011. An "As Found" Photographic Survey of the Midland Goods Shed, King's Cross Central, London Borough of Camden. Pre-Construct Archaeology Ltd. Unpublished Report.

Head, F.B. 1968. *Stokers & Pokers* Newton Abbot: David & Charles reprint (first published 1849).

Heald, H. 2010. *William Armstrong, Magician of the North* Newcastle upon Tyne: Northumbria Press.

Hill, F. 1974. *Victorian Lincoln* London: Cambridge University Press.

Holden, B. 1985. *The Long Haul- The Life and Times of the Railway Horse* London: J.A. Allen.

Hollingshead, J. 1861. *Ragged London* London: Smith Elder and Co.

Hughes 1921. *Hughes' London Business Directory* London: Hughes.

Humber, W. 1866. "On the design and arrangement of railway stations, repairing shops, engine sheds &c" *Minutes of the Proceedings of the Institution of Civil Engineers* Volume 25, 263–276.

Inwood, S. 1998. *A History of London* New York: Carroll & Graf.

Jackson, T. 2013. *British Rail, the Nation's Railway* Stroud: The History Press.

Jarrett, C. and Thompson. G. 2012. "A group of early 20th-century naval victualling finds from Royal Clarence Yard, Gosport, Hampshire" *Post-Medieval Archaeology* Volume 46 part 1, 89–115.

Jones, N. 2007. *The Plimsoll Sensation. The Great Campaign to Save Lives at Sea*. London: Abacus.

Kay, P. 2000a. "The First King's Cross – the 1850 GNR terminus in Maiden Lane and its subsequent fate" *London's Industrial Archaeology Journal* No. 7, 39–54.

Kay, P. 2000b. "Railway Archaeological Sites: King's Cross Goods Yard" *The London Railway Record*, No. 24, 361–370.

Kean, H. 1998. *Animal Rights: Political and Social Change in Britain since 1800* London: Reaktion Books.

Kelly & Co. 1861. *Post Office London Directory* London: Kelly & Co.

Kelly & Co. 1869. *Post Office Directory Huntingdonshire* London: Kelly & Co.

Kelly & Co. 1871. *Post Office London Directory* London: Kelly & Co.

Kelly & Co. 1882. *Post Office London Directory, Part 2* London: Kelly & Co.

Kelly & Co. 1892. *Post Office London Directory* London: Kelly & Co.

Kelly & Co. 1895. *Post Office London Directory, Part 3* London: Kelly & Co.

Kelly & Co. 1900. *Kelly's Directory of Lincolnshire* London: Kelly & Co.

Kelly & Co. 1904. *Kelly's Post Office London Directory* London: Kelly & Co.

Kelly & Co. 1912. *Kelly's Post Office Directory Derbyshire* London: Kelly & Co.

Kelly & Co. 1930. *Kelly's Post Office London Directory* London: Kelly & Co.

Kelly & Co. 1940. *Kelly's Post Office London Directory* London: Kelly & Co.

Kelly & Co. 1945. *Kelly's Post Office London Directory* London: Kelly & Co.

Kelly & Co. 1960. *Kelly's Post Office London Directory* London: Kelly & Co.

Kelly & Co. 1965. *Kelly's Post Office London Directory* London: Kelly & Co.

Kelly & Co. 1975. *Kelly's Post Office London Directory* London: Kelly & Co.

Knight, C. 1851. *Knight's Cyclopaedia of London*. London: George Woodfall & Son. Online at http://books.google.co.uk/books?id=iN49AAAAcAAJ&pg=PP9&dq=knight%27s+cyclopaedia+of+London [Accessed 10/06/2015].

Lake n.d. *The True John Rawlings*. Online at http://www.derynlake.com/rawlings.php [Accessed 26/11/2014].

Lewis, H. 2013/2014. "Somers Town Goods Yard: excavations at Brill Place, Camden NW1" *London Archaeologist* Volume 13 No. 11, 287–293.

Loft, C. 2006. *Government, the Railways and the Modernisation of Britain. Beeching's Last Trains* Abingdon, Oxon: Routledge.

Lovell, P. and Marcham, W. McB. (eds), 1938. *Survey of London: Volume 19, the Parish of St Pancras Part 2: Old St Pancras and Kentish Town.* Originally published by London County Council. Online at http://www.british-history.ac.uk/survey-london/vol19/pt2 [Accessed 10/06/15]

Lovell, P. and Marcham, W. (eds) 1938. *Survey of London: Volume 19: The Parish of St Pancras Part 2: Old St Pancras and Kentish Town*, pp. 60–2. Online at http://www.british-history.ac.uk/report.aspx?compid=64863 [Accessed 10/06/2015].

Marshall, G. 2013. *London's Industrial Heritage* Stroud: The History Press.

May, T. 2013. *The Victorian Railway Worker.* Princes Risborough: Shire Publications.

Mazurkiewicz, T. 2008. An Archaeological Watching Brief at Plots J, Q1, Q2, and R2 King's Cross Central, London Borough of Camden Pre-Construct Archaeology Ltd. Unpublished Report.

McNeil, I. 1968. *Joseph Bramah* Newton Abbot: David & Charles.

McNeil, I. 1972. *Hydraulic Power* London: Longmans.

Medcalf, J. 1900. "Railway Goods Depots: IV - King's Cross Goods Station" *Railway Magazine* Volume 6, April 1900, 313–20.

MoLA 2007. *Medieval and post-medieval brick and drain fabrics.* Online at http://www.mola.org.uk/resources/medieval-and-post-medieval-brick-and-drain-fabric-codes. [Accessed 9/12/2014].

MOLAS 1994. *Archaeological Site Manual* London: Museum of London Archaeology Service.

Morris, R. 2003. *The Archaeology of Railways* Stroud: Tempus.

National Telephone Company 1910a. *National Telephone Company Directory Vol. I* London: National Telephone Company.

National Telephone Company 1910b. *National Telephone Company Directory Vol. II* London: National Telephone Company.

National Telephone Company 1921. *National Telephone Directory* London: National Telephone Company.

Nock, O.S. 1974. *The Great Northern Railway* London: Ian Allan.

Nokes, G.A. 1938. *Locomotion in Victorian London* London: Oxford University Press.

O'Connell, S. 2007. "Motoring and modernity" in F. Carnevali and J. Strange (eds) *20th Century Britain. Economic, Cultural and Social Change* Harlow: Pearson Longman, 111–126.

Odlyzko, A. 2010. *Collective Hallucinations and Inefficient Markets: The British Railway Mania of the 1840s* Minnesota: University of Minnesota.

O'Gorman, T. 2007. An Archaeological Watching Brief on the Western Stables at King's Cross Goods Yard, Wharf Road, Off York Way, London Borough of Camden. Pre-Construct Archaeology Ltd. Unpublished Report.

Oliver, M.J. 2007. "The retreat of the State in the 1980s and 1990s" in F. Carnevali and J. Strange (eds) *20th Century Britain. Economic, Cultural and Social Change* Harlow: Pearson Longman, 262–278.

Orser (Jnr), C.E. 2002. *Encyclopaedia of Historical Archaeology* New York: Routledge.

Palmer, M. and Neaverson, P. 1992. *Industrial Landscapes of the East Midlands* Chichester: Phillimore.

Palmer, M., Nevell, M. and Sissons, M. 2012. *Industrial Archaeology: A Handbook* York: Council for British Archaeology.

Pigot & Co. 1839. *Pigot & Co's Directory of London* London: Pigot & Co.

Plimsoll, S. 1872. *Our Seamen: An Appeal* London: Virtue & Co.

Price, C. 2007. "Depression and Recovery" in F. Carnevali and J. Strange (eds) *20th Century Britain. Economic, Cultural and Social Change* Harlow: Pearson Longman, 145–61.

Public Health Act 1961. London: Her Majesty's Stationery Office.

Reynolds, T.S. 1983. *Stronger than a Hundred Men* Baltimore: The John Hopkins University Press.

Robbins, F.C. 1935. "Fish Traffic Dealt with at King's Cross Goods" *LNER Magazine* Volume 25, No. 1, 25. TNA ZPER 17/9.

Roberts, G A. 1938. "Remodelling of King's Cross Goods Station". *LNER Magazine* Volume 28, No. 11, 634–5. TNA ZPER 17/120.

Robertson, B. 1977. *Roman Camden* London: Camden History Society. Online at: http://archaeologydataservice.ac.uk/archiveDS/archiveDownload?t=arch-457-1/dissemination/pdf/vol02/vol02_10/02_10_250_255.pdf [Accessed 10/06/2015].

Robinson, H. 1887. *Hydraulic Power and Hydraulic Machinery* London: C. Griffin & Company Ltd.

Sabel, K. and Tucker, M. 2004. King's Cross Heritage Study Part 1: Historic Building Baseline Report: The Granary. ICHM Unpublished Report.

Sala, G.A. 1859. *Gaslight and Daylight: with some London Scenes they Shine Upon* London: Chapman & Hall.

Saunders, A. and Smith, N.J. 1970. "Hydraulic Cranes from Armstrong's Elswick Works, 1846–1936, a product life-cycle in capital goods" *Journal and Proceedings of the Industrial Market Research Association* 6 (No.2) May 1970.

Scott, P. 2007. *Triumph of the South: A Regional Economic History of Early Twentieth Century Britain* Aldershot: Ashgate Publishing.

Senatore, M. et al. 1982. "Great Northern Railway Offices, King's Cross, London" *Architects' Journal* 8 December, 31. LB Camden Local Studies & Archives Centre Jackdaw Box F23 (48.6626).

Simmons, D.A. 1983. *Schweppes: The First 200 Years* Washington D.C.: Acropolis Books.

Simmons, J. and Biddle, G. (eds) 1997. *The Oxford Companion to British Railway History* Oxford: Oxford University Press.

Smith, T. 1991. Hydraulic Power in the Port of London. *Industrial Archaeology Review* Volume 14 No. 1, 64–88.

Smith, T. 2008. Hydraulic Power at the Kings Cross Goods Yard. Unpublished Report commissioned by Pre-Construct Archaeology.

Smith, T. 2009. Turntables Delivered to the Kings Cross Goods Yard. Unpublished Report commissioned by Pre-Construct Archaeology.

Stamp, G. 1990. "From Battle Bridge to King's Cross: Urban Fabric and Change" in M. Hunter and R. Thorne (eds) *Change at King's Cross From 1800 to the Present*, 11–39.

Stratton, M. 1997. "A bibliographical overview of the railway heritage". In P. Burman & M. Stratton (eds) *Conserving the railway heritage* Abingdon: Routledge, 34–58.

Stretton, C.E. 1901. *The History of the Midland Railway* London: Methuen.

Swenson, S.P. 2006. *Mapping Poverty in Agar Town: Economic Conditions Prior to the Construction of St. Pancras Station in 1866*, 1–55. Online at http://eprints.lse.ac.uk/22539/1/0906Swensen.pdf [Accessed 10/06/2015].

Swindells, D. 2008. "The end of clubbing in King's Cross" *Timeout*, 2nd January 2008. Online at http://www.timeout.com/london/clubs/the-end-of-clubbing-in-kings-cross [Accessed 10/06/2015].

Talbot, O. 1974. "The evolution of glass bottles for carbonated drinks". *Post-Medieval Archaeology* Volume 8, 29–62.

The Labour Party 1945. *Manifesto for the General Election: Let Us Face the Future* London: The Labour Party.

The Railway Chronicle for 1844 London: J. Francis

Thomas, W.M. 1851. "A Suburban Connemara" *Household Words* Volume VI, pp. 308–15. Online at: http://books.google.co.uk/books?id=Ge5LAAAAcAAJ&pg=PA308&dq=suburban+connemara+household+words&hl=en&sa=X&ei=kCgSUf-pO8TStQaAnoBo&ved=0CD4Q6AEwAg#v=onepage&q=suburban%20connemara%20household%20words&f=false [Accessed 10/06/2015].

Thompson, G. and Gould, M. 2010. Historic Building Recording of the Midland Goods Shed, King's Cross Central, London Borough of Camden. Pre-Construct Archaeology Ltd. Unpublished Report.

Thompson, G., Gould, M. and O'Gorman, T. 2011. Historic Building Recording of the Granary Group of Buildings within the Eastern Goods Yard, King's Cross Central, London Borough of Camden. Pre-Construct Archaeology Ltd. Unpublished Report.

Thompson, G. and Matthews, C. 2011. Historic Building Recording of the Regent's Canal Walls, Excel Bridge and the Wall adjacent to Camley Street Natural Park, King's Cross Central, London Borough of Camden. Pre-Construct Archaeology Ltd Unpublished Report.

Thompson, G. and Matthews, C. 2012. Historic Building Recording of the Laser Building and the Retaining Wall for the Granary Group of Buildings, Kings Cross Central, London Borough of Camden. Pre-Construct Archaeology Ltd. Unpublished Report.

Thompson, G. and O'Gorman, T. 2009. Historic Building Recording of the Coal and Fish Offices, King's Cross Central, London Borough of Camden. Pre-Construct Archaeology Ltd. Unpublished Report.

Thornbury, W. 1878. *Old and New London: Volume 2* 296–8. Online at http://www.british-history.ac.uk/source.aspx?pubid=340 [Accessed 10/06/2015].

Thorne, R. 1990. "The Great Northern Railway and the London Coal Trade" in M. Hunter and R. Thorne (eds) *Change at King's Cross From 1800 to the Present* London: Historical Publications Ltd., 111–123.

Thorne, R. with Duckworth, S. and Jones, B. 1990. "King's Cross Goods Yard" in M. Hunter and R. Thorne (eds) *Change at King's Cross From 1800 to the Present* London: Historical Publications Ltd., 91–110.

Townend, P. N. 1975. *Top Shed: a pictorial history of King's Cross Locomotive Depot* London: Ian Allan Ltd.

Townley, A. 2013. *New Trent Brewery, Spen Lane – George Robinson Jun & Co – New Trent Brewery Co Ltd*, <http://crowle.org/?p=1929> [Accessed 26/11/ 2014].

Transport Act 1968. London: Her Majesty's Stationery Office.

Tuck, H. 1845. *The Railway Shareholders Manual*, Second Edition London: Effingham Wilson.

Vernon, T. 2007. *Archibald Sturrock: Pioneer Locomotive Engineer* Stroud: The History Press.

Wade, G.A. 1900. "The Horse Department of a Railway" *Railway Magazine* Volume 8, 208–213.

Walford, E. 1878. *Old and New London: Volume 5*. Online at http://www.british-history.ac.uk/report.aspx?compid=45241 [Accessed 10/06/2015].

Weale, J. 1851. *London Exhibited in 1851, elucidating its natural and physical characteristics, its antiquity and architecture, its arts, manufactures, trades and organizations, its social, literary and scientific institutions and its numerous galleries of fine art* London: John Weale. Online at https://play.google.com/books/reader?id=lCMLAAAAYAAJ&printsec=frontcover&output=reader&authuser=0&hl=en_GB [Accessed 10/06/2015].

Weinreb, B. and Hibbert, C. (eds). 1993. *The London Encyclopaedia*. London: Macmillan.

Whitehouse, P. and St John Thomas, D. 2002. *The London and North Eastern Railway A Century and a Half of Progress* Newton Abbot: David & Charles.

Williams, M. 2013. *Steaming to Victory: How Britain's Railways Won the War* London: Random House.

Williams, R. 1988. *The Midland Railway- A New History* London: David & Charles.

Wilson G. 2007. *London United Tramways: A History 1894–1933* Abingdon: Routledge.

Wolmar, C. 2007. *Fire & Steam: How the Railways Transformed Britain*. London: Atlantic Books.

Woodcroft, B. 1857. *Subject-Matter Index (made from titles only) of patents of invention, from March 2, 1617 (14 James I.) to October 1, 1852 (16 Victoriæ) Part II. (N. To W.)* London: The Great Seal Patent Office.

Wragg, D. 2012. *Wartime on the Railways* Stroud: The History Press.

Wrottesley, J. 1979. *The Great Northern Railway. Volume 1: Origins and Development* London: Batsford.

Cartographic Sources

Davies Map of 1834

Rocque Map of St Pancras Parish, 1769

Thompson Map of St Pancras Parish, 1801

Thompson, J. 1804. A map of the parish of Saint Pancras, situate in the county of Middlesex from a minute and correct survey / taken by J. Tompson ; engraved by I. Palmer, Store Street, Bedford Square.

Stanford Map of St Pancras Parish, 1834

St Pancras Parish Map, 1849

Parliamentary Plan of the Great Northern Railway's Proposed London Terminus, 1846

The Geological Survey of England and Wales (Sheet 260)

Goad Insurance Sheet 12/400 1891 Kings Cross Goods Yard

Goad Insurance Sheet 12/400 1921 King's Cross Goods Yard

Goad Insurance Sheet 12/400 1942 King's Cross Goods Depot

Index

compiled by Christina Reade

Illustrations are denoted by page numbers in *italics* or by *illus* where figures are scattered throughout the text.

accumulator 76–7, 119, 122, 143, 150, 188, 250, *288*, 288–95
    bases 151, *152*, 323
    building material 154
Accumulator House 150, 172, 188, 190, 192, 194; *see also* Accumulator Tower
Accumulator Tower
    original construction 30, 77, 79–80, 122, 126, 150, 190
    new expansion 150, 151–3, 190
    demolition 286
    derelict 142
    see also Hydraulic Station; Midland Shed Accumulator Tower
Agar Town 10, 14, 23, 27, 95, 121
Agar Town Depot 95, 121, 159, 299, 323
Agar, William 13–16
Air Raid Precautions Act 1937 245
air raids
    Gotha bombers 205–206
    precautionary measures 210, 211, 245–7, 248–50, 254, 256
    shelter later use 258, 266, 279
    Zeppelin 205
    see also London Blitz
Alfrichbury 9–10
Allenbury 9–10, 17
Ambergate Railway 93
Armstrong, William 55, 76–7, 80, 288–95; *see also* Armstrong engine; W.G. Armstrong & Co.; W.G. Armstrong, Mitchell & Co.
Armstrong engine 80, 150, 187, 192
Army Railway Council 203
Attlee, Clement 257, 317

Bagnigge Wells 10
Battle Bridge 10–13, 15, 17, 195, 212
    renamed to King's Cross 12–13
Beart's Patent Brick Company 120
Beeching, Richard 277
Benjamin Walker motor 148
Big Four 220, 222, 231, 239, 245–6, 256, 257, 266, 268, 314
Bishopsgate Goods Yard 222–3, 229

Boiler House
    disuse 227, 286, 295–6
    fuel delivery 72
    new buildings 150, 151–3, 187–8, 193, 293
    building material 196
    reconstruction and extension 190, 192, 192–3, 228, 295
    original building 72, 80–81, 153, 289
    see also Hydraulic Station
boiler supports 193, 196, 228, 326
bottle warehouse *see* Glass Bottle Warehouse
Boudicca 10
Brassey, Thomas 25, 28
brick
    fabrics 154, 196, 239, 333
        Blaes 154, 196, 333
        Fletton 140, 196, 239, 266, 333
        Kiln 87, 154, 196, 239, 285, 333
        Post-Great Fire 16, 36–7, 87, 103, 104, 154, 172, 196, 239, 333
        Staffordshire Blue 182, 185, 333
        Yellow Stock 36, 87, 103, 104, 122, 154, 196, 239, 333
        red /purple stock 16–17, 31, 33, 52, 75, 77, 78, 79, 80, 149, 151–3, 154, 190, 191, 192, 193, 234, 235
    forms 16–17, 31, 33, 36–7, 52, 48, 68, 75, 80, 87, 103, 104, 151–3, 154, 172, 190, 192, 193, 196, 234, 235, 239, 265, 267, 333
    bullnosed engineering 210, 236, 285
    stamps 36, 48, 87, 111, 196, 239, 333
    industry 11, 15, 87
    trade 37–8, 139–40, 179, 196–7, 225, 279, 298, 303
brickwork
    Flemish bond 38, 52, 56, 75, 84, 103, 104, 122, 125, 129, 163, 238
    English bond 17, 31, 33, 36, 56, 77, 79, 117, 122, 126, 129, 149, 191, 234
British Rail 67, 140, 177, 265
British Railways 229, 250, 257, 266, 268, 275, 277–80, 285, 287, 296, 317–18, 323–5
    denationalisation 278
British Railways Board 277–8, 279; *see also* Railway Executive Committee
British Road Services 279
British Transport Commission 257–60, 266, 277
Brown, Charles 206–207
Brydone, Walter 98–9, 120–1, 134–5, 144
building material 87; *see also* brick; tile; granite; stone

Caledonian Railway 159
Campbell Geddes, Eric 220
Cambridge & Lincoln 21
Cambridge & York 18

Cambridge Street Coal Drops   128–9, 132, *157*, 158, 171, 193, 300–303, 324
Camden Goods Yard   197, 277, 293, 298, 321–3, *321*, 325
Camden, London Borough of   278, 285–6
Camden Town   13, 27
Canal Branch
   construction   34
      building material   36
   design   29, 35, 133
canal tunnels
   design   29, *30*, 33–5
   construction   35–7
      building material   36–7
   see also Canal Branch; Regent's Canal
cap stones   17, 33, 35–6, 80
capstan   29, 51, 119, *145*, 293–5
   Capstan 1   149
   Capstan 2   149
   Capstan 3   147, 149
   Capstan 4   147, 149
   Capstan 5   147, 149
   Capstan 6   147–9, *148*
   see also Hydraulic Capstan Shunting
Carriage Shed   *25*, 73–5, 120–24; see also Midland Goods Shed
cattle   67, 140, *141*, 322
Churchill, Winston   230
Clarke, Herbert   126–8, 171
Clarke, Seymour   29, 47, 54–5, 72, 76–7, 94–5, 98–104, 105–110, 114–17, 121, 132–5, 157, 215, 310
Clerkenwell   10, 11
coal
   distribution networks   20, 88, 93–5, 98, 133, 156–7
   decline   222, 318
   economic importance   81, 88, 118, 136, 222
   handling facilities   67, 81–2, 85, 86, 88, 121, 139–40, 150, 173, 212, 299–301, 322, 324
   import   20, 37–8, 138
   private merchants   126, 128, 131, 171, 229; see also Plimsoll, Samuel
   sea trade   156–7, 311
   transshipment   33, 50, 213, 316
   see also Cambridge Street Coal Drops; Coal Drops; Coal Offices
Coal and Fish Offices   207, 217, 229, 285–6
Coal and Stone Basin   *25*, 31, 33–5, 37, 82, 140, 181, 197, 303, 312, 325
Coal Drops   *25*, 50, 77 81–2, 86, 126, 128–9, 133, 138, 171, 285, 299, 301; see also Eastern Coal Drops
Coal Offices *25*, 31, 81–6, *81*, 126–35, 171–2, 213–14, 228–301, 303; see also Coal and Fish Offices
Corn Laws   24, 65, 311
Crane Base 1   77–8, 100
Crane Base 2   100
Crockett, Edwin   70, 103, 108, 111, 114, 164–5, 185

Cubitt, Joseph   24, 27–9, 33–4, 47, 51, 55, 67 73–7, 81–2, 107, 114, 133–4, 213, 289, 325
Cubitt, Lewis
   canal arms   34
   Main Goods Shed design   37–8, 40, 45, 47, 48, 50, 51–2, 54–5, 56, 73–4, 76–7, 101, 114, 119, 213, 305, 322
      functionality   298–9
      schedule of prices   73–4
   paintings
      Granary exterior   *30*, 35, 36, 38, 77, 80
      Transit Shed interiors   291
   Passenger Station design   89, 164
   Pickford's goods shed   322
   roofing plan   35
   vision for King's Cross   29–30, 329
Cubitt, Thomas   11
Cubitt, William   20, 27, 29

Denison, Edmund   18–20, *18*
Dickens, Charles   10, 14
Direct Northern   18–21
Drakefield Estate   11, 23, 27

East Handyside Canopy   74, 162–4, 230, 260, 262, 265, 302
East London line   246
Eastern Midlands Railway   159
Eastern Coal Drops   129, 171, 209, 272, 311; see also Coal Drops
Eastern Goods Offices
   alterations   102, 111–16, 266, 286
      staff welfare   102, 186–7
      Transposition Scheme   232
   bomb damage   251
      repairs   255, 266
   construction   101–3, 104–107
   design   54, *55*, 101–102
   later use   279, 280
Eastern Goods Shed
   Air Raid Precaution measures   248–50
   alterations   210, 259–65, 268, 279–80, 285, 287; see also Transposition Scheme; Overhead Offices; Rail Parcels Depot
   development   181, 230–3
      archaeological evidence   233–34, 235
      built heritage evidence   235–6
      building materials   196–7, 239
   finds assemblage
      glass bottles   272–5
      iron weights   243
      wheelbarrows   176
   inefficiencies   231
   see also Eastern Transit Shed; Main Goods Shed; Train Assembly Shed; Western Transit Shed

Eastern Transit Shed
- alterations 107-9, 111-14, 161, 163, 230, 287, 305
  - conversion to Eastern Goods Shed 181-2, 230
  - conversion to Road Distribution Terminal 284-5
- bomb damage 251
  - repairs 254, 255, 268
- construction 38-43, 45-7, 67
- design 29
- finds assemblage 270-3
- function 37-38
- hydraulic machinery 76-7
- movement of goods 63, 67-8, 99, 136, 181, 230, 298, 302-303
- see also Eastern Transit Shed Stables; Goods Shed complex

Eastern Transit Shed Stables
- alterations 167, 169, 236, 250
- construction 47, 49, 51
- design 37-8, 51
- finds assemblage 214
- horse welfare 214-16
- see also Goods Yard; Main Goods Shed

engine beds 87, 154, *189*, 191, 196, 296, 326
- Engine Bed 1 80, 192
- Engine Bed 2 80, 190
- Engine Bed 4 152-3, 191-2, 228
- Engine Bed 5 152-3, 191-2, 228
- Engine Bed 6 191-2, *227*, 228

East Coast Joint Stock 95

Engine House
- demolition 286
- original construction 80, 150, 289
  - building material 333
- new construction 150, 151-3
  - further alterations 187-8, 190-92, 194, 208
- see also engine beds

Engineer and Railway Volunteer Staff Corps 203
Euston Allies *19*, 20, 74, 93-4, 159
Euston Road 10-12, 23; see also New Road
Euston Station 13, 17, 18, *19*, 93, 120, 205

Farringdon 98, 192, 208, 295
Ferme Park 98, 179, 204, 211
Fever and Smallpox Hospitals 10, 23-4, 27, 90
fish 37, 94, 140, *141*, 159, 222, 229, 230, 266, 274, 301, 303, *314*, 316
- Milk and Fish Depot 323
- Fish and Pay Office 161
- see also Coal and Fish Office
Freightliner terminal 277-8, 279

Gainsborough, Sheffield & Chesterfield 20

Gas Light and Coke Company 195; see also Imperial Gas Light and Coke Company
Gasworks Basin 16, 17, *25*, 133
Gasworks Viaduct 133, 195-6, 198
geology 6, 9, 28
Gibbs, Joseph 18, 21-2
glass
- bottles 175, (*illus*) 270-6
  - beer 270-1
  - brandy 273
  - gin 275
  - milk 270, 275-6
  - poison 274
  - sauce 274
  - scotch whisky 271
  - soft drinks 271-6
  - stopper 273, 274
  - wine 275
- manufacture see Joseph Bagley & Co.; Kilners
- panes 59, 125, 162, 260, 265
- trade 37, 270-5
Glass Bottle Warehouse 122-3, 124-5, 129, 172, 208-209; see also Kilners; Midland Goods Shed
Goods Shed complex see Main Goods Shed
Goods Yard
- bomb damage 251-3, 255, 268; see also London Blitz
- construction of
  - architect; see Cubitt, Lewis
  - building material 87-8, 154
  - contractor; see Jay, John
  - design 29
  - ground works 27-8, 31
  - human cost 90-2
  - location 13, 23-4
- canal infrastructure 96, 154, 298-9, 311, 316; see also Canal Basin; canal tunnels; Granary Basin; Regent's Canal
- finds assemblage 243, 273-5
- goods traffic 94-5, 136, 138-41, 154, 197, 211, 298-303, 314-18
- container traffic 223, 240
- see also Coal and Fish Offices; Coal Drops; Coal Offices; Eastern Goods Shed; Eastern Transit Shed; Goods Shed complex; Goods Warehouses; Granary; Main Goods Shed; Main Goods Station; Midland Goods Shed; Potato Market; Train Assembly Shed; Unclaimed Goods Warehouse
- horses 212-17
  - stabling 167; see also Eastern Transit Shed Stables; Horse Provender Store; Western Transit Shed Stables
- hydraulics 134, 143, 149, 154, 223, 250, 266, 288-95
- see also Hydraulic Metering Station; Hydraulic Station

importance of   297, 327–8
later re-use   285–6, 287
methodology   4–5, 325–6
modernisation   223, 228, 239, 240
communication   226, 239; *see also* Telegraph Office
electrical power   227, 239
post-nationalisation   257–8, 268, 279, 317
rail infrastructure   98–9, 114, 132, 133, 136, 173, 280, 298–9, 304–309, 311–12
road infrastructure   31, 86, 98–9, 136, 161, 173, 239, 298–9
recession   230–31
staff   160, 174–7, 206, 240, 257; *see also* labour force
welfare   160, 173, 176, 210, 239–40, 258, 265, 302; *see also* Inward Staff Messroom; Lamp Room and Coffee Shop; trade unions; Western Goods Offices
synopsis   297–328
topography   6
Goods Warehouses   25, 29–30, 35–6, 54; *see also* Eastern Transit Shed; Western Transit Shed
Goods Way and Excel Bridge   206–207, *220*
Government Works Schemes   231
grain   37–8, 90, 115, 154, 289, 316, 322; *see also* Granary
Granary
alterations   104, 107, 115–16, 239, 284–5, 286, 287, 310–312
fire prevention improvements   164–5
conversion to Eastern Goods Shed   181–2, 185, 318
Overhead Offices   183–6
modernisation   223–6, 260, 265
bomb damage   251, *252*
brickwork   104, 182, 185–6
built heritage record   56–64, 116, 164–5
cats   177
construction   35, 54–56, 88, 90–92
building materials   56, 59, 87, 196, 260
design   29, *30*, 37, 43, 51, 54–5, 298, 327
facilitation of goods handling   37, 55, 56, 59, 64–67, 88, 116
goods traffic   38, 55, 115–16, 139–40, 185, 223–5, 304, 316
hydraulics   55, 76–7, 80, 88, 207–208, 289, 292
modernisation   223–6, 227
later use   280, 286, 287; *see also* Pickfords
rail road 98–9
railway infrastructure   67–72, 98–9, 108, 232, 235, 279, 304
road infrastructure   146, 154, 161
see also Eastern Transit Shed; Main Goods Station; Overhead Offices; Train Assembly Shed; Western Transit Shed
Grand Junction Railway   13, 15, 18, 20

Granary Basin
bomb damage   253, *253*
construction   34–6
walls   31, 35–6
building material   36, 87
decline   160, *197*, 198, 206, 312, 317
goods handling   99–100
hydraulics   76–8, 99–100, 133
modernisation 223
infilling   206–208, 273, 312, 313–14, 317
car park   217, 229
garage   229, 239; *see also* RME Garage
railway infrastructure   99, 113–14
granite   43, 48, 59, 77–8, 98–9, 169
Aberdeen   17, 87–8,
Guernsey   98
Mountsorrel   98–9, 120
Markfield   120
Great Central   179, 181, 198, 222, 246, 271, 305; *see also* Marylebone Passenger Station and Goods Yard
Great Depression   222, 230–31, 240
Great Eastern   95, 222, 229, 246, 271
Great Exhibition   55, 93–4, 156
Great North of England Railway   18
Great Northern Railway Act   22, 23, 25, 27
Great Northern Railway Company
coal trade   93–5, 126–9, 132, 299; *see also* Coal Offices
comparative development   321–8
competition   13, 88–9, 93–5, 136, 159, 162, 173, 179–81, 198, 298
congestion   95, 98, 114, 154, 181, 197, 206, 211
development of King's Cross
community impact   27
compared to rivals   321–8
location   6, 23–4, 33
name   13
economic growth   94–5, 143, 159, 179–80, 198, 312–14
formation   18–22
goods yard facilities   98, 179; *see also* Farringdon; Ferme Park; Goods Yard; Holloway Road
government control   203, 211, 219–20, 313–14
importance of King's Cross   96, 327–8
investment in   18, 22, 25
London suburban network   97–8, 160
technological advances   2, 143, 154, 179–80, 226, 243; *see also* hydraulic power
track network   87–9, 93–4, 136, 196–7, 270–75
trains   159, 179, 222
see also London & York; London & North Eastern Railway
Great Western   197, 220, 224, 227, 289
Grouping Act   220, 228, 325

Harrison's Brickworks   11
Holloway Road   95, 98, 140, 167, 173, 248
    power station   227
Homer, Thomas   15
horses   6, 29, 86, 138, 143, 154, 162, 174–6, 212–17
    decline   223, 239, 228–9, 314, 317, 323
    horse-drawn cart   44, 66–7, 81–2, 147, 161, 228, 238, 300, 311, 322
    horse-drawn dray   138, 214, 300, 323
    welfare   167–8, 173, 213, 215–16; *see also* Eastern Transit Shed Stables ; Horse Provender Store/Stables; Western Transit Shed Stables
    shunting   28, 98–9, *212*, 293–4
Horse Department   83, 85, 171, 215, 228, 301
Horse Provender Store / Stables   72, 134–6, 167–8, 169
Hudson, George   18–22, *18*
Huish, Mark   93–4
Humber, William   95
    Goods Station report   56, 65, 70, 72
    Plan of 1866   *34*, 35–40, 43–4, 47, 52, 54, 56, 65, 67–8, 70, 72, 75–7, 79, 81, 86, 99, 102, 110–11, 114, 116, 122–4, 132
Hydraulic Metering Station   266–7
Hydraulic Station
    alterations   150–53, 187–90, *189*, 194, 197, 206–208, 228, 325–6
    building material   151–3, 154, 196
    electrification   313
    World War II   250
    construction   28, 31, 33, 77
        archaeological evidence   78–81, 323
        building material   79, 87
    design   29, 30, *30*
    power output   55, 78, 134, 150, 154, 187, 288–95; *see also* Armstrong engine; Benjamin Walker motor; steam pumping engine; Tannett Walker engine
    rail infrastructure   72
    water supply   77, 134, 187, 206–207, 208, 291–2, 295; *see also* Pump House
    see also Accumulator House; Accumulator Tower; Boiler House; Engine House; Hydraulic Metering Station
hydraulics
    impact on welfare   174–6
    machinery
        accumulator   122, 126, 150, 292
        capstan shunting   119, 143–50, 154, 162, 212, 231–2, 293–5, *293*, *294*, 305, 316
        crane   37, 44, 47, 55, 70, 76, 133–4, 223, 288–92; *see also* Crane Base
        hoist   168, 176, 207, 288-91
        jigger   176, 291–2
        lift   323
        winch   63–7, 311
    network   133

    extension programme   146, 294–5
    pipes   7, 55, 77–8, 162, 187, 206–208
technology
    electrification   223, 224, 227, 233, 239, 295–6, 313
    pioneering use   2, 29, 88, 288, 322, 327
    power   *see* Hydraulic Station

Imperial Gas Light and Coke Company   12, 15–17, 132, 300; *see also* Gas Light and Coke Company; St Pancras Gasworks
Inwards Staff Messroom   236–9
    building material   238
    conversion to Outwards Staff Messroom   239
Islington   10, 27
Ivatt, Henry   176, 187

Jay, John   27–8, 30, 33, 38, 52, 54, 79, 81, 87, 90–92, 132, 135
    brick stamp   36, 48, 87
Johnson, Richard   103, 115, 119, 132, 144, 146, 161–2, 165–8, 169, 172–3, 215
Joseph Bagley & Co.   140, 177; *see also* bottle warehouse

King George IV monument   12, *14*
King's Cross Goods Yard   *see* Goods Yard
King's Cross Passenger Station   *see* Permanent Passenger Station
Kilner Brothers   122, 124–5, 140, 162, 172, 208, 272, 311
Kirk Knight & Co.   169, 182, 286

labour force
Great Northern   90–92, 206
    impact of World War I   206
London & North Eastern   174–7, 242–4
    impact of World War II   246–7
    women   177, 206, 210–11, 238, 243–4, *244*, 246–7, 251, 258, 265, 266
    welfare   90–92, 101, 119, 160, 173, 240, 242; *see also* trade union; welfare facilities
Lamp Room and Coffee Shop   25, 72–3, 113–14, 116–17; *see also* Refreshment Club
Laser Building   2, 236, 253, 266, 286; *see also* Outwards Staff Messroom
Londinium   9
Leroux, Jacob   10
Lloyd-George, David   219
Locke, Joseph   20–21
locomotives   159, 179, 222
    Flying Scotsman   222
    Mallard   222
London & Birmingham   13, 18, 321, 322
    Camden depot   13
    see also Euston station

London & North Eastern Railway
    creation of   211, 220–22
    decline   313–16
    early years   217, 222–3, 239
    importance of King's Cross   138–40, 327–8
    modernisation   217, 223–9, 231, 239–240; see also
        Transposition Scheme
    nationalisation   257–9, 268, 317
London & North Western Railway Company   2, 19, 24,
        33, 88, 93–4, 120, 136, 159, 197, 275, 278, 325
    Camden depot   2, 13, 24, 33, 88, 298, 321–3, 327; see
        also Camden Goods Yard
London & York   18–22, 23; see also Great Northern
        Railway
London Blitz   106–7, 111, 248, 250, 251–3, 256
    death toll   251, 253
    repairs   251, 254–6
London Brick Company   196, 278
London Chatham & Dover Railway   96, 98
London Hydraulic Power Company   266–7, 295–6; see
        also Hydraulic Station
    Metering Station   see Hydraulic Metering Station
London, Midland & Scottish   220, 222, 275, 323–5
London Passenger Transport Board   245–6
Lynx Express Ltd.   280, 285

machine bed   191, 250
Maiden Lane Bridge   6, 15–17, 25, 33–4, 90, 161
Main Goods Shed
    alterations   144–9, 161, 310
        conversion to Eastern Goods Shed   181–4,
            230–6
    construction   303
        ground works   31
            archaeological evidence   68–72
            building material   87–8
    contrasted to other companies   323–5
    design   35, 37–38, 303, 322
    finds assemblage   175–6, 275
    goods traffic   297–8, 301–302, 304, 323
    railway infrastructure   67–72, 98–9, 136
    road infrastructure   161
    traffic management   88, 297–8, 303–5
    see also Eastern Goods Shed; Eastern Transit Shed;
        Granary; Train Assembly Shed; Western Transit
        Shed
Main Goods Station see Main Goods Shed
Manchester Sheffield & Lincolnshire   181, 288; see also
        Great Central
Marylebone Passenger Station and Goods Yard   181, 198,
        222, 271, 296, 299, 305
    bomb damage   248
Marylebone Road   10; see also New Road
Medcalf, Josiah   138–40, 159, 161, 165, 167, 174–7, 215, 243,
    messrooms   266; see also Inwards Staff Messroom

Metropolitan line   95, 96, 98, 181, 246
Middlesex County Hospital for Smallpox   10
Midland Goods Shed   75, 119, 121–6, 146, 161–4, 172–3,
        208–211, 226, 230, 250, 279–80, 291, 312, 314; see
        also Kilner's bottle manufacturers; Pump House;
        Unclaimed Goods Warehouse
Midland Goods Shed Office Block   125
Midland Railway   18, 19, 93–5, 98, 120–21 132, 136, 157,
        159, 277–8, 323–5
    telephones   179
    see also Somers Town Goods Depot; Agar Town
        Depot; St. Pancras Passenger Station
Midland Roundhouse   94–5
Midland Shed Accumulator Tower   162–4
Ministry of Munitions   204
Ministry of Transport   204, 220, 231, 239, 245–6, 313
Moscrop, Andrew   171–2, 228
motor vehicles   216–17, 223, 228–9, 238–9, 240, 285, 314,
        317, 323
Munich Crisis   245–6

National Carriers Ltd.   280, 285
National Freight Consortium   280; see also National
        Freight Corporation
National Freight Corporation   279, 280, 285, 287; see also
        National Freight Consortium
New Road   9, 10, 12, 13, 17, 23, 88, 321, 323,
North British Railway   87, 95, 159, 196, 271
    axle box cover   138, 139
    glass bottles   271
North Eastern   87, 93, 95, 138, 159, 196, 220
    glass bottles   270–71
North Union Railway   18

Oakley, Henry   103, 108 114, 116, 117, 119, 122, 136, 143,
        161–2, 164, 165–6, 182
Outwards Goods Shed   see Western Goods Shed
Outwards Staff Messroom   236–9, 253, 266; see also Laser
        Building
Overhead Offices   114, 169, 183–6, 254, 265, 286

Pentonville Road   10, 12; see also New Road
Permanent Passenger Station   6, 31, 33, 173, 226, 248,
        278–9
    development of   23–7, 74, 88, 119
    location   10, 23–5
Pickford's   280, 293, 322–3
pitch Macadam   99
Portland cement   16, 31, 185
Potato Market   98, 115, 118–19, 118, 119, 122, 140, 141,
        143, 149, 161–4, 172–3, 176–7, 230, 286, 293, 310, 318,
        323–4
    bomb damage   251, 251, 253

pottery
    institutional wares   200–201
    trade   37
Plimsoll, Samuel   128–9, 132, *156*, 156–8, 171, 300–301,
Plimsoll Viaduct   *96*, 132, 158; see also Cambridge Street
    Coal Drops
prehistory   9
Pump House   250

races to the north   159
Rail Parcels Depot   279, 280, 281–7, 318, *321*, 323; see also
    Road Distribution Terminal
Railway Act of 1921   *see* Grouping Act
Railway Executive Committee   203, 211, 217, 246–8, 250,
    256, 260, 313, 317; see also British Railways Board
Railway Mania   20, 23–5, 28
Refreshment Club   117, 146, 165–7, 210
    extension   198, 200–201
    finds   200
Regent's Canal
    community impact   17, 243
    construction   6, 12, 15–16
        North Wall   16, *25*, 31
        Retaining Wall   16, *25*, 31, 33, 48, 78–9, 82
        South Wall   86
    decline   95–6, 160
    goods traffic   95–6, 159–60, 298–9, 311
        beer   273
        brick   87
        coal   133
        stone   88
    importance of   12, 17, 24, 33, 88
    water provision   187, 291
        horses   215
    see also Canal Branch; London & Birmingham; St
        Bartholomew's Hospital; St Pancras Gasworks
Regent's Canal and Dock Company   195–6; see also
    Regent's Canal Company
Regent's Canal Company   33–4, 77, 96, 128, 132–4, 206,
    291; see also Regent's Canal and Dock Company
Reilly, Billy   285
Rennie, John   18, 21
Rickett Smith & Co.   127–8, 130, 171
RME Garage   229, 251, *253*, 267
Road Distribution Terminal   280, 281–5, 287
Roadline UK Ltd.   280; see also Lynx Express Ltd.
Rocque, John
    map of 1769   10, *12*

Saxon   9
Shoreditch   10
shunting capstans   *see* hydraulic shunting capstans

slate
    North Welsh Countess   209, 218–19
    replacement of   74, 281–2
    roofs   60, 73, 101, 125, 148, 184, 261
Smallpox Hospital   *see* Fever and Smallpox Hospitals
Smith, John   11
    Dust Ground   11, *11*
Somers Bridge   17, *25*, 31, 33–5, 82, 86, 161, 206
Somers Town   10–11, 13, 23, 27, 91, 266
Somers Town Goods Depot   159–60, 161–2, 173, 203,
    266, 278, 296, 299, 302, 321, 323–5
    coal drops   323–4
Southern Railway   220, 229, 246
Spitalfields Goods Station   222–3
St Bartholomew's Hospital   10, 15–16, 23–4, 27
St Chad's Well   10
St Pancras Gasworks   12, 13, 15–16, 17, 33, 195, 206–207,
    300; see also Gas Light and Coke Company;
    Imperial Gas Light and Coke Company
St Pancras Old Church   *1*, 9, 95
St Pancras Passenger Station   95, 121, 205–206, 278, 279,
    323
St Pancras Prebendary   10, 13
steam pumping engine   76, 80, 187, 191, 208, 288–95, *290*
Stirling, Patrick   159, 176, 187
stone
    trade   38, 99, 139–40, 197, 298, 303, *303*
    York stone   87, 129, 152, 154, 192, 196
    see also granite
stoneware
    ginger beer bottle   *274*, 275
sugar beet   222, 224–5, *225*, 240, 316

Tannett Walker and Co.   119, 143, 146–9, 187–8, 293–5;
    see also Tannett Walker engine
Tannett Walker engine   148, 150, 153, 188, 191–2, 295; see
    also Tannett Walker and Co.
Temporary Passenger Station   25, 52, 54, 73–5, 88–9, 119,
    136, 298, 302; see also Potato Market
Temporary Smiths' Shed   *see* Carriage Shed
Telegraph Office   122, 226
    modernisation   314
    pneumatic tube   226
    see also telephones
telephones   164, 226, 242
    use in signal boxes   179
textiles
    manufacture   93, 295
    trade   37, 138–40
Thatcher, Margaret   280
Thomas, W.M.   14
tile
    form
        Roman style   59
        stamp   117
    manufacture   11, 87

trade unions   180, 198, 206, 240
 General Strike of May 1926   222
Train Assembly Shed
 building modifications   114, 144, 146–9, 154, 232–5, 260, 265, 268, 280, 284–5, 287, 305, 318–19
  Air Raid Precaution measures   250
  later re-use   285–6
 construction   28–30, 33, 52–54, 67–72
 building materials   87, 196
 design   38, 51–54
  railway lines   67–8
 finds assemblage
  axle box cover   138, *139*
  glass   272–3
  milk bottles   275
  spanners   174, *175*
 see also Eastern Goods Shed; Goods Shed complex; Overhead Offices; Rail Parcels Depot; Road Distribution Terminal
Transport Act of 1919   220
Transport Act of 1962   277
Transposition Programme *see* Transposition Scheme
Transposition Scheme   231–3, 236–7, 239–40, 256, 260, 268, 314, 316–7
 reversal   317–18
turntable   38, 44, *46–7*, 47, 50–2, *53*, 54, 56, 63, 67–70, 72, 87, 108, 113–14, 135, 144, 146–9, 175, 231–2, 234–5, 279, 294, 305, 311
 Turntable A   *26*, 68–9, *69*, *97*, 100, 147, 149, 175, 275
 Turntable B   *26*, 68–9, *69*, 77, *97*, 100, 147, 149, 275
 Turntable C   114, *189*
 Turntable D   *145*, 146
 Turntable E   *145*, 149

Unclaimed Goods Warehouse   265

vegetable traffic   37, 118, 161, 173, 174, 230, 140, 301–2, 316, 324
 corn   55, 56, 65, 88, 93, 140
 potato   93, 115, 118, 119, 136, *141*, 159, 162, 173, 230, 301–302, 310, 323; *see also* Potato Market

wagon shunting   28, 148–9, *212*
Wakefield, Lincoln & Boston   20
Walker, Benjamin   148; *see also* Tannett Walker Engine
Walker, Herbert Ashcombe   203
War Railway Council   203
Watkin, Edward   181, 198
Way and Works Committee   103, 108, 116, 119, 132, 133, 143, 150, 162, 168, 172, 210, 226, 236, 251
Wedgewood, Ralph   231, 239, 240
weighbridges   86, 134, 213
welfare facilities   72–3, 91, 117–18, 130, 160, 166, 173, 176, 181, 198, 200, 210–11

 London & North Eastern   236–40, 251, 265, 266, 280
  female welfare   238, 258, 265, 266
West Handyside Canopy   119–20, 162–3, 169, 230, 253, 262, 303
West London Extension   96
Western Coal Drops   131, 171, 181, 291, 301, 311
Western Goods Offices
 alterations   102–103, 108, 116, 266, 286
  Transposition Scheme   233
  staff welfare   102–103, 186–7, 236, 279, 243, 251
 bomb damage   251
 construction   101–103, 104
 design   54, *55*, 101–102
 later use   279
Western Goods Shed   *182*
 alterations   260, 312, 314
  transposition   314
 bomb damage   251, 256
  repairs   258
 construction   181–2, 197–8, 304–305, 311, 324
 hydraulic system   294–5
 offices   186, 233, 243
Western Transit Shed   *25*
 alterations   107–10, 230, 268, 287
  conversion to Eastern Goods Shed   181–2
  conversion to Road Distribution Terminal   280–5
 construction   38–44
  brickwork   38–9, 43–4
  design   29
 finds assemblage
  glass   273–5
  stoneware   273
 function   37–38
 hydraulic machinery   76–77, 133
 movement of goods   67–68, 99–100, 136, 181, 298
 see also Eastern Goods Shed; Western Transit Shed Stables
Western Transit Shed Stables
 alterations   167, 169, 236, 250
 construction   33, 47, 48–51
 design   37–8, 47
 finds assemblage   214
 horse welfare   214–16
 see also Goods Yard; Main Goods Shed
W.G. Armstrong & Co.   55, 122, 288; *see also* Armstrong, William; W.G. Armstrong, Mitchell & Co.
W.G. Armstrong, Mitchell & Co.   150; *see also* Armstrong, William; W.G. Armstrong & Co.
Wharf Road Viaduct   *25*, 31, 81–86, 129, 135
whisky   138
 bottles   271–2

Wiggins, Edward 47, 82, 86, 127, 128, 134–5, 213
Wollstonecraft, Mary 10
World War I 203–206, 210–11, 312
    impact on labour force 206
    mobilisation 203–204
    post-war economics 219–23, 239, 313–14
World War II 245–8
    Air Raid Precaution measures 245, 248–50, 254, 256
        shelter later use 266, 279
    impact on Goods Station 316
    impact on labour force 246–7
    see also Air Raid Precautions Act; London Blitz

York & North Midland 18